Lecture Notes in Computer Science   5787

Commenced Publication in 1973
Founding and Former Series Editors:
Gerhard Goos, Juris Hartmanis, and Jan van Leeuwen

Choong Seon Hong Toshio Tonouchi
Yan Ma Chi-Shih Chao (Eds.)

Management Enabling the Future Internet for Changing Business and New Computing Services

12th Asia-Pacific Network Operations
and Management Symposium, APNOMS 2009
Jeju, South Korea, September 23-25, 2009
Proceedings

 Springer

Volume Editors

Choong Seon Hong
Kyung Hee University
Department of Computer Engineering
1 Seocheon, Giheung, Yongin, Gyeonggi 446-701, South Korea
E-mail: cshong@khu.ac.kr

Toshio Tonouchi
NEC Corporation
Service Platforms Laboratories
1753 Shimonumabe, Nakahara-Ku, Kawasaki, Kanagawa 211- 8666, Japan
E-mail: tonouchi@cw.jp.nec.com

Yan Ma
Beijing University of Posts and Telecommunications
Information Network Center
10 Xitucheng Road, Haidian District, Beijing 100876, China
E-mail: mayan@bupt.edu.cn

Chi-Shih Chao
Feng Chia University
Department of Communications Engineering
100 Wenhwa Road, Seatwen, Taichung 40724, Taiwan
E-mail: cschao@fcu.edu.tw

Library of Congress Control Number: 2009934160

CR Subject Classification (1998): C.2, D.2, D.4.4, K.6, H.3-4, I.6

LNCS Sublibrary: SL 5 – Computer Communication Networks
and Telecommunications

ISSN   0302-9743
ISBN-10   3-642-04491-3 Springer Berlin Heidelberg New York
ISBN-13   978-3-642-04491-5 Springer Berlin Heidelberg New York

springer.com

© Springer-Verlag Berlin Heidelberg 2009
Printed in Germany

Typesetting: Camera-ready by author, data conversion by Scientific Publishing Services, Chennai, India
Printed on acid-free paper   SPIN: 12761357   06/3180   5 4 3 2 1 0

Preface

We are delighted to present the proceedings of the 12th Asia-Pacific Network Operations and Management Symposium (APNOMS 2009), which was held in Jeju, Korea, during September 23–25, 2009.

Recently, various convergences in wired and wireless networks, and convergence of telecommunications and broadcastings, are taking place for ubiquitous multimedia service provisioning. For example, broadband IP/MPLS wired networks are actively converged with IEEE 802.11e wireless LAN, IEEE 802.16 Wireless MAN, 3G/4G wireless cellular networks, and direct multimedia broadcast (DMB) networks. For efficient support of service provisioning for ubiquitous multimedia services on the broadband convergence networks, well-designed and implemented network operations and management functions with QoS-guaranteed traffic engineering are essential. The converged network will open the way for a new world with emerging new businesses and computing services. The Organizing Committee (OC) selected "Management Enabling the Future Internet for Changing Business and New Computing Services" as the timely theme of APNOMS 2009.

Contributions from academia, industry and research institutions met these challenges with 173 paper submissions, from which 41 high-quality papers (23.7% of the submissions) were selected for technical sessions as full papers, and 32 papers were selected as short papers. In addition, we had nine papers in innovation sessions for on-going research. Diverse topics were covered, including Traffic Trace Engineering, Configuration and Fault Management, Management of IP-Based Networks, Autonomous and Distributed Control, Sensor Network and P2P Management, Converged Networks and Traffic Engineering, SLA and QoS Management, Active and Security Management, Wireless and Mobile Network Management, and Security Management.

The Technical Program Committee (TPC) Co-chairs would like to thank all those authors who contributed to the outstanding APNOMS 2009 technical program, and thank the TPC and OC members and reviewers for their support throughout the paper review and program development process. Also, we appreciate KICS KNOM, Korea, IEICE TM, Japan, and IEEE ComSoc for their sponsorship, as well as IEEE CNOM, IEEE APB, TMF and IFIP WG 6.6 for their support of APNOMS 2009.

September 2009

Choong Seon Hong
Toshio Tonouchi
Yan Ma
Chi-Shih Chao

Organization

Organizing Committee

General Chair

Young-Tak Kim   Yeungnam University, Korea

Vice Co-chairs

Young-Myoung Kim   KT, Korea
Hiroshi Uno   NTT, Japan
Victor Wu-Jhy Chiu   Chunghwa Telecom, Taiwan

Technical Program Committee Co-chairs

Choong Seon Hong   Kyung Hee University, Korea
Toshio Tonouchi   NEC, Japan
Yan Ma   Beijing University of Posts and
                             Telecommunications, China
Chi-Shih Chao   Feng Chia University, Taiwan

Members

Adarsh Sethi   University of Delaware, USA
Aiko Pras   University of Twente, The Netherlands
Akira Idoue   KDDI R&D Labs, Japan
Alexander Keller   IBM, USA
An-Chi Liu   Feng Chia University, Taiwan
Antonio Liotta   University of Essex, UK
Chung-Hua Hu   Chung Hwa Telecommunication, Taiwan
Chu-Sing Yang   National Cheng Kung University, Taiwan
Cynthia Hood   Illinois Institute of Technology, USA
Daniel W. Hong   KT, Korea
Deokjai Choi   Chonnam University, Korea
Fei-Pei Lai   National Taiwan University, Taiwan
Filip De Turck   Ghent University, Belgium
Gabi Dreo Rodosek   University of Federal Armed Forces, Germany
Gahng-seop Ahn   City University of New York, USA
Haci Ali Mantar   Gebze Institute of Technology, Turkey
Hanan Lutfiyya   University of Western Ontario, Canada
Han-Chieh Chao   National Ilan University, Taiwan
Hiroyuki Tanaka   NTT-West, Japan
Hwa-Chun Lin   National Tsing Hua University, Taiwan
Hyunchul Kim   Seoul National University, Korea
Ian Marshall   Lancaster University, UK

Tutorial Co-chairs

Hong-Taek Ju   Keimyung University, Korea
Masaki Aida   Tokyo Metropolitan University, Japan

Special Session Co-chairs

Jae-Hyoung Yoo   KT, Korea
Yuka Kato   Advanced Institute of Industrial Tec, Japan

DEP Co-chairs

Tae-Sang Choi   ETRI, Korea
Kohei Iseda   Fujitsu Labs, Japan

Exhibition and Patron Co-chairs

Dong-Sik Yoon   KT, Korea
Tadafumi Oke   NTT Comware, Japan
Chung-Hua Hu   CHT-TL, Taiwan

Poster Co-chairs

Wang-Cheol Song   Jeju University, Korea
Marat Zhanikeev   Waseda University, Japan

Innovation Session Co-chairs

Hoon Lee   Changwon National University, Korea
Kiyohito Yoshihara   KDDI R&D Labs, Japan

Publicity Co-chairs

Ki-Hyung Kim   Ajou University, Korea
Jun Kitawaki   Hitachi, Japan
Jiahai Yang   Tsinghua University, China
Han-Chieh Chao   NIL, Taiwan

Financial Co-chairs

Yoonhee Kim   Sookmyung Women's University, Korea
Shingo Ata   Osaka City University, Japan

Publication Co-chairs

Mi-Jung Choi   Kangwon National University, Korea
Seung-Yeob Nam   Yeungnam University, Korea

Local Arrangements Co-chairs

Kwang-Hui Lee   Changwon National University, Korea
Jae-Oh Lee   KUTE, Korea
Seok-Ho Lee   Metabiz, Korea

Secretaries

Young-Woo Lee	KT, Korea
Akihiro Tsutsui	NTT, Japan

International Liaisons

Ed Pinnes	Elanti Systems, USA
Raouf Boutaba	University of Waterloo, Canada
Carlos Westphall	Santa Catalina Federal University, Brazil
Marcus Brunner	NEC Europe Ltd., Germany
Rajan Shankaran	Macquarie University, Australia
Alpna J. Doshi	Satyam Computer Services, India
Teerapat Sanguankotchakorn	AIT, Thailand
Borhanuddin Hohd Ali	University of Putra, Malaysia
Rocky K.C. Chang	Hong Kong Polytechnic University, China

Advisory Board

Makoto Yoshida	University of Tokyo, Japan
Masayoshi Ejiri	Studio IT, Japan
Doug Zuckerman	Telcordia, USA
Qiliang Zhu	Beijing University of Posts and Telecommunications, China
Seong-Beom Kim	KTFDS, Korea
Young-Hyun Cho	KTH, Korea

Steering Committee

Nobuo Fuji	NTT Advanced Technology, Japan
Hiroshi Uno	NTT, Japan
James Hong	POSTECH, Korea
Kyung-Hyu Lee	ETRI, Korea
Yoshiaki Tanaka	Waseda University, Japan
Young-Tak Kim	Yeungnam University, Korea

Technical Program Committee

Co-chairs

Choong Seon Hong	Kyung Hee University, Korea
Toshio Tonouchi	NEC, Japan
Yan Ma	Beijing University of Posts and Telecommunications, China
Chi-Shih Chao	Feng Chia University, Taiwan

All OC members are also part of the Technical Program Committee.

Additional Paper Reviewers

Akihiro Tsutsui, NTT, Japan
Al-Sakib Khan Pathan, Kyung
Hee University, Korea
Cao Trong Hieu, Kyung
Hee University, Korea
Carlos Westphall, Santa Catalina
Federal University, Brazil
Dae Sun Kim, Kyung Hee University,
Korea
Georgios Exarchakos, TUE,
The Netherlands
Haci Ali Mantar, Gebze Institute of
Technology, Turkey
Hideaki Miyazaki, Fujitsu Laboratories
Ltd., Japan
Hiroshi Uno, NTT, Japan
Hiroyuki Tanaka, NTT, Japan
Hoon Lee, Changwon University,
Korea
Hoon Oh, University of Ulsan, Korea
Hyunseung Choo, Sungkyunkwan
University, Korea
Isabelle Chrisment, LORIA -
University of Nancy 1, France
Iwona Pozniak-Koszalka, Wroclaw
University of Technology, Poland
Jae-Oh Lee, University of Technology
and Education, Korea
James Hong, POSTECH, Korea
Jeng-Yueng Chen, Hsiu Ping Institute
of Technology, Taiwan
Jin Ho Kim, Kyung Hee University,
Korea
Joberto Martins, Salvador University -
UNIFACS, Brazil
John Strassner, Waterford Institute of
Technology, Ireland
Joon Heo, NTT, Japan
Jun Kitawaki, Hitachi, Japan
Ki-Hyung Kim, Ajou University,
Korea
Kiyohito Yoshihara, KDDI R&D Labs,
Japan

Kohei Iseda, Fujitsu Labs, Japan
Marat Zhanikeev, Waseda University,
Japan
Masaki Aida, Tokyo Metropolitan
University, Japan
Md. Abdul Hamid, Kyung Hee
University, Korea
Md. Abdur Razzaque, Kyung Hee
University, Korea
Md. Mustafizur Rahman, Kyung Hee
University, Korea
Md. Obaidur Rahman, Kyung Hee
University, Korea
Md. Shariful Islam, Kyung Hee
University, Korea
Muhammad Mahbub Alam, Kyung
Hee University, Korea
Muhammad Shoaib Siddiqui, Kyung
Hee University, Korea
Phlipp Hurni, University of Bern,
Switzerland
Ramin Sadre, University of Twente,
The Netherlands
Satoshi Imai, Fujitsu, Japan
Seungyup Nam, Yeungnam University,
Korea
Shingo Ata, Osaka City University,
Japan
Sungwon Lee, Kyung Hee University,
Korea
Syed Obaid Amin, Kyung Hee
University, Korea
Tadafumi Oke, NTT, Japan
Tae-Sang Choi, ETRI, Korea
Takashi Satake, NTT, Japan
Teerapat Sa-nguankotchakorn, AIT,
Thailand
Tomohiro Ishihara, Fujitsu Labs.,
Japan
Torsten Braun, University of Bern,
Switzerland
Tran Hoang Nguyen, Kyung Hee
University, Korea

Table of Contents

Autonomous and Distributed Control

Sensor Network and P2P Management

Converged Networks and Traffic Engineering

SLA and QoS Management

Active and Security Management

Wireless and Mobile Network Management

Security Management

Short Papers

Traffic Trace Engineering

Pham Van Dung[1], Marat Zhanikeev[2], and Yoshiaki Tanaka[1,3]

[1] Global Information and Telecommunication Institute, Waseda University
1-3-10 Nishi-Waseda, Shinjuku-ku, Tokyo, 169-0051 Japan
[2] School of International Liberal Studies, Waseda University
1-6-1 Nishi-Waseda, Shinjuku-ku, Tokyo, 169-8050 Japan
[3] Research Institute for Science and Engineering, Waseda University
17 Kikuicho, Shinjuku-ku, Tokyo, 162-0044 Japan
dungpv@fuji.waseda.jp, maratishe@aoni.waseda.jp,
ytanaka@waseda.jp

Abstract. Traffic traces captured from backbone links have been widely used in traffic analysis for many years. By far the most popular use of such traces is *replay* where conditions and states of the original traffic trace are recreated almost identically in simulation or emulation environments. When the end target of such research is detection of traffic anomalies, it is crucial that some anomalies are found in the trace in the first place. Traces with many real-life anomalies are rare, however. This paper pioneers a new area of research where traffic traces are engineered to contain traffic anomalies as per user request. The method itself is non-intrusive by retaining the IP address space found in the original trace. Engineering of several popular anomalies are shown in the paper while the method is flexible enough to accommodate any level of traffic trace engineering in the future.

Keywords: Traffic, Analysis, Trace, Replay, Simulation, Emulation, Anomaly.

1 Introduction

Many traffic traces captured from backbone links are available for network research [1]. In fact, there are many research papers and tools [2] that can analysis traffic content of such traces. On the practical side, one of the most popular uses of traffic traces is *replay* where conditions and states of the original traffic are recreated almost identically in simulation or emulation environments. When the *replay* aims to detect traffic anomalies, however, reality is such that it is not always easy to find traces which inherently contain anomalies one expects to find and detect. Having studied the content of a traffic trace, one might want to modify the trace in some way to increase the share of a certain traffic type or add traffic from malicious attacks.

On the other hand, several studies recently were dedicated to classification of traffic types [3], [4]. To the best of authors' knowledge, the common aim of such classification methods is to identify all traffic types present in a trace, namely to find out which applications generate which traffic type and what is the share of traffic held by each type in total traffic soup.

C.S. Hong et al. (Eds.): APNOMS 2009, LNCS 5787, pp. 1–10, 2009.
© Springer-Verlag Berlin Heidelberg 2009

This paper goes one step further by not only allowing to classify traffic in order to discover the traffic content, but also by facilitating various engineering routines among various traffic types. The paper uses IP aggregation to split a trace into traffic entities where each is composed of a traffic node and a number of connections coming from its communication parties. In this paper, traffic node is defined as a collection of IP addresses which share a common prefix. Traffic classification is done by using basic properties of a traffic entity. Six particular types of traffic each representing a popular traffic type in real life are defined while the method can also fit other traffic types if needed.

When properly classified into individual traffic types, traffic trace can also be subjected to engineering in order to obtain more traffic of a given type. In this paper, the keyword "traffic engineering" stands for changing traffic of some type(s) into another, a targeted type. The modification process is implemented within a well controlled environment with a set of rigid rules created in order to retain sanity in the morphing process.

Engineering of several traffic types are presented in this paper as an example, while the method itself is capable of dealing with other types as long as those are well defined. The paper describes in detail how to manipulate each possible type in turn as well as all of them at once to obtain more traffic of some typical types. Results in terms of number of major occurrences of each traffic type in the trace prove the fact that a traffic trace can be easily engineered to obtain more of target traffic.

The remainder of this paper is organized as follows. Six traffic entities (occurrences of traffic types in trace) are defined in Section 2. Section 3 stipulates a set of regulations used for the process of engineering. Section 4 presents the proposed algorithm of discovering communication parties and describes how to manipulate traffic in order to obtain more of target traffic. Results of engineering of three typical traffic types as well as evaluation of performance are shown in section 5. Finally, the conclusion and directions of future work are drawn in Section 6.

2 Traffic Entity

The concept of traffic entity is created when IP aggregation is used to cluster a number of IP addresses which share a common prefix into a *traffic node*. While traffic node is simply the prefix itself, traffic entity is characterized by a number of properties including the number of IP addresses inside the prefix, number of *destinations* in connections originated from the prefix, number of *sources* in connections terminated at the prefix, number of *incoming/outgoing connections* and incoming/outgoing *traffic volume*. As a major concept throughout the paper, *traffic entity* is defined as a traffic node and a number of incoming and outgoing connections to or from it. By applying the concept of traffic entities, it is possible to successfully split traffic into a number of traffic entities interacting with each other on a timeline.

The properties of a traffic node bring out an image of users inside a prefix by indicating how crowded they are and how actively they interact with the rest of network community. Correspondingly, this paper uses combinations of properties as criteria to classify traffic entities. After classifying entities, basic traffic units such as *packets*, *flows*, *connections*, and *nodes* are thereof labeled by the type of entity they belong to.

In addition, the remaining traffic which does not belong to any types defined above is considered as unclassified traffic. Among many types the method works with six particular entity types as illustrated in Fig. 1.

In Fig. 1, each traffic entity is composed of a traffic node depicted by a clouded shape on the left with one or several IP addresses inside and connections depicted by lines with arrows to or/and from other cluster of users on the right. The size of the cloud on the left indicates the density of a traffic node. Similarly, the arrows show the connection direction while the thickness of lines indicates how much traffic in bytes the connections transfer. In order to determine whether a property is small or large in the classification process, this paper sets two thresholds, *min* and *max* for each property. The thresholds are inferred from real values of the property of all the nodes. The thresholds are picked as follows. Given that there are one hundred different values of a property sorted in an ascending list, the *min* threshold is chosen as value of the 30th element in the list while the *max* is value of the 70th element. A property of a node is considered as very small if its value is less than the *min* and very large if its value is greater than the *max* threshold.

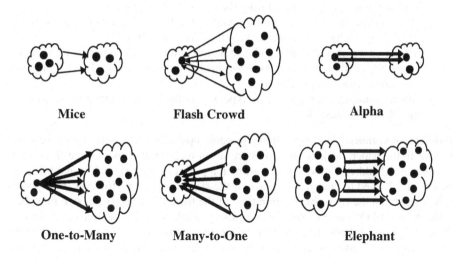

Fig. 1. Traffic entity definitions

Each traffic entity type in Fig. 1 maps closely to a popular type of traffic in real life. Mice entity represents the case where a small group of users creates a small amount of traffic and sends to few other users over Internet. Flash Crowd entity stands for a famous social phenomenon which occurs when a web server suffers a surge in number of requests within a short period of time due to large social events, thus, rendering the server unavailable [5]. Alpha entity represents high-rate transfers between a pair of a single source and a single destination [6]. One-to-Many and Many-to-One traffic entities are the cases where a large amount of traffic exchanged between a single server and a group of users. These might be content distribution where many users together download files from a server or content aggregation where users together upload files to a server. Finally, Elephant entity represents the case where a large group of users actively join and create much traffic in network [7].

3 Rules of Modification

The ultimate goal of this paper is to modify traffic among types that is to change one or several traffic types to a given target type. Obviously, before performing the modification, reasonable boundaries for the process have to be defined. This section stipulates which operations are permitted during modification as well as sets a number of related constraints. The modification process itself from this point on will be referred to as *morphing*.

Constraints: This paper performs morphing in order to get more of a given traffic type while trying to be as unobtrusive as possible. Below is the list of rules governing the morphing process:

1. It is not allowed to generate new IP addresses. Instead, the method should do within the IP space extracted from the original traffic trace.
2. The modification process must guarantee that the total number of packets and flows stay unchanged no matter how they are manipulated.
3. Based on a fact that source and destination of a packet in a packet trace are always physically located on opposite sides of the link, new packets, if any must also conform strictly to this rule.
4. When changing traffic type X to type Y only the traffic belonging to traffic entity X is to be modified.

Possible operations: This paper limits possible operations to modifying packets and flows. In the paper, packets and flows are seven-tuple units (time, source IP address, destination IP address, source port, destination port, protocol, size). The modification process is allowed to change the time and the size of a packet. However, new packet time must be in the range of duration time of the trace and new packet size must be a standard size of the current protocol. It is also possible to replace source or destination IP or even both of them. In case both are replaced, new source IP address must be the old destination IP address, and vice versa, to ensure that no new communication pairs are generated out of thin air. Depending on each target type, one or several operations will be used to manipulate to obtain expected results.

4 Traffic Type Morphing

The morphing process can be divided into two phases. The former is to discover all IP space related to the traffic node within the entity. For a given target traffic type of morphing, the goal of this phase is to determine sides for every IP address of the type. In plain words, all IP addresses must be attributed only to side A or side B provided that A and B are two sides of the backbone link.

The second stage is to select all traffic entities suitable for morphing and manipulate their connections by using the rules and guidelines defined above.

4.1 Pre-morphing IP Space Discovery

This paper uses a simple procedure to discover the entire IP space related to a given traffic entity. From this point on this process will be referred to as *Round Trip IP Space Discovery* (RTSD) as illustrated in Fig. 2. The paper exploits IP aggregation to diminish overhead of the algorithm by operating with traffic nodes instead of individual IP addresses. More specifically, the RTSD algorithm attempts to determine how many IP addresses can be used in morphing itself as well as which side relevant to the source traffic node they are found at.

This paper assumes that a traffic trace has the most popular traffic node, a node which has the largest number of total incoming and outgoing connections. General idea of the algorithm is to track out side for sources based on the communication pairs extracted from the original trace. The most popular node is selected as a starting point in attempt to make the overall process completed in the smallest number of round trips with the maximum number of nodes localized. The procedure is performed in round trips where each performs two similar tasks. The former is to discover nodes in side B from pairs where either source or destination node has already been known as located in side A. In a similar manner, the latter is to discover nodes in side A. The algorithm completes when no more nodes can be discovered, which normally happens when the entire IP space within the trace has been successfully located and attributed to the either side.

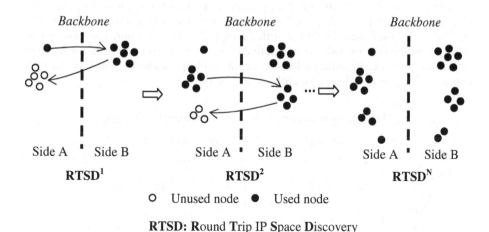

O   Unused node   ● Used node

RTSD: Round **T**rip IP **S**pace **D**iscovery

Fig. 2. Process of discovering side of sources

In regard to performance of the RTSD algorithm, although most sources are either directly or indirectly connected to the most popular node, there are always isolated communities where their member nodes only communicate with each others. In fact, these isolated nodes can be forced to get involved in the process of discovering by decreasing length of the prefix used in the aggregation. In addition, multiple round trips also help to improve the performance of the algorithm. The more round trips the algorithm takes to perform, the more traffic nodes are localized.

4.2 Traffic Morphing

Once the RTSD algorithm has completed its work, the morphing process can begin. Due to a fact that each entity type has its own characteristics, the process of morphing must be implemented individually for each target type. The differences among cases of different target types are indicated by operations used to morph the source entities. The set of operations is determined based on the definitions of both *src* (source) and *dst* (destination) entity types. However, there is a generic four-step algorithm which can be applied to every case of modification as follows:

1. Select all entities of *src* type from total traffic entities.
2. Use **RTSD**(*src*) to discover IP space that will be later used for morphing.
3. From the entities whose sources are localized in the step 2, select suitable ones based on the definition of the *dst* entity type.
4. Based on the definitions of both the *src* and *dst* type, determine proper operations and morph *n* source entities into *m* target ones using the operations to create a number of new *src* entities according to the user's requirement, where $n \mathrel{!=} m$ in most cases.

After selecting all entities of the *src* type, the generic algorithm invokes RTSD(*src*) to discover IP space of *src* entities. Note that only the *src* nodes are used in the run of RTSD. This certainly limits the performance of the RTSD, yet guarantees that the algorithm does not waste time on localizing side for an enormous number of superfluous sources. In Step 3, among localized entities only ones which have the *potential* to become *dst* entities are selected for further morphing. The algorithm compares properties of current entities with those in definition of *dst* entity to make decision for such selection. Next, the algorithm selects a certain set of operations to morph *src* into *dst* entities. This part is different for each *src* and *dst* types.

Table 1. Interchangeability with priorities among traffic types

MI: Mice	**AL:** Alpha	**FC:** Flash Crowd	
OM: One-to-Manny	**MO:** Many-to-One	**EL:** Elephant	**UN:** Unclassified

src / dst	MI	AL	FC	OM	MO	EL	UN
MI	0	2	0	0	0	0	1
AL	2	0	3	4	5	0	1
FC	2	0	0	4	3	0	1
OM	0	0	2	0	0	0	1
MO	0	0	0	0	0	0	1
EL	0	0	0	0	0	0	1

Not all traffic types can be subject for morphing. Based on experience, for each *dst* this paper recommends a priority list for *src* traffic types. Table 1 shows the interchangeability among types with priorities in range from 0 to 5 where 0 means it is impossible to change *src* to *dst* traffic, 1 means the *src* is the best choice for creating more *dst* traffic and 5 means *src* is the worst case. The table shows that unclassified traffic (UN) is the best choice for any *dst*. This is because changing unclassified traffic not only leaves other types intact, but also helps to convert an amount of abundant traffic to one of the types.

Because of space limitations, this paper offers the morphing examples for two cases of *dst* traffic, Elephant and Flash Crowd. Creating more Elephant entities is quite simple because no RTSD is required and only a single operation is needed while creating more Flash Crowds is the most complicated case in all target traffic types. Henceforth, the paper also uses shortened names of traffic types as in Table 1 for the purpose of limiting space.

More El: Table 1 indicates that only UN traffic can be used for creating more EL traffic. Morphing UN into EL does not require any operation that can violate the link-side constraint. Therefore, it is not necessary to run RTSD for this case. Next, to perform the selection in the step 3, only UN nodes that have many IP addresses aggregated inside and many outgoing connections are able to become EL entities therefore are chosen. Each of these entities easily becomes an EL entity by increasing sizes of packets of outgoing connections. More concretely, size of a selected UN packet is converted to a random size in the list of existing EL packets of the current protocol. Thus, changing packet sizes is the only operation used to manipulate in the step 4 of the general algorithm.

More FC: Unlike the case of EL, creating more FC traffic is complicated because of the nature of flash events. It has been stated that FC packets often arrive at servers with a rule of time [8]. This paper creates new FCs based on the flash crowd generation model proposed in [8]. According to the model, a FC has three phases in sequence, *ramp-up*, *sustained* and *ramp-down*, where each is accompanied by a function of time to calculate a number of requests to server. Hence, to create a new FC, this paper makes requests from selected sources to a selected server with request counts as well as packet arrival times are modified to conform to the model in [8]. In addition, for some cases of *src* types, many entities rather than one are used to construct a new FC. In those cases, the algorithm needs to invoke the RTSD algorithm for discovering sides for the *src* entities.

Let us turn back to the four-step procedure. Potential types in descending priority for making more FC traffic are UN, MI, MO and OM as in Table 1. Among the types, it is easiest to change MO traffic to FC. Certainly, not all MO nodes can become FC entities. Only those who have little outgoing traffic are usable. Each selected MO node can be modified to become a new FC by changing its incoming connections into incoming requests to its server. Besides, arrival times of requests must also be changed to conform to the rule in the model. Changing OM to FC can also be accomplished similarly to the case of MO, with an additional task of reversing the direction of connections. Here, reversing direction means replacing source IP address of every packet of a connection with destination address and vice versa.

Changing UN and MI traffic to FC involves more operations because an individual entity of each case is not eligible to become a new FC. Fortunately, the procedures of changing these two traffic types to FC are similar to each other. To change UN entities to new FCs, only UN nodes that have very few sources inside and little incoming traffic are eligible. Each new FC will include a server node and a list of user nodes. Then, a unique IP address inside a server node will be determined to become a real

FC server. Concerning the link-side constraint, user nodes must be selected from different sides of the server. Namely, RTSD is mandatory in these cases. Eventually to create a new FC, all outgoing connections of the server node are converted to incoming requests to the server address. Meanwhile, outgoing connections of user nodes are converted to requests destined to the server by replacing destinations of their packets and changing their packet sizes. Again, the packet arrival times need to be rearranged to conform to the time rule in the model.

5   Results and Performance

This paper uses 30 seconds of a *WIDE* packet trace [9] as a playground to analyze traffic content and modify traffic types. The modification results are shown in Table 2 in the form of traffic content distribution in terms of entity counts of traffic types before and after morphing. In regard to the original traffic content, the trace contains 3693 traffic entities. Table 2 shows that most entities are MI as its entity count is an order of magnitude greater than that of other entity types. Originally, there are 6 FCs in the input traffic. There are 4327 traffic nodes extracted from 5792 communication pairs of the nodes in the trace. Among those, 634 nodes are discovered as unclassified, corresponding to about 14% of total.

For each of three typical target types, EL, MI and FC, the table shows results of changing each of possible types separately as well as result of changing all of them at once. As mentioned before, unclassified traffic is not counted as an entity type. However, in Table 2, it is the first choice in all three cases of modification.

Table 2. Distribution of traffic content before and after morphing

Type	Before	More EL	More MI			More FC				
		UN	UN	AL	All	UN	MI	MO	OM	All
MI	3246	3017	3593	3328	3718	2929	1042	3018	3112	1124
FC	6	4	4	5	4	13	24	9	10	36
AL	157	151	145	2	1	152	148	162	146	138
OM	38	31	35	42	36	35	34	41	0	0
MO	22	21	20	23	22	22	21	0	25	1
EL	224	526	218	244	221	205	207	243	203	191

The Table 2 shows that 302 new EL entities are created by morphing UN into EL nodes. Surprisingly, not only EL traffic is affected by the modification process but all types of traffic are. In fact, morphing process affects properties of traffic nodes which results in changes in their thresholds. In case of EL, morphing UN into EL increases outgoing traffic of some UN nodes, which consequently increases incoming traffic of some other nodes as well. As a result, *max* threshold of outgoing traffic increases while its *min* threshold decreases. Similar reasoning can be applied to incoming traffic property. Eventually the changes of the thresholds explain why the entity counts of all other types decrease. This side effect, however, is diminished by the success of creating a larger number of additional traffic entities.

One can get more MI traffic by morphing UN into AL traffic. While the UN can be used to make more 347 MI entities, the AL can only help to make more 82 ones. Besides, changing UN makes entity counts of all other types decreased while changing AL only takes one FC away, yet increases OM, MO and EL counts. In case even more MI traffic is required, the maximum of 3718 entities can be created, among them 472 new entities, by choosing to morph both UN and AL into MI.

With 6 existing FCs in traffic, one can get more FCs by several morphing processes. UN traffic can be used to make 7 new FCs, MI helps to make 18 more, MO makes 3 and OM makes 4. However, if one individual source type alone does not make enough new FCs, one can pick entities from other types and use them for morphing. In case of 30 new FCs, all of possible types are operated at once. Especially, changing UN and MI both decrease all remaining types; changing MO makes more AL, OM and EL traffic but reduces MO one and changing OM reduces MI, AL and EL traffic but makes more MO.

With regard to the performance, the engineering method is implemented using database tables. For this reason, this paper evaluates the performance of the method in terms of amount of time required to complete the process, the number of times the algorithm required to read an entry or update/write it back to the database as well as the ratio of localized nodes to total nodes the RTSD algorithm can discover.

Table 3. Evaluation of performance in terms of database operations

Metrics	More EL		More MI		More FC	
	RTSD	Morphing	RTSD	Morphing	RTSD	Morphing
% Nodes picked	0	31%	0	34%	77%	37%
Reads	0	256	0	284	4048	792
Updates/Writes	0	1820	0	1488	2998	5413
Time (s)	0	51	0	47	154	143
Total Time (s)	51		47		297	

This paper uses 12-bit mask (20-bit prefix) for IP aggregation. Table 3 shows the evaluation of performance with various metrics for three cases of engineering. First, RTSD is only required in the case of FC where 77% of sources are discovered in the trace. As for the ratio of nodes selected in the step 3 of the generic algorithm presented earlier, 37% of *src* nodes are selected to create FC traffic while that of the EL is 31% and of the MI is 34%.

The morphing algorithm reads from database when it searches for nodes and updates (writes) whenever a new entity is created with a number of packets and flows are modified. From the table, number of updates is much greater than that of the reads indicating that the algorithm mainly spends time on updating data. Meanwhile, instead of updating, the RTSD algorithm writes new discovered nodes to database. With total 4327 traffic nodes, using 4 round trips to run, the RTSD can discover side for 2998 nodes among all nodes of possible types as seen in Table 3.

Finally, as for the processing time, without running RTSD, the processes of engineering EL and MI traffic only took less than 1 minute of actual time to complete. On the contrary, in the case of FC, the algorithm needs 154 seconds for the RTSD and 143 seconds for the morphing, totaling to almost 5 minutes to complete.

6 Conclusion

This paper proposes a new method which can be used to morph one or more types of traffic into one or more other types within the commonly known traffic types. The model enables to engineer traffic trace by manipulating entities under predefined regulations. The morphing process happens at packet lever but guarantees that the original IP space of the traces remains intact. In order to deliver on this promise, the process conforms to number of rules presented in this paper and used both as guidelines and as constraints in the morphing process. The results of using the proposed method in practice were also shown for a few well understood traffic types.

Results of morphing in terms of entity count for each type and its share in total count indicated that target traffic of a given type can be easily created as per user request with the help of the proposed method. The evaluation of performance showed that the engineering method can be completed in a reasonably short amount of time on fairly large traffic traces. For future work, authors plan to define new types of entities in order to allocate the share of unclassified traffic present in each trace. Secondly, while traffic entity count is the only metric mentioned in the discussion, other basic metrics like IP address count, packet count, flow count and byte count will be included in future research works in order to get a deeper insight into traffic types prior to performing the morphing process proper.

References

1. Internet Traffic Archive (ITA), http://ita.ee.lbl.gov/html/traces.html
2. Pham Van, D., Zhanikeev, M., Tanaka, Y.: Effective High Speed Traffic Replay Based on IP Space. In: ICACT, Phoenix Park, Korea, Feburary 2009, pp. 151–156 (2009)
3. Karagiannis, T., Papagiannaki, K., Faloutsos, M.: BLINC: Multilevel Traffic Classification in the Dark. In: ACM SIGCOMM, Philadelphia, USA, August 2005, pp. 229–240 (2005)
4. Szabó, G., Szabó, I., Orincsay, D.: Accurate Traffic Classification. In: IEEE WOWMOM, Finnland, June 2007, pp. 1–8 (2007)
5. Jung, J., Krishnamurthy, B., Rabinovich, M.: Flash Crowds and Denial of Service Attacks: Characterization and Implications for CDNs and Web Sites. In: WWW Conference, Hawaii, USA, May 2002, pp. 532–569 (2002)
6. Lakhina, A., Crovella, M., Diot, C.: Characterization of Network-Wide Anomalies in Traffic Flows. In: Internet Measurement Conference, Italy, October 2004, pp. 201–206 (2004)
7. Estan, C., Varghese, G.: New Directions in Traffic Measurement and Accounting. In: ACM SIGCOMM, San Francisco, USA, November 2001, pp. 75–80 (2001)
8. Pham Van, D., Zhanikeev, M., Tanaka, Y.: Synthesis of Unobtrusive Flash Crowds in Traffic Traces. In: IEICE General Conference, Ehime, Japan, March 2009 BS-4-19 (2009)
9. WIDE Project, http://tracer.csl.sony.co.jp/mawi/

Understanding Web Hosting Utility of Chinese ISPs

Guanqun Zhang[1,2], Hui Wang[1,2], and Jiahai Yang[1,2]

[1] The Network Research Center, Tsinghua University
[2] Tsinghua National Laboratory for Information Science and Technology (TNList)
Beijing, China, PA 100084
zgq07@mails.tsinghua.edu.cn, hwang@cernet.edu.cn,
yang@cernet.edu.cn

Abstract. By the end of 2008, China has 298 millions of Internet users served by 214 ISPs. But little is known to researchers about the details of these ISPs. In this paper, we try to understand web hosting utility of Chinese ISPs. After a brief introduction to ISPs in China, we use a simple and general method to uncover the web hosting information of ISPs including websites and web pages they host. We present a metric to evaluate the web hosting utility based on a new model which relies on public available data. Finally, a ranking of Chinese ISPs is given according to their web hosting utility. We analyze this ranking from the point of view such as IP allocation and geographical features. We believe that results of this work are not only helpful for researchers to understand networks in China, but also beneficial to interdomain capacity planning in China.

1   Introduction

The Internet in China is growing faster in the recent years. In 1997, there were only 620,000 Internet users according to the 1st Statistical Survey Reports on the Internet Development in China [1] published by CNNIC (China Internet Network Information Center). At the end of 2008, the number of users raised to 298 millions [2] which is almost equal to one-quarter of the whole population in China. By the end of 2008, China had become the top one country in the number of Internet users in the world.

Internet service providers (ISPs) play an important role in China Internet development. But we know little about the details of Chinese ISPs. By now, few research papers have been published on this topic due to ISP's privacy policy and huge difficulties in AS-level measurement. Most research works [3, 4] related to China Internet only introduced the status of Chinese ISPs briefly according to some published statistical reports. With some exception, Zhou [5] presented an AS-level topology of China Internet based on traceroute data probed from servers of major ISPs in China.

Researches on all aspects of Chinese ISPs are very interesting and significant because information about Chinese ISPs is desired by not only researchers but also common Internet users. For example, web hosting utility information and user accessing utility information of one ISP are needed by Internet users to decide which ISP to

C.S. Hong et al. (Eds.): APNOMS 2009, LNCS 5787, pp. 11–20, 2009.
© Springer-Verlag Berlin Heidelberg 2009

access. Companies also need this information to choose an ISP where their websites can be hosted with lower cost and fast access by their customers.

In this paper, we try to uncover the web hosting utility of Chinese ISPs. Web hosting [6] is a type of Internet hosting service that allows individuals and organizations to provide their own websites accessible via the World Wide Web. In this paper, we assume ISPs which host a large amount of popular web contents are considered high utility in web hosting. Based on this idea, a new model is presented to quantify ISP's web hosting utility in Section 3. This new model relies only on public available measurements and flexible enough to incorporate ISP's private data. The quantified value is called *WHU* (i.e. *Web Hosting Utility*). Statistically, the *WHU* value of one ISP is proportional to the average amount of web traffic sending to outside Internet users within a period of time. This value is significant for inter-AS traffic matrix estimation and peering relationship inference between ISPs. Besides the WHU value, the number of websites and web pages one ISP hosted can also be used to complementally describe the web hosting utility of that ISP.

In order to avoid huge difficulties in data measurement, similar with paper [7], we use a measurement method to uncover the web hosting utility of ISPs based on search engines and hot search keywords. However, our method pay more attention to the actual Internet conditions in China, and it is easier to understand and realize. Two search engines are employed in our method: *Baidu*, the leading Chinese language search engine in China, and *Google*, the most famous search engine in the world. Hot keywords are consulted from search engines' websites, and most of them are Chinese characters.

With the results obtained from our experiment, we rank the Chinese ISPs according to the number of websites and web pages they hosted as well as the WHU value. We also analyze this ranking from the point of view such as IP allocation and geographical features. To our best of knowledge, this is the first report about the web hosting utility information of Chinese ISPs. We believe that the results of this work are not only helpful for researchers to understand networks in China, but also beneficial to interdomain capacity planning in China.

The paper is organized as follows. In Section 2 we introduce the basic information about Chinese ISPs. In Section 3, we describe our new model and the measurement method used. The results of our experiment are presented in Section 4 followed by the result analysis in Section 5. We conclude our work in Section 6.

2   ISPs in China

2.1   AS Numbers Allocated to China

APNIC provides IP resource allocation and registration services which support the operation of the Internet in Asia-Pacific area. According to our statistics of AS allocation related data published by APNIC [8], 433 AS numbers (ASNs for short) have been allocated to China. But 55 of 433 AS numbers are not in use at present. In addition, 33 of 433 ASNs are allocated to CNGI (China Next Generation Internet) project which is a set of interconnected IPv6 experimental networks. As CNGI is still in experimental and research stage instead of a commercial Internet and few websites are built on it. For this reason, we will not take these 88 (55+33) ASes into account in inferring the web hosting utility information.

2.2 Status of Chinese ISPs

The traditional description of an ISP is a company that offers its customers to access to the Internet. In this paper, we define ISP as an organization which has at least one AS number and provides at least one of three Internet service, including user access, lower tier ISP transit and web hosting.

We match ASNs with their corresponding ISPs by collecting AS descriptions from APNIC WHOIS database. According to our statistics, there are 214 ISPs in China. This number is far beyond our imagination. With a little surprise, we find only 108 of 214 are companies which mainly make profit from providing Internet service. The rest 106 ISPs are allocated to broadcasting & TV companies, oil companies, manufacturing companies, sports companies and so on. Obviously, these 108 ISPs are expected to have high utility as web service providers.

Unfortunately, little is known about the details of ISPs. The majority of Internet users in China are only familiar with the four most famous ISPs (so-called *big four*): China Telecom, China Mobile, China Unicom and China Education and Research Network (CERNET). They are all state-owned and the business scope of which cover the whole China. However, the rest of 210 ISPs are strange to us. So it is very interesting and significant to uncover the information of small ISPs in China.

2.3 Imbalanced Allocation

At present, 214 ISPs in China totally have 378 ASNs in use. But this allocation is not balanced. Most of ASNs are allocated to several famous ISPs. For example, the *big four* totally have 112 ASNs which is nearly 30% of the total ASNs of China. In contrast, 86% of ISPs only have one AS number. Table 1 presents the number of ASNs and IP addresses allocated to *big four*.

Table 1. AS numbers and IP address numbers allocated to *big four*

ISPs	# of AS	Percentage (%)	# of IP address [3]	Percentage (%)
China Telecom	24	6.3	65,490,944	36.1
China Mobile	15	4.0	21,331,968	11.8
China Unicom	23	6.1	37,471,232	20.7
CERNET	50	13.2	13,560,320	7.5
total	112	29.6	137,854,464	76.1

With no doubt, the *big four* occupy most of AS and IP resources in China. This imbalanced resource allocation will certainly lead to an imbalance in web contents distribution of ISPs. What we concern is how this imbalance affects the web hosting utility of ISPs and whether this is the only influence factor. We will give answers to these questions in section 6.

3   Evaluation Model and Measurement Method

3.1   Model to Quantify Web Hosting Utility

The number of websites and web pages can be used to describe the web hosting utility of one ISP, but it is infeasible to obtain the two numbers metrics and accurately. In addition, to evaluate the web hosting utility of ISP, it is better to use a single metric than two. Therefore, we define WHU as the only metric instead of websites or web pages. The WHU values of ISPs are derived from our new model described below. To our best knowledge, this is the first criterion to quantify the web hosting utility of ISPs accurately. The basic idea of our model is: ISPs that host a large amount of hot web contents are considered to have high utility in web hosting. To determine popular web contents on the Internet, we use search engines and hot keywords to help us. In our model, hot web contents are represented by search results returned by a search engine. Specifically, $P(k,u)$ is defined to denote a piece of hot web content determined by keyword k and URL u returned by a search engine.

Before presenting the quantitative model, we first define a set of related notations. We use K to denote the keywords set collected. $N(k)$ is defined as the number of searches for keyword k. This value reflects the popularity of a keyword. We define $U(k)$ as a set of matched URLs returned by search engines when performing a query using keyword k. $C(u)$ denotes the clicked rate of the web page represented by URL u. This value reflects the popularity of a web page. $S(u)$ is the size of web page represented by URL u. $W_{sa}(u)$ denotes the web server address of URL u.

The hotness of web content $P(k,u)$ is defined by the following equation

$$H(P(k,u)) = N(k) \times C(u) \quad \text{where } k \in K, u \in U(k) \tag{1}$$

where $H(P(k,u))$ is the hotness of web content. We use $W(P(k,u))$ to denote the weighted size of web content $P(k,u)$.

$$W(P(k,u)) = H(P(k,u)) \times S(u) = N(k) \times C(u) \times S(u) \quad \text{where } k \in K, u \in U(k) \tag{2}$$

With the definition above, we define $WHU(x)$ in equation (3) to denote the web hosting utility of ISP x.

$$WHU(x) = \sum_{k \in K, u \in U(k)} W(P(k,u)) \quad \text{where } W_{sa}(u) \in I_P(x) \tag{3}$$

$I_P(x)$ is an IP address set allocated to ISP x. With equations (1) ~ (3), we summarize the steps taken to compute the web hosting utility $WHU(x)$ for every ISP x in Algorithm 1.

Algorithm 1. Computation of $WHU(x)$

---

1. initialize $WHU(x)$ to 0 for every ISP x
2. for each keyword k in keywords set K

3.	for each URL u in URLs set $U(k)$
4.	$whu = N(k) \times C(u) \times S(u)$
5.	extract web server address wsa from u
6.	map wsa to ISP isp
7.	$WHU(isp) = WHU(isp) + whu$

3.2   Measurement Method

In order to compute the WHU value of ISPs described in Algorithm 1, we use a simple and general measurement method. There are four steps in our method:

1. Collecting hot keywords. Hot keywords and number of searches are collected from *top.baidu.com*, a website of *Baidu*.
2. Obtaining matched URLs. For each keyword collected in step 1, we query *Google* and *Baidu* to retrieve a set of most closely matched URLs.
3. Extracting web server address from URLs in step 2 by querying DNS servers.
4. Mapping URLs (web content) to ISP hosted according to web server address. We obtained the IP addresses allocated to each ISP in China according to AP-NIC data. Differently, Chang [7] achieved this mapping using BGP tables.

Compared with Chang's method in [7], the measurement method we used in this paper is simpler and more feasible in China. Employing two search engines, we collected more matched URLs. This may help us to uncover small ISPs with fewer web contents. We also avoided difficulties in obtaining BGP routing tables.

3.3   Discussion

There are five unknown parameters in Algorithm 1. Three of them can be straightly obtained from the measurement described above. These three parameters are: K, $N(k)$ and $U(k)$. The two parameters left are $S(u)$ and $C(u)$.

$S(u)$ consists of two parts: the size of web page and the size of embedded objects. Embedded objects cannot be ignored as they contribute a large amount of web traffic. We first select ten famous video sharing websites which totally occupied over 90% market in China [9]. Then we crawl the keyword-retrieved URLs individually and extract the embedded objects from the web pages which belong to the top 10 websites.

$C(u)$ represents the click rate of URL u. In general, search results with high rank are more frequently clicked. This is a user search behavior habit. We download a 30-day user query logs from the website of Sogou [10], a famous search engine in China. In order to find the relationship between clicked rate and rank of an URL, we plot a log-log graph using 30-day Sogou data in Figure 1(a). The horizontal axis represents the rank of URL, the vertical axis represents the log-value of clicked rate: the ratio of clicked times on that rank to the total clicked times.

Figure 1(a) shows that the clicked rates of URLs on the first result page (10 results per-page default) are much higher than results on other pages. In Figure 1(b), we use two straight lines to fit the log-log relationship shown in Figure 1(a). Points in Figure 1(b) present the average values of 30-day data. For the two straight lines,

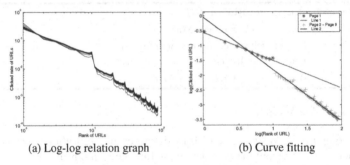

(a) Log-log relation graph   (b) Curve fitting

Fig. 1. Relationship between clicked rate and rank of an URL

Table 2. Fitting results

	a	b	SSE	R-square	RMSE
Line 1	-0.9282	-0.5779	0.0107	0.9866	0.0365
Line 2	-1.7390	-0.0732	0.1732	0.9879	0.0471

we use equation $y=ax+b$, where x and y stand for the horizontal axis and vertical axis in Figure 1(b) respectively. Line 1 fits for the top 10 results while Line 2 fits for the rest 80 results. The fitting results are listed in Table 2 where parameters a, b stand for the slope and the ordinate respectively.

$C(u)$ is computed with two equations listed in Table 2. So far, all the inputs needed in Algorithm 1 can be derived from public data and public measurements which mean our model can work without any ISP's proprietary data.

4   Experiment and Results

In our experiment, we totally collected 1158 hot keywords. These keywords cover most of areas including politics, economics, business, sports, entertainments, and et al. We selected top 50 matched URLs for each hot keyword from *Baidu* and *Google*, totally 115,800 URLs. With the measurement method described in section 3.3, we identified 73 different ISPs and 96 ASes in China.

4.1   Websites and Web Pages Hosted by ISPs

In our experiment, we obtained 28591 websites and 107175 web pages distributed on 73 ISPs. Table 3 lists the top 10 ISPs by number of websites and web pages. From Table 3, we can see that the top 3 ISPs have huge advantages in number of websites and web pages compared with other ISPs. With no doubt, they are the top 3 ISPs providing web hosting service in China.

4.2   Web Hosing Utility of ISPs

Using Algorithm 1, we quantified the web hosting utility for 73 ISPs in China we identified. Table 4 lists the top 5 and the bottom 5 ISPs as well as their WHU values.

Table 3. Top 10 ISPs by number of websites and web pages

Rank	ISP	# of Websites	Rank	ISP	# of Web-Pages
1	China Telecom	12936	1	China Telecom	37236
2	China Unicom	6011	2	China Unicom	31691
3	CERNET	2312	3	CERNET	16089
4	China Networks Inter-Exchange	877	4	China Abitcool	7628
5	China Mobile	820	5	China Networks Inter-Exchange	2664
6	Beijing DianXinTong	408	6	China Mobile	2213
7	China Abitcool	388	7	Beijing DianXinTong	1760
8	ZhengZhou GIANT	208	8	Beijing Net Infinity	1219
9	Alibaba	169	9	Beijing Zhongguancun	1028
10	Beijing Jingxun	160	10	Beijing Jingxun	1003

Table 4. Top 5 ISPs and bottom 5 in China according to their WHU values

Rank	ISP	WHU Value
1	China Unicom	80050.34
2	CERNET	56862.35
3	China Telecom	42657.64
4	China Abitcool	26600.40
5	China Networks Inter-Exchange	10382.40
69	China Ministry of Science and Technology	0.0283
70	Great Wall Broadband	0.0165
71	China State Post Bureau	0.0127
72	Shenzhen Information and Network Center	0.0062
73	China Cultural Heritage Information and Consulting Center	0.0022

Table 5. Top 10 websites in total page size

Rank	Website	Type	Page Size (KB)	ASN	ISP
1	tudou.com	video sharing	2867452	4837	China Unicom
2	sina.com.cn	general	2760460	4538	CERNET
3	youku.com	video sharing	1916139	9308	China Abitcool
4	56.com	video sharing	1047313	4837	China Unicom
5	sohu.com	general	834272	4538	CERNET
6	qq.com	general	275602	4538	CERNET
7	baidu.com	search engine	273854	9308	China Abitcool
8	ku6.com	video sharing	42422	23724	China Telecom
9	amazon.cn	network shopping	40735	9298	Beijing Net Infinity
10	6.cn	video sharing	39652	4134	China Telecom

From the ranking list, we can see that there is a huge distance between the top 1 and the bottom 1 ISP on web hosting utility. The top 5 ISPs are all networking companies which mainly make profit from providing Internet service. In contrast, 3 of the bottom 5 ISPs belong to government departments. Although China Telecom hosts more websites and web pages than China Unicom and CERNET shown in Table 3, it has lower WHU value than China Unicom and CERNET. In order to explain this puzzle, we list the top 10 websites in Table 5 ranked by total page size they host. We also show the ASNs and their ISPs that host the websites.

From Table 5 we can see that the top 10 websites belong to 5 different ISPs respectively. Three top general websites locate in CERNET and two big video sharing websites locate in China Unicom. In contrast, China Telecom only has two small video sharing websites. Some frequently-accessed websites deploy web servers in several ISPs to speed up user access. In this paper, the DNS server we used to extract web server IP addresses from URLs locate in CERNET. This may leads to the missing of some web server IP addresses for some large websites.

5   Discussions

From the results listed in Table 3 and Table 4 we notice that the distribution of web content is imbalanced among ISPs in China. The reason for this imbalanced distribution may be various. In this section, we try to interpret the imbalance from the point of IP allocation and geographical features.

5.1   IP Allocation

We draw the WHU values and the number of IP addresses of 73 ISPs in Figure 2. Here, the amount of IP is not the total IP addresses allocated to an ISP, because some ASs of an ISP do not host any website or they are not identified by our experiments. In other words, we only count in IP addresses allocated to 96 ASs.

Though the number of IP addresses is not decreasing with WHU values proportionally, the correlation coefficient of two lines in Figure 2 is 0.76. It means that IP allocation may be a possible cause for the imbalanced distribution.

We define the density $D(x)$ of an ISP x as the following equation and draw the density curve of 73 ISPs in Figure 3.

$$D(x) = 1000 \times WHU(x) / |I_P(x)|$$

(4)

In Figure 3, some ISPs have extremely high density value such as ISPs ranked in 7 and 11. We call them the dense ISPs as they host larger amount of hot web contents with fewer IP addresses. We list the top 5 dense ISPs in Table 6.

Table 6. Top 5 dense ISPs in China

Rank	WHU rank	ISP	Density
1	7	Beijing Zhongguancun	130.81
2	11	ZhengZhou GIANT	107.15
3	8	Alibaba	106.50
4	4	China Abitcool	43.42
5	15	Beijing Gu Xiang	22.29

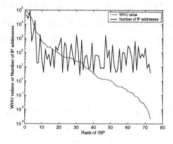

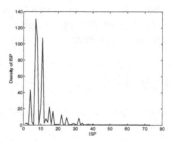

Fig. 2. Number of IP address vs. WHU values   **Fig. 3.** The density line

The top 1 ISP in Table 6 hosts an IT products information website www.zol.com.cn. And Alibaba hosts an online shopping website www.taobao.com. Nowadays, IT products and online shopping are warmly welcomed in China. These two websites contribute most of the WHU value of Zhongguancun and Alibaba. From this point of view, the WHU of an ISP is also related to the kind of websites it hosts.

5.2 Geographical Features

In China, regional economic development and Internet development are imbalanced [12]. Corresponding to this imbalance, China Telecom and China Unicom allocate ASNs for some economically developed areas to promote the Internet development in these areas. We classify AS numbers which belong to the above two ISPs with their geographical position. We try to find the effect of regional difference in imbalanced distribution of hot web contents. Table 7 lists the top 5 regions in China as well as ASNs they have.

From Table 7 we notice that the distribution of hot web contents is quite related to geographical features. Beijing, Guangdong and Shanghai are not only developed in

Table 7. The WHU value of different regions in China

Rank	Region	ASN	ISP	WHU
1	Beijing	4808, 17620, 23724	China Unicom, China Telecom	24992.15
2	Guangdong	17623, 17816,17622	China Unicom	7935.85
3	Shanghai	17621,4812	China Unicom, China Telecom	4572.08
4	Hebei	17672	China Telecom	249.68
5	Tianjin	17638	China Telecom	135.87

Table 8. GDP data and Internet resource occupancy of the top 3 regions

Region	Per Capita GDP (¥) [9]	Popularity [3]	Domain Names [3]	Websites [3]
Beijing	58204	60.0%	21.4%	12.9%
Guangdong	33151	48.2%	11.3%	15.0%
Shanghai	66367	59.7%	6.5%	6.2%
Nation average	18934	22.6%	3.3%	3.3%

economy but also strong in Internet construction. We list the statistical data on Internet and economy of the top 3 regions in Table 8. Popularity in Table 8 stands for penetration rate of Internet in each region. Numbers in column 4 and 5 is occupancy.

6   Conclusions

The detailed information of Chinese ISPs is significant not only for researchers but also for operators of individual ISPs. In this paper, we focus on the web hosting utility of ISPs. We present a new metric WHU to evaluate the web hosting utility of an ISP based on a simple and general measurement method. We rank Chinese ISPs according to the WHU values as well as number of websites and web pages. The problem of imbalanced distribution of hot web contents among ISPs is serious. The imbalance of IP allocation and regional Internet development in China are partial causes for this problem. In this paper, web server IP addresses for some large websites may miss because we only use the DNS servers located in CERNET. As part of future work, we try to add more measurement points in different ISPs to resolve this problem.

References

1. The First Statistical Survey Reports on the Internet Development in China,
   http://www.cnnic.net.cn/download/manual/en-reports/1.pdf
2. The 23rd Statistical Survey Reports on the Internet Development in China,
   http://www.cnnic.net.cn/uploadfiles/pdf/2009/3/23/153540.pdf
3. Randolph, K., Chen, Y.: The Internet in China: A Meta-Review of Research. The Information Society 21(4), 301–308 (2005)
4. Jonathan, Z., Enhai, W.: Diffusion, Use, and Effect of the Internet in China. Communications of the ACM 48(4), 49–53 (2005)
5. Shi, Z., Guo, Z., Guo, Z.: Chinese Internet AS-level Topology. IET Communications 1(2), 209–214 (2007)
6. Borka, J.: Web Hosting Market Development Status and Its Value as an Indicator of a Country's E-Readiness. Telecommunications Policy 32(6), 422–435 (2008)
7. Hyunseok, C., Sugih, J., Morley, M., Walter, W.: An Empirical Approach to Modeling Inter-AS Traffic Matrices. In: IMC 2005, ACM, Berkeley (2005)
8. APNIC Reports and Statistics,
   http://www.apnic.org/info/reports/index.html
9. China Websites Ranking,
   http://main.chinarank.org.cn/statistics/hot_vid.html
10. Sogou User Query Logs, http://www.sogou.com/labs/dl/q.html
11. 2008 China Statistical Yearbook,
    http://www.stats.gov.cn/tjsj/ndsj/2008/indexeh.htm
12. He, L., Gui, L., Le, Q.: Spatial-Temporal Analysis of Regional Disparities of Internet in China. Chinese Geographical Science 14(4), 314–319 (2004)

Internet Application Traffic Classification
Using Fixed IP-Port

Sung-Ho Yoon, Jin-Wan Park, Jun-Sang Park, Young-Seok Oh,
and Myung-Sup Kim

Dept. Of Computer and Information Science, Korea University, Korea
{sungho_yoon,jinwan_park,junsang_park,youngsuk_oh,
tmskim}@korea.ac.kr

Abstract. As network traffic is dramatically increasing due to the populariza-
tion of Internet, the need for application traffic classification becomes important
for the effective use of network resources. In this paper, we present an application traffic classification method based on fixed IP-port information. A fixed
IP-port is a {IP, protocol, port} triple dedicated to only one application, which
is automatically collected from the behavior analysis of individual applications.
We can classify the Internet traffic accurately and quickly by simple packet
header matching to the collected fixed IP-port information. Therefore, we can
construct a lightweight, fast, and accurate real-time traffic classification system
than other classification method. In this paper we propose a novel algorithm to
extract the fixed IP-port information and the system architecture. Also we prove
the feasibility and applicability of our proposed method by an acceptable experimental result.

Keywords: Traffic monitoring and analysis, Traffic classification, Application
identification, Fixed IP-port.

1   Introduction

Nowadays, network traffic is rapidly increasing because of the increase of Internet
users and high speed network links. It is caused by not only traditional Internet service such as WWW, FTP, and e-mail, but also development of various multimedia
services: for example, multimedia streaming, P2P(peer-to-peer) file sharing, gaming,
and so on. Consequently, the total Internet traffic volume has rapidly grown, which
makes the need of traffic monitoring and classification inevitable for efficient network
management [1, 2].

Network traffic classification is a series of processes to capture packets from a target network link and determine the identity of packets from the perspective of categorization. For example, it is to figure out the corresponding application names of the
captured network packets. The result of classification can be effectively utilized on
network management and control. We can accurately block or shrink the traffic from
applications individually. Also, it can improve the understanding of network user's
activity. Traffic classification can be carried out under various categorization perspectives such as application-layer protocol, traffic types, application programs, and so on.

C.S. Hong et al. (Eds.): APNOMS 2009, LNCS 5787, pp. 21–30, 2009.
© Springer-Verlag Berlin Heidelberg 2009

Previously, many methodologies have been proposed for accurate traffic classification such as well-known port-based, signature-based and machine learning-based, and so on. All above methods, however, have some limitations in applying them to classifying the traffic on a real operational network. Well-known port-based method using IANA port information cannot classify some applications which use dynamic or unknown ports [3]. Signature-based [4] method requires formerly created payload signatures and cannot be applied to the encrypted payloads. Machine learning-based method [5] also has limitation such as requisite of prior learning and difficult of adaptation to newly emerging application. The weakest point of last two methods is much processing overhead in applying to real-time classification system.

This paper proposes an application traffic classification method based on fixed IP-port information. A fixed IP-port is a {IP, protocol, port} triple dedicated to only one application, which is automatically collected from the behavior analysis of individual applications. A fixed IP-port for an application is absolutely accurate because it is extracted from end hosts where the application process is actually running to create traffic. It can overcome the limitations of the currently existing methods. We can classify the Internet traffic effectively and quickly by simple packet header matching to the fixed IP-port information. Therefore, we can construct a lightweight, fast, and accurate real-time traffic classification system than other classification method. In this paper we propose a novel algorithm to extract the fixed IP-port information and the system architecture. Also we prove the feasibility and applicability of our proposed method by an acceptable experimental result.

The rest of the paper is organized as follows. Section 2 describes the definition of the fixed IP-port. The methodology to automatically extract the fixed IP-port for each application is described in section 3. Section 4 describes the traffic classification system based on the fixed IP-port. In section 4, an experiment and result are presented. Section 5 summaries this paper and shows possible future work.

2   The Fixed IP-Port, What Is It?

A fixed IP-port is defined as a {IP address, port number, transport protocol} triple which is dedicated to a single application only. Actually it is server socket information which is open to the public for a service delivery. For example, in case an application requires a user authentication before providing service, the login server's {IP, protocol, port} triple is to be one of the fixed IP-ports of that application.

2.1   Motivation of the Fixed IP-Port

The port number based traffic classification is no longer accepted on current highly dynamic network environment. However we believe that {IP, protocol, port} based method might be acceptable, because most of the current popular Internet applications have at least one dedicated server and port for the application. The best example of this is the login server and port for user authentication. In some circumstances, we have to divide the traffic with the same application protocol into each specific service or application. In this case the {IP, protocol, port} information can be utilized effectively, because each service is provided from different servers with different IPs even though they use same protocol and port number.

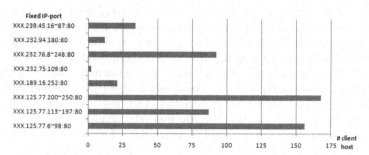

Fig. 1. The Fixed IP-ports of the gom.exe from Gretech Inc

To convince ourselves we made a simple investigation on our campus network. We selected one application called Gom (gom.exe) [7] from Gretech Inc. The Gom is a multimedia player which is popular in Korea and Japan. We collected all TCP and UDP connections from the Gom during 7 days from our campus network using KU-MON system and analyzed the remote IPs and ports. The KU-MON is a real-time traffic monitoring system developed by our research group.

Fig. 1. shows the number of local hosts having a connection to the remote {IP, port} pairs when the host plays the Gom application. All of the remote {IP, port} pairs listed in Fig.1 are not used by other applications. The {IP, port} pairs in Fig. 1 are used only Gom applications. The result shows that there are some {IP, port} pairs used by Gom application only. Some of them appear every time the Gom application is running while other appears irregularly. This result makes us confirm the existence of fixed IP-port. BLINC [3] explains these {IP, port} pairs as "farm".

2.2   The Fixed IP-Port Based Traffic Classification

If we have a collection of fixed IP-port triple for every application, by simple packet header matching we can easily identify the application name of every traffic data. But it is impossible because of the dynamically assigned port numbers and P2P applications.

However, if we can find at least one fixed IP-port triple for each application, we can determine that an application is running on the host or not. We know all the applications running on a host by this simple matching. For the non-determined traffic flows we can easily develop a method to determine the application name, because that should be one of the applications on the host and we know all of the applications running on the host.

How to collect the fixed IP-port triple for every application? In this paper we propose a novel method to automatically extract the fixed IP-port information by the flow behavior analysis of individual applications.

3   Fixed IP-Port Extraction Algorithm

In this chapter we describe the proposed fixed IP-port extraction algorithm. The Input to this algorithm is a set of flows determined the target applications, and the output is a set of fixed IP-port triples. The output could be nothing if the target application does not have one. The proposed algorithm is applied to each application separately.

The key of the proposed algorithm is to select the fixed IP-port triples among all the IP-port triples appear in the flow records of a target application. To achieve this goal, we distinguished IP-port triples of an application into three types: permanent IP-port, temporal IP-port, and dynamic IP-port. The permanent IP-port is an IP-port triple which is permanently used for the target application only. The best example of this is the login server's IP-port triple. The dynamic IP-port is an IP-port triple which is dynamically selected one and never shared by other flows. An example of this type is the client's IP-port triple in HTTP connection. The temporal IP-port is an IP-port which is a dynamic one, but is shared by other flows for certain time interval. An example of this type is the temporal P2P host IP-port, which is used only small amount of time for the P2P application and used by other application later.

The dynamically selected IP-ports should be eliminated while the permanently selected IP-ports should be defiantly selected as the fixed IP-port. The temporally selected IP-ports should be effectively un-selected, which is not the fixed IP-port in our definition.

3.1   Considerations

The fixed IP-port Extraction methodology proposed in this paper should consider the follows three performance metrics.

Coverage: We have to find the fixed IP-port triples as many as possible. Our methodology should find at least one fixed IP-port for each application.

Accuracy: The selected fixed IP-port triples should be accurate enough which minimize the FP(false positive) and FN(false negative) [6].

Overhead: We have to minimize the memory and processing overhead of the fixed IP-port extraction system. Among huge amount of IP-port triples it is important to eliminate the dynamic IP-port triples as quickly as possible to reduce memory space and processing burden.

3.2   The Fixed IP-Port Extraction Algorithm

Fig. 2 describes the overall algorithm for the proposed fixed IP-port extraction as a state transition diagram. These are the most important in this paper.

All IP-port triples exist in one of the five states (Discard, New, Candidate, Fixed, Time_wait). The IP-port triple in the Fixed state is the fixed IP-port for the target application. As aforementioned this extraction algorithm executes separately for each target application. The input to the algorithm for an application is all the flow records whose application names are determined.

Initially all possible IP-port triples exist in the Discard state, which means there is no fixed IP-port for a target application. Actually no information is stored in the Discard state. If an IP-port appears in a flow record, the IP-port moves to the New state. If an IP-port satisfies some specific conditions, it moves to the Candidate state and the subsequent Fixed state. The IP-ports in the Fixed state are the fixed IP-port for the corresponding application, which is used for the traffic classification.

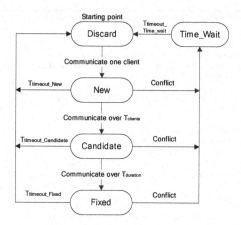

Fig. 2. The State Transition Diagram for Fixed IP-port Extraction

In the New, Candidate, and Fixed states, we prevent that an IP-port remains in the states forever using conflict and timeout check. The Time_wait state exists for a conflicted IP-port with other application to reenter the New state again just after eliminated. This causes high system burden because the reentered conflicted IP-port might be eliminated again soon by another conflict flow record.

We have to consider several conditions for the state transition. We use three threshold values for state transition conditions as follows.

$T_{clienthost}$ It is number of different client hosts to a single IP-port triple. If one IP-port is fixed IP-port, all client flows using the IP-port belong to same application. Therefore, an IP-port triple having over $T_{clienthost}$ clients can be determined as the fixed IP-port. If $T_{clienthost}$ is increase, accuracy, also, increase, but, completeness decreases.

$T_{duration}$ It means usage time (last usage time – registered time). If we use only $T_{clienthost}$ to determine the fixed IP-port, we can make a mistake to determine a temporal IP-port such as P2P application as fixed IP-port. However, we can distinguish these temporal IP-ports by using usage time. The feature of P2P application is short usage time. Therefore, we make $T_{duration}$ and use it to extract real fixed IP-port.

$T_{timelimit}$ Initially, all IP-ports wait in memory until excess $T_{clienthost}$ and $T_{duration}$. Unfortunately, we cannot store all IP-port triples forever due to the limit of physical memory. It, also, causes fail of real time classification system because of system overhead. Therefore, some IP-ports which over time limit without status change should be removed.

These thresholds make our extracting algorithm flexible and efficient in various environments. The main point in Fixed IP-port Algorithm is the transition among states in the state transition diagram. The proposed algorithm implements according to Fig. 3.

The code is divided in two parts. First part is to register new IP-port to the IPT as a New state or to update the condition of the existing IP-ports according to the new flow records: (conflict, number of client, and last usage time). If new flow record does

1:	**procedure** Fixed IP-port Extraction for an Application A
2:	IPT ← IP-port Table
3:	
4:	**for** each input flow record **do**
5:	search the corresponding IP-port record in IPT
6:	**if** found in IPT **then**
7:	update the status of the record;
8:	**else** add a new record with the state as **New**;
9:	**end for**
10:	
11:	**for** each flow record in IPT **do**
12:	Check the state of the record
13:	**if** state == **New then**
14:	**if** conflict **then** move to **Time_Wait**;
15:	**else if** over **Tclienthosts then** move to **Candidate**;
16:	**else if** over **Ttimelimit then** move to **Discard**;
17:	**else if** State == **Candidate then**
18:	**if** conflict **then** move to **Time_Wait**;
19:	**else if** over **Tduration then** move to **Fixed**;
20:	**else if** over **Ttimelimit then** move to **Discard**;
21:	**else if** State == **Fixed then**
22:	**if** conflict **then** move to **Time_Wait**;
23:	**else if** over **Ttimelimit then** move to **Discard**;
24:	**else if** State == **Time_Wait then**
25:	**if** over **Ttimelimit then** move to **Discard**;
36:	**end for**
37:	
38:	**end procedure**

Fig. 3. The Pseudo Code for the Fixed IP-port Extraction Algorithm

not exist in the IPT then a record is added to the IPT. If the existing record and new record are belongs to different application, the conflict value is checked. The new and existing record belong to the same application, the usage time is updated.

The Second part is to transfer the state of IP-port record in the IPT according to status changed in the previous step in the status transition diagram. If a record in any state is not accessed over the $T_{timelimte}$ time, the record moves to the Discard state. If a record is conflicted with other application then it moves to the Time_Wait state. If a record is accessed over $T_{clienthost}$ during over $T_{duration}$ time, then the record moves to the Fixed state. The IP-port triples in the Fixed state are the fixed IP-ports of the application as aforementioned.

4   The Traffic Classification System Based on the Fixed IP-Port

Fig. 4. shows the fixed IP-port based traffic classification system we developed in our campus network. This system consists of three subsystems: the fixed IP-port extraction system, the application traffic classification system, and the traffic verification system. It is a part of our KU-MON system.

The fixed IP-port extraction system is the realization of the proposed fixed IP-port extraction algorithm, which is described in chapter 3. The traffic classification system classifies Internet traffic into corresponding applications by simple header matching using the fixed IP-port records. The traffic verification system verifies the classification results in various points of view.

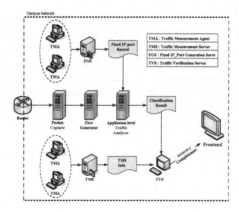

Fig. 4. The Fixed IP-port Extraction System Structure

4.1 Fixed IP-Port Extraction System

The fixed IP-port extraction system uses the TMA(Traffic Measurement Agent) [6] records and flow data as its input data. TMA is a tiny agent program which is installed in end-hosts, gathers network socket information and periodically sends it to the gathering server. Whenever a new socket is created by an application process, the TMA gathers the socket information, and sends them to the FGS and TMS. We installed TMAs at about 300 end-hosts in our campus network. TMA records from about 100 end-hosts are used by the fixed IP-port extraction system and other TMA records are used by the traffic verification system.

The fixed IP-port extraction system generates a ground truth flows from the TMA records and flow records. Based on the ground truth flows the system extract fixed IP-port according to the aforementioned algorithm.

4.2 Traffic Classification System

This system is simple. We can classify traffic with flow-based real-time classification because fixed IP-port is lightweight. Firstly, packets are captured from target network link (Packet Capture). And then, we make them into flows (Flow Generator). In the Application-level Traffic Analyzer, we classify traffic according to fixed IP-port by compare with just flow header which contains 5-tuples. Therefore, it is very lightweight and simple.

4.3 Traffic Verification System

The verification is the measurement of accuracy of classification by comparing ground truth with the result of the classification algorithm. So, the most important is to create correct ground truth. Previous papers used various ground truth data for verification of their algorithm. BLINC[3] uses result of payload-base signature classification; and Constantinou et al.[8] used result of well-known port-based classification. These ways to create ground truth data have some weakness. They used uncertain ground truth, result of other classification method, for valid their algorithm.

These are not suitable for correct verification. To make up previous uncertain way for creating ground truth, Szabó et al [9] proposed better way that installs driver in end host, creating traffics, and checks identification of all user traffic. It seems be cover the shortage of ground truth. However, it cannot actually create enough ground truth because of the need of installation on all end-hosts. It causes large overhead to end-hosts, so, they would be reluctant to this. In this paper we propose verification using TMA. For correct verification, we divide TMA group into two groups, one is for creation of Fixed IP-port, and another is for verification of our algorithm because we use TMA on both extraction and verification.

Correct ground truth is important, in a same manner, it is also important to define evaluation parameters for algorithm verification. This paper shows accurate evaluating parameters for verification of classification algorithm. Previously, [10] defines above some parameters, but we add more to make accurate and detail.

The parameters consist of three parts such as coverage, accuracy, and completeness. First of all, Coverage means the number of application identified by the algorithm. Second parameter is accuracy. It means how accurate the result comparing with ground truth. Accuracy might have various ranges of verification according to the range of ground truth data. Accuracy consists of overall accuracy and individual accuracy. Especially, individual accuracy has FP(False Positive) and FN(False Negative). The FP of application X means that the algorithm classifies that some traffic is application X, but it is false. The FN of application X means that the algorithm classifies that some traffic is not application X, but it is false. Especially, FN divides FN-Unclassified and FN-Mis_Classification. The former is that the algorithm cannot classify application X. The latter is the algorithm classifies application X to other, incorrectly. [10] states that the latter (FN-Mis_Classification) is more harmful in network management. The last one is completeness which means the rate of traffic classified by the algorithm.

The Frontend of the verification system provides the following information to network managers. First of all, it shows us about basic information of the traffic. Secondly, it also shows evaluation of the algorithm such as completeness. Last is result of verification. All three parts are reported on Web every minute. So, we can check the algorithm.

5   Experiment and Results

We applied the proposed the Fixed IP-port extraction algorithm in our campus network traffic. We extract the fixed IP-ports during 12 days, and then classified traffic during the next whole day.

We determined the threshold values during the first 6 days of our 13 day experiment. To get efficient threshold values, we checked the state remaining time and P2P server usage time. Fig. 6. represents cumulative distribution function (CDF) of NC (candidate register time – new register time), New state remain time, and usage time of IP-port using P2P application. According to CDF, to eliminate dynamic and temporal IP-ports, $T_{duration}$ is set to 24 hours. To reduce the overhead of remaining records, $T_{timelimit}$ is set to 48 hours. $T_{clienthost}$ is set to 2 for high coverage.

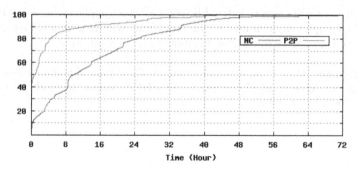

Fig. 6. CDF graphes for state usage time to get $T_{timelimit}$ and $T_{duration}$

Table 2. The result of traffic classification using Fixed IP-port

Completeness	28.8%(flow)
Accuracy	95.39%(flow)
Overhead	231,705(fixed)/238,543(total)
Coverage	346(fixed)/354(total)

We examined total 354 applications to extract fixed IP-ports. As a result, we extracted fixed IP-ports in 346 applications (about 98%). The total number of fixed IP-port extracted was 231,705.

Table 2. also shows the result of traffic classification using these Fixed IP-port and its accuracy. It shows the high accuracy but somewhat low completeness of the result, which means that most of applications use dynamically selected ports which should be determined using other classification method.

We extract the fixed IP-ports in most of all application. But some applications like a traditional client-server applications (ftp, ssh, mail) does not have fixed IP-port while most of modern applications does. Concerning the overhead, most IP-port remained in the Candidate state moved to the Fixed state in our system, which indicates the low overhead of our system.

6   Conclusion and Future Work

Real-time Internet traffic classification is inevitable for the efficient traffic management of network. Many of the currently existing methodologies for traffic classification have some difficulties in adopting in real operational network because of its complexity. So, this paper proposed a fixed IP-port based traffic classification approaches. A fixed IP-port is a {IP, protocol, port} triple dedicated to only one application, which is automatically collected from the behavior analysis of each applications. In this paper we proposed a novel algorithm to extract fixed IP-ports using flexible threshold values which make this algorithm adapt to various environment. Therefore, we can construct a lightweight, fast, and accurate real-time traffic classification system than other classification method.

In the future, we will extract the fixed IP-ports for more applications, and utilize it in the very first step of our real-time traffic classification system. In addition, we are interested in verification for traffic classification based on application.

Acknowledgments. This work was supported by the Korea Research Foundation Grant funded by the Korean Government (MOEHRD, Basic Research Promotion Fund) (KRF-2007-331-D00387).

References

1. Kim, M.-S., Won, Y.J., Hong, J.W.-K.: Application-Level Traffic Monitoring and an Analysis on IP Networks. ETRI Journal 27(1) (February 2005)
2. Sen, S., Wang, J.: Analyzing peer-to-peer traffic across large networks, Internet Measurement Conference (IMC). In: Proc. Of the 2nd ACM SIGCOMM Workshop on Internet measurement, pp. 137–150 (2002)
3. Karagiannis, T., Apagiannaki, K.P., Aloutsos, M.F.: BLINC: Multilevel Traffic Classification in the Dark. In: Proc. of ACM SIGCOMM (August 2005)
4. Sen, S., Spatscheck, O., Wang, D.: Accurate, Scalable In-Network Identification of P2P Traffic Using Application Signatures. In: WWW 2004, New York, USA (May 2004)
5. Zander, S., Nguyen, T., Armitage, G.: Automated Traffic Classification and Application Identification using Machine Learning. In: LCN 2005, Sydney, Australia, November 15-17 (2005)
6. Park, B.-C., Won, Y.J., Kim, M.-S., Hong, J.W.: Towards Automated Application Signature Extraction for Traffic Identification. In: Proc. of the IEEE/IFIP Network Operations and Management Symposium (NOMS) 2008, Salvador, Bahia, Brazil, April 7-11, pp. 160–167 (2008)
7. GOM Player: Multimedia Player Service, http://www.gretech.com/
8. Constantinou, F., Mavrommatis, P.: Identifying known and unknown peer-to-peer traffic. In: IEEE International Symposium on Network Computing and Applications (NCA), Cambridge, MA, USA, July 2006, pp. 93–102 (2006)
9. Szabó, G., Orincsay, D., Malomsoky, S., Szabó, I.: On the validation of traffic classification algorithms. In: Claypool, M., Uhlig, S. (eds.) PAM 2008. LNCS, vol. 4979, pp. 72–81. Springer, Heidelberg (2008)
10. Risso, F., Baldi, M., Morandi, O., Baldini, A., Monclus, P.: Lightweight, Payload-Based Traffic Classification: An Experimental Evaluation. In: Proceeding of Communications, ICC 2008, IEEE International Conference (2008)

Accuracy Improvement of CoMPACT Monitor by Using New Probing Method

Kohei Watabe, Yudai Honma, and Masaki Aida

Tokyo Metropolitan University, 6–6, Asahigaoka, Hino-shi 191-0065, Japan
`watabe-kouhei@sd.tmu.ac.jp`

Abstract. CoMPACT monitor that we have proposed is new technique to measure per-flow quality of service (QoS). CoMPACT monitor provide scalable measurement of one-way delay distribution by the combination of active and passive measurement. Recently, a new probing method for improving the accuracy of simple active measurement have been proposed. In this paper, we adopt the new probing method as active measurement of CoMPACT monitor, and show that the combination improves the accuracy of measurement remarkably for UDP flows.

1 Introduction

As the Internet spreads, various new applications (telephony, live video etc.) appeared. On the other hand, the quality of service (QoS) that is required by these applications has become diversified, too. Moreover, it is expected that the diversification of application and requirement will continue in the future.

In order to meet such varied requirements for network control, we need a measurement technology to produce detailed QoS information. Measuring the QoS for each of multiple flows (e.g., users, applications, or organizations) is important since these are used as key parameters in service level agreements (SLAs). One-way packet delay is one of the most important QoS metrics. This paper focuses on the measurement of one-way delay for each flow.

Conventional means of measuring network performance and QoS can be classified into two types: passive and active measurements.

Passive measurement monitors the target user packet directly, by capturing the packets, including the target information. Passive measurement is used to measure the volume of traffic, one-way delay, round-trip time (RTT), loss, etc. and can get any desired information about the traffic since it observes the actual traffic. Passive measurement can be categorized into two-point monitoring with data-matching processes and one-point monitoring.

Passive measurement has the advantage of accuracy. However if we perform passive measurement in large-scale networks, the number of monitored packets is enormous and network resources are wasted by gathering the monitored data at a data center. Moreover, in order to measure delay, it is necessary to determine the difference in arrival time of a particular packet at different points in the network. This requires searching for the same packet pairs monitored at the

C.S. Hong et al. (Eds.): APNOMS 2009, LNCS 5787, pp. 31–40, 2009.
© Springer-Verlag Berlin Heidelberg 2009

different points in the monitored packet data. This packet matching process lacks scalability, so passive measurement lacks scalability.

Active measurement monitors QoS by injecting probe packets into a network path and monitoring them. Active measurement can be used to measure one-way delay, RTT, loss, etc. It cannot obtain the per-flow QoS, though it is easy for the end user to carry out. The QoS data obtained by active measurement does not represent the QoS for user packets, but only QoS for the probe packets.

By complementary use of the advantages of active and passive measurements, the authors propose a new technique of scalable measurement called change-of-measure-based passive/active monitoring (CoMPACT monitor) to measure per-flow QoS [1,2,3,4].

The idea of CoMPACT monitor is as follows. The direct measurement of the one-way delay distribution of the target flow by passive measurement is difficult due to the scalability problem. So, we try to obtain the one-way delay distribution of the target flow by using a transformation of one-way delay data obtained by active measurement. The transformation can be determined by passively monitored traffic data for the target flow of the measurement. The problem of scalability does not arise, because the volume of traffic can be measured by one-point passive measurement without requiring data-matching processes.

We have believed Poisson arrivals (intervals with exponential distribution) is appropriate to a policy of probe packets arrivals since we can apply PASTA (Poisson Arrivals See Time Averages) property to it.

However, recent work [5] indicates that many distributions exist that are more accurate than an exponential distribution if a non-intrusive context (ignoring the effect of probe packets) can be assumed. Moreover, according to [5] we can find a distribution that is suboptimal in accuracy by selecting an inter-probe time according to the parameterized Gamma distribution.

This study has applied this Gamma-probing to the active measurement part of CoMPACT monitor and tried to improve CoMPACT monitor's in accuracy. The effectiveness of this Gamma-probing was verified in case of the delay/loss process monitored by simple active measurement in [5]. However, CoMPACT monitor observes a process that depends on a value for traffic that is obtained by passive measurement. This process obviously differs from delay/loss processes that are influenced by all flows on the network. Therefore, it is necessary to confirm whether Gamma-probing is effective as a part of CoMPACT monitor.

This paper confirms that Gamma-probing is appropriate when measuring the complementary cumulative distribution function (CDF) of individual flows by simulation, and shows that the effect is remarkable for the flows that use the User Datagram Protocol (UDP).

2   Summary of CoMPACT Monitor

CoMPACT monitor estimates an empirical QoS for the target flow by converting observed values of network performance at timing of probe packet arrivals into a measure of the target flow timing. Now, let $V(t)$ denote the network process under observation (e.g. the virtual one-way delay in the network path at time t), and X_k denote a random variable which is observed $V(t)$ with a certain timing (e.g. the timing of user k's packet arrivals). The probability for X_k to exceed c is

$$P(X_k > c) = \int 1_{\{x > c\}} dF_k(x) = E_{F_k}[1_{\{x > c\}}]$$

where $F_k(x)$ is the distribution function of X_k and 1_A denotes the indicator function that takes the value 1 if the event A is true and 0 otherwise.

If we can directly monitor X_k, its distribution can be estimated by $\sum_{n=1}^{m} 1_{\{X_k(n) > c\}}/m$ for sufficiently large m, where $X_k(n)$ ($n = 1, 2, \cdots, m$) denote the nth observed value. Now, let us consider the situation that X_k cannot be directly monitored. Let Y denote a random variable that is observed $V(t)$ at a different timing (e.g. timing of probe packet arrivals) independent of X_k. Then we consider the relationship between X_k and Y.

Observed values of X_k and Y are different if their timing is different, even if they observe a common process $V(t)$. X_k and Y can be related by each distribution functions $F_k(x)$ and $G(y)$, and $P(X_k > c)$ expressed by measure of X_k can be transformed into measure of Y as follows.

$$P(X_k > c) = \int 1_{\{x > c\}} dF_k(x)$$

$$= \int 1_{\{y > c\}} \frac{dF_k(y)}{dG(y)} dG(y) = E_G \left[1_{\{Y > c\}} \frac{dF_k(Y)}{dG(Y)} \right]$$

Therefore, $P(X_k > c)$ can be estimated by

$$\frac{1}{m} \sum_{n=1}^{m} 1_{\{Y(n) > c\}} \frac{dF_k(Y(n))}{dG(Y(n))}, \quad \text{for sufficiently large } m. \tag{1}$$

where $Y(n)$ ($n = 1, 2, \cdots, m$) denote the nth observed value. Note that this estimator does not need to monitor the timing of X_k, if we can get $dF_k(Y(n))/dG(Y(n))$. This means the QoS of a specific flow (as decided by k) can be estimated by just one probe packet train that arrives with a timing of Y.

This subsection briefly summarizes the mathematical formulation of the CoM-PACT Monitor [4]. We assume the traffic in the target flow can be treated as a fluid. In other words, we assume packets of the target flow are more numerous than active probe packets.

Let $a(t)$ and $V(t)$, respectively, denote the traffic in the target flow at time t and the virtual one-way delay on the path that we want to measure. $a(t)$ and $V(t)$ are a nonnegative deterministic processes assumed to be right-continuous with left limits and bounded on $t \geq 0$. Considering to measure the empirical one-way delay distribution, the value we want to measure is the ratio to all traffic of the target flow of traffic for which the delay to exceeds c, which is given by

$$\pi(c) = \lim_{t \to \infty} \frac{\int_0^t 1_{\{V(s) > c\}} a(s) ds}{\int_0^t a(s) ds} \tag{2}$$

This can be estimated through m times monitoring by

$$Z_m(c) = \frac{1}{m} \sum_{n=1}^{m} 1_{\{V(T_n)>c\}} \frac{a(T_n)}{\sum_{l=1}^{m} a(T_l)/m} \tag{3}$$

for sufficiently large m (see [4] for details), where T_n $(n = 1, 2, \cdots, m)$ denotes the nth sampling time, and each time of sampling corresponds to a time of probe packet arrival. Active and one-point passive measurement are used respectively to observe $V(T_n)$ and $a(T_n)$. Note that one-point passive measurement can be conducted very easily here, compared with two-point passive measurement for measuring the one-way delay.

If we extract the quantity $\sum_{n=1}^{m} 1_{\{V(T_n)>c\}}/m$ from (3), this quantity is a simple active estimator that counts the packets for the delay to exceeds c. However, (3) is weighted by $a(T_n)/(\sum_{l=1}^{m} a(T_l)/m)$, which is decided by the traffic in the target flow when probe packets arrive. This means that the one-way delay distribution (measured by active measurement without bias) is corrected to the empirical one-way delay distribution by the bias of the target flow (observed by passive measurement). $a(T_n)/(\sum_{l=1}^{m} a(T_l)/m)$ in (3) corresponds to $dF_k(Y(n))/dG(Y(n))$ in (1).

3 Suboptimal Probe Intervals

Since the PASTA property is good for non-biased measurement, Poisson arrivals (intervals with an exponential distribution) have been widely used as policy of probe packets arrivals for active measurement. However, if arrival process of the probe packets is stationary and mixing, under non-intrusive conditions, the following equation holds and we can also ignore the effects of probe packets under non-intrusive conditions.

$$\lim_{m \to \infty} \frac{1}{m} \sum_{n=1}^{m} f(X(T_n)) = \lim_{t \to \infty} \frac{1}{t} \int_0^t f(X(t)) dt$$
$$= E[f(X(0))] \qquad \text{a.s.,} \tag{4}$$

where f is an arbitrary positive function and the second equality follows from the stationary and ergodicity of the target process $X(t)$. [6] proved (4) and named this property NIMASTA (Non-Intrusive Mixing Arrivals See Time Averages).

For example, there are processes whose intervals obey the Gamma distribution, the uniform distribution, etc [6]. Note that periodic-probing, with determinate intervals is not a mixing process, and does not satisfy (4).

The recent study [5] also reported that NIMASTA-based probing is suitable for measurement. That provides an improvement in the accuracy of the measurement. We can select a suboptimal probing process in terms of accuracy under the specific assumption by using an inter-probe time given by the parameterized Gamma distribution.

If we estimate the mean of $X(0)$ by using active measurement, estimator $\hat{p}$ is

$$\hat{p} = \frac{1}{m} \sum_{n=1}^{m} X(T_n). \tag{5}$$

It is assumed that the autocovariance function $R(\tau) = \text{Cov}\,(X(t), X(t - \tau))$ is convex. It can be proven that under the foregoing assumptions, no other probing process with an average interval of μ has a variance that is lower than that of periodic-probing (see [5]). A lower variance of the estimator is connected with accuracy. So, periodic-probing is the best policy if we focus only on variance.

On the other hand, periodic-probing does not satisfy the assumptions of NI-MASTA due to non-mixing, so periodic-probing is not necessarily the best. This is because a phase-lock phenomenon may occur and the estimator may converge on a false value when the cycle of the target process corresponds to the cycle of the probing process.

To tune the tradeoff between traditional policies obeying Poisson arrivals and periodic-probing, [5] proposes a suboptimal policy that gives an inter-probe time that obeys the parameterized Gamma distribution. The probability density function that is used as the intervals between probe packets is given by

$$g(x) = \frac{x^{\beta-1}}{\Gamma(\beta)} \left(\frac{\beta}{\mu}\right)^{\beta} e^{-x\beta/\mu} \qquad (x > 0), \tag{6}$$

where $g(x)$ is the Gamma distribution whose shape and scale parameters are β and μ/β, respectively. μ denotes the mean, and β is the parameter. When $\beta = 1$, $g(x)$ reduces to the exponential distribution with mean μ. When $\beta \to \infty$, the policy reduces to periodic-probing because $g(x)$ converges on $\delta(x - \mu)$.

If the autocovariance function is convex, it is proven that the variance of estimator $\hat{p}$ sampled by intervals according to (6) monotonically decreases with β. We can achieve near-optimal variance of periodic-probing, since (6) corresponds to periodic-probing towards limit $\beta \to \infty$. The problem of incorrectness due to phase-lock phenomenon can be avoided if we tune β to a limited value (a probing process that has intervals as set by (6) is mixing). Solving the tradeoff between a traditional policy obeying Poisson arrivals and periodic-probing, we can get a suboptimal probing process if we give β an appropriate value.

The effectiveness of the Gamma-probing has been verified in the cases of simple virtual delay and loss processes using data from large-scale passive measurement and simulations [5]. However, these results do not directly mean that Gamma-probing can effectively be applied to CoMPACT. Comparing (5) with (3), we can consider that $X(t)$ as observed by CoMPACT monitor is

$$X(t) = 1_{\{V(t)>c\}} \frac{a(t)}{\overline{a}(t)} \tag{7}$$

where we let $\overline{a}(t)$ denote the time average of the traffic $a(t)$. Since this process is weighted by the traffic $a(t)$ of a specific flow, its propertyies are different from simple virtual delay or loss processes that is effected by all flows on the network. In this paper, we apply Gamma-probing to CoMPACT monitor and confirm its effectiveness.

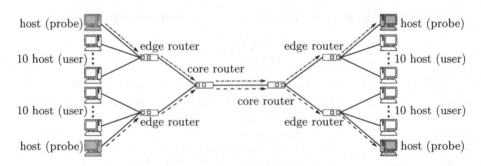

Fig. 1. Network model

4   The Effectiveness of Suboptimal Probe Intervals

4.1   Simulation Model

We investigated the effectiveness of Gamma-probing in the framework of CoM-PACT monitor, through NS-2 based simulations [7].

The network model we used in the simulation is shown in Fig. 1. There are 20 pairs of source and destination end hosts. Each end host on the left in Fig. 1 is a source and transfers packets by UDP to the corresponding destination end host on the right. User flows are given as ON/OFF processes and categorized into the four types listed in Table 1, with there are five flows in each type.

Probe packet trains are categorized into the five types listed in Table 2. Mean probe intervals of each type is 0.5 s. Note that *Exp* and *periodic* in Table 2 are special cases of Gamma distribution. 300 trains of each type are streamed on the two routes shown in Fig. 1, so the total number of probe packet trains in the network is 3000. To analyze the variance of the estimator, we streamed a large number of probe packet trains. Of course we can estimate the empirical delay from only train. Note that parameters of *Exp* and *periodic* in Table 2 are parameters of the Gamma distribution corresponding to each probing.

User flow packets and probe packets are 1500 bytes and 64 bytes, respectively. Link capacities are identical at 64 Mbps. Delay occurs mainly in the link between the core routers, since it is a bottleneck, but no loss occures, because there is sufficient buffering.

We ran the simulation for 500 s. The non-intrusive requirement was satisfied, since the ratio of the probe stream to all streams is 0.00197%. We observed the traffic by passive measurement of queue for the edge router on the source side.

4.2   One-Way Delay Distribution

In this subsection, we show that CoMPACT monitor can estimate the empirical one-way delay when using Gamma-probing. The estimation of the complementary CDF of the one-way delay experienced by user flows will beshown below.

To estimate the complementary CDF of the one-way delay experienced by flow #1, we use probe packet trains with parameter $\beta = 1, 25$ and $\beta \to \infty$

Table 1. Type of user flows

Flow ID	Mean ON/OFF period	Distribution of ON/OFF length	Shape parameter	Rate at ON period
#1-5	10s/5s	Exp	-	6 Mbps
#6-10	5s/10s	Exp	-	6 Mbps
#11-15	5s/10s	Parete	1.5	9 Mbps
#16-20	1s/19s	Parete	1.5	9 Mbps

Table 2. Type of probing

Distribution of probe intervals	Parameter of Gamma distribution
Exp	$(\beta = 1)$
Gamma	$\beta = 5$
Gamma	$\beta = 25$
Gamma	$\beta = 125$
Periodic	$(\beta \to \infty)$

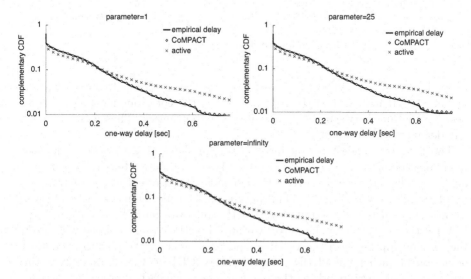

Fig. 2. The estimation of comprementary CDF (flow #1)

respectively. Each result, with 95% confidence intervals, is shown in Fig. 2. Note that the horizontal axis, which is the one-way delay, corresponds to c in (2). To compare the empirical delay with the estimate from CoMPACT monitor, we include the estimate from active measurement in the plot.

In Fig. 2, we can see that the CoMPACT monitor gives good estimates of the true value. We cannot judge the superiority or inferiority of any type of probe packet trains. To represent each flow type, we have plotted for flows #6, 11 and 16, getting results similar to Fig. 2.

4.3   Verification of Variance

In this subsection, we verify the relationship between the parameter of Gamma distribution that is used as the inter-probe time, and the variance of estimator. If we can apply the theory of [5] (explained in section 3) to CoMPACT monitor, the variance of the estimator decreases as the parameter increases.

Since we assume that the autocovariance function of the target process is convex, we will discuss its convexity. The target process observed by CoMPACT monitor is given by (7). We compute $V(t)$, which is the virtual one-way delay, by

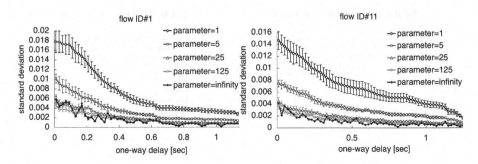

Fig. 3. Standard deviation of estimator

using the queue size (byte) of the core router on the source side. In our simulation model, the greatist part of the delay occurs at the core router on source side, because the link between the core routers is a bottleneck. Thus we can ignore any delay at other routers.

The autocovariance function for $c = 0.1$ for flow #1 with 95% confidence intervals is depicted in Fig. 4. To plot Fig. 4, we used data where the queue size and the traffic are recorded every 0.01 seconds. To represent at each flow type, we plotted flows #6, 11 and 16. This permitted the conclusion that none of these results contradicted the assumption that the autocovariance function is convex.

With the convexity of the autocovariance function confirmed, we will verify the relationship between the variance and parameter β. We plot the standard deviation of each point of the complementary CDF in Fig. 3. Note that the horizontal axis corresponds to the one-way delay, which is c in (2). Error bars indicate the 95% confidence interval when the standard deviation calculated from 30 probe packet trains is considered to be a single data point.

The standard deviation clearly decreases as β increases from $\beta = 1$ to $\beta = 125$. In periodic-probing corresponding to $\beta \to \infty$, the standard deviation is often larger than that for $\beta = 125$ and $\beta = 25$. This reversal is a sign of incorrectness due to the phase-lock that occurs when the cycle of the target process corresponds to the cycle of the probing process.

Consequently, for the case of UDP flow, it is confirmed that we can obtain adequate accuracy with a suboptimal probing process if we tune the parameter of Gamma distribution that we use as the inter-probe time.

4.4   Simulation in the TCP Case

We used UDP flows in the above simulation, but we also simulated the network for the case of TCP flows and obtained an interesting result. We replaced UDP flows with TCP flows in the above simulation model, but left other conditions the same. 300 packets was specified as the maximum window size.

In the case of TCP, user traffic is strongly affected by TCP traffic control. Therefore, the observed traffic of TCP flow may differ greatly depending on the timing of the sampling, compared with UDP. A condition of this simulation is

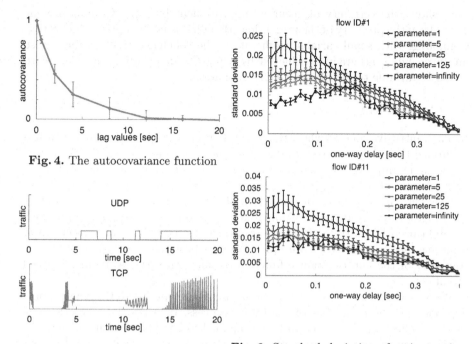

Fig. 4. The autocovariance function

Fig. 5. Traffic process

Fig. 6. Standard deviation of estimator in the TCP case

that user traffic is very unstable (see Fig. 5). Comparing these results with those from UDP simulation, we can confirm whether Gamma-probing can be applied to flows with a unstable traffic, or not.

In fact, we were able to confirm that the estimator converged to the true value and that the autocovariance function of the target process was convex, as in the UDP simulation.

However, the standard deviation of the estimator was different from that in the UDP simulation. Fig. 6 shows a plot of the standard deviation of each point versus the corressponding CDF point. The error bars idndicate 95% confidence intervals with the standard deviation calculated from 30 probe packet trains considered to be a single datum, as in the UDP simulation.

Now let us focus on $\beta \rightarrow \infty$ (corresponding to periodic-probing) in Fig. 6. The phase-lock phenomenon can be observed quite clearly by comparing these results to the UDP simulation. A clear reversal of the standard deviation is observed only with periodic-probing. This is the same as in the UDP simulation, though there the reversal does not necessarily happen. This result proves that periodic-probing is not optimal.

Moreover, we can confirm that there is no significant difference. From $\beta = 1$ to $\beta = 125$, there is a tendency to decrease. But confidence intervals overlap each other, which suggests that the reversal may be due to chance.

In short, the certainty of accuracy improvement by using Gamma-probing depends on the property of the target flow, though it is not completely ineffective. Comparing with small variable traffic, it can be concluded that no guaranteed improvement in accuracy is achieved by using Gamma-probing in a case with seriously variable traffic like this TCP simulation.

5   Conclusion

In a non-intrusive context where the effect of probe packets can be ignored, it was confirmed that the accuracy of estimating the complementary CDF of one-way delay can be improved by using Gamma-probing as part of applying CoMPACT monitor estimates.

However, the autocovariance function of the target process must be convex, obeying the assumption in [5]. Compared with simple delay/loss processes which are influenced by all traffic in the network, the traffic process in a specific flow some has periodicity. When we apply Gamma-probing to a real network, the convexity of the autocovariance function of the target process requires special attention.

We were able to show that the accuracy improvement due to Gamma-probing depended on the properties of the target process. In particular, there is a remarkable effect when gamma-probing is applied to UDP flows with stable traffic volumes, but there is no guaranteed improvement in accuracy when the same technique is applied to unstable traffic flows.

References

1. Ishibashi, K., Kanazawa, T., Aida, M., Ishii, H.: Active/passive combination-type performance measurement method using change-of-measure framework. Computer Communications 27(9), 868–879 (2004)
2. Aida, M., Ishibashi, K., Kanazawa, T.: CoMPACT-Monitor: Change-of-measure based passive/active monitoring–Weighted active sampling scheme to infer QoS. In: Proc. IEEE SAINT 2002 Workshops, Nara, Japan, Feburary 2002, pp. 119–125 (2002)
3. Aida, M., Miyoshi, N., Ishibashi, K.: A scalable and lightweight QoS monitoring technique combining passive and active approaches–On the mathematical formulation of compact monitor. In: Proc. IEEE INFOCOM 2003, San Francisco, CA, April 2003, pp. 125–133 (2003)
4. Aida, M., Miyoshi, N., Ishibashi, K.: A change-of-measure approach to per-flow delay measurement systems combining passive and active methods: On the mathematical formulation of CoMPACT monitor. IEEE Transactions on Information Theory 54(11), 4966–4979 (2008)
5. Baccelli, F., Machiraju, S., Veitch, D., Bolot, J.: On optimal probing for delay and loss measurement. In: Proc. Int. Measurement Conf. 2007, San Diego, CA, October 2007, pp. 291–302 (2007)
6. Baccelli, F., Machiraju, S., Veitch, D., Bolot, J.: The role of PASTA in Network Mesurement. In: Proc. ACM SIGCOMM 2006, Pisa, Italy, September 2006, pp. 231–242 (2006)
7. The Network Simulator – ns-2, http://www.isi.edu/nsnam/ns/

Proposal and Evaluation of Data Reduction Method for Tracing Based Pre-patch Impact Analysis

Kenji Hori and Kiyohito Yoshihara

Kddi R&D Laboratories Inc. 2-1-15, Ohara, Fujimino-shi, Saitama 356-8502 Japan
{hori,yosshy}@kddilabs.jp

Abstract. Patching of applications on servers and PCs is a significant source of expense and frustration for IT staff. Organizations are concerned that patches will negatively impact the stability of their existing applications, and this concern results in significant delays in applying patches to publicly announced vulnerabilities of the applications. While Tracing based Pre-patch Impact Analysis method (T-PIA) is an effective way of measuring the actual impact of recent patches on the applications, it causes large storage and bandwidth usage of servers and PCs running the applications. We propose and evaluate a method to reduce such usage caused by T-PIA. Compared to the existing method, our method reduces storage and bandwidth usage by 83 and 89%, respectively. This verifies the ability of our method to reduce such usage.

Keywords: Software Patch Management, Kernel-Tracing.

1 Introduction

Software patches update applications or OS's component (e.g., kernels or drivers) of the servers and PCs. Major objective of applying the patches (i.e., patching) is fixing the security vulnerabilities and fixing the bugs. However, patching is a significant source of expense and frustration for IT staff. Organizations are concerned that patches will have impacts on their existing applications' behaviors, and this concern results in significant delays in applying released patches. This delay causes the outbreak of disruptive worms (e.g., Code Red) or the system crashes lead to data loss.

The greatest bottleneck in the patching process is the time required to test a patched subset of systems for stability before rolling out the patch to all systems [1]. Organizations deploying a patch typically want to test patches even after the corporation releasing the patch has tested them. One common reason for this second round of testing is the existence of legacy, third-party and in-house applications that have not previously been tested. Even when the patch is known to only update one or two libraries, organizations often cannot take advantage of this knowledge to reduce the testing load because they do not have an exhaustive list of the library dependencies (e.g., Microsoft's Internet

C.S. Hong et al. (Eds.): APNOMS 2009, LNCS 5787, pp. 41–50, 2009.

Explorer and Office depends on a library *mshtml.dll*) on their machines. They may not even have an exhaustive list of the softwares that is in regular use on each machine, even though keeping an accurate inventory is a recommended best practice in systems management [2].

An existing work [3] measures the actual impact of recent patches on the applications before applying them, using the dependency information (e.g., all dynamic library loads that occur in a running system) directly captured via the dedicated kernel driver. The information is uploaded to the central server, and the server analyzes the impact of patches using the information. Therefore [3] works with legacy, third-party and in-house applications for which no explicit dependency information is available. Hereafter, we call this method Tracing based Pre-patch Impact Analysis (T-PIA). However, because a number of the duplicate dependency information is captured, T-PIA requires significant storage and bandwidth usage for the machines where the dependency information is captured. Such usage make the applications slower and unstable.

In this paper, we propose and evaluate a method to reduce such usage in T-PIA. In our proposed method, the duplicate dependency information is detected and eliminated while capturing. Therefore, such usage can be reduced.

We organize the remainder of the paper as follows: In Section 2, we give a brief explanation of existing T-PIA along with its problem. In Section 3, we present a method to detect and eliminate the duplicate dependency information. In Section 4, we describe the implementation based on our proposed method for evaluation. In Section 5, we evaluate our method using the implementation. Finally, in Section 6 we conclude with our planned future work.

2   Existing T-PIA

[3] provides the T-PIA method, which can measure the actual impact of recent patches on the applications before applying them. Figure 1 shows an overview of T-PIA. Note that we use the terms that we defined, in order to abstract the framework of T-PIA.

In order to analyze the impact of patches, T-PIA uses the *DI (Dependency Information)* between applications and objects (e.g., libraries or files), as mentioned later in this section. For an example of DI, in Fig. 1, an application AP1 reads a configuration file F1 and loads a library L1 in its bootstrap process. Then these accesses to F1 and L1 should be recorded as DI of AP1. DI is captured by the *kernel-tracing*. Kernel-tracing is a method of tracing (or capturing) the activity (e.g., loadings of the libraries and accessing to the files) of any applications via the dedicated kernel driver. Dtrace[4], systemtap[5], and the one used in Flight-Data Recorder (FDR) [6] are the typical implementations of such driver. Kernel-tracing applications such as FDR [6] can obtain the capturing output by accessing a dedicated API (e.g., standard outputs of a dedicated shell command). A typical capturing output consists of the application's name (such as shown in the *AP* column of Fig. 2 (b)) and associated name of the object (such as shown in the *Object* column of Fig.2 (b)). *TDI (Traced DI)* is a kind of DI that is obtained by kernel-tracing. For example, if the above-mentioned DI among AP1, F1 and L1 are obtained by kernel-tracing, then we refer to this type of DI as TDI.

TA (Target for Analysis) is the machine where we want to collect DI. And *DC (Data Collector)* is the user-mode daemon on TA that is responsible for capturing TDI, writing it into log, and uploading these logs to the central server. In order to configure DC and to start the data collection, the operator makes an installer package of DC for a particular TA. The package should contain a DC, installer and the DC's configuration file. *PIAM (Pre-patch Impact Analysis Manager)* is central server that is responsible for receiving TDI sent from a DC, analyzing the patch impact using TDI, and presenting the result to an operator. PIAM also provides a dedicated GUI tool for the operators to make an installer package of DC.

Stored TDI Table is the log data for collected TDI. It exists in the DC's memory or storage. Figure 2 (a) is an example of the table. *"AP"* and *"Object"* denote the name of the applications on a TA (e.g., AP1 and AP2 in Fig. 1) and the name of the objects (e.g., F1 and L1 in Fig. 1), respectively.

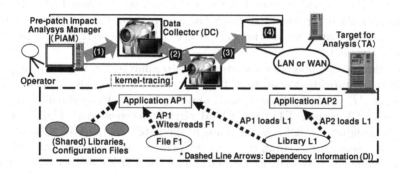

Fig. 1. Basic Operation of T-PIA

(a) Stored TDI Table

AP	Object	row #
/usr/bin/httpd	/lib/libcrypto.so.4	1
/usr/bin/sendmail	/etc/sendmail.cf	2
/usr/bin/sendmail	/var/spool/queue/mail	3
/usr/bin/mysqld	/etc/my.cnf	4

(b) A Newly Captured TDI

AP	Object
/usr/bin/sendmail	/var/spool/queue/mail

Fig. 2. Example of Stored TDI Table

The following configuration parameters should be pre-determined by an operator and should be designated in the DC's configuration file; otherwise hardcoded ones in DC's binary are used: 1) *Data Collection Period*, DC collects TDI until the elapse of this time period that begins from the start of collection, and

2) *Upload Server Address*, this is the destination ip address (or hostname) which the log (i.e., content of the Stored TDI Table) is uploaded.

Then the basic operation of T-PIA is follows:

Step 1. (Fig. 1 (1)): An operator designates required parameters to the DC's configuration file, and then makes an installer package of DC for a particular TA using the GUI tools on PIAM. Usually, an Upload Server Address is set to PIAM's IP address.

Step 2. (Fig. 1 (2)): PIAM installs a DC onto a TA remotely.

Step 3. (Fig. 1 (3)): DC collects TDI in the Data Collection Period using kernel-tracing. The newly captured one will be added to the end of the Stored TDI Table. DC uploads collected TDI to the Upload Server Address.

Step 4. (Fig. 1 (4)): PIAM performs pre-patch impact analysis using the received TDI. After that, the operator can confirm the result of the analysis on PIAM's GUI tools.

In the Step 4, the operator designates the name of the patch via the GUI. Then PIAM seeks the provided content of the patch from the received TDI. If the one is found, PIAM determines the patch has a impact on the TA. If not, PIAM determines the patch has no impact on the TA. The determination result is shown in the GUI.

For example, suppose the operator want to analyze the impact of a patch *Patch1* which provides the updated file "/etc/sendmail.cf". Also suppose a TDI consists of AP = "/usr/bin/sendmail" and $Object$ = "/etc/sendmail.cf" is captured at a TA named *TA1*, and PIAM already received that TDI. Now the operator designates *Patch1* as the name of the patch via the GUI. Then PIAM seeks "/etc/sendmail.cf" from the received TDI. Because the received TDI contains above-mentioned TDI, PIAM determines the *Patch1* has a impact on the TA and presents the result to the GUI.

Because the applications access to the same libraries or files repeatedly, The duplicate TDI is captured. Hereafter, we call these TDI as *DTDI (Duplicate TDI)*. However, [3] does not have a method to detect and eliminate the DTDI. So a number of the duplicate information is captured, and is uploaded to the central server. This causes the significant storage and bandwidth usage for the machines where the dependency information is captured.

3 Method to Eliminate Duplicate Dependency Information

In this section, we propose a method to eliminate DTDI entailed by T-PIA. With our method, in addition to above mentioned operations of existing T-PIA, DC detects for duplicate information in the Stored TDI Table, then eliminates DTDI. DC performs such process during the collection. In order to achieve this, we introduce an additional configuration parameter of DC, *DTDI Search Algorithm*.

This is the name of the algorithm that is used to seek a duplicate TDI from the Stored TDI table. One of the following well-known searching algorithms should be designated: 1) *sequential search*, 2) *binary tree search*, or 3) *hash table search*. Then, we modified Step 3. of Sect. 2 as follows:

Modified Step 3. DC collects TDI in the Data Collection Period using kernel-tracing. If a TDI as that shown in Fig. 2 (b) is newly captured, then DC looks up the newly captured TDI in the Stored TDI Table by using the designated DTDI Search Algorithm. If is found, the newly captured one will be discarded. Otherwise, the newly captured one will be added to the end of the Stored TDI Table. DC uploads collected TDI to the Upload Server Address.

3.1   DTDI Elimination Details

In the above-mentioned modified Step 3, the elimination (i.e., looking up and discarding) of the DTDI is performed as follows. Note that this process is performed every time, immediately after the new one is collected. The designated DTDI Search Algorithm is used for searching for both *AP* and *Object*.

Step 1. DC looks up the *AP* of the newly captured one ("/usr/bin/sendmail" in Fig. 2 (b)) in the first column ("*AP*") of the Stored TDI Table.

Step 2. If a TDI that has the same *key1* as the newly captured one is found, then DC compares the *Object* of the newly captured one ("/var/spool/queue/mail" in Fig. 2 (b)) with the second column ("*Object*") of the rows that has the same *AP* as the newly captured one.

Step 3. If the *Object* matches, then DC considers that the newly captured one is duplicated, and discards it. If the *Object* does not match, and if there are two or more TDI that has the same *AP* as the newly captured one, Step 2 is performed again for this *AP*.

Step 4. If there are no more TDI that has the same *AP* as the newly captured one, this DTDI elimination process is ended (no DTDI is found).

Example of DTDI Elimination. For example, if a TDI (Fig. 2 (b), *AP* = /usr/bin/sendmail, *Object* = /var/spool/queue/mail) is newly captured, then DC looks up the *AP* of the newly captured one ("/usr/bin/sendmail" in Fig. 2 (b)) in the first column ("*AP*") of the Stored TDI Table.

DC finds it ("/usr/bin/sendmail") in the row #2 of the Stored TDI Table. Then DC compares the *Object* of the newly captured one ("/var/spool/queue/mail") with the *Object* of the row #2 of the Stored TDI Table ("/etc/sendmail.cf").

Because it does not match "/var/spool/queue/mail", and there is one more TDI that has the same *AP* as the newly captured in the row #3 of the table, DC compares the *Object* of the newly captured one ("/var/spool/queue/mail") with the *Object* of the row #3 of the Stored TDI Table ("/var/spool/queue/mail"). Because the *Object* matches, DC considers that the newly captured one is duplicated, and discards it.

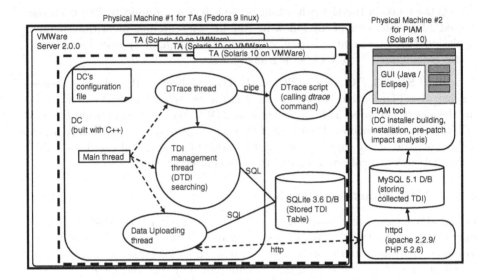

Fig. 3. Example Implementation for Evaluation

Effects on Pre-patch Impact Analysis. The DTDI elimination does not disturb the pre-patch impact analysis. This is because PIAM use only the information of *AP* and *Object* in TDI as mentioned in Sect. 2, and these information are not omitted in the DTDI elimination.

4 Example Implementation for Evaluation

We have implemented our method as a part of our T-PIA tools. Figure 3 is the illustration of our implementation. Both PIAM and TA runs on the Solaris 10 platform; however, a virtual machine environment (VMWare Server 2.0.0, VMWare Inc.) is used to accommodate a number of TAs into a single physical machine, which runs Fedora Core 9 Linux. Each of the virtual machines is assigned 1GB of RAM. All virtual machines use the bridge-mode of the virtual NIC. The virtual NICs are set to 100Mbps/Full Duplex. All of them are connected to the same ethernet segment/IP subnet where the physical machines are directly connected. No virtual machine environment is used for PIAM. We used two physical machines: one for PIAM and the other for nine TAs running on VMWare. All physical machines use the same hardware platform (SC1435, Dell Inc., dual quad-core AMD Opteron 2350 2.0GHz, 12GB of RAM and 1GbE NIC). All programs except PIAM's GUI are built using the gcc version 3.4.3 and libstdc++.so.6.0.3. PIAM's GUI is built with Java/Eclipse. We used dtrace[4] as the kernel driver for the kernel-tracing. Dtrace can be called from the command shell, and returns the result (TDI) through the standard output.

We implemented all of the three DTDI Search Algorithms mentioned in Sect. 3: 1)*sequential search,* 2)*binary tree search,* and 3)*hash table search.* We implement the *sequential search* from scratch. The *binary tree search* is implemented using *std::map* in the standard C++ libraries (libstdc++). The *hash table search* uses *unordered_map* in the Boost 1.37 C++ libraries. The initial hash table size is 10000. We used SQLite 3.6 as the database engine for the Stored TDI Table.

We used the PUT method of the http (Hyper Text Transfer Protocol) as the protocol for sending TDI from the TA to PIAM. In PIAM, apache 2.2.9/PHP 5.2.6 is used to receive the data. No data compression/encryption is used. DC uploads the entire contents of the Stored TDI Table after the Data Collection Period.

5   Evaluation

We evaluated our proposed method as follows. The objectives of this evaluation are: 1) to evaluate how much storage and bandwidth usage on TA are reduced compared to the case where no DTDI elimination is performed, and 2) to compare the computational overhead incurred by the three DTDI Search Algorithms mentioned in Sect. 3.

Evaluation Method. We installed a DC onto one TA and measured the following: 1) the database size on TA at the end of the Data Collection Period, as a measure of storage usage, 2) the transmitted payload bytes in the http session of Step 3 in Sect. 2, as a measure of bandwidth usage and 3) DC's processing time for database searching, as a measure of computational overhead incurred by the DTDI elimination. The DC's processing time was measured as the time taken for Step 1 through Step 4 in Sect. 3.1.

The implementation described in Sect. 4 were used. We used only one TA and one PIAM for this evaluation. We used 1, 5, 10, 20 and 30 minutes as the Data Collection Period.

In order to simulate the activity of an actual server on TA, ten commonly used server applications, as shown in Table. 1, were used. The server side of these applications ran on TA (i.e., on top of a virtual machine on the physical machine #1), and the client side applications ran directly on top of the Fedora Core OS of the physical machine #1. Clients were initiated from the shell scripts at five second intervals.

For each test, only one server-client pair of the application in Table. 1 was used. One of the three DTDI Search Algorithms (*sequential search*, *binary tree search*, and *hash table search*) was used for each test. We also tested the case where no DTDI elimination is performed.

We ran three tests for each combination of a) a pair of the server-client applications, b) a DTDI Search Algorithm, and c) a length of the Data Collection Period, and the results are averaged over three tests. Then these averaged results are averaged over all pairs of the server-client applications.

Evaluation Results. The results are shown in Fig. 4, Fig. 5, and Table. 2. With our method (The lines noted as "w/ DTDI elimination" in Fig. 4 and Fig. 5),

Table 1. Applications used on TA

Pair#	Server	Client
1	snmpd ver 5.0.9	snmpwalk ver 5.0.9
2	named ver 9.3.5, sendmail ver 8.13.8	perl ver 5.10.0, Net::SMTP ver 2.31[1]
3	in.dhcpd ver 1.4	dhclient ver 4.0.0
4	radiusd ver 2.0.2	radclient ver 1.118
5	ipop3d ver 2006e.96	fetchmail ver 6.3.8
6	imapd ver 2006e.378	fetchmail ver 6.3.8
7	httpd ver 2.0.63	http_load ver 12mar2006
8	mysqld ver 5.0.67	mysql ver 5.0.67
9	nfsd ver 1.8	mount ver 2.13.1, umount ver 2.13.1[2]

[1] For each initiation of the client, a mail was sent by the Net::SMTP. Then the sendmail looked up the DNS MX records.

[2] For each initiation of the client, an NFS volume was mounted. Immediately after that, the volume was unmounted.

the database size and the transmitted payload bytes, as an average for all the Data Collection Periods, are reduced by 83 and 89%, respectively. Because the number of the DTDI increases with the length of the Data Collection Period, the effectiveness of the DTDI elimination is greater the longer the period.

In Fig. 4, if the DTDI elimination is not performed, the database size rises to about 0.75 [megabytes] in 30 minutes. However, it should be noted that this is the average number. In the worst case, this size rises to about 2.7 [megabytes] in 30 minutes (the case ipop3d and fetchmail was used). Assuming this worst case increase rate is lasting, the database size rises to about 130 [megabytes] in 24 [hours]. This number is not necessarily negligible, especially for the relatively old TAs, which is likely to run the legacy applications. With our method, we can safely use T-PIA on these machines because the size is reduced by about 83% (108 [megabytes]) compared to the case where no DTDI elimination is not performed. Though, it is apparent that the effectiveness of the DTDI elimination is smaller the fewer DTDI.

The results of the DC's processing time for database searching are shown as the average time for an entire DTDI elimination process (Step 1 through Step 4 in Sect. 3.1). The average time is determined for all searches that occurred during a test run for a particular DTDI Search Algorithm. Compared to the processing time for the *sequential search* and the *binary tree search*, the one for the *hash table search* is only 28 and 68% for the former two, respectively. While it is generally accepted that the *hash table search* is faster than the others, our results confirmed that fact in an implementation of T-PIA. Because the difference in the results for the *binary tree search* and for the *hash table search* is not necessarily large (0.037 vs. 0.025 [msec]), the algorithm should be chosen considering the memory footprint for these algorithms.

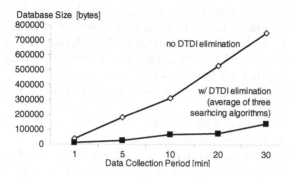

Fig. 4. Comparison of database size

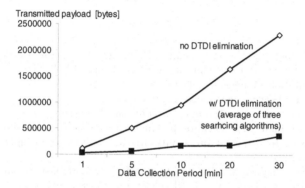

Fig. 5. Comparison of transmitted payload bytes

Table 2. Comparison of DC's processing time

Algorithm name	DC's processing time [msec]
sequential search	0.090
binary tree search	0.037
hash table search	0.025

6   Conclusion

We proposed a method to reduce storage and bandwidth usage of T-PIA. Our method achieves this by detecting and eliminating duplicate dependency information during the dependency information collection. We provided an example of implementation for evaluation. We presented the implementation evaluation results, which verified the ability of our method to reduce such usage. In our evaluation, compared to the existing method, our method reduces storage and bandwidth usage by 83 and 89%, respectively. We also compared the required processing time for the searching algorithms of the duplicate dependency information. The processing time of the hash table search algorithm is only 28 and 68% of the ones for the sequential search and the binary tree search, respectively. Our results verified the theoretical advantage of the hash table search quantitatively in actual implementation.

For future work, in order to mitigate peak bandwidth usage, we plan to examine the details of the data uploading algorithm for the dependency information. Comparison of the memory footprints for the algorithms of the duplicate dependency information is also desirable.

It is not clear how much TA's CPU time is spent by dtrace. Therefore evaluation of individual processing time of dtrace is necessary. It is also desirable that further evaluation in more sophisticated case, such as the case where the two or more server applications are used at one time.

References

1. Beatie, S., Arnold, S., Cowan, C., Wagle, P., Wright, C.: Timing the Application of Security Patches for Optimal Uptime. In: Proc. Usenix LISA (2002)
2. ITIL Configuration Management Database,
   http://www.itil-itsm-world.com/itil-1.htm
3. Dunagan, J., Roussev, R., Daniels, B., Johnson, A., Verbowski, C., Wang, Y.M.: Towards a self-managing software patching process using black-box persistent-state manifests. In: Proc. of Int. Conf. on Autonomic Computing (ICAC 2004), pp. 106–113 (2004)
4. Cantrill, B.M., Shapiro, M.W., Leventhal, A.H.: Dynamic instrumentation of production systems. In: Proc. of. USENIX 2004, pp. 15–28 (2004)
5. Eigler, F.C.: Problem solving with systemtap. In: Proc. of the Ottawa Linux Symposium 2006, pp. 261–268 (2006)
6. Verbowski, C., Kiciman, E., Kumar, A., Daniels, B., Lu, S., Lee, J., Wang, Y.M., Roussev, R.: Flight Data Recorder: Monitoring Persistent-State Interactions to Improve Systems Management. In: OSDI 2006, pp. 117–130 (2006)

High-Speed Traceroute Method for Large Scale Network

Naoki Tateishi, Saburo Seto, and Hikaru Seshake

NTT Network Service Systems Laboratories, NTT Corporation 9-11,
Midori-Cho 3-Chome, Musashino-shi, Tokyo, 180-8585 Japan
{tateishi.naoki,seto.saburo,seshake.hikaru}@lab.ntt.co.jp

Abstract. As the scale of IP networks becomes larger, network management becomes more complex and the turn-around time of fault recovery is longer. This is because there are problems in adopting current essential tools for network management, such as a ping, Traceroute, and Simple Network Management Protocol (SNMP) to large-scale IP networks. We have been investigating ways to improve the performance of these tools, and proposed a high-speed ping and showed its advantage. For this paper, we propose and evaluate a high-speed traceroute method for collecting topology information from large scale networks. Our method is used to increase the speed of network topology collection to reduce topology-searching packets, which are sent in excess using the current traceroute method, as well as adopting our high-speed ping technology. In terms of reducing packets, our method can be used to send these searching packets to end points first (with the current traceroute method, they are sent last), detect nodes on the shared routes to those end points, and send searching packets to these nodes only. By implementing the above, we show that our proposed method can be used to reduce searching packets by 70% and the time to collect topology information by 95% compared with the current traceroute.

1 Introduction

As the scale of IP networks become larger, fault detection and diagnosis require more turn-around time and are more complex. ISPs and telecom carriers test their networks using traditional operation tools, such as ICMP [1], SNMP [2], and Traceroute [3]. However, these essential tools do not perform well in large-scale networks, and this low performance causes complicated fault diagnosis and increases the cost of network equipment.

We proposed a high-speed ping method, and the experimental results show that this method can be used to shorten fault detection time by over 90% compared with the current ping in large-scale IP networks by parallelizing packet-sending and - receiving threads [4].

We propose and evaluate a high-speed traceroute method for specifying failure points and the extent of the failure. The current traceroute method does not perform well in large-scale IP networks because it requires a large amount of time to collect network topology information. This method is used for sending many packets, so the network load tends to increase. To solve these problems, we focused on a method of

C.S. Hong et al. (Eds.): APNOMS 2009, LNCS 5787, pp. 51–60, 2009.

parallelizing packet-sending and -receiving threads, and optimizing the number of topology-collecting packets.

We introduce applications of traceroute in Section 2, discuss the problems of the current traceroute method in Section 3, propose a high-speed traceroute method in Section 4, and discuss the results of our evaluation of the method in Section 5.

2 Applications of Traceroute

In this section, we show two examples of applications.

First example is to check route settings of routers or switches (Fig.1a). Some routes of traffic are often changed due to network fault or reconstruction. Operators some-time set unintended route settings (solid line shown in Fig.1a) instead of normal route settings (dashed line shown in Fig.1a) to routers, and they can only check validity of the settings by executing traceroute from source node and checking routes.

Second example is to check routes of video traffic from video contents delivery server (Fig.1b). Video traffic (solid line shown in Fig.1b) is very to affect other traf-fics and the routes affects quality of the traffic, so collecting tree-shape traffic routes is important to qualify video services by executing traceroute from source node.

However, in both examples, if a lot of routes are set to be target, it takes a lot of time and effort to execute traceroute due to low performance.

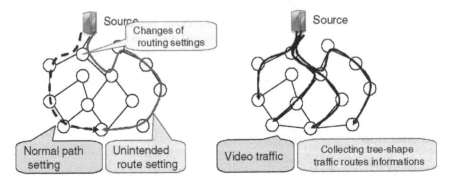

Fig. 1a. Validating changed routes   **Fig. 1b.** Collecting video traffic routes

3 Current Traceroute Method

The current traceroute method is used to send packets in serial (Fig. 2), so delay of reply from some nodes affects the topology-collecting time of the entire target net-work. Sending excessive packets to a target network increases the load of network equipment, such as a router (Fig. 3).

For collecting network topology from the source to node C, The current traceroute method is used to send packets by setting destination header C and increasing the Time To Live (TTL) header by 1, then it detects all receiving interface addresses on the path to C. First, the source sends a packet by setting the destination addresses C

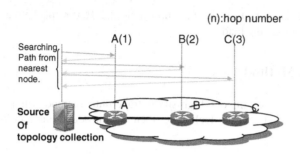

Fig. 2. Topology collection using current traceroute method

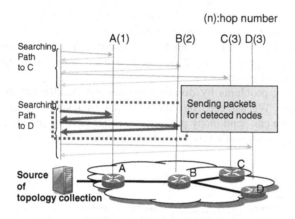

Fig. 3. Sending and receiving of excess packets

and TTL 1 to the packet header. Then, the source receives a reply from A as the first hop on the path to C. Second, the source sends packets by setting the same destination and TTL 2 to the packet header, then the source receives a reply from B as the second hop on the path of C. In the latter, the source sends packets by setting TTL 3 to the packet header, then it receives a reply from C, and we can see that the path to C consists of A-B-C.

Next, we show how to collect other topology when a network is constructed with source-A-B-C and node D connected to node B. For collecting network topology from the source to node D, a traceroute sends packets by setting destination header D and increasing the TTL header, then, the current traceroute method is used to detect all nodes on the path to D by sending packets. However, nodes A and B are already detected in the process to collect the topology from node C, so we can say that the current traceroute method is used to send verbose packets. This example is simplified so the verbosity of sending packets is at most a few packets. However, with large-scale networks consisting of 100,000 nodes and one node has three child nodes, the amount of sending packets for collecting the entire network topology is 1.04 million, and the topology-collecting time is 1 hour and 55 minutes by setting the packet-sending rate to 150 packets/sec. The current traceroute method is used to send packets (TTL=1) to all target nodes for detecting the first hop node, so 10,000 packets are sent to the node, which increases the load.

4 Proposed Method

In this section, we explain our high-speed traceroute method. First, we introduce our high-speed packet-sending technique, then we introduce our topology-collecting technique.

4.1 Parallelization of Packet Sending and Receiving Threads

In the current traceroute method, one process sends packets to one target. Figure 4 (a) shows an example. However, this method is problematic in large-scale networks. We can only execute hundreds of processes simultaneously because many operating systems only permit the execution of up to this amount. Also, we cannot shorten the intervals for sending pings because this method must wait for a reply of sending packets until a timeout occurs.

We solved these problems by dividing threads and parallelization. Figure 4 (b) shows an example. We assigned one thread to send packets and another thread to receive packets. We can send many packets at a faster pace by parallelizing the sending and receiving threads so that they are simultaneous.

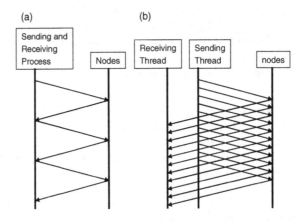

Fig. 4. (a) Current traceroute method   (b) Proposed method

4.2 Topology-Collection Method

As we discussed in Section 2, the current traceroute method is used to send excessive packets so making the topology-collecting time longer in proportion to the network scale. To solve this problem, we propose a method for reducing excessive topology-searching packets, topology-collecting time, and the load of network equipment.

The current traceroute method is used to send packets to the nearest node from the source first in searching the paths to the end points; therefore, the source node sends at least as many nearest node-searching packets as the number of end nodes. Therefore, if the shared routes on the paths to end nodes can be detected using the topology-collecting techique, it can be used to reduce excessive packets. In terms of reducing packets, our method can be used to send these searching packets from the end points, which is opposite of the current traceroute method. Our method then can be used to detect common routes on the path to the end points and send packets by setting the destination as a node on the common route. Figure 5 shows the flow of our method. If the source sends packets by setting destination C, D and TTL 3 to the IP header, the source receives replies from nodes C and D. It then sends packets with setting destination C, D and TTL 2 to the IP header, receives two replies from B, and determines that C and D are the children nodes of B. Then, the source sends packet to search the parent nodes of B with setting destination B and TTL 1 to the IP header and source receives a reply from A. The repetition of this sequence reduces excessive packets.

If we adopt this method for collecting network topology, the number of hops of all target nodes must be given. Therefore, we must set an initial assumption hop number of all target nodes at the first execution. Operators know the structure of their network briefly so we can set an initial assumption hop number. We can use the results from previous topology collections after the second execution. For example, if the packets are sent by setting destination A, B, C, D and TTL 2 to the IP header, the source receives replies from A and B with hop numbers below 2. Then, the source sends packets by subtracting 1 from TTL in the packet header. In contrast, the source receives packets from B as a reply of node C and D, with hop numbers over 2. Then, source sends packets by adding 1 to TTL in the packet header until the source receives replies from C and D. In this way, this method can be used to find the hop number of all target nodes. In addition, this method can be used to deal with the addition of new nodes and changes in network topology.

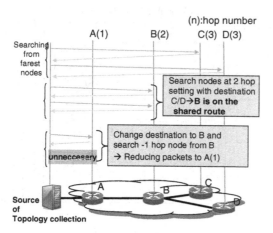

Fig. 5. Proposed topology-collection method

5 Evaluation

In the previous section, we discussed two high-speed traceroute methods. We evaluated these method in a large-scale IP network with tens of thousands of nodes. In this section, we report the results of evaluation and discussions.

5.1 Evaluation Procedure

We collected network topology of a large-scale network similar to the network discussed in Section 2 (NW-reference), then we measured the number of sending packets and the time of collecting network topology. Our proposed method will be affected by the network structure. Compared with network B (NW-B), network A (NW-A) has many divaricated points at nodes near the source, and network B (NW-B), which has many divaricated points at nodes far from source, searching topology packet will further decrease under NW-B because the number of nodes on the common route will increase. Thus, we also measured the effect on the two networks, NW-A and NW-B.

Chart.1 Evaluation patterns

	Parallelization pf packet sending and receiving threads	topology collecting method	
		Reduction of sending packets	Using previous information of network topology
Current Traceroute	×	×	×
Pattern 1	O	×	×
Pattern 2	O	O	×
Pattern 3(proposed)	O	O	O

O : enabled, × : disabled

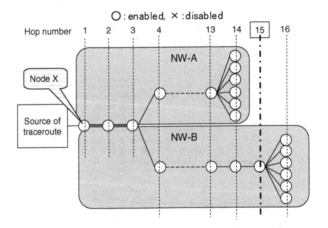

Fig. 6. Simple structure of NW-A and NW-B

More specifically, NW-A has divaricated points with an average hop number of about 13, and NW-A has divaricated points with an average hop number of about 15 (shown in Fig. 6). We evaluated four patterns, as shown in Table 1, the current traceroute method, our proposed method (Pattern 3), with only the high-speed packet-sending function of our proposed method (Pattern 1), and with the high-speed packet-sending and search-reducing packet functions with no information of previous topology collections and setting the initial assumed hop number to 15 (Pattern 2).

We evaluated the sending-packet number to node X and the CPU load of node X, and the topology collecting time and sending packet number of the entire network. This is because the current traceroute method is used to send the largest number of packets to node X under the four patterns mentioned above.

We used a server with a Xeon X3070 (2.66 GHz) CPU and 2-GB memory as the source node.

5.2 Results

5.2.1 Reduction of Sending Packets

The number of searching packets in NW-A and NW-B is shown in Fig. 7, and the number is normalized by the number of packets that are sent by the current traceroute. The same number of packets is sent by Pattern 1. Pattern 2 reduces searching packets by 73% in NW-A and 81% in NW-B, and Pattern 3 reduces searching packets by 74% in NW-A and 83% in NW-B.

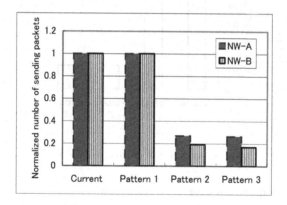

Fig. 7. Packet reduction rate

Figure 8 shows the number of packets with the setting of TTL 1 in the packet header and the number is normalized by the number of packets that are sent by the current traceroute. The same number of packet is sent by Pattern 1. Pattern 2 reduces searching packets by 90% in NW-B, and Pattern 3 reduces searching packets by 99.7% in NW-B.

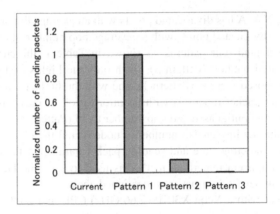

Fig. 8. Packet reduction rate (TTL=1)

Figure 9 shows the CPU load of node X collecting network topology of NW-B. The number is normalized with Pattern 1, and Patterns 2 and 3 reduce the load by about 80 %.

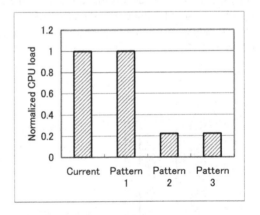

Fig. 9. Normalized CPU load at node X

5.2.2 Topology Collection Speed

Figure 10 shows the topology-collecting time, which is normalized by the time taken by the current traceroute. Pattern 1 reduces the time by 96.4%, and patterns 2 and 3 reduce the time by 97.3%. We obtained similar result on NW-A .

5.3 Discussion

5.3.1 Reducing Amount of Sending Packets

The rate of searching-packet reduction in NW-B, which has many divaricated points at the nodes far from the source, is larger than that in NW-A, which has many divaricated

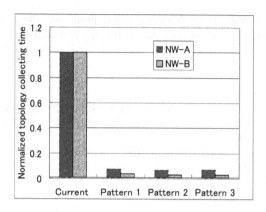

Fig. 10. Topology-collecting time

points at the nodes near the source(shown in Fig. 6). These results suggest that the rate of searching-packet reduction is large when many paths to the end point are shared and many nodes are on the shared route. This is because the more far divaricated points increase, the more far aggregated points increase, so searching packets sent from the source to the aggregated points are reduced. In addition, the results of pattern 2 show that the proposed topology-searching method enables us to reduce searching packets even by using the assumption initial hop value, and the results of pattern 3 show that packet reduction rate is improved considerably by using the results of previous topology collections. However, the initial hop value in pattern 2 had an error of at most 2 or 3 hops compared with the real topology, so the difference in the packet reduction rate between patterns 2 and 3 converged within only a few percentage points.

Our proposed topology-collecting method also enables us to reduce the number of packets sent to the nearest node to the source (shown in Fig. 7) and to decrease the CPU load. From these results, we can say that our proposed methods is efficient for load mitigation of a target network.

5.3.2 Topology-Collection Speed

From Fig. 9, parallelization of packet sending and receiving threads enables to reduce the time to collect topology information by 95% compared with the current traceroute. On the contrary, the sending-packet reduction method has little effect on topology-collecting time compared with thread parallelization. This is because the packet sending time is reduced instead of increasing the topology calculation time to detect a shared route.

6   Conclusion

We conclude from our evaluations that our proposed traceroute method enables us to collect network topology by reducing packets in a short time.

In the future, we will investigate high-speed SNMP with a high-speed packet-sending and receiving technique that we applied to our high-speed traceroute method, and we will develop methods for detecting failure points and the extent of the failure using data from a high-speed network-state collecting technique.

References

1. Postel. J.: Internet Control Message Protocol, STD 5, RFC 792 (September 1981)
2. Case, J., Fedor, M., Schoffstall, M., Davin, J.: The Simple Network Management Protocol, STD 15, RFC 1157 (May 1990)
3. Malkin, G.: Traceroute Using an IP Option, RFC1393 (January 1993)
4. Tateishi., N., et al.: Methods for rapidly testing node reachablity with congestion control and evaluation. In: Ma, Y., Choi, D., Ata, S. (eds.) APNOMS 2008. LNCS, vol. 5297, pp. 491–494. Springer, Heidelberg (2008)

Fault Diagnosis for High-Level Applications Based on Dynamic Bayesian Network[*]

Zhiqing Li, Lu Cheng, Xue-song Qiu, and Li Wu

Networking and Switching Technology State Key Laboratory,
Beijing University of Posts and Telecommunications, Beijing 100876
zhiqing323@bupt.cn

Abstract. In order to improve the quality of Internet service, it is important to quickly and accurately diagnose the root fault from the observed symptoms and knowledge. Because of the dynamical changes in service system which are caused by many factors such as dynamic routing and link congestion, the dependence between the observed symptoms and the root faults becomes more complex and uncertain, especially in noisy environment. Therefore, the performance of fault localization based on static Bayesian network (BN) degrades. This paper establishes a fault diagnosis technique based on dynamic Bayesian network (DBN),which can deal with the system dynamics and noise. Moreover, our algorithm has taken several measures to reduce the algorithm complexity in order to run efficiently in large-scale networks. We implement simulation and compare our algorithm with our former algorithm based on BN (ITFD) in accuracy, efficiency and time. The results show that our algorithm can be effectively used to diagnose the root fault in high-level applications.

Keywords: fault diagnosis, dynamic Bayesian network, approximate inference, noisy environment.

1 Introduction

In order to ensure the availability and effectiveness of the network and improve the quality of services, it is important to quickly and accurately diagnose the root fault from the observed symptoms and knowledge. Because High-level applications are implemented in distributed systems, any fault in software or hardware would probably cause problems of availability or quality of services. Anyway the more complex of the service, the more uncertain of the relationship between symptoms and root faults [1]. Therefore in order to diagnose these faults, some probabilistic inference techniques are present, such as techniques based on Bayesian networks.

At present, people pay much attention on probabilistic fault diagnosis based on Bayesian networks which mostly include IHU [2], Shrink [3], Maxcoverage [6] and so on. All these algorithms assume that the state of the managed system does not

---

[*] **Foundation Items:** Sub-topics of 973 project of China (2007CB310703), Fok Ying Tung Education Foundation(111069).

C.S. Hong et al. (Eds.): APNOMS 2009, LNCS 5787, pp. 61–70, 2009.

change during the time interval of fault diagnosis. They can complete fault localization and perform relatively well using the current observations. However, they do not consider the changes of node status during the time of observing and inference. In distributed systems, routing and states of one component between end-to-end services may change dynamically caused by link congestion, dynamic routing or other reasons. If the network is large, the observation period must be set long, therefore the status of the same node in one observation period may change, we call these nodes contradiction nodes which are incorrectly considered to be caused by noises in the algorithms based on Bayesian networks. This increases the possibility of diagnostic errors.

Historical faults in management information base reflect the information of the system state through which we can analysis and get the probabilistic message and the state transition message of each node. If there is one node which fails more frequently, the prior probability of this node should be updated and become larger. However in BN the network is assumed to be static, hence the prior probability (characteristic) of each node is assumed to be fixed. So fault diagnosis techniques based on dynamic Bayesian networks attract our attention.

1.1 Related Work

Dynamic Bayesian network (DBN) [4] extends static Bayesian network (BN) by introducing the notion of time, namely, by adding time slices and specifying transition probabilities between these slices. DBN use the Markov assumption that the future system state is independent of its past states given the present state. As a result, a DBN is defined as a two-slice BN, where the intra-slice dependencies are described by a static BN, and inter-slice dependencies describe the transition probabilities. This feature just describes the dynamic systems. It is for this reason that diagnosis techniques based on dynamic Bayesian network become more and more important in probabilistic fault diagnosis techniques for dynamic system. The latest research results about exact inference include forwards-backwards algorithm, frontier algorithm and the interface algorithm [4] etc. Because the computational complexity of exact algorithms based on DBN is so high that we will lose diagnostic significance. Then we convert to approximate algorithms which include The Boyen-Koller (BK) algorithm [5], the factored frontier (FF) algorithm [4] and PF algorithm [7]. BK algorithm takes in a factored prior, does one step of exact updating using the jtree algorithm, and then projects the results back down to factored form. Unfortunately, for some models, even one step of exact updating is intractable. FF algorithm and PF algorithm has a lower diagnosis accuracy although their computational complexity is not so high.

1.2 Contributions in This Paper

In this paper, we propose a new probabilistic inference technique based on DBN, which is able to localize root causes in noisy and dynamic environments with high accuracy and efficiency. The observed symptoms could be probing results by active measurement [8] [9], or alarms which are received by the management station. Based on the symptoms, the algorithm can locate faults of high-level applications in dynamic system and show that it has a better performance than the inference based on static Bayesian model. Because characteristic of each node in dynamic system changes influenced by network environments, we add time information to DBN model to handle this dynamics and propose a simple tool to deal with noise. Moreover, our algorithm has taken several measures to reduce the algorithm complexity in order to diagnose in large-scale networks. The contributions made by this paper are:

1) We use DBN model to deal with system dynamics. In other words our algorithm uses transition probabilities between time slices to update the prior probability of the nodes which can improve the accuracy greatly.

2) We propose an approximate algorithm to reduce the computational complexity in large-scale networks. First, in adjacent time slices, we only update the probability of the nodes whose statuses are faulty, and the transition probability of normal nodes is approximately considered to be $P(F^t=0|F^{t-1}=0)=1$. Thus, if one node fails more frequently, the prior probability of it increases more rapidly. Second, if the observations between adjacent time slices are same, we hold that the state of each node does not change, so we do not perform inference and then we output the results diagnosed in the former time slice.

3) We implement the algorithm based on DBN and ITFD[10] and then compare them in accuracy, efficiency and time at 0% noise level and 5% noise level. Experiments show that the algorithm based on DBN outperforms that based on BN.

2 DBN Model for Diagnosis

As DBN is an extension of BN in time slices, first of all, we construct the static BN (Fig.1. (a)) using EM algorithm according to symptom nodes, fault nodes and the dependences between them. The BN model is a two layer directed graph, and then we extend the model by introducing the notion of time. Each node has two states: 0(normal) or 1(faulty). For a fault F_i, if it happens we denoted $F_i=1$ and $F_i=0$ otherwise. The general inference problem in Bayesian networks is NP hard, and the running time is exponential with the number of nodes. If the probability $P(|F|>k)$ is very small, ignoring such low probability events would make a good trade-off between increasing the running time and lowering the accuracy.

The DBN model is as Fig.1.(b). The first time slice is the initial part, each fault node has a prior $P(F_i^1)$, and each symptom node has a conditional probability table (CPT) $P(S_i^1|F_i^1)$ where S_i^1 is i'th symptom node at time 1 and F_i^1 is i'th fault node at time 1. The second part is a two-slice temporal Bayesian net (2TBN), for each fault node, it also has a transition probability $P(F_i^t|F_i^{t-1})$ which reflects the transfer characteristics of each node as the time slice goes. The value of it also depends on the characteristics of the network. For example, if the visits of one component is too large, it is more likely to be faulty in the next time slice, so the value of both $P(F_i^t=1|F_i^{t-1}=1)$ and $P(F_i^t=1|F_i^{t-1}=0)$ are larger than $P(F_i^t=0|F_i^{t-1}=0)$ and $P(F_i^t=0|F_i^{t-1}=1)$. Our algorithm aims to calculate the posterior probability $P(F^t|F^{t-1},S^t)$, where F^t is the set of root causes, S^t is a set of observations. Our algorithm performs fault diagnosis based on this model.

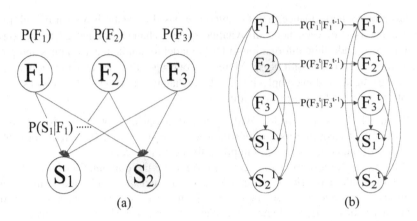

Fig. 1. (a) BN with 5 nodes (b) DBN with 5 nodes by extending BN in time slice. The model is a two-slice DBN (B_1, B_t), where B_1 is for initial time slice and B_t for each subsequent time slice, we perform inference literately in B_t. In the intra-slice both B_1 and B_t are static BN, in the inter-slice, we perform approximate inference by only updating the nodes with state changing.

3 Fault Localization Using DBN Model

In this section, we will introduce our approach for fault localization in High-level applications using dynamic Bayesian network.

The fault nodes of DBN correspond to the logical links or physical links. The probability of each fault node which reflects the property of the node changes with the dynamic network environments. The symptom nodes correspond to the end-to-end services or some applications. When failures take place in some of these nodes in one time slice, malfunction or decrease of quality of services will be observed. The prior probability of these nodes should be updated and increased. Therefore if the failure of one node takes place more frequently, the probability of this node should become larger, which is great important for us to diagnose the root fault. For example, if the database denies to be accessed, the root can be that the host which the database belongs to is down. It also can be that some of the links congest or the packages are lost. All these reasons which are corresponding to the fault nodes in DBN happen independently with a probability. If the bandwidth of network is relatively small, link congestion may often take place in some time slices. For this reason, the probability of nodes corresponding to link congestion should be updated more than others and become larger.

It can be seen that the characteristic of each node in DBN will change with the dynamic network environments. However in BN the network are assumed to be static, hence the probability(characteristic) of each node are assumed to be fixed and does not change. In the realistic application, if the failure of one node takes place more frequently, the probability of this node should become larger. It is very important for us to diagnose the root and improve the performance of the algorithm.

3.1 Algorithm Description

We utilize active probing method to detect the failures. In order to simulate the dynamics of High-level applications, when injecting faults, we produce the faults according to the prior probability in the first time slice but in the subsequent time slices we produce faults according to the transition probability and the state of this node in the last time slice. Then we produce symptoms and send them to the diagnosis engine. When the diagnosis engine receives the suspected nodes F_{sus}^t and observations S_o^t, first we determine the time slice information.

If it is in the first time slice, we use an observation rate $Ratio_{Fi}$ for each node to filter the spurious faults and observations. After filtering spurious nodes which are mostly caused by noise, we obtain a new set of suspected nodes which will be used to filter the whole DBN model. In other words in DBN model we only reserve the symptoms which can be explained by the fault nodes in the new set of suspected nodes. Then we start the diagnosis algorithm based on BN. During the inference, we obtain one or more sets of nodes $H_i=\{F_i^t=1,F_j^t=1......F_k^t=1\}$ called Hypothesis, each of which can explain all the symptoms independently. For each Hypothesis H_i, we calculate a brief $B(H_i^t,S_o^t)$ as the equation (1),where S_o^t is the observations, $pa(S_j^t)$ is a set of nodes which can explain S_j^t, F^t is the set of all fault nodes in the whole DBN model. We obtain the most likely Hypothesis H^{*t} as equation (2) and output H^{*t} as results of the algorithm. Meanwhile, the posterior probability of each node in H^{*t} should be calculated as equation (3). Finally we update the prior probability for each node in the next time slice in H^{*t} using the posterior probability and transition probability as equation (4).

$$B(H_i^t,S_o^t) = \prod_{F_i^t \in F^t} P(F_i^t) \prod_{S_j^t \in S^t} P(S_j^t \mid pa(S_j^t))$$

$$= \prod_{F_i^t \in H_i^t} P(F_i^t = 1) \prod_{F_i^t \in F^t/H_i^t} P(F_i^t = 0) \prod_{S_j^t \in S_o^t} P(S_j^t = 1 \mid pa(S_j^t)) \prod_{S_j^t \in S^t/S_o^t} P(S_j^t = 0 \mid pa(S_j^t))$$

(1)

$$H^{*t} = \arg\max_{H_i^t} B(H_i^t \mid S_o^t) = \arg\max_{H_i^t} B(H_i^t,S_o^t)$$

(2)

$$Postprior(F_i^t \mid pa(F_i^t)) = p(F_i^t = 1) \prod_{F_j^t \in F^t/F_i^t} p(F_j^t = 0) \prod_{s_j^t \in s_o^t} p(s_j^t = 1 \mid F_i^t = 1)$$

(3)

$$p(F_i^t) = \sum_{F_i^{t-1}} p(F_i^t \mid F_i^{t-1}) * Postprior(F_i^{t-1} \mid pa(F_i^{t-1}))$$

(4)

If it is not in the first time slice, we first compare the symptoms in the current time slice S_o^t with the symptoms in the last time slice S_o^{t-1}. If they are exactly the same($S_o^t = S_o^{t-1}$),we believe that none of nodes have changed, that is the state of the system does not change. Therefore we output the most likely Hypothesis H^{*t-1} in the last time slice as the results of the current time slice. If the symptoms are not the same ($S_o^t != S_o^{t-1}$), we get every node $F_i^{t-1} \in H^{*t-1}$ and add it to the set of suspected nodes in the current time slice $F_{sus}^{t} = F_{sus}^{t} \cup H^{*t-1}$ because the fault nodes in H^{*t-1} may still be faulty in the current time slice. An observation rate Ratio$_{Fi}$ for each node is used to filter spurious nodes and observations caused by noise. In order to reduce the complexity in DBN model we only reserve the symptoms which can be explained by the fault nodes in the new set of suspected nodes. Then the inference based on BN will be started and output one or more Hypothesis H_i^t. We calculate the brief of each Hypothesis $B(H_i^t,S_o^t)$ as equation(1) and output the most likely Hypothesis H^{*t} as equation(2).At the same time the posterior of each node in H^{*t} can be calculated as equation(3). Finally we update the prior probability for each node in the next time slice in H^{*t} using the posterior probability and transition probability as equation (4).The workflow of our algorithm is shown as Fig. 2.

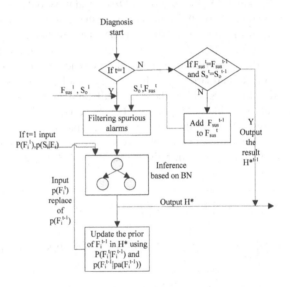

Fig. 2. The workflow of the algorithm based on DBN

In the distributed system, when an end-to-end service failure takes place, each node in its path can cause it probably. So we call these nodes suspected nodes F_{sus}. For each node in F_{sus}, if another end-to-end service which is successful passes through it, we consider this node to be ok and remove it from F_{sus}. If there are no nodes in F_{sus}, we hold that the end-to-end service failures are caused by noise which should be filtered using a mechanism before inference.

We propose a simple but effective tool as equation (5)[11] to deal with spurious faults and symptoms caused by noise. Experiment show that it can work well and improve the accuracy with less running time. For a real fault, it is highly unlikely that all symptoms can be seen. However if the number of observations is lower than a threshold, we consider it to be spurious fault. For example, if a node failure F_i^t can arouse five symptoms, but the observations are only two, suppose all the conditional probability are sane, so Ratio$_{Fi}$=0.4 which is lower than a threshold 0.5. This node will be considered to be spurious and be removed from $F_{sus}{}^t$. If no node in $F_{sus}{}^t$ can explain a symptom, we hold that this symptom is spurious and remove it from the observations.

$$Ratio_{F_i{}^t} = \frac{\sum\limits_{S_i \in S_o} P(S_i{}^t \mid F_i{}^t)}{\sum\limits_{S_i \in S} P(S_i{}^t \mid F_i{}^t)} \tag{5}$$

3.2 Analysis of Complexity

The complexity of algorithms based on DBN at present is so high that it is difficult to diagnose in large-scale networks. Because in each time slice we should perform update for each node and do inference in the whole DBN. Generally, the number of nodes is so large that even performing one time update and inference is intolerable. To reduce the computational complexity, we improve the algorithm as follows:

1) If the observations received in the current time slice are the same as observations in the last time slice ($S_o{}^t = S_o{}^{t-1}$), the status of each node will be considered not to change. The diagnosis results in the last time slice will be outputted as results of the current slice. The algorithm only needs to do one time comparison and not to do inference, so the time is $O(|S_o|)$.

2) If the observations are different between adjacent time slice ($S_o{}^t != S_o{}^{t-1}$), we add H^{*t-1} to $F_{sus}{}^t$ and perform diagnosis because the nodes in H^{*t-1} may be still faulty in the current slice. The algorithm only update the prior probability of the node in H^{*t}, thus if a node becomes faulty more frequently, the prior of it is larger than others, which is great important to improve the accuracy. Here we consider the transition probability $p(x_i{}^t = 0 \mid x_i{}^{t-1} = 0) = 1$ approximately.

3) After filtering we only reserve the symptoms which can be explained by the fault nodes in the new set of suspected nodes. The DBN model is simplified so much that the algorithm complexity reduces highly. If the number of real fault is $K(K < |F_{sus}|)$, the maximum number of Hypothesis $|F_{sus}|^k$, so the time is $O(|F_{sus}|^k)$ not $O(|F|^k)$, where $|F|$ is the whole number of fault nodes and $|F_{sus}| << |F|$.

4 Simulation

In this section, we simulate the network environment and adopt active probing method to obtain the symptoms and detect suspected fault. Then we implement ITFD and our algorithm with java and compare them in accuracy, efficiency and time.

4.1 Simulated Networks

We use network simulation tool INET[11] to generate the network topology. The connection between each pair of nodes is considered as an end-to-end service which corresponds to the symptom node in DBN model whose status could be possibly observed. The faults of DBN correspond to network edges. If a service passes though an edge, then we add dependence relationship between them. Finally we extend BN by adding time information and obtain DBN model.

For each fault node, we produce a prior probability which can be used to inject fault in the first time slice. For each symptom node, we produce a conditional probability which represents the influence from fault node to symptom node. In the first time slice, for each fault node we randomly generate a number range such as $[0,1]$ referring to prior probability. If the prior probability falls into the range, this node will be considered as faulty and the symptom node which traverses this fault node will be fail as observed. In the subsequent time slice, for each fault node whose status is 0, we randomly generate a number range referring to transition probability $P(F_i^t=1| F_i^{t-1}=0)$, otherwise referring to $P(F_i^t=0| F_i^{t-1}=1)$. If the transition probability fall into the range, the status of this node will be changed from 0 to 1 or from 1 to 0. Because the transition probability $P(F_i^t=1| F_i^{t-1}=1)$ and $P(F_i^t=0| F_i^{t-1}=0)$ are close to 1, each node can most probably keep its status in adjacent time slice. For each real fault, all symptom nodes which traverse this fault node will fail as observed.

For each observation, all fault nodes in its path can be added into F_{sus}. For each fault node in F_{sus}, if at least one symptom node whose status is 0 passes through it, this node will be considered as normal and removed from F_{sus}. Both the suspected nodes F_{sus} and observations are sent to diagnosis engine.

In distributed systems, the status of symptom nodes may change probably during propagation such as loss of packet or spurious symptoms. If some observations are lost, the diagnosis engine may feel incorrectly that there is no such observation. If the engine receives some spurious symptoms, it is useless to do such inference. For this reason, we inject spurious symptoms with a noise level 0% or 5%. In other words, we change the status of symptom nodes from 0 to 1 or from 1 to 0 with the noise level.

4.2 Metrics

We utilize two metrics named accuracy and false positive to evaluate the performance of the two algorithms. H is the set of diagnosis results, F is the set of real faults.[2]

$$Accuracy = \frac{H \cap F}{F}, \qquad falsepositive = \frac{H - F}{F}$$

4.3 Simulation Results

We implement our algorithm based on DBN (ADBN)) and the ITFD based on BN, where the time slice in DBN is T=50. All the figures show results averaged 50 runs.

Without noise: Fig.3.(Left) compares the diagnosis accuracy of the algorithm based on DBN and the algorithm based on BN(ITFD). As the number of nodes becomes larger, the accuracy of ADBN remains higher than 90% but that of ITFD declines slightly. When the number of nodes is more than 200, the accuracy of ITFD drops sharply but that of ADBN declines elegantly. With the number of nodes up to 500, the accuracy of ITFD is worthless. The false positive of ADBN is lower than ITFD shown in Fig.3(Right), in other words, the precision of ADBN is higher than ITFD. As number of nodes is more than 200, the precision of ITFD drops sharply but that of ADBN drops slightly.

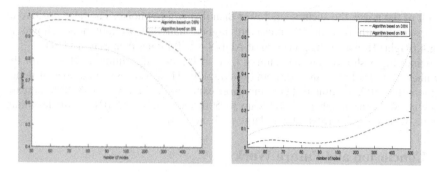

Fig. 3. Comparison of the two algorithms with noise=0% (Left) Accuracy (Right) Precision. The falsepositive of ADBN is lower than that of ITFD,that is the precision of ADBN is higher than that of ABN.

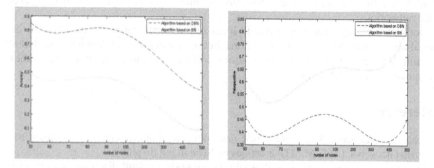

Fig. 4. Comparison of the two algorithms with noise=5% (Left) Accuracy (Right) Precision

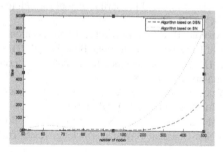

Fig. 5. Comparison of the two algorithms in fault localization time. Both of the two algorithms runs quite rapidly with the number of nodes less than 200. However, as the number of nodes is more than 200, the running time of ITFD increases exponentially but that of ADBN is still less than 300ms.

With 5% noise: As the number of nodes becomes larger, the accuracy of both of the two algorithms declines but ADBN is higher than ITFD by 40% shown in Fig.4.(Left). We can propose that the ITFD cannot work in noise environment. As the result of the noise, the accuracy is lower than that without noise. The false positive of the two algorithms with 5% noise are higher than that without noise shown in Fig.4.(Right).However the precision of ADBN is still higher than that of ITFD.

Time: Fig.5 shows that the running time of the two algorithms is close with the number of nodes less than 200, both quite rapidly. However, as the number of nodes is more than 200, the time of ITFD increases exponentially but that of ADBN increase slightly. With the number of nodes up to 500, the time of ADBN is still less than 300ms which is quite acceptable in large-scale networks.

5 Conclusion and Future Work

We have presented a technique to diagnose and locate end-to-end services faults in dynamic system and showed that it has a better performance than the inference based on static Bayesian model. We used dynamic Bayesian model to handle system dynamics and a simple tool to deal with noise. Our approximate algorithm has taken several measures to reduce the algorithm complexity in order to diagnose in large-scale networks.

Our new technique can perform fault localization for the current time using the historical alarms and the current observations, but it cannot predict when and which node will fail. We believe that the fault prediction will become more and more important in the management of IP network.

References

1. Steinder, M., Sethi, A.S.: Probabilistic fault localization in communication systems using belief networks. IEEE/ACMTransactions on Networking (in press)
2. Steinder, M., Sethi, A.S.: Non-deterministic event-driven fault diagnosis through incremental hypothesis updating. In: Goldszmidt, G., Schoenwaelder, J. (eds.) Integrated Network Management VIII, Colorado Springs, CO, pp. 635–648 (2003)
3. Kandula, S., Katabi, D., Vasseur, J.P.: Shrink: A tool for failure diagnosis in IP networks. In: Proc. ACM SIGCOMM (2005)
4. Murphy, K.P.: Dynamic Bayesian Networks: Representation, Inference and Learning, PHD degree thesis of University of California, Berkeley (2002)
5. Boyen, X., Koller, D.: Tractable inference for complex stochastic processes. In: Proc. of the Conf. on Uncertainty in AI (1998)
6. Kompella, R.R., Yates, J., Greenberg, A., Snoeren, A.: Detection and Localization of Network Black Holes. In: Proc. IEEE INFOCOM (2007)
7. Arulampalam, S., Maskell, S., Gordon, N.J., Clapp, T.: A Tutorial on Particle Filters for On-line Nonlinear/Non-Gaussian Bayesian Tracking. IEEE Tran. on Signal Proc. 50(2), 174–188 (2002)
8. Natu, M., Sethi, A.S.: Probe Station Placement for Fault Diagnosis. In: Proc. IEEE GLOBECOM 2007 (2007)
9. Natu, M., Sethi, A.S.: Probabilistic fault diagnosis using adaptive probing. In: Clemm, A., Granville, L.Z., Stadler, R. (eds.) DSOM 2007. LNCS, vol. 4785, pp. 38–49. Springer, Heidelberg (2007)
10. Cheng, L., Qiu, X.-s., Meng, L., Qiao, Y., Li, Z.-q.: Probabilistic Fault diagnosis for IT Services in Noisy and Dynamic Environments. In: Proc. IM 2009 (2009)
11. Winick, J., Jamin, S.: Inet-3.0: Internet topology generator, Technical Report CSE-TR-456-02, University of Michigan (2002)

Novel Optical-Fiber Network Management System in Central Office Using RFID and LED Navigation Technologies

Masaki Waki, Shigenori Uruno, Yoshitaka Enomoto, and Yuji Azuma

NTT Access Network Service Systems Laboratories, NTT Corporation
1-7-1, Hanabatake, Tsukuba, Ibaraki 305-0805, Japan
{waki,uruno,enomoto,azuma}@ansl.ntt.co.jp

Abstract. Optical IP services such as FTTH have been growing rapidly, leading to an urgent need to manage the optical fiber network from the central office. We have developed an optical-fiber distribution facility called an IDM. However, the changing FTTH environment means we need further efficiency and cost-effectiveness for network construction and operation work. This paper proposes a novel optical-fiber network management system for the IDM that realizes reducing working time, less skilled work and eliminating human error. This system consists of two technologies with control software, namely optical distribution identification technology using RFID, and automatic navigation technology using an LED and DB. In this paper, we describe optimized work flows for installation and removal work. Moreover, we achieved a reduction of up to 48 % in working time with less skilled workers and accurate DB management by using our proposed system.

Keywords: FTTH, human error, RFID, automatic navigation system.

1 Introduction

Optical IP services such as fiber to the home (FTTH) have been growing rapidly throughout the world. In Japan, the number of FTTH customers has increased tenfold in the last 5 years [1]. In addition, NTT plans to achieve 20 million customers by 2010 [2]. In this kind of environment, we have to reduce optical fiber network costs related to construction, maintenance and operation. At the same time, our goal is to spread the use of FTTH services and related components that link an optical line terminal (OLT) in an NTT central office to optical network units (ONU) in a customer's home. Therefore, we have developed optical fiber network facilities for central offices that consist of an integrated distribution module (IDM), optical fiber cables and a management system [3][4]. These facilities realize increased work efficiency and high-density connectors.

Figure 1 shows the optical access network topology for typical FTTH services in Japan. There are many kinds of access network line and the type depends on the FTTH service being provided. In recent years, FTTH in Japan has provided an access network bandwidth of 100 Mb/s. Figure 1 (a) shows the basic FTTH service with a

C.S. Hong et al. (Eds.): APNOMS 2009, LNCS 5787, pp. 71–81, 2009.

gigabit Ethernet-passive optical network (GE-PON) [5]. This network serves a maximum of 32 customers per optical fiber by using two kinds of splitter in the central office and near the customer's house. The other access network topology shown in Fig. 1 (b) is a single star (SS) topology, which is used for businesses and common carrier leased lines. To manage these services in the access network, there is a strong need to be able to change a service easily and rapidly, and to realize further efficiency and cost-effectiveness as regards network construction and operation.

In this paper, we propose a novel optical-fiber network management system for the IDM that reduces working time, requires less skilled work and prevents human error. This system consists of two technologies with control software, namely optical distribution identification technology using radio frequency identification (RFID), and automatic navigation technology using a light emitting diode (LED) and a database (DB). We describe optimized work flows for installation and removal work. As a result, we achieved a working time reduction of up to 48 % with less skilled workers, and accurate DB management by using a novel system.

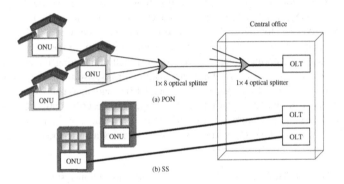

Fig. 1. Access network topology

2   Installation and Removal Work on IDM

In this section, we describe actual distribution facilities with a management system in a central office and the service order flow.

2.1   Distribution Facilities in Central Office

The progress made on such aspects of the optical access service as operation and maintenance technologies has been inadequate compared with the growth in service demand and the development of optical line and transmission technologies [6]. We have developed an IDM to increase the work efficiency and to solve the problem of optical fiber cable congestion. In addition, it is likely that there will be a shortage of available floor space. We realized a high-density connector and a space-saving module by optimizing the architecture and design of the module. Figure 2 shows the optical fiber distribution facility in a central office using the IDM. The IDM realizes simple optical fiber distribution, and integrates the following functions; the accommodation of splitters for splitting optical signals, the termination and the connection of feeder and central office cables, the selection of an optical fiber to be tested and the launching of a maintenance signal. The IDM consists of IDM-A, which terminates and connects feeder cables from outside the central office, and IDM-B, which terminates and connects all the optical fiber cables that run from the OLTs.

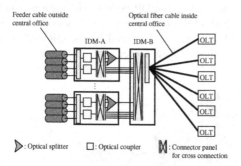

Fig. 2. Optical fiber distribution facilities in central office

This configuration makes it possible to minimize the additional distribution of optical fiber cables even if an IDM-A is newly installed. Therefore, the configuration reduces the number of cables and simplifies the cable configuration.

Figure 3 shows a photograph of the IDM-A. We use an MU connector with a 1.1 mm diameter optical fiber cord. Although the cord is temporarily housed in a plug holder, it can be extracted from the holder and connected if needed. The jumper unit connects the optical fiber cord and the feeder cable outside the central office and different types are used depending on the service.

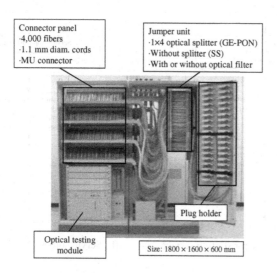

Fig. 3. Integrated distribution module (IDM-A)

2.2 Management System in Central Office Using QR Code

Figure 4 shows the configuration of a system designed to manage central office optical distribution facilities using a two-dimensional identification code, called a QR
code (ISO/IEC 18004). The registration and accessing of fiber information are facili-
tated by attaching an information tag printed with the QR code to the optical fiber
cord used in the IDM, and reading the QR code on the information tag with an identification code reader, called a QR code reader. The QR code includes information for
facility identification. Moreover, the data read by the QR code reader can be output
from a mobile terminal to the database server.

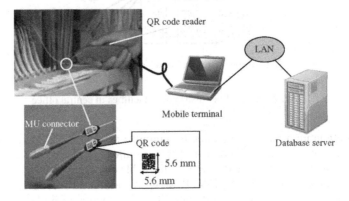

Fig. 4. Management system using QR code

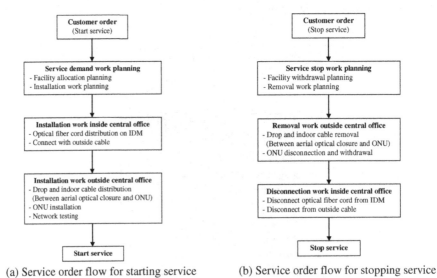

(a) Service order flow for starting service   (b) Service order flow for stopping service

Fig. 5. Service order flow from the customer's service demand

Figure 5 shows the service order flow from the customer's request for a service to its provision. When starting a service as shown in Fig. 5 (a), the order flow is as follows; after receiving the customer's service request, the operator undertakes service demand work planning, which consists of facility allocation and installation work planning. Next, a worker in a central office undertakes distribution work such as fiber cord distribution on the IDM according to the work plan. Finally, after completing the distribution work inside the central office, the worker undertakes distribution work outside the central office such as drop and indoor cable distribution, ONU installation and network testing. The FTTH service starts once all the distribution work and testing are finished. On the other hand, when stopping a service as shown in Fig. 5 (b), the order flow is as follows; after receiving the customer's order to discontinue the service, the operator undertakes service stop work planning, which comprises facility withdrawal and removal work planning. Next, a worker outside the central office performs removal work such as drop and indoor cable removal, ONU disconnection and withdrawal. Finally, after finishing the removal work outside the central office, the worker performs disconnection work such as disconnecting the fiber cord from the IDM. After all the removal and disconnection work is finished, the FTTH service is discontinued. When changing a service, the worker would perform the service stop and start procedures in turn.

3 Architecture Design of Novel Optical-Fiber Network Management System

In the previous section, we described actual distribution facilities with a management system in a central office and the service order flow. However, as mentioned above, increasing FTTH service demand, and the introduction of novel FTTH services allows us to further improve efficiency and cost-effectiveness. Therefore, we focused on work at the IDM, and investigated how to reduce both working time and the required level of skill, and to eliminate human error.

Figure 6 shows our novel optical-fiber network management system. It consists of optical distribution identification technology and an automatic navigation system, a connection tool optimized for this system, a PC with control software referred to as a control terminal, and a DB server that provides service and access network information. The features of this system are as follows:

(1) ID discrimination function

We mounted RFID chips on connectors and connector adaptors, and installed an RFID reader, an ID discrimination device and a solenoid as a controlling trigger in the connection tool. Each RFID-chip-mounted adaptor has a unique ID. An RFID-chip-mounted connector has a unique ID and stores customer information such as the services the customer uses. The ID discrimination device in the connection tool linked to a control terminal uses a USB cable that supplies DC power.

(2) DB autonomous installation

Once a connector has been connected or disconnected, control software immediately updates the DB. Moreover, it is possible to crosscheck the records in the DB when a worker needs to confirm the difference between order information and the current state of the optical fiber distribution.

(3) Alarm notification and connector holder locking in discriminating different IDs
If a worker selects the wrong connector when disconnecting an optical fiber cord, the control software activates an alarm tone and posts a message on the PC display. Moreover, the connection tool trigger that controls the on/off connector remains locked.

(4) Automatic navigation using LED
LEDs mounted on both sides of the connector adaptor are illuminated when a installation or removal order is input.

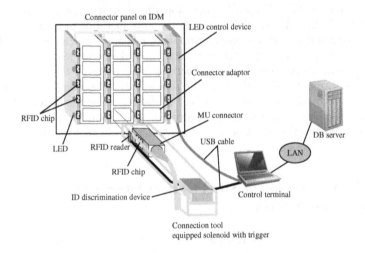

Fig. 6. IDM mounted proposed system

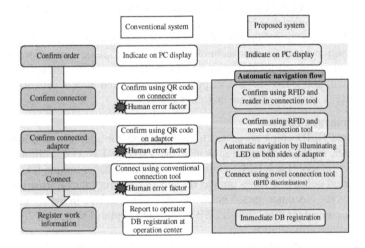

Fig. 7. Conventional and proposed installation work flow

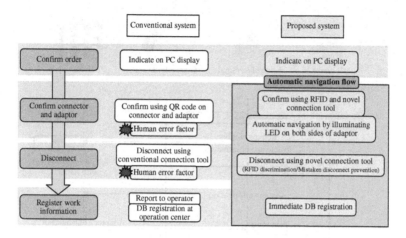

Fig. 8. Conventional and proposed removal work flow

Figures 7 and 8, respectively, show the conventional and proposed installation and removal work flow for an optical fiber network facility. It can be seen that a conventional IDM is subject to human error because it depends on manual work and visual confirmation. In addition, after confirming that the connector and connector adaptor are correct using the QR code, the worker must look away in order to remove the QR reader and attach the MU connector to the connection tool. For these reasons, it is possible for the worker to make a mistake. Therefore, a person working on an IDM with the currently used system must perform many crosschecks and this is time-consuming. To solve this problem, we investigated a way of eliminating human error from IDM work. Then, we developed an IDM work support system using the RFID and LED navigation technologies. This system employs an LED and sound navigation, and it enables the worker to continue concentrating on the target and thus prevents any errors as regards connector disconnection.

Figures 9 and 10 show the information flow when installing and removing a network. This flow involves four devices, namely a DB server in the management office, a control terminal with a connection tool, and an IDM equipped with an RFID and an LED navigation system. The information flow during installation is as follows; first, a worker in a central office downloads operation orders using the control terminal. Then, the worker selects the operation order using control software, and the details of the order including the adaptor number and connector information are input into the control software. Next, control software illuminates the appropriate LEDs that are mounted on both sides of connector adaptor on the IDM. The worker then places an appropriate connector in the connection tool, and the RFID reader in the connection tool confirms this by reading the RFID mounted connector. The control software determines whether the connector is correct, and informs the worker by both sound and display. Next, the worker connects the connector to the correct connector adaptor using the connection tool, and then the RFID reader in the connection tool confirms this by reading the RFID mounted connector adaptor. The control software makes a determination and informs the worker in the same way as that mentioned above. Finally, the control software uploads the installation data to the DB server.

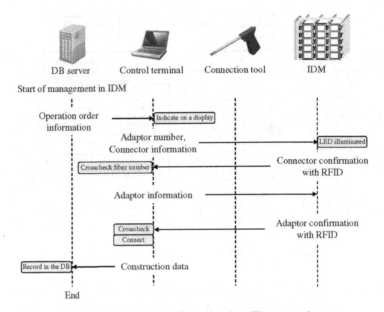

Fig. 9. Information flow when installing network

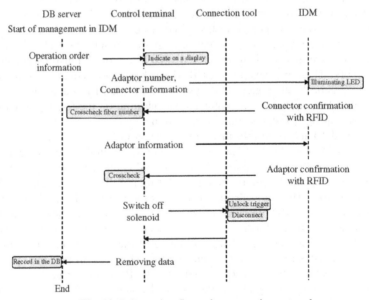

Fig. 10. Information flow when removing network

The information flow when removing a network is almost the same as the flow described above. However, the most important point in the work flow is to ensure that the wrong connector is not disconnected. If this were to happen, customers using

FTTH services would be cut off from the FTTH network. To prevent such a occurrence, we installed a solenoid beside the control trigger in the connection tool. Therefore, the connection tool trigger remains constantly locked. The trigger is released only if the control software determines that the connector adaptor is correct.

We believe that human error can be eliminated by using the proposed system including the solenoid-control connection tool, LED navigation and automatic crosscheck functions. Moreover, IDM workers can perform without having to undertake many crosschecks. Therefore, all workers who have any level of skill can perform the above work accurately and quickly.

4   Prototype Manufacture and Evaluation Result

We have manufactured a prototype based on the design concept described in the previous section. Figure 11 (a) shows the connector adaptor mounted with 16 RFIDs and LEDs on both sides. Figure 11 (b) shows an MU connector equipped with an RFID, and Fig. 11 (c) shows a connection tool equipped with an RFID reader and a solenoid to control the locking of the connector holder.

(a) Connector adaptor
    with RFIDs and LEDs

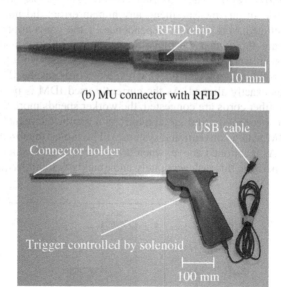

(b) MU connector with RFID

(c) Connection tool optimized for proposed system

Fig. 11. Prototype manufacture based on proposed system

Figure 12 shows the IDM equipped novel optical fiber network management system. Figure 12 (a) shows the appearance of the connector panel on an IDM and the connection tool and an enlarged view of connector adaptors with illuminated LEDs. Figure 12 (b) shows the control software window when a network is being installed.

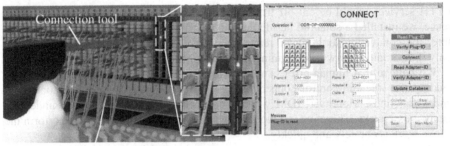

(a) Enlarged view and appearance of IDM   (b) Control software window

Fig. 12. Enlarged view and appearance of IDM, and control software window

This shows that the worker can confirm the operation order and the current work according to the work flow shown in fig.7 or 8.

Next, we investigated the workability of the IDM mounted in conventional and proposed management systems. We measured the time needed for installation and removal work. The participants were five people who were unfamiliar with IDM work. In this measurement, any human errors did not happen. Figure 13 shows the averaged working time dependence of the conventional and proposed systems. The figure shows that the times needed for installation and removal using the proposed system were respectively 28 and 48 % less than with the conventional system. Additionally, a connector panel with fiber cords on the IDM used for this measurement was exactly aligned. As the currently used IDM is providing the FTTH services and the fiber cords are congested, the worker spends more time crosschecking and looking for a connector and an adaptor. As a result, the actual working time when using an IDM with a conventional system will be longer than in this measurement.

As mentioned above, when using the proposed system, it is possible for all workers to perform this task accurately and quickly without frequent crosschecking.

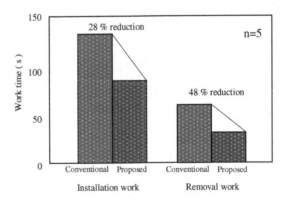

Fig. 13. Working time dependence of conventional and proposed systems

5 Conclusion

We investigated a novel optical fiber network management system designed to reduce working time and the required level of skill, and to eliminate human error. We achieved a working time reduction of up to 48 % with less skilled workers and accurate DB management by using our proposed system. These results show that the proposed system is a promising approach for achieving increased efficiency and cost-effectiveness for access networks.

References

1. Online,
   http://www.soumu.go.jp/main_sosiki/joho_tsusin/eng/Releases/
   NewsLetter/Vol19/Vol19_19/Vol19_19.pdf
2. Online, http://www.ntt.co.jp/news/news07e/0711zrmh/
   bqyt071109d_06.html
3. Tachikura, M., Mine, K., Izumita, H., Uruno, S., Nakamura, M.: Newly developed optical fiber distribution system and cable management in central office. In: Proc. 50th IWCS, pp. 98–105
4. Arii, M., Enomoto, Y., Azuma, Y., et al.: Optical Fiber Network Operation Technologies for Expanding Optical Access Network Services. NTT Technical Review 5(2), 32–38 (2007)
5. Shinohara, H.: Broadband Access in Japan: Rapidly Growing FTTH Market. IEEE Communications Magazine 72 (2005)
6. Kawataka, J.: Study of improvement in operation and maintenance of optical access network facilities. In: 11th APNOMS, innovation session, pp. 31–34 (2008)

Mobility Management Using Virtual Multi-parent Tree in Infrastructure Incorporated Mobile Ad Hoc Networks*

Trung-Dinh Han and Hoon Oh**

School of Computer Engineering and Information Technology, University of Ulsan,
(680-749) San 29, Mugeo 2-Dong, Nam-gu, Ulsan, South Korea
trungdinhvn@yahoo.com, hoonoh@ulsan.ac.kr

Abstract. Mobile nodes in the integrated network of infrastructure networks and mobile ad hoc networks need an efficient registration mechanism since they register with Internet gateway via wireless multiple hops. Tree-based mobility management is highly dependable in that mobile nodes register with Internet gateway along tree paths from itself up to a gateway without using flooding. In this paper, we introduce a new method using a virtual multi-parent tree in which a node maintains multiple parents whenever possible. A mobile node can still retain its upstream path to a gateway even though it loses connectivity to one of its parents. The new scheme is evaluated against the tree-based method and the other traditional methods - proactive, reactive, and hybrid by resorting to simulation. The result shows that the new method outperforms the other ones and also is very robust against the change of topology.

Keywords: Mobile IP, Mobility Management, Multiple Wireless Hops, Multi-Parent Tree, Registration.

1 Introduction

Wireless mobile nodes can communicate with each other via wireless multiple hops by building a mobile ad hoc network (MANET) without the help of infrastructure base stations. However, they often need to have an access to the Internet or communicate with mobile nodes (MN) in other MANETS. In this case, a mobile node has to register with Internet gateway and remain registered so that other nodes can find easily where it is located; this registration activity causes a lot of control overhead since it is done via wireless multiple hops unlike the original Mobile IP or Cellular IP that only manages the mobility of nodes with a wireless one hop [1]. Therefore, the Mobile IP that deals with mobile nodes in MANETs requires an efficient registration or mobility management scheme.

Recently, a lot of researches have been conducted regarding mobility management; however, the existing schemes suffer from a high mobility management

---

* This work was supported by 2008 Research Fund of University of Ulsan.
** Corresponding author.

overhead due to control messages flooded over the networks, curtailing effective bandwidth. Broch et al. [8] assume that IG is equipped with two network interface cards, where one that executes the standard IP mechanism acts as a bridge to Internet nodes while another does as a bridge to mobile nodes; this protocol uses DSR protocol to discover IG that rely on flooding. Jonsson et al. [4] focus on when to register with a foreign agent using MIPMANET Cell Switching algorithm in order to reduce control overhead. In [12], Sun et al. discuss how Mobile IP and AODV can cooperate to discover multi-hop paths between mobile nodes and IGs. Prashant et al. [11] propose a hybrid mobility management scheme that combines the techniques such as agent advertisements, TTL scoping and caching of agent advertisements, eavesdropping, and agent solicitation. In [5], Ammari et al. propose a three-layer approach in which a mobile gateway is used as an interface between mobile nodes and IG that executes Mobile IP and DSDV [3]. Some other studies on gateway discovery were conducted based on AODV [6, 10]; however, they still use flooding.

The traditional approaches for node registration are in general categorized into three types: Proactive scheme [6], reactive scheme [6, 8, 11], and hybrid scheme [6, 11]. In the proactive scheme (Proactive), an IG floods advertisement message periodically to inform its presence to the mobile nodes that are visiting its coverage area. Upon receiving the advertisement message, the mobile nodes set up the forward paths to IG and can register with the IG that has issued the advertisement message along the paths. Since this method allows all mobile nodes to register with a gateway regardless of whether they participate in active communication or not, remote nodes can make connection request to the registered mobile nodes all the time. Nevertheless, the repeated broadcasting of IG advertisements and solicitations can have a negative impact on the MANET due to excessive flooding overhead. In the reactive scheme (Reactive), only the mobile node that wants to participate in communication floods IG solicitation message to explore an IG, setting up a reverse path to IG. This approach also suffers from the flooding of IG solicitation message. The hybrid approach (Hybrid) was proposed to complement the shortcomings of two approaches, where connectivity is maintained in such a way that IG floods advertisements, but only within a limited number of hops, say k-hop, from the IG while the mobile nodes residing outside k-hop range flood IG solicitations only if they want to communicate with any mobile nodes. This approach still uses flooding for registration and solicitation.

Meanwhile, the tree-based approach (Tree-Based) does not use an inefficient flooding for registration, and instead, trees are formed and maintained such that every node in a tree keeps track of information of all its descendent nodes. However, it has to find a new parent if it loses connection to its parent, requiring another join process. The new approach proposed in this paper is basically based on the tree-based approach [7] where every node registers with the gateway along tree paths; however, it is different in that a node maintains multiple parents. Thus, each node in a tree maintains multiple paths virtually to gateway. That is, a mobile node can still retain its upstream path to a gateway even though it

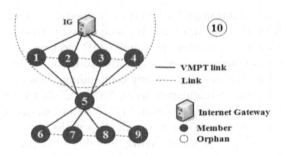

Fig. 1. Network Model

loses connectivity to one of its parents without going through a tree maintenance process. The proposed method is evaluated against the tree-based approach and the traditional mobility management schemes for different network scenarios in terms of control overhead, registration ratio, registration latency and registration jitter. The simulation results show that the new approach greatly outperform the previous approaches and is highly robust against the variations of network scenarios.

The remainder of the paper is organized as follows. Section 2 describes network model with useful definitions. The formal description of a new mobility management method is given in Section 3. In Section 4, we evaluate different methods by resorting to simulation and finally, we make concluding remarks in Section 5.

2   Preliminary

2.1   Network Model

Mobile nodes (MN) form the multiple trees, each of which stems from an Internet gateway (IG) being a root as in Fig. 1 in which the solid lines indicate Virtual Multi-Parent Tree (VMPT) links and the dashed lines do links that do not belong to the trees. A node is said to be a *member* if it belongs to any tree, and is said to be an *orphan* if it does not have any parent. We define a *primary parent* as a node that an orphan node joins for the first time. An IG acts as both home agent (HA) and foreign agent (FA). We assume that a node can overhear packets that are transferred among the other nodes [9]. We further assume that all nodes have to be ready to receive communication request from other distant nodes regardless of the schemes, proactive, reactive or hybrid.

2.2   Message Definitions

We define some control messages used to build VMPT. We denote the type of control message as *CT*.

- IG-Hello = (CT), IG broadcasts this message periodically.
- MN-Hello = (CT, HopToIG), MN sends this message to its primary parent periodically.
- J-REQ = (CT), MN sends this message to its primary parent to join.
- CR-REQ = (CT), an orphan responds with this message to its child which sends MN-Hello.

2.3 VMPT Information Structure (VTIS)

2.3.1 VTIS Definition

We define some notations for convenience with respect to node i as follows. $i.IG$ indicates IG to which node i belongs. $i.NS$ indicates a set of neighbors of node i. $i.P$ and $i.C$ denotes a set of parents and children of node i, respectively. $d(i, j)$ stands for the distance in hops from i to j; if j is IG, it is simply referred to as $HopToIG$.

   Definition 1: $VTIS(i) = (i.P, i.C, d(i, i.IG))$ where $i.P = \{x|\ x \in i.NS, d(i, i.IG) = d(x, x.IG) + 1 \}$, $i.C = \{x|\ x \in i.NS, d(i, i.IG) = d(x, x.IG) - 1 \}$.

2.3.2 VTIS Maintenance

A node can get a child when it either receives J-REQ or overhears MN-Hello from a node which has HopToIG greater than its own one by one. A node can get a parent when it either receives ACK to J-REQ or overhears MN-Hello from a node which has HopToIG less than its own one by one.

   Every member in a VMPT sends MN-Hello periodically to its primary parent. The parents update their VTIS whenever they receive or overhear MN-Hello from their children. If a node does not receive or overhear MN-Hello from its child for the specified interval, it deletes the child. If a node fails to send MN-Hello to its primary parent, it selects another parent to send MN-Hello and if it succeeds, it changed its primary parent. In case that there does not exist a parent to select, it becomes an orphan node. Thus, every node in a VMPT can maintain its VTIS as indicated in Definition 1.

3 VMPT Mobility Management

3.1 VMPT Establishment

3.1.1 MN-IG Join Process

Let us consider the procedure that MN joins IG. If a MN i receives IG-Hello from an IG and $d(i, i.IG) > 1$, it sends J-REQ to join. If it receives ACK from the IG, it becomes a child of the IG. If an IG receives J-REQ, it takes the sender as its child.

3.1.2 MN-MN Join Process

Consider the procedure that a MN joins a MN being a member of a VMPT. When an orphan overhears a MN-Hello from a neighbor member, it sends J-REQ to the neighbor to join. Upon receiving ACK, it becomes a child of the neighbor. A member that receives J-REQ takes the sender as its child.

3.2   VMPT Maintenance

3.2.1   VMPT Maintenance Process

A mobile member node *unicasts* MN-Hello to its primary parent while an IG *broadcasts* IG-Hello periodically. By overhearing MN-Hello, a node can update its neighbors. We describe how VMPT links are maintained dynamically below.

   A. Children Maintenance: If a parent either receives or overhears MN-Hello from its child, it updates its children. If the parent overhears MN-Hello from its neighbor that has HopToIG less than its own HopToIG by one, it takes the node as a new child. If it does not receive MN-Hello from a certain child for the specified interval, it deletes the child.

   B. Multi-Parent Maintenance: A node that overhears MN-Hello from a neighbor which has HopToIG equal to the HopToIG of its primary parent takes the node as one of its parents. If a node does not overhear MN-Hello from one of its parents for the specified interval, it deletes the parent.

3.2.2   Parent-Change Process

Suppose that a node x overhears MN-Hello or receives IG-Hello from its neighbor y. If $d(x, x.IG) > d(y, y.IG) + 1$, x joins y by initiating the MN-MN Join Process or following the MN-IG Join Process. If it changes its parent, it updates its VTIS.

3.2.3   Children-Release Process

If an orphan node receives MN-Hello from its children, it replies with CR-REQ to ask them to leave. The child that either fails to send MN-Hello or receives CR-REQ will select another parent as a primary parent from its VTIS to send MN-Hello. In case that there does not exist a parent to select, it becomes an orphan and then selects a neighbor member y that gives the minimum $d(y, y.IG)$ among its neighbor list and initiates the MN-MN Join Process or IG-MN Join Process.

4   Performance Evaluation

Three scenarios S1, S2, and S3 are used according to IG locations (center, top center, and corner). If IG is located in the center, mobile nodes will have a shorter average distance to the IG, resulting in a decreased control overhead. In the other two cases, nodes will have a relatively higher distance, increasing control overhead. We use four metrics *control overhead, registration ratio, registration latency* and *registration jitter* to evaluate the different mobility management schemes which are *Reactive, Proactive, Hybrid, Tree-Based*, and *VMPT*.

   We performed comparison in the QualNet version 3.9 by changing IG location (at center (S1), top center (s2), and corner (s3)), varying number of nodes (*nNodes*) as 50, 75, and 100 or maximum speed (*mSpeed*) as 0, 10, 20, 30, 40, and 50 (m/s) in the terrain of 1000 x 1000(m^2). Figure 2 shows an S1 with a different number of nodes. The used parameter values are given in Table 1.

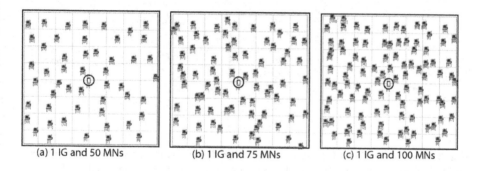

(a) 1 IG and 50 MNs   (b) 1 IG and 75 MNs   (c) 1 IG and 100 MNs

Fig. 2. Three deployment scenarios (A circled one indicates IG)

4.1 Performance Evaluation of Mobility Management Schemes

In this section, we study performance of the Tree-Based and the traditional mobility management methods. Fig. 3 - Fig. 6 compare Tree-Based approach against the traditional registration methods - Reactive, Proactive, and Hybrid. According to Fig. 3, Hybrid always generates more overhead than the simple Proactive unlike we have expected. One reason is that it is very hard to set a right value of k in Hybrid. Another reason is that Hybrid and Reactive use a ring-expanding search algorithm to find a IG or some registered node and in this process, if it fails to find any, it increases TTL by 2 and then reinitiate a search. On the contrary, Tree-Based shows the smallest overhead of all even though it pays some cost to maintain trees. This is because Tree-Based does not use a flooding in maintaining trees and managing node mobility.

Referring to Fig. 4, Proactive shows the lowest registration ratio of all. This is because when a node receives advertizement message, it waits until the registration interval timer expires, leading to the change of topology that frequently

Table 1. Simulation parameters and values

Parameters	Value
Mobility Pattern	Random Waypoint
Pause Time	30
Number of Nodes	Varying density (1 fixed IG)
Dimension	1000 x 1000
Transmission Range	250 m
Wireless Bandwidth	2 Mbps
Registration Interval	5 seconds
Advertisement Interval	2 seconds
Advertisement zone (IG)	2 hops
Increment in the expanding ring search (MN)	2 hops
Simulation Time	600 seconds

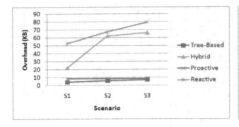
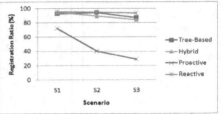

Fig. 3. Control overhead for different scenarios (nNodes = 50, mSpeed = 10 m/s)

Fig. 4. Registration ratio for different scenarios (nNodes = 50, mSpeed = 10 m/s)

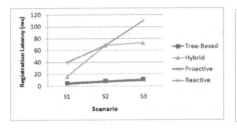
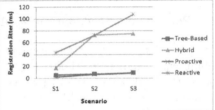

Fig. 5. Registration latency for different scenarios (nNodes = 50, mSpeed = 10 m/s)

Fig. 6. Registration jitter for different scenarios (nNodes = 50, mSpeed = 10 m/s)

invalidates the established routes. This is even severe as the average distance of a node to IG becomes longer as in S2 and S3. Note that S3 has longer distance to IG on average than S2 since the former has the IG located at the corner.

In Fig. 5, Proactive shows the lowest registration latency since it does not perform any route discovery and it just sends registration message if the route is valid. In case of Reactive, it discovers the route to IG if the route is not valid and then sends registration message, causing the highest registration rate and the highest registration latency. Fig. 6 shows the curve pattern of registration jitter similar to the curves of registration delay.

Proactive is not recommended overall since it shows the lowest registration ratio. So, we may say that Hybrid is one of dependable solutions among the traditional approaches. However, the Tree-Based is always very stable over the variations of IG location scenarios. According to simulation results, it is conjectured that the message flooding affects very negatively network performance.

4.2 Performance Evaluation of VMPT

We found out that performance of theTree-Based scheme was highly dependable over the traditional schemes according to simulation study. Now, we evaluate the effectiveness of the newly proposed method, VMPT against the Tree-Based.

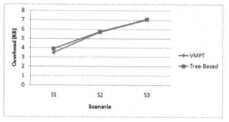

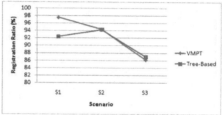

Fig. 7. Control overhead for different scenarios (nNodes = 50, mSpeed = 10 m/s)

Fig. 8. Registration ratio for different scenarios (nNodes = 50, mSpeed = 10 m/s)

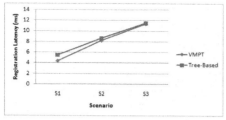

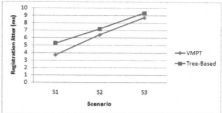

Fig. 9. Registration latency for different scenarios (nNodes = 50, mSpeed = 10 m/s)

Fig. 10. Registration jitter for different scenarios (nNodes = 50, mSpeed = 10 m/s)

In Fig. 7 Fig. 10 show a comparision between VMPT and Tree-Based schemes for different scenarios. Referring to Fig 7, VMPT and Tree-Based have almost the same overhead. However, it is worth noting that VMPT has more parents in S1 rather than in S2 or S3 since tree depths in S1 are smaller. Therefore, VMPT has slightly lower overhead in S1 because it maintains more parents and thus it initiates Join Process less frequently which involves J-REQ message. Consequently, the two graphs in Fig. 8 show a relatively big gap in S1. Referring to Fig. 9 and Fig. 10, VMPT shows a slight improvement over Tree-Based for the same reason that is explained by multiple parents.

Again VMPT shows a highly stable performance for the variations of the other parameters as shown in Fig. 11 to Fig. 14. Paths from MN to IG In the

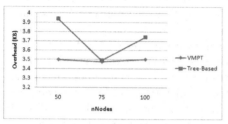

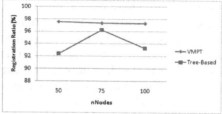

Fig. 11. Control overhead with varying nNodes (S1, mSpeed = 10 m/s)

Fig. 12. Registration ratio with varying nNodes (S1, mSpeed = 10 m/s)

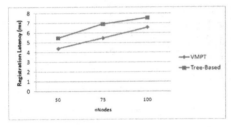

Fig. 13. Registration latency with varying nNodes (S1, mSpeed = 10 m/s)

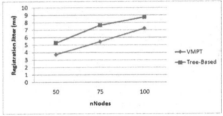

Fig. 14. Registration jitter with varying nNodes (S1, mSpeed = 10 m/s)

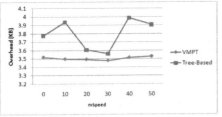

Fig. 15. Control overhead with varying mSpeed (S1, nNodes = 50)

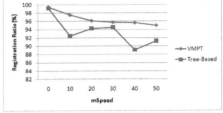

Fig. 16. Registration latency with varying mSpeed (S1, nNodes = 50)

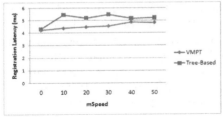

Fig. 17. Registration latency with varying mSpeed (S1, nNodes = 50)

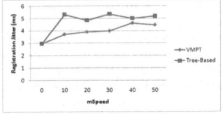

Fig. 18. Registration jitter with varying mSpeed (S1, nNodes = 50)

sparse network are more likely to be corrupted and may have the longer average length. Nevertheless, VMPT remains stable over the wide range of *nNodes* since it maintains multiple parents and thus can easily adapt to the change of network topology. Latency and jitter increase slightly for the two approaches; however, the gap between two graphs indicates that VMPT is more stable than Tree-Based.

Fig. 15 Fig. 18 show proformace curves over the change of maximum node speed from 0 to 50 m/s. VMPT that enables a quick repair of broken tree paths based on multiple parents shows a very robust performance over the Tree-Based against the high mobility of nodes. On the contrary, the Tree-Based becomes unstable since it pays more cost to handle the broken paths as node mobility increases.

5   Conclusions

In this paper, we proposed a new VMPT mobility management method which maintains multiple parents whenever possible. We compared the new method with Tree-Based and the traditional approaches - Proactive, Reactive and Hybrid. The simulation results show that the VMPT method outperforms the other ones and also are very robust against the change of IG location scenarios and the change of the parameter values that characterize the considered network.

References

1. Perkins, C.: IP Mobility Support, Request For Comments(Standard) 2002, Internet Engineering Task Force (October 1996)
2. Perkins, C.E., Royer, E.M.: Ad-hoc on-demand distance vector routing. In: Second IEEE Workshop on Mobile Computing Systems and Applications, February 1999, pp. 90–100 (1999)
3. Perkins, C.E., Bhagwat, P.: Highly dynamic destination-sequenced distance-vector routing (DSDV) for mobile computers. ACM SIGCOMM: Computer Communications Review 24(4), 234–244 (1994)
4. Jonsson, D., Alriksson, F., Larsson, T., Johansson, P., Maguire, J.G.: MIPMANET - Mobile IP for mobile ad hoc networks. In: Proceedings of IEEE/ACM Workshop on Mobile and Ad Hoc Networking and Computing (MobiHoc 2000), August 2000, pp. 75–85 (2000)
5. Ammari, H., El-Rewini, H.: Integration of mobile ad hoc networks and the Internet using mobile gateways. In: Proceedings of 18th Parallel and Distributed Processing Symposium 2004, April 2004, pp. 218–225 (2004)
6. El-Moshrify, H., Mangoud, M.A., Rizk, M.: Gateway discovery in ad hoc onde-mand distance vector (AODV) routing for Internet connectivity. In: Radio Science Conference, NRSC 2007, March 2007, vol. 1-8 (2007)
7. Oh, H.: A tree-based approach for the Internet connectivity of mobile ad hoc networks. Journal of Communications and Networks 11(3), 261–270 (2009)
8. Broch, J., Maltz, D.A., Johnson, D.B.: Supporting hierarchy and heterogeneous interfaces in multi-hop wireless ad hoc networks. In: Proceedings of International Symposium on Parallel Architecture, Algorithms, and Networks, June 1999, pp. 370–375 (1999)
9. Jubin, J., Tornow, J.D.: The DARPA packet radio network protocols. Proceedings of the IEEE 75(1), 21–32 (1987)
10. Ruiz, P., Gomez-Skarmeta, A.: Enhanced Internet connectivity for hybrid ad hoc networks through adaptive gateway discovery. In: Proceedings of the 29th Annual IEEE International Conference on Local Computer Networks (LCN 2004), November 2004, pp. 370–377 (2004)
11. Prashant, R., Robin, K.: A hybrid approach to Internet connectivity for mobile ad hoc networks. In: Proceeding of IEEE WCNC 2003, March 2003, vol. 3, pp. 1522–1527 (2003)
12. Sun, Y., Belding-Royer, E.M., Perkins, C.E.: Internet connectivity for ad hoc mobile networks. International Journal of Wireless Information Networks special issue on Mobile Ad Hoc Networks (MANETs): Standards, Research, and Applications 9(2), 75–88 (2002)

Lightweight Traffic Monitoring and Analysis Using Video Compression Techniques

Marat Zhanikeev[1] and Yoshiaki Tanaka[2,3]

[1] School of International Liberal Studies, Waseda University
1-6-1 Nishi-Waseda, Shinjuku-ku, Tokyo, 169-8050 Japan
[2] Global Information and Telecommunication Institute, Waseda University
1-3-10 Nishi-Waseda, Shinjuku-ku, Tokyo, 169-0051 Japan
[3] Research Institute for Science and Engineering, Waseda University
17 Kikuicho, Shinjuku-ku, Tokyo, 162-0044 Japan
maratishe@aoni.waseda.jp, ytanaka@waseda.jp

Abstract. Traffic analysis based only on IP address is a new research area where traffic anomalies can be detected by studying clusters of IP addresses extracted from traveling packets. Such analysis is normally spatial and needs IP addresses to be put in a multi-dimensional map. This paper proposes a novel method that converts such maps to 2-dimensional graphical form and applies video compression techniques to create MPEG-2 VBR movies where frames are individual snapshots of IP space in time. The paper proves that this combination is suitable for traffic monitoring and detection of DDOS attacks as well as large-scale traffic anomalies caused by social phenomena.

Keywords: traffic analysis, traffic monitoring, anomaly detection, IP space, video compression.

1 Introduction and Research Target

Traditional traffic analysis and especially anomaly detection relies on traffic flow information exported by NetFlow. There are several other standards and software products which compare to NetFlow, but the underlying concepts are the same, - packets are sampled and joined into flows, which are exported to a remote location where the traffic analysis takes place.

It is common for traffic analysis methods to "unwrap" flows in order to extract other information. IP space extracted from flows could be used to study communication patterns or one might need to count how many sources communicate to a single destination.

Whenever NetFlow-like technology is used as basis for traffic analysis, performance overhead becomes important. First, it takes time to aggregate raw packets into flows. Secondly, it takes time and traffic overhead to collect flow records from a network device for analysis. Overhead also occurs when almost any information it extracted from IP flows.

On the other hand, many traffic anomalies exhibit themselves in IP space. The term "IP space" throughout this paper is used to describe the totality of

C.S. Hong et al. (Eds.): APNOMS 2009, LNCS 5787, pp. 92–101, 2009.

IPv4 addresses found in trace. IP space is normally a temporal properly because a limited time interval is used to collect IP addresses before they are analyzed by graphical methods introduced in this paper.

The two immediate examples of traffic anomalies which "show up" in IP space are DDOS and Flash Crowds. The former often uses spoofed source IP addresses which have strong effect on IP space created by otherwise normal traffic. Flash Crowds, on the other hand, are local by nature, - Flash Crowds are created as the result of temporarily increased network activity directed to a single destination. Sources of these communication pairs are not completely random but rather "social" and tend to create local clusters in IP space.

There is a new emerging kind of social traffic phenomena which is generated by distributed botnet attacks. Botnets are zombie computers which are controlled by hidden attackers and tend to become active at roughly the same time, thus, resembling Flash Crowds. There are, however, subtle differences between the two.

If IP space is converted to a 2-dimensional graphical form, traffic analysis can be enhanced by a very well developed area of video compression. As will be shown in the paper, traffic analysis and video compression have common goals as long as the data in question comes in graphical form. Specifically, MPEG-2 was found to be the most suitable compression format given that it uses variable bitrate (VBR), which can be used to represent the degree of change occurring in graphical data.

This paper uses only IPv4 addresses [1]. Although it is predicted that at the current level of consumption, IPv4 address pool will run out by the end of year 2011 [2], it is still unclear as to whether IPv6 is the only alternative. In fact, a few scenarios are considered in research, most of which are planning to extend the use of IPv4 addresses. On the other hand, conversion of IPv4 addresses into graphics images is not new. There are several research projects in this area, one of which is CAIDA IPv4 WHOIS Map project [5]. However, to the extent of authors' knowledge no methods have yet been proposed to explore these graphical images for the purpose of detecting traffic anomalies.

This paper proposes a method which converts IP addresses collected from raw packet headers to a graphical form, stores them as individual graphics image files and finally compresses them into a MPEG-2 variable bitrate stream where each frame corresponds to graphics images of IP space. The proposed method exploits the fact that MPEG-2 encoding process achieves high compression ratio by performing interframe compression where it leaves in intermediate frames only the parts of graphical image which are different compared to the graphical image from the previous frame. The effectiveness of interframe compression translates almost directly into the notion of video stream bitrate, where the larger the difference between frames the higher is the bitrate.

When IP address space is converted into graphical image, stream bitrate of encoded still images will directly represent change and will make it possible to detect several important traffic anomalies. Additionally, the paper proves that the proposed method of traffic monitoring is effective especially in cases when traffic anomalies are rooted in social phenomena. With the Internet rapidly becoming a fully distributed and decentralized endeavor, traffic anomalies will exhibit more social features in the future.

Apart from the proposed method, the paper also presents details about its practical implementation, which was made possible with several open source software products. This in turn facilitates the discussion of practical applications of the proposed method at the end of this paper.

2   Construction of 2D Graphical IP Maps

The trick with video compression is to keep in mind that each frame is analyzed by the encoder based on 8x8 or in some cases 16x16 blocks of pixels. This block size imposes rigid requirements on the frame size in pixels. Normally, values are required to be divisible by 8 or 16. Specifically, MPEG-2 requires frame dimensions to be divisible by 16.

The numbers 8 and 16 also make perfects sense in the world of traffic analysis. When prefixes are used to analyze IP addresses, the two most commonly used prefixes are 8/ and 16/. The "forward slash" notation will be used throughout this paper to represent the length of the prefix. 8/ means that only 8 bits from the head of IP address are used. Naturally, 16/ is twice longer.

The 8/ prefix also bears geographical meaning in most of its 256 values. Originally and still sometimes today, IANA used to distribute the 8/ addresses based on geographical location. When it comes to 16/ prefix, address distribution there is not so regular. Statistically, only from 5 to 20 percent of each 8/ address is occupied permanently. Some of the rest are either dynamic or reserved addresses which are active much less often. In any case, 16/ addresses are normally distributed among large private or public organizations including ISPs.

The method to position IP address in 2-dimensional space is shown in Fig.1. First, 8/ and 16/ of an IP address are converted to simple integer values and used to position the block in 2048x1536 space. If to denote 8/ as a and 16/ as b, then $x = 8a$ and $y = 6b$.

The reason for allocating 8 pixels for each bit in first byte of each IP address is simple two-fold:

- if only one pixel were allocated for each value in 8/ prefix, the resulting image would be only 256 pixels wide, which is too small;
- horizontal cluster of 8 pixels can be used to present additional statistics about each address in 8/; in this paper it is used to create visual difference among active IP addresses with high packet counts and those which send/receive only a few or even one packet.

The vertical structure positioned by the value of 16/ prefix also contains 6 pixels for each value in 16/. However, here each pixel is mapped to a bit in 22/ prefix map, thus, visualizing activity in addresses within 6 more bits of each 16/ prefix. So, position of each pixel vertically is $y = 6b + c$ where c takes values from 0 through 6 and represents bits between 16/ and 22/. Prefixes from 20/ to 22/ are

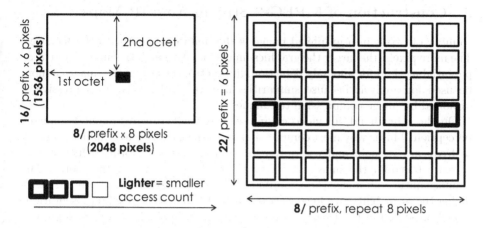

Fig. 1. Method used to position a single IPv4 address on a 2-dimensional map using 8/, 16/ and 22/ prefixes as integer values

at the longest prefixes found in global BGP advertisements and can be used in packets traveling through both intra- and inter-ISP backbone links.

In addition to the above positioning rules, some coloring rules are used as well. The proposed method does not actually use color, so, instead, the level of lightness is used on the scale from black to white colors. All graphics and video processing algorithms are much more sensitive to changes in brightness than in color. The word "color" hereinafter always means "brightness" of white color. Rules are as follows:

1. Default color of each block is black. The first packet that carries the IP address of the bock will color its horizontal line in white color. All successive packets with the same IP address will gradually make the color "lighter" until a threshold of 128 (half-grey) is reached where the color saturates with no further change.
2. When 2 or more bits between 16/ and 22/ are active, they affect each other's colors by making them brighter. Each neighbour doubles the count of all other 22/ bits in the block.
3. The coloring of each horizontal line is exponential where ends are colored in the color defined by packet count directly while the color of each inner pixel is defined by the number remaining after a right bitwise shift operation (division by two). In short, small packet counts fade out faster than large ones.

The above coloring rules exist with a sole purpose of creating graphical diversity. Given that video compression treats all frames as visual images, the more diversity exists in them the more effective is the monitoring technique based on the bitrate of the output video stream. Naturally, other methods to graphically present IP address blocks in 2-dimensional maps may exist, but for the sake of simplicity this paper only presents one.

3   Construction of MPEG-2 Stream from IP Maps

Number of pixels in a graphical image is the main contributor to the file size. The more pixels, the larger the graphics image file. Although this may sound as a potential method for anomaly detection, in practice the size of the image cannot be used for analysis because similarities in strings of pixels are aggressively exploited by image compression techniques.

On the other hand, all video compression technology cares about is human perception. This is why most compression techniques transform graphical images to frequency domain and use only a portion of first transform coefficients to recreate (decode) the image at receiver side. In plain words, the more local "abnormalities" an image (or computed difference between frames) contains, the higher the probability this local area is "noticed". Taken the pixel representation model presented in Fig.1 earlier, detectability of individual IP addresses placed in graphics image can be estimated by the following equation:

$$D = p\ var_{j=i-1,i,i+1}(2^{n_j} e^{-(c_j-1)}),\ i = 0..255. \tag{1}$$

In (1), c stands for packet count to/from the target IP address, n is the number of "vertical" neighbours, i.e. IP addresses which share the 22/ prefix, var stands for mathematical variance among values for the three horizontal neighbour 8/ addresses, and p represents the probability that the previous or the next IP address on 16/ prefix is active at the same time. According to statistics [2], p is normally between 0.8 and 0.95 and can be dropped from the equation for its insignificance.

The scope of this paper cannot accommodate elaboration on (1), but a few notes are in place. First, $c = 1$ guarantees highest detectability unless the 8/ neighbours on both sides are the same. Secondly, increasing values of c will exponentially decrease detectability unless 8/ neighbours exhibit very different trends. In general, detectability is higher when a particular IP address has packet counts and 22/ neighbours different from those of its 8/ neighbours.

The visualization in Fig.2 is called "mirror bar plot" and is used when two sequences are synchronized in time and have to be displayed together both for visual clarity and to enable visual comparison. Both upper and lower sequences in such plots contain only positive values. Values in all such plots in this paper are also normalized since real values convey no meaning.

The notion of bitrate comes directly from the processing stage in video compression called quantization. In quantization, a frame is analyzed and compared to previous (key frame in most cases) in order to find differences in image. If there are no differences, bitrate remains low since it requires little throughput to transmit little change. However, major changes in graphical content will result in larger encoded frames which in turn will require high bitrate (throughput) to transmit them. In plain words, video compression looks into the graphical content of frames rather than frame size and can detect changes in content even when frame size remains roughly the same.

The dynamics of bitrate are clearly shown in Fig.2 where the start-up phase of traffic monitoring is shown. In start-up, frames have no pixels in them in

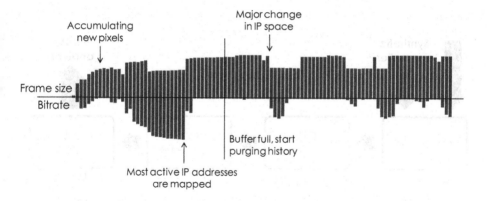

Fig. 2. Example of a start-up phase. History is set to 50 frames, each frame accumulates 1s of traffic.

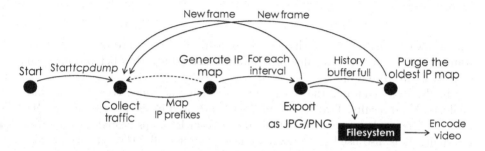

Fig. 3. The program logic starting from raw IPv4 packet collection and up to export of individual frames as JPEG images

the beginning which is why the frame size will gradually increase as pixels are accumulated. Bitrate also reacts accordingly by increasing until a relatively balanced state is achieved. In Fig.2, the balanced state is reached even before frame buffer was filled and started to purge old frames. From that point on, bitrate is relatively low and reacts only to local changes in graphical content.

The program logic used to convert IP space into graphical form is shown in Fig.3. The process starts at *tcpdump* which is used for collecting raw packets from network interfaces and is constantly looping between traffic collection and IP map generation as each packet header has to be inspected and either its source of destination address is mapped to the image. Only source addresses are mapped in the proposed method, while destination addresses may be a better choice in other cases.

At the end of each collection interval, graphical image is saved to a JPEG file and is later used to encode video. If frame buffer is full at the time, oldest frames are discarded at the end of each collection interval. The reason for the buffer is simple. Without the buffer, graphical images would be too random and

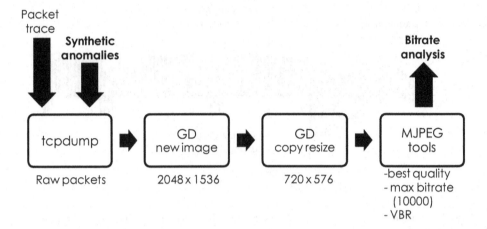

Fig. 4. Overall process from packet collection to analysis of MPEG-2 stream generated from IP maps

would always contain major changes in each frame. To counteract this, an n-frame history is retained at all times and are merged together before graphical content is saved to a file.

Finally, Fig.4 contains the overall process including video compression and analysis of end results. Two open source tools are used in the process. First, GD library [7] is used to generate JPEG images from IP maps stored by program internally. It is also used to resize the image from its full 2048x1536 frame size down to 720x576, which is standard for MPEG-2 streams. Resulting JPEG images are used as input into *mjpegtools* software [8] which converts multiple still JPEG images into a MPEG-2 variable bitrate stream.

Traffic monitoring itself is performed only on the resulting video stream. Although bitrate is not the only metric which can be used to monitor traffic, for the sake of simplicity only bitrate is used in this paper. In networking terminology, bitrate is directly equal to throughput.

4   Evaluation Results

This section considers practical uses of the proposed traffic monitoring framework. Specifically, DDOS attacks and FlashCrowds [4] are evaluated and compared.

When DDOS attacks happen, it is common to encounter IP addresses which are deliberately spoofed by attackers. This scenario can easily be synthesized. As per Fig.4 above, the process starts from packet collecting using tcpdump. However, to have a controlled and reiterable environment, packet traces from the WIDE traffic archive [3] were used instead. On top of using a recent trace from F-samplepoint of the WIDE network (average throughput 70 to 120Mbps), DDOS was synthesized and fed into the tcpdump stream. A simple DDOS synthesis

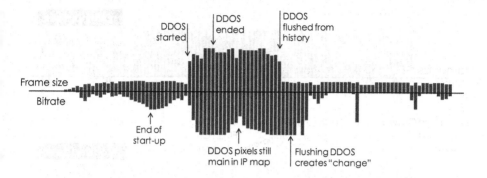

Fig. 5. Example of how DDOS attacks reveal themselves in video stream

was used where about 30% of IP address pool at the selected time in trace was artificially randomized.

Fig.5 clearly detects DDOS both in frame size and in bitrate. DDOS synthesis was started past the start-up phase so that it is not affected by it. Although DDOS attacks normally have short lifespan, they will affect video stream as long as they remain in the frame buffer. Bitrate will also be affected by it the whole time. Additionally, when DDOS is finally flushed from frame buffer, it will generate abrupt change in IP space of the monitor and will cause bitrate to rise again. In plain words, DDOS will reveal itself in 2 pulses, - one when the attack itself takes place and the other when it it flushed from the buffer.

Fig.6 shows how a Flash Crowd anomaly would reveal itself in stream bitrate. Synthesis of a Flash Crowd is much more complicated given that one has to mimic social behaviour. In the above example, the number of 8/ random prefixes gradually grew up to 100 with exponential rate of 1.18, while each 8/ prefix hosted up to 100 random 16/ prefixes growing with the rate of 1.15, each containing up to 200 random 32/ prefixes (IP addresses) growing at the rate of 1.18. The exponential growth rate helps Flash Crowd to grow, reach its peak, and exponentially fade out. The rates were tuned over several experiments and resulted in around 40 frames in the synthetic trace.

As can be seen from Fig.6 Flash Crowd-like behaviour does not exhibit drastic changes which is why frame sizes both with and without Flash Crowd presence are almost identical. However, the change in bitrate can still be detected and comes from the fact that many local clusters of IP addresses light up in the image and pass change detection thresholds when encoded in the video stream.

As far as the overhead (time consumption) in the overall process from Fig.4 goes, clearly, the packet collection phase is the most performance-hungry. Once the IP address is extracted from the packet, the process is relatively easy and is entirely based on bitwise operations in software. Bitwise operations cause only mild computational overhead since they happen directly in low-level registers in CPU instead being the result of complex computations. Also, since

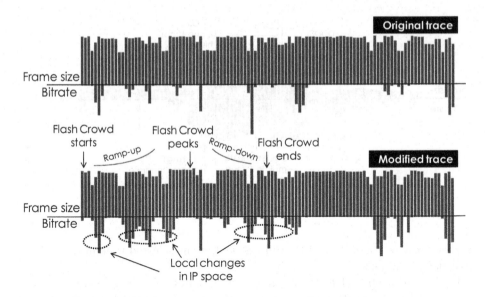

Fig. 6. Example of how Flash Crowd anomaly can reveal itself in stream bitrate

8/, 16/, and 22/ prefixes are treated as integers and used directly to address an element of an array in memory, overhead in positioning IP address in graphical image is minimal.

The use GD library and *mjpegtools* is also beneficial for the overall process since these tools have been developed for many years and are optimized for best performance. This means that these parts of the overall process from Fig.4 are also optimal in their performance. In general, the implementation of the proposed method is limited only by the speed with which packets can be collected from interface, which means that *tcpdump* is the only performance bottleneck in the process.

5   Conclusions

This paper proposed a novel method of traffic monitoring based entirely in graphics. IP addresses are extracted from raw packets and mapped to a 2-dimensional graphical image. Series of images are then encoded into MPEG-2 video stream using variable bitrate and highest possible quality. The use of bitrate which is directly affected by changes in graphical content of images and, thus, the original IP space, facilitates many traffic monitoring tasks at the client end. Although this paper only analyzed the bitrate of the video stream, MPEG-2 streams contain other information that can be used for traffic monitoring as well. For example, quantization coefficients for each intermediate frame are also stored in the stream and can be used to localize the change within the image.

The implementation of the proposed method was made possible using existing open source tools. In fact, the two tools which were used to generate graphical and video content are optimized to the best of performance due to high demands on the part of compression of video content. This area of research has been actively developed for many years and leaves little space for improvement.

The performance of the process used by the proposed method is very lightweight compared to NetFlow-based traffic monitoring. One particular performance boost comes from the fact that the proposed process does not have store IP addresses in hash tables as NetFlow does. Instead, bits from IP addresses are used directly to index elements within permanent array structures. In case of NetFlow, the necessity to contain a hash table of all flows is the biggest performance challenge.

The use of MPEG-2 streams can also have practical uses in network monitoring as well. Since bitrate is called throughput in networking world, when traffic experiences little change, video stream will require little throughput. On the other hand, drastic changes in traffic will raise bitrate of video stream and will require more throughput between network element and NMS. In whole, this is the description of a real-time monitoring system with very modest demands for the volume in meter-NMS communications. This makes it possible to implement the proposed system directly on network devices and stream monitoring video over the web.

In the future work, authors plan to look further into other models of translating IP addresses into blocks of pixels. Depending on the monitoring target, models other than the one proposed in this paper may offer better performance. Additionally, authors plan to address the need for benchmark traffic traces which can be used to test the performance of a given detection target. Such benchmarks already exist in the area of video compression where MPEG-2 encoding is tested on various graphical content. Similar benchmarks can be created using sample traffic traces with anomalies already inside. Development of such benchmark test cases is necessary for the future development of the proposed method.

References

1. Rekhter, Y., Moskowitz, B., Karrenberg, D., de Groot, G.J., Lear, E.: RFC 1918. Address Allocation for Private Internets (1996)
2. IPv4 Address Report, http://www.potaroo.net/tools/ipv4/
3. MAWI Working Group Traffic Archive, http://tracer.csl.sony.co.jp/mawi/
4. Jung, J., Krishnamurthy, B., Rabinovich, M.: Flash Crowds and Denial of Service Attacks: Characterization and Implications for CDNs and Web Sites. In: WWW Conference, Hawaii, USA, pp. 532–569 (2002)
5. IPv4 WHOIS Map,
   http://www.caida.org/research/id-consumption/whois-map/
6. Lakhina, A., Crovella, M., Diot, C.: Characterization of Network-Wide Anomalies in Traffic Flows. In: Internet Measurement Conference, Italy, pp. 201–206 (2004)
7. GD Graphics Library, http://www.boutell.com/gd/
8. MJPEG Tools, http://mjpeg.sourceforge.net/

NETCONF-Based Network Management System Collaborating with Aggregated Flow Mining

Tomoyuki Iijima, Yusuke Shomura, Yoshinori Watanabe, and Naoya Ikeda

Alaxala Networks Corp.,
890 Kashimada, Saiwai-ku, Kawasaki, Kanagawa 212-0058 Japan
{tomoyuki.iijima,yusuke.shomura,yoshinori.watanabe,
naoya.ikeda}@alaxala.com

Abstract. As the speed of the Internet getting higher and accommodating a broader range of applications' traffic, it becomes more difficult to monitor, analyze and manage the network traffic. To realize automatic network management of high-speed networks, we have developed two systems. One is a high-speed per-flow basis traffic analyzer using aggregated flow mining (AFM) that works with links of 10Gbps or higher. The other one is a network management system (NMS) using NETCONF as configuration protocols between NMS and network elements. For automatic network management, it is important that the NMS has flexibility to control a wide variety of network elements and also capabilities so that it can collaborate with flow collectors that capture and analyze network traffic data flows. To meet these requirements, a new application programming interface (API) that enables an NMS to control network elements, which include flow collectors, was developed.

Keywords: Aggregated Flow Mining, NETCONF, Flow Collector, API.

1 Introduction

The expanding scale and ubiquity of the Internet is increasing the total volume of traffic that it carries. Moreover, the Internet now has to accommodate traffic from a broader range of applications. In spite of the rapid innovation involved in expanding the Internet, less attention has been paid to network operation and management. For example, there is still no innovative technology that can control every element on the network in a unified manner. Such control is presently done in an outdated manner and depends on the know-how and experience of individual operators.

Network management is divided into two categories: monitoring and controlling. Monitoring and then analyzing a flow at high-speed is the first step towards network operation and management. Since it is impossible to monitor all the data passing through the Internet, various flow collectors have been developed to perform high-speed traffic analysis on a per-flow basis by using packet-sampling technology. These flow collectors receive sampled packets from network elements such as routers and switches that support sFlow [1] or NetFlow [2] and then

C.S. Hong et al. (Eds.): APNOMS 2009, LNCS 5787, pp. 102–111, 2009.

analyze their behaviors per flow. The second step is controlling the network elements by a network management system (NMS) according to the flow analysis results. In this step, the NMS is required to flexibly and dynamically obtain flow analysis results from the flow collectors. This collaboration between the NMS and flow collectors is the basic technology for automating network operation and management.

We have been developing a high-speed per-flow-basis traffic analyzer as well as a NETCONF-enabled NMS. The analyzer uses aggregated flow mining (AFM), which can be applied to links of 10Gbps or higher. The NMS uses NETCONF as a configuration protocol between the NMS and network elements, and it can configure network functions such as VLAN, filter, network interface, node, and route. As a solution to the above-described missing link between flow collectors and an NETCONF-enabled NMS, this paper shows a new application programming interface (API) for an NMS to control flow collectors, which leads the way towards automatic network management of the future high-speed Internet.

2   Issues

2.1   Lack of Interface between NMS and Flow Collectors

In the case of flow analysis technology, there are several technologies available such as sFlow [1] and NetFlow [2]. Standardization of flow analysis technology, such as IPFIX (IP Flow Information eXport) protocol [3] and Packet SAMPling (PSAMP) protocol [4], is also ongoing in the Internet Engineering Task Force (IETF). These protocols are aimed at just defining the way of transferring basic statistical information per flow from network elements to flow collectors. These collectors are able to store and analyze that flow statistical information. Network operators can thus look at the information on flow statistics stored on collectors, grasp the traffic behavior, and configure network elements accordingly. It is, however, troublesome for network operators to do these steps. It is preferable that an NMS dynamically obtains the flow analysis results, interprets the traffic behavior, and configures network elements automatically. This process would reduce the operator's workload significantly.

When developing such an NMS, however, two issues need to be taken into consideration: first, it is impractical for an NMS to directly receive and process flow-statistics information sent from network elements because of their heavy loads; second, there is presently no means for an NMS to obtain flow-statistics information stored in collectors. As for the first issue, owing to the characteristics of the push model in packet-sampling technology, an NMS becomes overloaded with a huge volume of data if network elements keep sending flow-statistics information directly to the NMS. As for the second issue, flow collectors today do not provide an interface for an NMS to obtain flow-statistics information stored in collectors.

Given these considerations, it is desirable to make collectors with a common interface that enables an NMS to obtain the necessary flow-statistics information from them.

2.2   Lack of Interface for Configuring Network Elements

In addition to the interface provided by flow collectors, an interface for configuring network functions of network elements is necessary. By incorporating a self-management algorithm using those interfaces into an NMS, automatic network management is possible. For example, even if an anomaly in traffic is detected, a self-management algorithm incorporated in the NMS can solve the problem quicker than a human operator. This will help to significantly reduce the manpower needed for configuring and operating networks.

The Simple Network Management Protocol (SNMP) is one step in realizing automatic network management in terms of having defined a common standard and common framework in the control plane. Some efforts have been made to achieve automatic network management by going through processes consisting of monitoring, analyzing, planning, and configuring via the SNMP. Such efforts are successful in terms of monitoring, analyzing, and planning. In terms of executing configuration, however, the SNMP is not good enough [5]. The reason for this deficiency is that data managed by the SNMP is not object-oriented but variable-oriented. This state is troublesome for network operators in terms of setting each variable of the management information base (MIB) tree. In addition, there are no mechanisms for preventing the conflicts that occur when more than one NMS tries to modify the MIB tree simultaneously [6]. As a result, even though the Internet is progressing continually, there is as yet no action on establishing a working group for standardizing SNMPv4 [7].

3   Development of an Interface for Configuring Network Elements

3.1   Proposed Technologies

To provide an interface that enables an NMS to configure network elements while solving SNMP's shortcomings, we previously proposed using NETCONF, a data model, and Java [8][9]. Figure 1 shows the relationship between each technology.

NETCONF is standardized at IETF and defines the operation messages exchanged between the NMS and network elements. It is specified to use Extensible Markup Language (XML) for its description language [10] and Secure Shell (SSH), Simple Object Access Protocol (SOAP), or Blocks Extensible Exchange Protocol (BEEP) as its transport protocol. When both the NMS and network elements use NETCONF, they can communicate with each other. Since NETCONF has a lock/unlock capability, conflicts are prevented even when multiple NMSs access a network element in order to configure a network element simultaneously.

3.2   Architecture of NETCONF-Enabled NMS and Network Element

By applying the technologies described in section 3.1 and selecting SOAP as the transport protocol for NETCONF, it is possible to generate a Java application

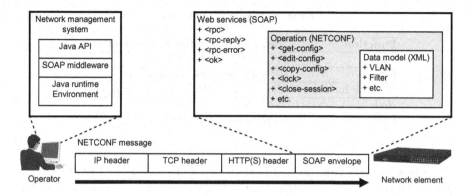

Fig. 1. Relationship between NETCONF, data model, and Java

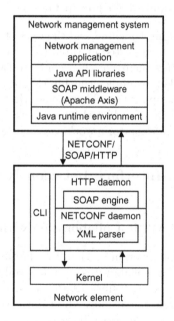

Fig. 2. Architecture of an NMS and a network element

programming interface (API) in a Web-services framework [11][12][13]. This API is easy to use compared to SNMP when a developer incorporates a self-management algorithm in an NMS. We developed a Java API and named it an "open-networking API" (ON-API).

Figure 2 shows the architecture of the NMS that implements ON-API and that of a network element. ON-API is installed on a PC. By importing ON-API into the Java development environment, developers can develop an NMS that

Table 1. ON-API's functions

Function	Data to be configured/collected
VLAN	VLAN ID, VLAN Name, etc
Filter	IP Address, MAC Address, etc
Network Interface	Slot ID, Port ID, Interface Name, etc
Node	Node Name, Location, etc
Route	Destination, Next Hop Address, etc

can configure network elements' functions according to the data model via the NETCONF protocol. Because the NETCONF daemon in a network element has an XML parser in it, NETCONF messages are parsed and instructions sent from the NMS are recognized.

3.3 Functions Implemented So Far in ON-API

ON-API's functions so far implemented and the type of data for configuring and collecting with each function are listed in Table 1. These functions are selected in consideration of the benefits incurred when configuration is made by the NMS instead of an operator. For example, the filter's rules tend to be input into network elements in large quantities. This means that these functions tend to be repeated many times when conducted by network operators. A huge benefit is therefore gained when an operator's workload is automated by an NMS.

4   Development of Interface between NMS and Flow Collectors

4.1 Architecture of NMS in Collaboration with Flow Collectors

Based on the consideration given in section 2.1, the proposed architecture of a NETCONF-based NMS that collaborates with flow collectors is shown in Fig. 3. A flow collector provides Java API libraries that enable an NMS to obtain flow-statistics information stored in collector. These Java API libraries are based on the same technologies as those described in section 3.1, namely, Web services. By utilizing these Java API libraries, it is possible to develop an NMS that can obtain flow-statistics information and configure network elements via NETCONF. In addition, a flow collector that can receive SOAP messages generated from the Java API was developed.

4.2 Data Model Exchanged between NMS and Flow Collectors

For providing the API, data manipulated by the API in the NMS, exchanged between the NMS and flow collectors, and used for obtaining flow analysis results needs to be modeled. Accordingly, we developed the data model shown in Fig. 4.

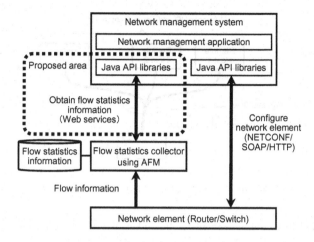

Fig. 3. Architecture of NMS collaborating with flow collectors

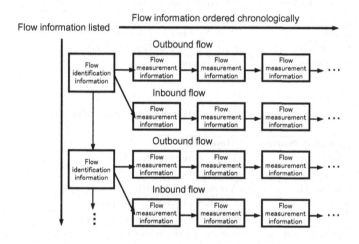

Fig. 4. Data model of flow statistics information

The data model of flow-statistics information is conceptually divided into two types of information: flow-identification information and flow-measurement information. Flow-identification information is used to distinguish a flow by a so-called "5-tuple," that is, source IP address, destination IP address, protocol, source port number, and destination port number. Moreover, flow-measurement information is data such as flow observation time and number of packets and number of bytes observed in a certain period of time. On top of that, the concept "variety per aggregated flow" (shown in Fig. 5) is incorporated into the flow-measurement information. Variety per aggregated flow indicate a variance in one of the pieces of information in the 5-tuple with some of them being equal. This

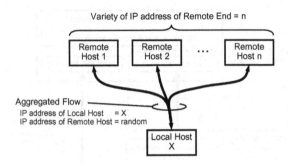

Fig. 5. Variety of IP address per aggregated flow

Table 2. Some elements of flow-identification information

Element Name	Meaning
localAddress	IP Address of Local End
remoteAddress	IP Address of Remote End
protocol	Protocol (TCP, UDP, etc.)
localPort	Port Number of Local End
remotePort	Port Number of Remote End
flowName	Flow Name

Table 3. Some elements of flow-measurement information

Element Name	Meaning
timestamp	Time obtaining statistics
duration	Measurement Time
bandwidth	Average Bandwidth
packets	Number of packets in Flow
vLocalAddress	Variety of IP Address of Local End
vRemoteAddress	Variety of IP Address of Remote End
vLocalPort	Variety of Port Number of Local End
vRemotePort	Variety of Port Number of Remote End

information is useful when a flow collector mines a flow that shows an abnormal behavior. Aggregated flow mining (AFM) applied in flow collectors can aggregate up to 1,000 packets together per aggregated flow, and it enables flow collectors to analyze traffic at 10-Gbps-and-higher links [14]. Multiple flow-measurement information is concatenated chronologically under flow-identification information. Tables 2 and 3 list elements of flow-identification information and flow-measurement information, respectively.

From the data model rearranged in the style of an XML Schema Definition (XSD) and then imported to a Web Services Description Language (WSDL) file, a Java API is generated automatically by a development tool such as Apache Axis. By using this Java API, network developers can easily develop an NMS that can processes the data listed in Tables 2 and 3.

5   Evaluations

An NMS that uses the API described in section 4.2 was developed. Figure 6 shows a request message sent from the NMS to flow collectors in order to obtain flow analysis results. The message is sent in XML format. This message requests the first thirty flows that have large variety of destination IP addresses.

```
<getSnapshot time="2008-03-18T03:00:00.000Z">
  <filter>
    <condition operator="eq" target="flowName" value="clientFlow" />
  </filter>
  <sort>
    <key order="descending">
      <target>out_vRemoteAddress</target>
    </key>
    <truncate offset="0" size="30" />
  </sort>
</getSnapshot>
```

Fig. 6. Request message for obtaining flow analysis results

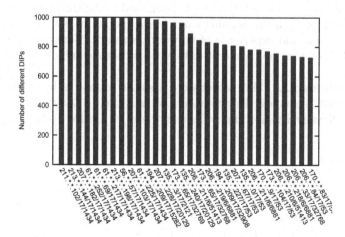

Fig. 7. Display of flow analysis results

A flow collector that can deal with messages generated from the API and sent from the NMS was also developed. Since the flow collector uses AFM technologies and the data model incorporates the concept of aggregated flow, the NMS can obtain statistics information already analyzed by flow collectors. The analyzed information can be obtained by just searching and sorting on the aggregated flow basis. The NMS does not have to do any analysis by itself. After the NMS obtained the flow analysis results, it can display a graph like that shown in Fig. 7.

The developed flow collectors can aggregate up to 1,000 packets per aggregated flow by AFM. When an aggregated flow with 1,000 different destination IP address is observed, it indicates that destination IP addresses of each packet inside an aggregated flow are different. And when this behavior is observed, it is reasonable to infer that a malicious user is scanning a particular port number of many hosts randomly. Figure 7 shows that port scanning of UDP/1434 is being conducted. The NMS can configure network elements by using ON-API according to the flow analysis made by flow collectors. In this case, the NMS can configure filter functions of network elements and make network elements drop packets sent from this source IP address. According to reference [8], a configuration of the filter function made by ON-API is quantitatively as good as that made by existing technologies.

6 Summary

An API that enables the NMS to obtain flow analysis results from flow collectors was developed. Utilizing these APIs makes it possible to develop an NMS that can obtain flow analysis results from flow collectors and configure network elements. When self-management algorithms are incorporated into the NMS, the NMS can achieve automatic network management by going through processes consisting of monitoring, analyzing, planning, and configuring network elements. An NMS developed in this manner can replace operations done so far according to the know-how and experience of network operators. Moreover, the workload of network operators–and the cost of their operations–can be reduced.

Acknowledgments. This research work has been supported by the New Energy and Industrial Technology Development Organization (NEDO) under the Development of Next-generation High-efficiency Network Device Technology Project.

References

1. Phaal, P., Panchen, S., McKee, N.: InMon Corporation's sFlow: A Method for Monitoring Traffic in Switched and Routed Networks, RFC3176 (2001)
2. Claise, B.: Cisco Systems NetFlow Services Export Version 9, RFC3954 (2004)
3. IP Flow Information Export (ipfix). In: International Engineering Task Force (IETF), http://www.ietf.org/html.charters/ipfix-charter.html
4. Packet Sampling (psamp). In: International Engineering Task Force (IETF), http://www.ietf.org/html.charters/psamp-charter.html
5. Pavlou, G., Flegkas, P., Gouveris, S., Liotta, A.: On Management Technologies and the Potential of Web Services. IEEE Communications Magazine 42(7), 58–66 (2004)
6. Munz, G., Antony, A., Dressler, F., Carle, G.: Using Netconf for Configuring Monitoring Probes. In: 10th IEEE/IFIP Network Operations and Management Symposium (2006)

7. Pras, A., Schonwalder, J., Festor, O.: XML-Based Management of Networks and Services. IEEE Communications Magazine 42(7), 56–57 (2004)
8. Iijima, T., Kimura, H., Kitani, M., Atarashi, Y.: Development of NETCONF-based Network Management Systems in Web Services Framework. IEICE Trans on Communications E92-B(4), 1104–1111 (2009)
9. Network Configuration (netconf). In: International Engineering Task Force (IETF), http://www.ietf.org/html.charters/netconf-charter.html
10. Enns, R.: NETCONF Configuration Protocol, RFC4741 (2006)
11. Choi, M.C., Choi, H.M., Hong, J.M., Ju, H.T.: XML-Based Configuration Management for IP Network Devices. IEEE Communications Magazine 42(7), 84–91 (2004)
12. Iijima, T., Atarashi, Y., Kimura, H., Kitani, M., Okita, H.: Experience of Implementing NETCONF over SOAP, RFC5381 (2008)
13. W3C, Web Services Activity, http://www.w3.org/2002/ws/
14. Shomura, Y., Watanabe, Y., Yoshida, K.: Analyzing the number of varieties in frequently found flows. IEICE Trans on Communications E91-B(6), 1896–1905 (2008)

Policy-Based Monitoring and High Precision Control for Converged Multi-gigabit IP Networks

Taesang Choi, Sangsik Yoon, Sangwan Kim, Dongwon Kang, and Joonkyung Lee

BcN Division, ETRI
Daejeon, Republic of Korea
{choits,ssyoon,wanni,dwkang,leejk}@etri.re.kr

Abstract. High speed real-time precise traffic monitoring and control has gained significant interests by researchers in this field recently. This is mainly due to the fact that precise traffic measurement and analysis in a high-speed IP network environment is not a simple task. Let alone, control of such traffic in real-time is another big challenge. This paper addresses such complicated issues and proposes a novel solution which can precisely measure traffic in high-speed IP networks, classify them per application, create detailed flow-aware traffic information, and control per application basis. It includes motivation, architecture, and mechanisms. We have embedded capabilities of lossless packet capturing, deep packet inspection, and flow generation into hardware level. We describe major features, design concepts, implementation, and performance evaluation result.

Keywords: policy-based monitoring and control, hardware-based flow record generation, traffic measurement.

1 Introduction

Recently, there have been various efforts such as research, development, and standardization to address issues for true convergence services and its corresponding infrastructure. Notable examples are 3GPP's efforts of defining IMS specifications, ETSI TISPAN"s efforts of fixed mobile convergence, ATIS's efforts of NGN, and finally ITU-T's efforts of NGN. Among them, ITU-T's NGN efforts are most comprehensive which coordinates all the rest activities into a single convergence platform for NGN. Convergence efforts were addressed in three perspectives: transport, service, and application aspects. In terms of transport aspect, wireless, mobile, and fixed transport network technology were converged into a common infrastructure and transport control. For service aspect, common service capabilities are defined and a service control framework is built over such capabilities. Applications take advantage of such underlying common capabilities and framework so that a variety of converged services and applications can be developed. True convergence in terms of requirements, architecture, and capabilities is still undergoing process. Cooperation between policy-based resource control and monitoring is one good example that is under study among experts in the relevant groups. QoS-aware mobility management is another example.

C.S. Hong et al. (Eds.): APNOMS 2009, LNCS 5787, pp. 112–121, 2009.
© Springer-Verlag Berlin Heidelberg 2009

Convergence is an essential part of the next generation networks in the current market. The real requirements for convergence applications and services are coming from the NGN service providers. Mergers among fixed telecommunications service providers and mobile service providers are happening in the global market. Vertical applications, therefore, no longer satisfy new customer's needs. Service providers themselves are actively seeking new service opportunities by leveraging such mergers. This introduces other complexities in their management. For example, traffic monitoring and QoS resource control were separate issues before. However, integrated management of the two becomes essential to provide high quality converged services. Real-time feedback of the monitored results to QoS resource control is no longer optional but mandatory capability for the fixed mobile convergence environment. This requires convergence of two technologies: traffic measurement and management and traffic resource control. Up to now, these fields have been developed independently in terms of research, development, and standardization. To meet the newly emerging requirements, however, tightly-coupled coordination among them is very important. Efforts on the research and development of these fields have been quite active and various solutions were introduced in the relevant market. Traffic measurement solutions which support upto 40 Gbps speed and volume are introduced. Resource admission control solutions handling upto 10 Gbps are available from various vendors. However, research and development on the convergence of the above mentioned technology are still pre-mature stage. Noting such requirements, we have conducted research and development recently. Our research focus was on the improvement of both hardware and software capabilities to deal with high speed and volume traffic measurement and real-time traffic control.

For measurement of high speed and volume traffic, there were various research efforts to address this issue with software improvement in the host OS kernel I/O driver level. It showed significant improvements but it has limitations which can't be solved without the help of hardware acceleration for the high-speed measurement. Thus, dedicated hardware has to be designed to meet such high speed and volume measurement requirements. Major capabilities that have to be considered in the hardware are high speed reliable packet processing including no loss/duplication packet capture, multi-layer filtering, and various sampling methods (fixed, probabilistic, or flow sampling). None of the currently available hardware whether it is a card or a standalone device can meet all the requirements mentioned above. Especially when it has to deal with very high speed links such as OC-48 or above.

The control of such traffic in real-time with high accuracy is another challenge. This is because high precision control requires support of accurate traffic measurement and classification of applications. Real-time traffic control especially in-line situation can cause serious problems to application services if it is not administered appropriately. Due to the diversity of the current and newly emerging converged applications such as various P2P-based overlay applications, UCC, and IPTV, the main difficulties come from highly dynamic nature of the development and the use of the applications. They are port number independent and asymmetric nature of application transactions. This means that distinguishing flows based on a port number and other header properties is not safe and accurate enough. Also an application transaction can consist of multiple sub-transactions, a series of Requests and Replies, which may follow different routing paths. Accurate flow identification for such a case requires distributed monitoring and correlation of sub-transactions which appeared in different paths. In this paper, we propose novel mechanisms for such challenges. It consists of hardware and software methodologies. For performance and scalability, packet in-spection, filtering/sampling, traffic anomaly handling, flow generation, and traffic management for control are all conducted in our novel hardware. Various traffic analysis including multi-path correlation and high precision applications recognition is the job of our software. We have incorporated the proposed mechanisms into our proof-of-concept system called Wise^{HITMAS}.

The paper is organized as follows. Section 2 examines related work. Section 3 describes our novel hardware and software methodology and implementation archi-tecture to meet the above challenges. A proof-of-concept system and deployment ex-periences are explained in Section 4. Finally, section 5 concludes our effort with potential future work.

2 Related Work

There have been many research and development efforts in the field of traffic meas-urement and analysis for the past decade. As a result, many tools were introduced to meet various objectives such as traffic profiling, traffic engineering, attack/intrusion detection, QoS monitoring, and usage-based accounting.

Various researches were conducted for high-speed traffic measurement based on Network Processors.[1] Their main focus is usually on a specific problem domain like performance study of efficient sampling and filtering algorithms, especially flow sampling.[2] Such algorithms can be important to study but is one part of large set of problems for solving high-speed traffic measurement. We have adopted some of the efficient algorithms (e.g., multi-stage filtering [3]) in our system.

For the precise recognition and classification of applications, most existing solutions are targeted for P2P, streaming applications identification, or security anomaly detec-tion. Cisco Systems' NBAR (Network-Based Application Recognition) provides basic application recognition facilities embedded in their Internet Operating System (IOS) for the purpose of traffic control. Most intrusion detection systems are also equipped with basic application recognition modules which usually function on a signature matching-basis. For streaming and P2P application identification, Mmdump[4] and SM-MON[5] used payload inspection method. Many other methods for P2P traffic characterization [6] based on port number matching were also published. Cisco's Service Control Engine (SCE) series and Netintach's Packetlogic provide more general purpose applications recognition solutions. These systems, however, exhibit shortcomings of performance, scalability, scope of coverage, and accuracy. For resource control solutions, there are several RACF implementations available in the market from vendors such as Huawei, Alcatel-Lucent, ZTE, and Operax[7] to name a few. These solutions are specifically targeted for complying with ITU-T Y.2111[8] Recommendation.

3 Novel Methodology for High-Speed Measurement and High-Precision Control

3.1 High Level System Overview

Based on the motivation described above, we have designed our system with chal-lenging objectives such as no loss/duplication packet capturing up to 40Gbps speed,

L2/L3/L4 filtering, packet and flow sampling, deep packet inspection, application flow generation at the hardware level, and traffic management capabilities such traffic shaping, policing, dropping, and scheduling. It is based on NP and its associated co-processors for DPI, regular expression engine, etc. Before we describe them in details, we briefly explain high level system architecture depicted in Fig. 1. The system is built as ATCA(Advanced Telecom Computing Architecture) compliant board with Network Processors and Co-processors interconnected by SPI 4 switch module and upto 40 Gbps fabric interface and flexible I/O modules which can adapt various combinations such as multiple 1G Ethernet, 2.5G POS, 10G and 40G Ethernet.

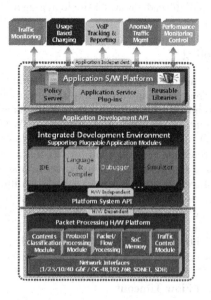

Fig. 1. High-level System Architecture

It consists of packet processing H/W platform, integrated software development environment, Application S/W platform, and platform system/application development APIs. The main design goal of the system is to provide hardware-based reliable performance and flexible software development environment so that various applications can be built in a timely manner to meet the real market requirements. For this we provide hardware dependent API and independent API which relieves hardware dependencies for the software development. It also aims to provide easy to develop IDE including high-level language and an associated compiler for system software programmers. Besides the ease of development, it allows multiple instances of applications running at the same time on the same platform. Duo to such feature, an integrated application described in this paper can be built. Finally, faster customization to meet the customer requirements is possible with the platform. In the following subsections, we describe the major novel methodology to address the above mentioned challenges.

3.2 Traffic Filtering

Traffic filtering is used in our system as a front-end module for various purposes. Our system is targeted to accommodate as many flows as possible in order to meet high-precision requirement. However, lossless traffic capturing on the link of OC-48 or higher speed is practically almost impossible especially when abnormal situation (e.g., syn flooding, DDoS attack, etc.) occurs. In fact, we don't need to capture all flows that may include traffic usage of less interest in terms of flow analysis such as virus traffic and miscellaneous control traffic (e.g., keep alive messages). Our traffic filtering functionality handles such situation. It filters out traffic which does not need flow record generation and, instead, creates a single aggregate flow record for keeping usage statistics information.

3.3 Smart Flow Sampling

Besides one-packet flow aggregation, we need more intelligence for collecting flows in extreme situations. There are many different ways of sampling traffic: simple periodic packet sampling to intelligent flow sampling. We defined a smart flow sampling based on a extended multi-stage filtering algorithm [3]. We use multiple stages of hash tables to keep tract of flows of a certain size. If a flow meets the criteria, it is selected in a stage one filter and continues until it satisfies all stages of filters. If succeeded, it is finally stored in a flow table. This reduces possibility of false judgment. When the decision on flow selection is made, we add additional criteria. Since the signature detection is performed before filtering, particular flows which meet signature criteria but does not exceed size threshold still be selected it is important flows to account its usage. The number of such flows can have a certain limitation to the threshold of flow table entry. We call it smart flow sampling. It is smart in the sense that the decision of flow selection is not simply depends on its size.

3.4 Time Bucket Based Flow Timeout

Typical flow timeout rule (e.g., Netflow rule) is FIN arrival, inactive timeout, active timeout, or flow memory full. Main drawback of this rule is that flow doesn't faithfully reflect application behavior. Application session or service may span multiple

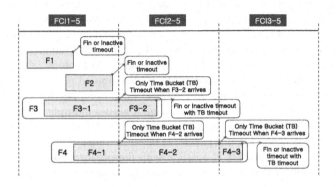

Fig. 2. Time bucket based Flow Timeout Mechanism

flows but there is no particular way of finding such relationship under current timeout mechanism. Other issues are the fact that analysis of flows may delay up to maximum active timeout period or inaccuracy of flow analysis which spans multiple analysis intervals can occur. These problems can't be solved by simply reducing the active timeout value because it causes unnecessary division of long-term flows. We have solved this problem with "time bucket based flow timeout" mechanism. Fig. 2 illustrates how it works. We define time bucket as flow creation interval (FCI). The default value is 5 minutes and can be flexibly adjusted. If a flow like F1 completes within a FCI by FIN arrival or inactive timeout, they are timed out normally. Also a flow like F2 completes within a FCI1-5 although FIN arrives or Inactive timeout occurs at FCI2-5 within inactive timeout period can normally time out within FCI1-5. However, a flow like F3 or F4 which spans multiple FCIs, they are forced to be timed out by a time bucket timeout. All the flows now contain an indicator that tells which FCI it belongs. This mechanism allows not only solving the above mentioned problems but providing accurate long-lived flow analysis and predictable performance of flow export and its associated memory usage.

3.5 Hardwired Flow Record Generation

We have defined a flow as extension of a normal flow. Our flow has a flow record and one or more associated packet records. Fig. 3 depicts the flow schematic. Flow record list is two hash tables kept in SRAM and actual values are stored in DRAM with packet records. The figure also shows that the how fragmented packets are handled for reassembly.

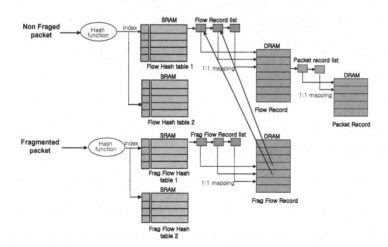

Fig. 3. Flow Data Structure

The extension of flow concept is defined for the purpose of providing more information to an analysis process for precise applications analysis. The packet record contains not only packet specific information but also has a part of payload which includes an application signature attached when it is found. Thus our flow contains just enough information about accurate application classification without keeping the entire payload. The generation of flow records in this detail level at hardware is one of the most important features of our system. We have met system performance objective of supporting upto 4 million flows in a hash table per second and actual export rate of 600,000 flows per second for 10Gbps link.

3.6 Two-Step Flow Merge

When flows are generated normally, it considers a single direction only. However, applications are usually asymmetric. It is up to the analysis process which is responsible for identifying the bi-directional relationship of application flows. We have designed our system to conduct such flow merge in two steps. During 1st step, it checks whether two same direction flows belong to one original flow. This situation can occur when traffic is load-balanced at flow level. If such flows are found then they are merged and aggregated. As a 2nd step, it checks two opposite direction flows which has generated by the same application. If such flows are found, they are merged. Such merged flows then can be transferred to the application process for further analysis. This makes the application process job much simple and scalable. Fig. 4 illustrates how two step merge can take place.

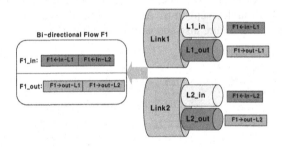

Fig. 4. Two-step Flow Merge

3.7 Application Classification Methodology

Other important novel features in our system are: general purpose Internet applications traffic classification algorithm, accurate application accounting, adaptability and expendability for newly emerging applications, and auto-detection mechanism of the new applications.

The objective of our application classification method is high-precision IP application identification. Since we are using several novel methods, the accuracy of application traffic accounting is much higher in comparison with most other currently available solutions. Since IP applications lifecycle is very dynamic, it is important to design the monitoring system to adapt such a characteristic. We designed a run-time configuration language, called Application Recognition Configuration Language (ARCL). When a new application appears, manual detection is very time-consuming and labor intensive and thus requires automation. We are currently working on such automation.

Our software methodology consists mainly of a precise application identification method, extensible applications recognition language, and flow definition extension.

Our approach is unique in that we classify the applications into the four distinctive types: Type-FP (Fixed Port-based Recognition), Type-PI (Payload Inspection-based Recognition), Type-DP (Dynamic Port-based Recognition), Type-RR (Reverse Reference-based Recognition). In the first method, recognition is performed on the basis of a predefined port number to application mapping. In the second method, recognition is performed on the basis of both port numbers and signatures. In the third method, recognition is performed on the basis of port numbers obtained by inspecting other flows' payloads. Lastly, recognition is performed on the basis of referential information obtained by recognizing a type-PI flow on the other links. We determined the types by inspecting, generalizing, formulating, and classifying more than 100 major IP applications. We gathered flow and entire packet specimens from four major networks including a large-scale campus network, an enterprise network, a major Internet exchange point, and an ISP.

The recognition language is simple and effective in that the system can be swiftly reconfigured to detect and recognize unprecedented or modified applications without the developer's writing and distributing extension modules for processing the newer applications, which usually results in fairly long period of shrunk recognition coverage even after detecting and analyzing the new applications. The basic hierarchy of ARCL semantics and syntax possesses three levels. The highest category is an application (application), the next is a representative port (port_rep_name), and the last is a subgroup (decision_group). There is a one-to-many relationship between a higher and a lower category. The basic concept can be explained by an example. Although most WWW services are provided through port 80, there are still many specific web services that are provided via other ports, such as 8080. All these representative ports pertain to the application "WWW" and have their own distinctive names; port_rep_names – "HTTP" for 80, "HTTP_ALT" for 8080, etc. Although a port_rep_name is tightly coupled with a port, the relationship between them is not one-to-one, nor fixed; a single port can be used by a number of different applications under different representative names. Packets in a flow can further be classified into a number of subgroups. For example, an HTTP flow subdivides into "HTTP_REQ" packets, "HTTP_REP" packets, ACK packets of HTTP_REP, etc. Each of these elementary subgroups constituting the entire context of a flow is a decision_group.

4 Implementation Experience and Results

4.1 Implementation and Deployment Experience

The proposed requirements and architecture is in the process of implementation as a proof-of-concept system, called Wise[*HITMAS]. It consists of a ATCA-compliant packet processing board, FM(Flow Mediator), AS(Application Server), DB server, and GUI. We are currently conducting various function and performance testing on our new system in a Lab environment with traffic generators, AX/4000 and IXIA IxLoad.

4.2 Performance Evaluation

As mentioned, our system performs real-time traffic statistics calculation. It generates top-N statistics per N number of source and destination hosts. It can handle up to 1.8Gbps 60 Byte packets without loss which matches with around 4 million pps. For 1500Byte packets, MA can fully support up to 10 Gbps even with real-time Top-N feature turned on. On the other hand, MA can support up to 2.4 Gbps with around 50 million pps without any packet loss in case that real-time Top-N statistics function turned off.

Measurement Data Accuracy Table by Monitoring Interval(1Day: 2006-0316 00:00 ~ 2006-03-16 24:00)					
Monitoring Interval	Measurement Data				
	Bps		pps		fps
5Min	169,338,875,356		53,474,742		8,806,233
1Min	169,854,435,559		53,587,637		8,826,019
7.5Sec	171,737,532,549		54,873,333		9,106,379
	5Min-X(%) Data				
5Min	–	–	–	–	–	–
1Min	-515560.202	-0.30%	-112.894	-0.21%	-19.785	-0.22%
7.5Sec	-2398657.192	-1.42%	-1398.590	-2.62%	-300.145	-3.41%
	1Day Total Sum				
Unit	Mbytes		Kpackets		Flows
5Min	–		–		–
1Min	-5,310.106		-9,525.431		-1,709,424,000
7.5Sec	24,705,408		-118,006,031		-25,932,528,000

Fig. 5. Measurement Data Accuracy by Monitoring Interval

Fig. 5 shows one interesting performance statistics which has been acquired from measurement in a real operational environment in one major ISP in Korea. Since our system can collect statistics data up to two seconds level, we conducted traffic measurement in different time intervals with the same raw data set. It shows that measurement analysis accuracy by intervals varies. The accuracy increases as the measurement time interval becomes more granular. This testing was conducted in a 1 Gbps International link. We can easily guess that the same testing with higher speed links will provide more differences. This result justifies that real-time measurement of very granular level will increase the accuracy of traffic usage statistics. Of course, this testing doesn't take other application classification algorithms into account. The accuracy will increase much more with them.

4.3 Deployment Experience in Real Field

We have deployed our system in one of major ISPs in Korea. The system is installed on the 10Gbps link that connects to another major ISP. The link is utilized in 6 – 7 Gbps in total bi-directionally. The average flow per second is less than 10,000 because the link does not carry much aggregated traffic. We have been deploying the system over 3 months now. We are testing various system functionality: packet capturing and flow generation in hardware level, flow filtering, two-step flow merging, DPI-based application signature matching, analysis server's performance, etc. The detailed analysis work hasn't been completed similarly as shown in the section 4.2. We will update in our future version.

5 Conclusion and Future Work

Based on the novel mechanisms we introduced, we have successfully implemented our policy-based monitoring and control system supporting up to 10Gbps speed and are currently working for enhancing it to support much higher speeds upto 40Gbps. It is very challenging to support such capabilities upto 40Gbps especially at the hardware level. As far as we understand, such attempt hasn't been made before by other research. Our short-term goal is to verify our system's functionality in a real-time situation. For longer-term, we are looking for alternative hardware design which can improve the current performance limitations capable of supporting upto 40Gbps such as the lack of main memory for more flow processing, less flexible programming environment for application developers, and integration of more functionality like flow merge into hardware. Such work is for further study.

References

1. Donghua, Chuang, R., Zhen, L., Jia, C., Ungsunan, N., Peter, D.: Handling High Speed Traffic Measurement Using Network Processors. In: Proc. of International Conference on Communication Technology 2006, Beijing (November 2006)
2. Estan, C., Varghese, G.: New Directions in Traffic Measurement and Accounting: Focusing on the elephants, ignoring the mice. ACM Transactions on Computer Systems (TOCS), 270–313 (2003)
3. Bloom, B.H.: Space Time Trade-Offs in Hash Coding with Allowable Errors. Comm. ACM 13(7), 422–426 (1970)
4. van der Merwe, J., Caceres, R., Chu, Y.-h., Sreenan, C.: mmdump- A Tool for Monitoring Internet Multimedia Traffic. ACM Computer Communication Review 30(4) (October 2000)
5. Kang, H.-J., Kim, M.-S., Hong, J.W.-K.: A method on multimedia service traffic monitoring and analysis. In: Brunner, M., Keller, A. (eds.) DSOM 2003. LNCS, vol. 2867, pp. 93–105. Springer, Heidelberg (2003)
6. Sen, S., Wang, J.: Analyzing peer-to-peer traffic across large networks. In: Proceedings of the second ACM SIGCOMM Workshop on Internet Measurement Workshop (November 2002)
7. Operax Resource Controller series,
   http://www.operax.com/products/default.asp
8. ITU-T Recommendation Y.2111, Resource and admission control functions in Next Generation Networks (2006)

The Design of an Autonomic Communication Element to Manage Future Internet Services

John Strassner[1,2], Sung-Su Kim[1], and James Won-Ki Hong[1]

[1] Department of Computer Science and Engieering, POSTECH, Pohang, Korea
{johns,kiss,jwkhong}@postech.ac.kr
[2] Telecommunications Systems & Software Group, Waterford Institute of Technology, Ireland
jstrassner@tssg.org

Abstract. Future Internet services will have vastly different requirements than the current Internet. Manageability, which has been largely ignored, will have the dual role of controlling capital and operational expenditures as well as enabling agile reorganization and reconfiguration of network services and resources according to changing business needs, user demands, and environmental conditions. Autonomic systems have the potential to meet these needs, but important architectural extensions are required. This paper examines the core building blocks of an autonomic system, and compares our approach to existing approaches with respect to Future Internet networks and networked applications. A special emphasis is placed on business driven services.

Keywords: Autonomic Architecture, Autonomic Element, Business Driven Device Management, FOCALE, Future Internet, Management.

1 Introduction

The current success of the Internet architecture has spurred advances in business, social, and technical communications. However, that simplicity is also the source of many of its inherent limitations [1][2][3]. Two of the most important are its architectural limitations and its inability to relate business needs to network services and resources offered. If a design for the Future Internet is to be *sustainable*, then it must not just be technically correct; it is far more important for it be economically motivating and extensible, so that it can support new business models and services.

There is currently a great deal of discussion concerning what type of design approach to take. There are three fundamentally different approaches for the design of the Future Internet that are now being pursued [4]. The first is an incremental, or "evolutionary" approach, and is evidenced by many solutions currently being applied to the current Internet that violate its architectural principles, such as Network Address Translators, Firewalls, and Virtual Private Networks. This approach is no longer sustainable [1]. The second is a revolutionary, or "clean slate", approach [5], which eliminates existing commitments, restraints, and assumptions, and starts with a new set of ideas. The third is a compromise between the above two approaches, and enables new ideas to evolve while simultaneously emphasizing backwards compatibility with the

C.S. Hong et al. (Eds.): APNOMS 2009, LNCS 5787, pp. 122–132, 2009.

existing Internet. This is very important to certain stakeholders, such as Internet service providers (ISPs), who have invested billions into their equipment and want to leverage those investments.

Unfortunately, most communities ignore the management aspects of the Future Internet. For example, many researchers consider the Simple Network Management Protocol (SNMP) [6] or current command line interfaces [7] appropriate for Future Internet management, but such network management protocols are not even suitable for the current Internet [8].

In our previous work, we summarized the management requirements of the Future Internet [9]. In this paper, we continue this work and define architectural components for managing the Future Internet. Our work is motivated by the fact that current and future networks and networked applications have vastly different requirements; this means that a single architecture for the Future Internet cannot simultaneously meet these different needs. Therefore, we need a modular approach that is distributable, so that a single programmable and reconfigurable system can be built that can be re-purposed to suit different application needs.

The rest of the paper is structured as follows. Section 2 briefly describes the salient principles of autonomic communications, and reinforces why this approach is well suited to meeting the management needs of the Future Internet. Section 3 concentrates on the *autonomic element* abstraction, and compares three different implementations of it. Section 4 compares the autonomic elements, and Section 5 presents our conclusions and outlines our future work.

2 Autonomic Communications and the Future Internet

This section first describes the components of an autonomic communications architecture, and then explains why this approach is well suited to meeting the management needs of the Future Internet.

2.1 Autonomic Communications Primer

The purpose of autonomic computing is to *manage complexity*. The name was chosen to reflect the function of the autonomic nervous system in the human body. By transferring more manual functions to involuntary control, additional resources (human and otherwise) are made available to manage higher-level processes.

The fundamental management element of an autonomic computing architecture is a control loop, as defined in [10][11]. Sensors retrieve data, which are then analyzed to determine if any corrections to the managed resource(s) being monitored are needed (e.g., to correct "non-optimal", "failed" or "error" states). If so, then those corrections are planned, and appropriate actions are executed using effectors that translate commands back to a form that the managed resource(s) can understand. If the autonomic network can perform manual, time-consuming tasks (such as configuration management) on behalf of the network administrator, then that will free the system and the administrator to work together to perform higher-level cognitive functions, such as planning and network optimization.

2.2   Future Internet Requirements

The current Internet design emphasized simplicity, and has inherent manageability, scalability, mobility, flexibility, security, trust and other limitations. The European FP7 program is addressing these shortcomings. Challenge 1, called Pervasive and Trustworthy Network and Service Infrastructures, recognizes that the current Internet architecture was not designed to cope with a wide variety and ever increasing number of networked applications, business models, and environments. Europe is advocating four *different* Internets: an Internet of Things, Services, Media, and Communities [12]. While the number of foci of Internets is not important, the splitting of different services that require different networks proves our hypothesis that a *single* architecture is not optimal for managing the Future Internet and its applications.

2.3   Meeting the Needs of the Future Internet Using Autonomics

However, just because four Internets are specified does not mean that four different architectures are required. First, this would greatly complicate sharing resources and services between these networks. Second, it would also make their management very difficult. Service and content providers, network operators, and many other constituencies would be adversely affected, because each would want to offer multiple services from two or more Internets to their subscribers. Hence, we believe that building a single framework that can support each of these needs, as well as future needs, is an important approach.

This is the reason that autonomics are inherently aligned with the Future Internet: autonomic technologies provide a means to reduce complexity. A typical scenario for the Future Internet is one in which a user requires *context-aware services*. This means that the devices that are being used should adapt the services and tasks that they are performing in accordance with changing user needs and environmental conditions. This adaptation should ideally be done with as little manual intervention as possible, which is the focus of Ubiquitous Computing. Mark Weiser said "In such a world, we must dwell with computers, not just interact with them....Interacting with something keeps it distant and foreign...Dwelling with computers means that they have our place, we have ours, and we co-exist comfortably [13]." That paper epitomizes many of the fundamental goals of ubiquitous systems, but especially the goal that the person now has to *think less* about his or her tasks and the environment because *unnecessary work has already been done or summarized by the computer(s)*.

Components of network management applications are typically chosen because they are the "best of breed" in performing a particular type of function. The obvious problem is that many different additional applications all share different views of the same object. Therefore, any new management approach must deal with the inherent heterogeneity of programming models and data representation for that domain. The FOCALE [14] autonomic architecture was designed to meet these and other requirements. In particular, FOCALE solves the above heterogeneity problem by defining an internal vendor- and technology-neutral language that can be mapped to vendor- and technology-specific languages (and vice versa) [11][15].

3   The Autonomic Element Abstraction

An *Autonomic Element* (AE) is a fundamental abstraction that describes how autonomic systems manage resources and services. This section will compare and contrast three very different AE designs.

3.1   The Autonomic Element of IBM

Fig. 1 shows IBM's AE [10]. The AE provides a common management façade to be used to control the resource by using a standardized set of interfaces. The autonomic manager implements the control loop. Sensors monitor the resource, while effectors send commands to the managed resource. The set of sensors and effectors forms the management interface that is used for communication between autonomic elements and the environment. The monitor stage collects, aggregates, filters, manages and reports on sensor data. The analyze stage examines the collected data to determine if its goals are being met, as well as to learn about its environment and help to predict future situations. The plan stage organizes the actions needed to achieve its goals. The execute stage controls the execution of a plan.

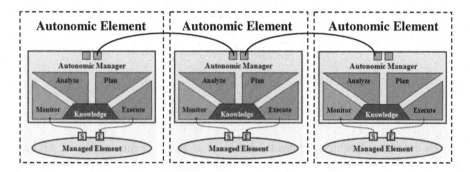

Fig. 1. Conceptual Diagram of IBM's Autonomic Element

3.2   The AE of CASCADAS

Fig. 2 shows the control loop from the FP7 CASCADAS program [16], which defines AEs as *distributed components* that can *autonomously self-organize* and provide *specific user communication services*. This flexibility enables them to self-adapt according to both social and network contexts. AEs have a "common" part that communicates using a set of standard interfaces that is also exposed to the outside world, and a "specific" part that uses dedicated interfaces to provide component-specific functionality (hence, only other AEs that need those functions will implement these dedicated interfaces). AEs use message-based communication to collaborate and provide aggregated and/or composited functions. The former defines a set of AEs that loosely collaborate to provide a service but ensure that each AE remains independent; the latter constructs a new AE that is the composition of each component AE.

Specific interfaces contain semantic descriptions of the set of features that a given AE can perform, and the conditions and actions required to accomplish each feature. The use of semantic features is not provided in the IBM approach. This enables the CASCADAS AE to use semantic routing (i.e., a recipient is addressed using its semantic properties) instead of using its logical or physical address. Each AE also has a "self-model", which defines the possible states and their associated transitions; this self-model is *published* to other components using a specific protocol that can describe the semantics of each state. The self-model can be *dynamically adapted* using the Facilitator, in response to (for example) context changes. Such changes are analyzed using the Reasoning Engine, and the Facilitator will adjust the model using one or more actions.

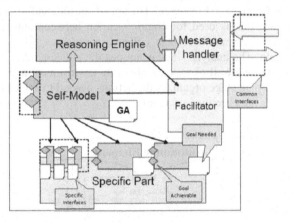

Fig. 2. The AE of CASCADAS

AEs use "plans", which defines the behavior of an AE. Features that an AE can provide are "achievable goals", while functionality that an AE requires are "needed goals". The Reasoning Engine implements the self-model, and the Facilitator is responsible for changing the self-model state machine when a feature exhibits different behavior than what was expected.

A unique approach in the CASCADAS architecture is that it does *not* continuously monitor the in-depth state of the entities that it is managing, but rather uses a restricted set of signals that provide an abstract view of the current state of the managed system. This enables the CASCADAS AE to take action only when needed.

3.3   The Autonomic Communication Element of FOCALE

Fig. 3 shows a simplified version of the FOCALE AE, which is made up of a set of *distributable components* connected by an enhanced enterprise service bus (ESB) [17] that supports simple as well as semantic queries. An ESB is an event-driven message broker. Our implementation is a distributed *content-based* message and retrieval broker; the difference between it and standard ESBs is that it can be used to orchestrate content, whereas standard ESBs are limited to orchestrating *messages*. Hence, the FOCALE ESB is really an enterprise *content* bus (ECB). The FOCALE Autonomic

Manager uses the ECB to orchestrate behavior. It can support different types of knowledge acquisition and distribution (e.g., push, pull, and scheduled) and perform common processing (e.g., semantic annotation, filtering and storage) before content is delivered to components. This enables components to register interest in knowledge in a more precise fashion, and thus reduce messaging overhead.

Sensor data is analyzed using the Observe component, which translates vendor-specific data into a vendor-neutral form. Translation is necessary, because in general, each device uses its own language to express its management data; hence, vendor-specific data must be translated into a common vendor-neutral form. This is done using a combination of information/data models and ontologies [15], where input data is matched to structures and patterns defined in the models and ontologies. This approach is unique to FOCALE.

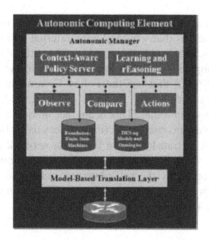

Fig. 3. Simplified Version of the FOCALE AE

The Compare component determines the current state of the managed entity from the normalized data and then compares it to its desired state. If the two are equal, the loop continues; otherwise, appropriate actions are computed and executed. This state-based approach is similar to that used in CASCADAS; however, in FOCALE, a set of adaptive control loops are implemented. An outer, or macro-control loop, is used to ensure that only policy rules applicable for the context at hand are used; an inner, or micro-control loop, is a variant of the Observe-Orient-Decide-Act (OODA) control loop [18], and is used to govern the functionality of the managed entity. Adaptation uses context-aware policy rules to determine both the specific components of the control loop as well as how each component functions. For example, a simple threshold comparison could be changed to a semantic relatedness test depending on the type of data, state, and context. This is shown in Fig. 4.

Since the action may be directed at a different managed entity (which, for example, could have been the root cause of the observed problem), the control loops could be changed (e.g., the original loop could be suspended or terminated and a new loop created). This can be done programmatically as well as through the use of different

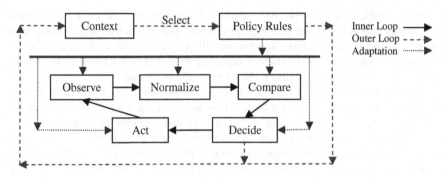

Fig. 4. FOCALE Control Loops

types of learning and reasoning algorithms. This is unique to FOCALE. FOCALE does not publish its model, as that would require external entities to fully understand the complex relationships between different managed entities that FOCALE governs.

Business goals and objectives can be directly related to the system using context-aware policy models [19][20] that use the Policy Continuum [21][22] to relate the needs of different constituencies (e.g., business, network, and programming people) that use *different terms* to each other. This is another unique aspect of FOCALE. For example, it enables business people to describe services using business concepts, such as revenue; this can then be translated to a form that network engineers can use and implement (e.g., traffic classification and conditioning configuration commands). This adaptation is reinforced by the self-governing nature of FOCALE, in that the system senses changes in itself and its environment, and determines the effect of the changes on the currently active set of business policies.

4   Autonomic Element Comparisons

This Section compares the strengths and weaknesses of the three AEs described; a summary is provided in Table 1.

The control loops of the three approaches are very different, and are summarized in the first 3 rows of Table 1. IBM has a single, pre-defined control loop, whereas CAS-CADAS and FOCALE both use adaptive control. FOCALE is the only approach that uses multiple control loops, and does so to separate the effect of large changes, such as those from applying new policy rules, from small changes, such as fine-tuning of parameters controlled by a set of policy rules. Both FOCALE and IBM use policy rules to govern control loop functionality.

IBM and FOCALE use a well-specified form of policy, which is lacking in the CASCADAS approach. As seen in the next two rows of Table 1, FOCALE uses a *set* of policy languages, which enables different constituencies to express policies in terms that are natural to them. IBM is limited to a single policy language.

The next 5 rows of Table 1 describe the respective knowledge bases of each AE. Both CASCADAS and FOCALE use dynamic knowledge bases, which IBM doesn't. Data scalability is limited in the IBM approach by the monitoring stage – if it cannot

Table 1. Comparison of IBM, CASCADAS, and FOCALE AE Approaches

Aspect	IBM	CASCADAS	FOCALE
Static or Adaptive Control	Static	Adaptive	Adaptive
Multiple Control Loops	Single	Single	Multiple
Policy Driven Control Loop	Yes	No	Yes
Policy Languages	One	None	Multiple
Policy Languages for Different Constituencies	No	No	Yes
Dynamic Knowledge Base	No	Yes	Yes
Data Scalability	Limited	Scalable	Scalable
Accommodates Heterogeneous Data	Limited by Common Base Event	Specific plugins can be added for new data	Yes, through model-based translation
Data Complexity	Limited to data that can be instrumented	Specific plugins can be added for new data	Uses patterns, models, ontologies to parse data
Data Semantics Encoded	No	Yes	Yes
AE Complexity	High	Low	Low
AE Components Distributed	No	Yes	Yes
Uses Ontologies	No	Could in Future	Yes
Semantic Matching	No	Yes	Yes
Self-Organizing	No	Yes	Yes
Uses Self-Model	No	Yes	Yes
Publishes Self-Model?	No	Yes	No
Model-Based Supervision	No	Yes	Yes
State-Driven	No	Yes	Yes
Adaptable State Machine	No	Yes	Yes
Autonomic Behavior - Loosely or Tightly Coupled	Tightly Coupled	Both	Both
Context-Aware	No	Yes	Yes
Supports Emergent Functionality	No	Yes	Yes
Communication Mechanism	Pre-Defined Interfaces	Messages and interfaces	ECB messaging and interfaces

keep up with the data that is being sent to it, the entire system suffers. In addition, the IBM approach is limited to data that is pre-defined in its static knowledge base, and offers no way to encode data semantics. This is one of the key limitations of the IBM architecture, as the knowledge repository is populated with data defined at development time, and cannot accept new or changed data discovered at runtime. In contrast, CASCADAS supports plugins for accommodating new data. Specific interfaces contain semantic descriptions of the set of features that a given AE can perform, and the conditions and actions required to accomplish each feature. FOCALE uses a combination of models and ontologies to generate software patterns to understand and learn from data.

In the IBM architecture, the AE combines the resource to be managed with the autonomic manager. Thus, self-awareness (of the AE) and awareness of the external environment of the AE are each limited. In contrast, both CASCADAS and FOCALE were designed as distributed systems from the start.

The IBM AE does not use semantics. CASCADAS uses semantics to determine decisions and also to route knowledge. FOCALE is similar, except that it uses both models and ontologies to represent different types of knowledge.

The IBM AE is not self-organizing; it cannot adapt to change, as its behaviors are controlled by the plan and execute parts using statically defined knowledge. Both FOCALE and CASCADAS use models and semantics, along with machine-based learning and reasoning, to organize their components to adapt to change. This enables them to support *emergent functionality.*

IBM does not provide state automata for management, whereas both FOCALE and CASCADAS do. In addition, both FOCALE and CASCADAS enable parts of their models to be changed. A unique feature of the CASCADAS AE is the publishing of its "self-model", which defines the possible states and their associated transitions. The self-model can be *dynamically adapted.* FOCALE does not publish its model, since it uses a complex governance model.

Contrary to both the IBM and FOCALE approaches, CASCADAS does not have a well-developed notion of policy; rather, it discovers deviations from the desired state of an entity and takes corrective actions to return that entity to its desired state. FOCALE is similar to CASCADAS, except it uses context-aware policies, and has a set of control loops that adjust the functionality of the autonomic manager according to changing context.

5   Summary and Future Work

This paper describes the key role that autonomic systems can play in managing the Future Internet and its applications. An AE enables a communications façade to be used to control managed resources and services that may or may not be autonomic using a common set of governance mechanisms.

Three different types of AEs were described and compared. CASCADAS and FOCALE offer significant advantages over the IBM approach in terms of scalability, ability to handle complex data, self-organization of the components of the AE, and communications flexibility. FOCALE has several unique features that provide advantages over CASCADAS, such as its ability to use different types of policy rules authored by different constituencies (e.g., network as well as business people), ability to accommodate heterogeneous data, use of an ECB, and more powerful control mechanisms that provide additional functionality.

FOCALE uses a model-based approach and emphasizes dynamic code generation. This reduces operational expenditures by removing the human from the configuration and monitoring processes in most circumstances, as changes to devices are dynamically constructed and applied under FOCALE's supervision. This emphasis in code generation drives the use of models and ontologies, from analysis and design through implementation and deployment. As such, a related goal of FOCALE is to build a knowledge framework that enables *reusable libraries of behavior* to be built that are then *matched based on their semantics*. This combination of extensions helps FO-CALE based applications realize the important vision of Weiser's *invisible manage-ment* in Ubiquitous Computing systems.

Future work will concentrate on building different FOCALE applications as speci-fied in the World Class University research program for Ubiquitous Health and Ubi-quitous Environment applications. We will report our results as soon as possible.

Acknowledgments. This work is sponsored in part by the WCU (World Class Univer-sity) program through the Korea Science and Engineering Foundation funded by the Ministry of Education, Science and Technology (Project No. R31-2008-000-10100-0). This work is also partially sponsored by Science Foundation Ireland under grant num-ber 08/SRC/I1403 (FAME).

References

1. Clark, D., Sollins, K., Wroclawski, J., Katabi, D., Kulik, J., Yang, X., Braden, R., Faber, T., Falk, A., Pingali, V., Handley, M., Chiappa, N.: NewArch: Future Generation Internet Ar-chitecture, NewArch Final Technical Report, http://www.isi.edu/newarch/
2. Blumenthal, M., Clark, D.: Rethinking the design of the Internet: the end to end arguments vs. the brave new world. ACM Transactions on Internet Technology 1(1), 70–109 (2001)
3. Feldmann, A.: Internet clean-slate design: what and why? ACM SIGCOM Computer Communication Review 37(3) (2007)
4. Kim, S., Won, Y., Choi, M., Hong, J., Strassner, J.: Towards Management of the Future Internet. Accepted for publication at the 1st IEEE/IFIP Workshop on Management of the Future Internet, Long Island, NY, USA, June 5 (2009)
5. McKeown, N., Girod, B.: Clean slate design for the internet, April 2006 Whitepaper (2006), http://cleanslate.stanford.edu
6. Harrington, D., Preshun, R., Wijnen, B.: An Architecture for Describing Simple Network Management Protocol Management Frameworks, RFC3411 (December 2002)
7. Cisco, http://www.cisco.com/warp/cpropub/45/tutorial.htm
8. Schonwalder, J., Pras, A., Martin-Flatin, J.-P.: On the future of Internet management tech-nologies. IEEE Communications Magazine 41(10), 90–97 (2003)
9. Kim, S.-S., Choi, M.-J., Hong, J.W.: Management requirements and operations of Future In-ternet. In: Ma, Y., Choi, D., Ata, S. (eds.) APNOMS 2008. LNCS, vol. 5297, pp. 156–166. Springer, Heidelberg (2008)
10. Kephart, J., Chess, D.: The Vision of Autonomic Computing. IEEE Computer 36(1), 41–50 (2003)
11. Strassner, J.: Autonomic Networking – Theory and Practice. In: 2008 IEEE Network Op-erations and Management Symposium (NOMS) Tutorial, Salvador Bahia, Brazil (2008)
12. Future Networks and Services – Developing the Future of the Internet through European Research, ftp://ftp.cordis.europa.eu/pub/fp7/ict/docs/futint-book_en.pdf
13. Weiser, M.: Open House. In Review, the web magazine of the Interactive Telecommunica-tions Program of New York University (March 1996)
14. Strassner, J., Agoulmine, N., Lehtihet, E.: FOCALE – A Novel Autonomic Networking Architecture. ITSSA Journal 3(1), 64–79 (2007)

15. Strassner, J.: Enabling Autonomic Network Management Decisions Using a Novel Semantic Representation and Reasoning Approach, Ph.D. thesis (2008)
16. Manzalini, A. (ed.): Deliverable 1.1, Report on state-of-art, requirements and AE model, CASCADAS Project, January 9 (2007)
17. Christudas, B.: Service-Oriented Java Business Integration: Enterprise Service Bus Integration Solutions for Java Developers. Packt Publishing (August 2008)
18. Boyd, J.R.: The Essence of Winning and Losing, June 28 (1995)
19. Strassner, J., de Souza, J., Raymer, D., Samudrala, S., Davy, S., Barrett, K.: The Design of a Novel Context-Aware Policy Model to Support Machine-Based Learning and Reasoning. Journal of Cluster Computing 12(1), 17–43 (2009)
20. Strassner, J., de Souza, J., van der Meer, S., Davy, S., Barrett, K., Raymer, D., Samudrala, S.: The Design of a New Policy Model to Support Ontology-Driven Reasoning for Autonomic Networking. Journal of Network and Systems Management 17(1) (March 2009)
21. van der Meer, S., Davy, S., Davy, A., Carroll, R., Jennings, B., Strassner, J.: Autonomic Networking: Prototype Implementation of the Policy Continuum. In: 1st IEEE International Workshop on Broadband Convergence Networks, pp. 1–10 (2006)
22. Davy, S., Jennings, B., Strassner, J.: The Policy Continuum – A Formal Model. In: 2nd International IEEE Workshop on Modelling Autonomic Communications Environments, Multicon, Berlin. Multlicon Lecture Notes, vol. 6, pp. 65–78 (2007)

Adaptive Grid Resource Selection Based on Job History Analysis Using Plackett-Burman Designs

Cinyoung Hur and Yoonhee Kim*

Department of Computer Science,
Sookmyung Women's University, Korea
{hurcy,yulan}@sm.ac.kr

Abstract. As large-scale computational applications in various scientific domains have been utilized over many integrated sets of grid computing resources,
the difficulty of their execution management and control has increased. It is
beneficial to refer job history from many application executions, in order to identify application's characteristics and to decide grid resource selection policies
meaningfully. In this paper, we apply a statistical technique, Plackett-Burman
design with fold-over, for analyzing grid environments and execution history of
applications. It identifies main factors in grid environments and applications,
ranks based on how much they affect. Especially, the effective factors could be
used for future resource selection. Through this process, application is performed
on the selected resource and the result is added to job history. We analyzed job
history from an aerospace research grid system. The effective key factors were
identified and applied to resource selection policy.

Keywords: Resource Selection, Job History, Grid Computing.

1 Introduction

A grid computing[4] is used effectively to support applications of various fields by
aggregating computing, network and storage resources. Especially, research capabilities of scientific domains have been increased with grid technology. Although many
research grids have been developed, mostly these are limitedly configured for particular domains. To increase the utilization of grid technology by sharing it with various
domains, it requires deep understanding of application's characteristics, as well as
large-scale computing resources for the applications [7]. Typically, scientific applications take a lot of time to solve problems with enormous size of data, required grid
environments based on characteristics of applications in a specific domain. For example, as a CFD(Computational Fluid Dynamics) [9, 13] application usually requires
numerical analysis and its visualization of the process, high performance computational grids are needed. If resources are configured without considering these characteristics, grid utilization could be lowered and the efficiency of research be degraded
as well.

---

* Corresponding author.

C.S. Hong et al. (Eds.): APNOMS 2009, LNCS 5787, pp. 133–142, 2009.
© Springer-Verlag Berlin Heidelberg 2009

It is important to understand and utilize thoroughly about job history [5] of applications executions on grid, as well as characteristics of the applications. Analysis of job history from various angles can help in diverse areas such as research in grid resource management, grid maintenance and operation and grid design. Also research applying various statistical methods on job history analysis is actively conducting. Key features of the methods help to define characteristics of applications and properties of grid as factors and to identify effects of each factor and interaction between factors. Execution of an application considered these factors can use grid efficiently. However, if it is too many, that makes difficult to identify effects and interaction of factors - Statistical approach is popularly used to find out these factors and their interaction from job history [10, 11]. Specifically, we use a Plackett-Burman design [3, 12] which is easy to quantify effects of factors which are significant in grid environments and the execution of applications. In addition, the quantified factors and its effects is useful for resource selection.

To apply job history for resource selection polices, each job execution is identified as a job profile, which includes requirement of an application and resource, the result of the execution, and other features understanding characteristics of applications and status of grid. We apply Plackett-Burman designs for screening the factors' impacts on execution of an application as conditions in resource selection policies. Resource selection policies use the effective factors for choosing resources, reflect gap between actual execution time and referred execution time in job history. Newly executed result is added to job history as a new job profile, and used for next resource selection.

Concludingly, we allocate resource efficiently according to characteristics of applications and the status of grid environment. As job profiles are accumulated to job history, we also understand applications deeply more and guarantee resource selection with better resource stability [1, 2].

The rest of this paper is structured as follows. In Section 2, we explain background and assumption. In Section 3, we present our resource selection algorithm. In Section 4, we describe its design and implementation. In Section 5, we evaluate our approach in many grid scenarios. In Section 6, we address the related work. Finally, in Section 7, we conclude and describe future work.

2 Background and Assumption

2.1 Characteristics of Target Application

Simulation of CFD(Computational Fluid Dynamics)[9], scientific discipline which solves a flow variation around target geometry using numerical methods, consists of three main steps. A mesh targeted to solve is constructed, then numerical analysis is applied on the mesh, and finally, resultant data are analyzed using a visualization software. In CFD simulation, the step of solving a mesh requires huge computing resources because it iteratively computes fluid dynamic equations on a mesh and processes resultant data for visualization. Before starting to solve a mesh, multiple simulations are dynamically created with various conditions and diverse ranges [8]. Since execution time of these simulations depends on the conditions and ranges, we define these as an application's characteristics and the numerical analysis on a mesh as a CFD application in this paper.

In general, a CFD application first transmits input files of simulations such as a mesh, information of parameters for analysis, and solvers of a mesh, to computing resources. Actual systems for CFD application is e-AIRS(e-Science Aerospace Integrated Research System) 2.0 [14], which is an aerospace integrated research system, currently running over Korea VOs(Virtual Organizations) and PRAGMA VOs. Korea VOs include clusters of KISTI(Korea Institution of Science and Technology Information) [15]. PRAGMA(Pacific Rim Application and Grid Middleware Assembly) [16] VOs provides large-scale grids composing of 35 institutional resources.

2.2 Effect of Factors and Statistical Methodology

P. Shivam [11] defines rigorously the notion of factors and their effect on system behavior. Based on his study, we define main causes in the performance of grid applications as factors and then analyze how much each factor impacts on the performance. For example, characteristics of computational applications, CPU speed and memory size could be key factors deciding performance of grid applications (See Table 1). Equation (1) represents factors and the most effective factor on an application is represented as shown in Equation (2).

$$F = \{f_1, f_2, ..., f_k\} \tag{1}$$

$$f_{max} = \{ f \mid most\ effective\ f \in F \} \tag{2}$$

Before starting to solve a mesh, multiple simulations are dynamically created with various conditions and ranges of parameters for solving. Since execution time of these simulations depends on the conditions and ranges on it, we define these as an application's characteristics and we name this numerical analysis on mesh as CFD analysis in the following

Table 1

Domains	Factors	
Application	Mach numbers (MA)	Local flow-speed divided by local speed of sound
	Reynolds numbers (RE)	Relative importance of inertial and viscous forces in a flow
	Angle of Attack (AOA)	Angle between a reference line and the oncoming flow
	Total iteration (ITMAX)	Number of iteration in application
Network	Bandwidth, Propagation delay, Location of Clusters	
Resource	CPU capability, Memory, Stability	

2.3 Plackett-Burman Designs [3, 12]

Consider a situation of a design with N factors, each of which having one of M values, it is possible to simulate a exhustive experiment consisting of M^N runs for a complete analysis. In case of each factor have only two values, simulating all the possible combinations would require 2^N simulations. Since this number of simulations is extremely high, designs of logically minimal number of simulations is required to estimate quickly the effect of each of the N fators.

Plackett-Burman(PB) designs, a well-established approach of this type, provide a method to logically minimize the number of simulations in cases where interactions of factors are not presented. The PB design measures the effects of k=N-1 factors in N runs, where N is multiple of 4. A design matrix is used for PB designs and each rows in the design matrix corresponds to different experiment configurations. For N = 8, the first rows of the matrixfor PB design, which is used in this paper, is acquired from [3]. The second and further rows are found by rotating the previous row to the right; the final (X-th) row is a series of low values.

An improvement on the PB design is PB designs with fold-over(PBDF). Using the PBDF, we can determine the effect of all of main parameters and selected interactions with only 2N experiments. In the case of PBDF, a further N rows having switched values is included, therefore total number of required experiments equals to 2N.

The columns of the matrix correspond to factors, and the rows correspond to different simulation configurations. There are a series of (+) and (-) values corresponds the high and low values of a factor which should be selected to be just outside the range of normal values. After running 2N simulations, the effect of each factor is computed by multiplying the result for each configuration by the value of the entry for that congifuration and factor combination. Then the results of all those multiplications are summed together to determine the overall effect for that factor. When comparing effects of different factors, only the magnitude is used.

3   Resource Selection Optimization Algorithm Based on Job History

3.1   Job Profiles and Job History

Job profiles contain application's characteristics, required conditions for executing job and information of resources. After the execution of an application, which can be defined as interactions among elements in profiles creates several results such as job execution time, completion ratio, the results are also included in a job profile that is used as a record unit of a executed job. Equation (3) represents a job profile and its information in it. A job profile is defined as j, one of a requested job j is allocated on particular resource(r) and executed during T_j.

$$P = \langle j, r, T_j \rangle \text{ where } j \in J = \{j_1, j_2, \dots, j_n\}, \ r \in R = \{r_1, r_2, \dots, r_m\}, \ T_j > 0 \tag{3}$$

Job history is a set of separate job profiles accumulated for a long time. Job history gets richer as jobs are executed more. From the job profiles, we can deduce common information, execution tendency, and patterns of target execution environment. Not all job profiles are applied for job history. Screening a job profile which is not acceptable due to exceptional execution is carried out with caution. Significant job profiles in job history are chosen for reference as PR(Profile Reference) set. A Plackett-Burman design is applied in the screening procedure.

$$P \in JobHistory \tag{4}$$

$$PR = \{P|\ r \in R_{fast}, PR \subseteq JobHistory\} \tag{5}$$

3.2 Resource Selection Optimization Algorithms

Objective of below algorithm is to get most effective factor and reference job profiles and optimal resource for job (See Figure 1). If effective factors does not exists, it assumes that consider PBDF(Plackett-Burman with Fold-Over) is never conducted and then estimates impact of factors on system with Plackett-Burman designs (2~6 lines). Rating the impact in decreasing order and f_{max} is defined the most effective factor related to resources. Using JobHistory(F,j) function, set of reference job profiles which contains information of candidate resources is obtained from job history, giving the first consideration to effective factors of applications (7, 11~16 lines). In each reference job profile, resources elements where job j was executed are gathered in set R(8 line). Estimation of resources based on f_{max} is required (9 line). For example, if f_{max} is stability, the stablest resource is selected. Finally, job is submitted to selected resource (10 line).

Given $j \in J, T_j$

1) f_{max} = null, PR = $\emptyset$, $r_{selected}$ = null
2) if f_{max} not exists then
3)   PBFD(F, T_j)
4)   Rating F in deceasing order
5) f_{max} = one of F related to resource
6) endif
7) PR = JobHistory(F,j)
8) R = PR(r)
9) $r_{selected}$ = r of f_{max}(R)
10) Dispatch j on $r_{selected}$

JobHistory(F,j)

11) PR = null
12) repeat each Profile
13)   Find P where C_j == f_{max}
14)   PR = PR ∪ P
15) until P == empty
16) return PR

Fig. 1. Algorithm for Referring Job Profiles

After finish the j on $r_{selected}$

1) P_{new} = $(j, r_{selected}, T_j)$
2)   JobHistory = P_{new} ∪ JobHistory
3)   if P_{new} == Failed then
4)   decrease Stability of $r_{selected}$
5)   else
6)   Err = $|T(P_{new}) - T(P_{selected})|$
7)   if Err < Threshold Then
8)   Increase Credit f_{max}
9)   else.
10)   Decrease Credit f_{max}
11)   Repeat PBFD(F, T_j)
12)   endif
13)   endif

Fig. 2. Adaptive Algorithm

Job executed on the selected resource could be succeeded or failed. Above algorithm provides adaptive approach with considering such situations (See Figure 2). New job profile produced after job execution is added to job history (2 lines). In case of job failed, stability of resource where the job is executed on is adjusted (3~4 lines). On the other hand, new job profile is compared with reference profile (5~11 lines). As gap between time elements of two profiles is smaller, the algorithm for referring job profiles is reliable (6 line). Credit of factor is increased when gap is smaller than threshold (7~8 lines), otherwise credit is decreased and PBDF is repeated (10~11 lines). The threshold depends on the user's requirement such as maximum tolerance of factors.

4 Design and Implementation of Job History-Based Resource Selection Optimization System

In this system, users are allowed to apply various policies for their resource selection by identifying characteristics of applications and properties of resources. Heterogeneous distributed resources are constructed as VO(Virtual Organization) through Virtualization. As shown in figure 3, Virtualization provides resource sharing service, resource and workload monitoring service, dynamic resource allocation service. Based on these services, System Middleware supports main features of resource selection with job history.

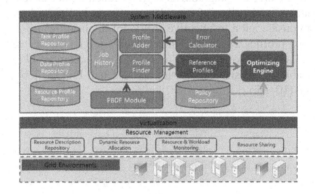

Fig. 3. Architecture of Resource Selection System based on Analysis of Job History

In System Middleware, job profiles are aggregated according to profile's schema which defines characteristics of applications and properties of resources. Aggregated profiles are managed in Profile Repositories (Task ·Data·Resource Profile Repository) and are utilized to analyze information of a particular application and resource with assistance of Profile Finder module. PBDF module, providing Plackett-Burman designs, helps to choose many reference job profiles from Job History component. Resource selection policies in Policy Repository decide resource by Optimizing Engine. Newly generated job profile is added to Job History by Profile Adder, and Error Calculation module periodically conducts estimation for maintaining accuracy of reference job profiles.

5 Experiments and Evaluation

5.1 Analysis of Factors

Prior to evaluation, main factors and job history defined above are as follows. In Plackett-Burman designs, the high and low values of factor should be selected to be just outside the range of normal values. We analyzed high and low values of each factor from job history. Values of MA, AOA are determined by quartile 1 and 3 of its distribution. On the other hand, values of RE and ITMAX is defined by mean.

   Among the factors in chapter 2, we select factors of CPU clocks, stability, and network for evaluation and categorize clusters in PRAGMA grid according to selected factors. The information of clusters is referred to PRAGMA grid monitoring service [17]. Values of Network factor are determined with bandwidth and location of cluster. Stability represents ratio of succeed job over whole executed job. Clusters categorized in boundary have values of factor near dividing line, so we utilize both high and low category.

5.2 Preliminary Experiment and Results

In this section, our experiments show that our approach looks for reliable resources for applications using statistical methods and previous job execution history. PBDF

Table 2. Simulation Result of PBDF based on Actual Job History

	Parameters							Execution Time (s)
	CPU Clocks	Stability	Network	RE	AOA	MA	ITMAX	
1	+	+	+	-	+	-	-	1507
2	-	+	+	+	-	+	-	2401
3	-	-	+	+	+	-	+	47889.5
4	+	-	-	+	+	+	-	898.33
5	-	+	-	-	+	+	+	279.5455
6	+	-	+	-	-	+	+	128
7	+	+	-	+	-	-	+	6786
8	-	-	-	-	-	-	-	195.52
9	-	-	-	+	-	+	+	14373
10	+	-	-	-	+	-	+	25267
11	+	+	-	-	-	+	-	1266
12	-	+	+	-	-	-	+	7504
13	+	-	+	+	-	-	-	625
14	-	+	-	+	+	-	-	26517
15	-	-	+	-	+	+	-	363.8
16	+	+	+	+	+	+	+	1811.083
Effect	-61235	-41668.5	-13353	64790.05	71254.74	-94770.3	70264.48	
Rank	1	2	3	4	2	1	3	

design matrix in Table 2 was comprehensively generated from analysis of real job history from the e-Science environment. It indicates that "CPU clocks" is the most effective factor among factors related to resource. Also, we can find out "MA" mainly decides execution time of application. These are used in the experiments.

The first scenario was a situation of submitting an application with specific parameters and completing successfully without any problems. This scenario allowed us to confirm the capability of our algorithms. The other scenario demonstrated cases of adjusting resource selection when a failure occurred. For each scenario, we compared the execution time of an application and computed the gap between actual execution time and referred one.

Following is an example for better understanding for the first scenario. Consider a job with parameters of specific values, e.g. RE:5, AOA:7, MA:0.7 and IT-MAX:10000. To select a resource for the job, we first gathered reference job profiles from job history with effective factors identified by PBDF. As a result, Table 3 was obtained through this process and we obtained information of candidate resources of {rocks-52, rocks-153, sakura, nucleus}. When we consider the minimum execution time, sakura met this requirement. However, sakura didn't provide acceptable reliability than other resources would have. Consequently, nucleus is selected as the best candidate resource for satisfying execution time and reliability requirements.

Table 3. An Example of Reference Job Profiles

JobID	RE	AOA	MA	ITMAX	Status	Resource	Execution Time
1000000233	5	7	0.7	10000	DONE	rocks-153	1701
1000000236	5	7	0.7	10000	DONE	rocks-52	2364
1000000238	5	7	0.7	10000	DONE	rocks-52	2289
1000000239	5	7	0.7	10000	DONE	sakura	1458
1000000240	5	7	0.7	10000	DONE	sakura	1491
1000000241	5	7	0.7	10000	DONE	nucleus	1321
1000000245	5	7	0.7	10000	DONE	sakura	851
1000000246	5	7	0.7	10000	DONE	sakura	900

In second scenario, we considered the situation after execution of a job. If a job were completed successfully, information of execution including execution time on selected resource would be added to job history. However, if a job were failed, we would regard the selected resource as unavailable and decrease its stability. For example, nucleus, the selected resource of above example, was crashed and led the job to be failed. Then stability of nucleus was adjusted to lower value than 100.

6 Related Works

NIMO [10] system that automatically learns cost models for predicting the execution time of computational scientific applications executing on large-scale computational grids. Accurate cost models helps to select candidate resources of good performance. NIMO generates training samples of biomedical application to learn fairly-accurate cost models quickly using statistical learning techniques. NIMO's approach actively deploys and monitors the application under varying conditions with no changes to the operating system or application. It obtains training data of applications from passive instrumentation streams.

GWA(The Grid Workloads Archive) [5, 18] is a study of gathering and utilizing job history. It presents a place for exchanging workload data and meeting point for grid users. Also, GWA defines requirements of building a workload archive and describes the approach taken to meet these requirements with the GWA. For example, a format for sharing grid workload information, and tools associated with this format are suggested as these requirements. Using these tools, GWA collects and analyzes data from nine well-known grid environments. Finally, several cases of utilizing GWA are presented in critical grid research and practical areas such as research in grid resource management, grid design, operation, and maintenance.

The theory of *design of experiments*(DOE) [3] is a rigorous theoretical field of statistics that studied planned experiments of factors affecting system behavior. Recently, grid research has applied DOE at improving performance of meta-scheduler [6] on grid environments.

7 Conclusion

In this paper, we investigated the problem of resource selection for scientific grid applications. Also, we presented that resource for a job is selected based on analysis of job history that encapsulates knowledge of an application and grid environments. Under these objectives, we apply Plackett-Burman designs with Fold-Over for identifying characteristics of CFD applications in an aerospace research grid and properties of large-scale networked grids. Since these characteristics are considered as factors that impact on CFD applications execution time, resources are chosen based on these factors. Our approach also allows adaptive resource selection by estimating accuracy of job history after every job execution. Also, since job history gets rich as jobs are executed, reliability of resource selection is improved.

Future work will involve extending our approach to utilize grid with various applications. Algorithms in this paper will be refined by identifying realistic weight of effective factors and making a comparative study of real job history. Also, further research is needed to guarantee efficient execution of various applications by supporting a service for easily defining characteristics of each application. Also, we will apply our strategy on various requirements of users to expand the base of study for improvement of resource utilization and management. These objectives will be evaluated with more precise experiments in the future.

References

1. Hur, C., Kim, Y., Kim, J., Cho, K.W.: A Prediction Strategy of Resource Selection with Pattern of Job History in e-AIRS 2.0 System. In: KIISE HPC Winter Conference (February 2009)
2. Hur, C., Kim, Y.: Prediction Scheduling with Patterns of Job History in e-AIRS System. In: KIISE Fall Conference, October 2008, vol. 35(B), pp. 520–525 (2008)

3. Montgomery Douglas, C.: Design and Analysis of Experiments, 5th edn. John Wiley & Sons, Inc., Chichester (2000)
4. Krauter, K., Buyya, R., Maheswaran, M.: A taxonomy and survey of grid resource management systems for distributed computing. Software—Practice & Experience 32(2), 135–164 (2002)
5. Losup, A., Li, H., Jan, M., Anoep, S., Dumitrescu, C., Wolters, L., Epema, D.H.J.: The Grid Workloads Archive. Future Generation Computer Systems 24(7), 672–686 (2008)
6. Vanderster, D.C., Dimopoulos, N.J., Sobie, R.J.: Improved Grid Metascheduler Design using the Plackett-Burman Methodology. In: Proceedings of the 21st International Symposium on High Performance Computing Systems and Applications (May 2007)
7. Wrzesinska, G., Maassen, J., Bal, H.E.: Self-adaptive applications on the grid. In: Proceedings of the 12th ACM SIGPLAN symposium on Principles and practice of parallel programming, San Jose, California, USA (March 2007)
8. Casanova, H., Obertelli, G., Berman, F., Wolski, R.: The AppLes Parameter Sweep Template: User-level middleware for the Grid. In: IEEE Supercomputing 2000, Dallas, Texas, USA, November 4-10 (2000)
9. Kim, J., Kim, Y., Kim, B.S., Ahn, J.W., Kim, J.-H., Lee, J.-S., Choi, J.G., Lee, G.-B., Cho, J.H., Kim, J., Hur, C., Kang, S.-H., Rye, G.-Y.: Development of e-Science technology for fluid dynamic research in various fields report I-08-GG-05-01R-1. Korea Institute of Science and Technology Information (2008)
10. Shivam, P., Babu, S., Chase, J.: Active and Accelerated Learning of Cost Models for Optimizing Scientific Applications. In: International Conference on Very Large Data Bases (VLDB), Seoul, Korea (September 2006)
11. Shivam, P.: Proactive Experiment-Driven Learning for System Management. Department of Computer Science. Duke University (2007)
12. Park, S.H.: Design of Experiments, Minyoungsa (2005)
13. Computational Fluid Dynamics, http://www.cfd-online.com/
14. e-AIRS 2.0,
    http://repository.kisti.re.kr:8080/gridsphere/gridsphere
15. Korea Institution of Science and Technology Information,
    http://www.kisti.re.kr/
16. Pacific Rim Application and Grid Middleware Assembly,
    http://www.pragma-gird.net
17. PRAGMA Grid Monitoring home page,
    http://pragma-goc.rocksclusters.org/scmsweb/
18. The Grid Workloads Archive, http://gwa.ewi.tudelft.nl

Automated and Distributed
Network Service Monitoring

Giovan Germoglio, Bruno Dias, and Pedro Sousa

Centro de Ciências e Tecnologias da Computação, Departamento de Informática,
Universidade do Minho, Campus de Gualtar, 4710-057 Braga, Portugal
{gicage,bruno.dias,pns}@di.uminho.pt

Abstract. Decentralized architectures have been proposed using functional hierarchies of several middle-level managers with delegation of management activities. However, no management applications have been implemented and
widely used that could sanction this approach for real use. Also, current frameworks or mechanisms in network management do not directly address automation. More efficient network management systems with higher levels of
availability and automation are needed. In this paper we propose a design and
specify an automated and distributed network services monitoring system for
Internet services that should be capable to effectively gather and calculate
relevant operational parameters that will be used to determine the availability
level of a network or application service. Based on this availability level, the
system recommends or automatically triggers a service reconfiguration or a total or partial network service replication.

Keywords: Network Services Management, Automation, Distributed
Monitoring.

1 Introduction

The last twenty years have been prolific on standards, protocols, technologies and as-
sorted mechanisms on the computer communications field, resulting from an intensive
and increasing research activity and driven by user demands for better devices, services
and applications. The same applies for the specific area of network management.

In general, the technologies used to implement management systems are based on an
overly centralized agent-manager architecture, which puts too much effort on the manager, or application, side. This classic client-server paradigm uses manager's invocations
of remote procedures that implement a management activity or function. Within these
frameworks, management functions should be classified using the old and classic Open
Systems Interconnection Network Management Framework (OSI/NMF) [13] taxonomy:
fault, configuration, accounting, performance and security management activities. More
decentralized architectures have been proposed using functional hierarchies of several
middle-level managers or delegation of management activities (or management code) to
the agents (or management servers), but no real management applications have been
implemented and widely used. Furthermore, automation is still an open issue in network

C.S. Hong et al. (Eds.): APNOMS 2009, LNCS 5787, pp. 143–150, 2009.
© Springer-Verlag Berlin Heidelberg 2009

management, which is not efficiently addressed by current frameworks or mechanisms. The only form of automation pursued by standards are the policy management approaches defined by the Internet Engineering Task Force's (IETF) Policy Framework working group, that has defined the Policy Core Information Model [14] extensions to the Distributed Management Task Force (DMTF) Common Information Model (CIM) [15] specification for exclusive usage on policies definition, and the IETF's work on policy provisioning specified as the Common Open Policy Service Protocol (COPS) [16]. Again, these approaches had, until now, no significant impact on available network management products in respect to automation or distribution.

Internet Services availability is of great importance for many user-distributed applications and is completely dependent on the correct operation and availability of all major standard network services. Our overall purpose is to design and specify an automated and distributed network services monitoring and configuration system for Internet services focused on the calculation of availability levels. In this article we discuss the architecture of the monitoring sub-system that should be able to effectively calculate the availability level of a network or application service and instruct the configuration sub-system on further actions (backup, replication, etc).

2 Related Work

Several research and development projects already exist for Internet services monitoring, although the most relevant were deployed for DNS monitoring, either using active or passive techniques.

The RIPE NCC DNS Monitoring Service project (DNSMON) [23] uses a direct approach and tries to actively test the availability of the DNS system using a globally distributed active measurement service. This is, probably, the only network services project deployed on a global scale and uses injected DNS test traffic to measure QoS information. On the other hand, the dnslogger [24] is a Weimer's DNS passive logging software. In this approach, sensors capture DNS packets on the network and forward them to an analyzer module integrated on its architecture. This solution is bounded to a specific network service and does not use a real-time monitoring mechanism, although non real-time availability studies could be performed from the logged data.

The mobile agent platform for Customer-based IP Service Monitoring (CSM) that exploits the unique capabilities of mobile agents is implemented by Günter et. al [6]. Agents are deployed by ISPs and by customers to perform measurements tailored to individual monitoring needs and use filter objects to describe protocol packets in to be analysed. This is a more generic concept since it could permit, depending on the specification of the filter objects, monitoring of several network services.

The MonALISA project [25] defines a distributed architecture for service monitoring and is based on a scalable dynamic and distributed architecture. This model used a higher-level component to decide which monitoring mechanisms and tools to use depending on the overall monitoring goal, which is also one of the requisites of our monitoring system.

Choi and Hwang [26] focused their approach on in-service end-to-end flow monitoring by utilizing packet flows to gather information about QoS parameters. Test packets are generated periodically and circulate on an application flow so to allow collection of important statistical information such as network throughput, packet loss, delay and jitter. References [1] to [3] also provide mechanisms for calculation of network or end-to-end application performance parameters.

It is clear that the monitoring concepts behind these projects are not universal and they are overly dependent on the specific needs of the monitoring process of the specific network service availability they intent to monitor. These tools concentrate their efforts on monitoring some direct or indirect connectivity or functional parameters of a specific network service.

3 Automated Network Services Management

In the context of our project, we are interested in define a generic mechanism for measuring and calculate availability levels for any network service. Furthermore, the monitoring sub-system should integrate an automated decision process and the capability to interact with a configuration sub-system, as depicted on Figure 1.

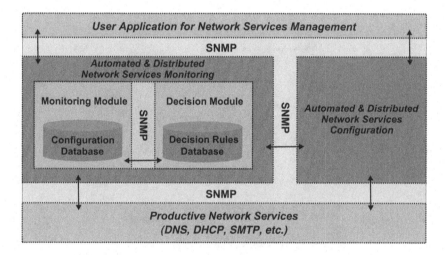

Fig. 1. The Automated Network Management Framework

The internal design of the monitoring sub-system comprises two types of modules: the Monitoring and the Decision Modules. The deployment of these modules in separate network devices/hosts enhances the resilience capabilities of this sub-system. On a typical management domain there will be a much larger number of monitoring modules than decision modules.

3.1 Decision Module

An Active Decision (AD) module , which could be implemented like a Lightweight Policy Management Server (LPMS), must derive decisions from a pre-defined set of strategic, administrative and operational management policy rules when applied to a set of network services QoS metric values that should be continuously evaluated by the monitoring modules. When this metrics imply reduced availability levels an Automated and Distributed Network Services Configuration sub-system (ADNSC), such as depicted on Figure 1, should be automatically notified for an immediate network services reconfiguration procedure. The ADNSC sub-system will be responsible for the reconfiguration process. In the present project we are only interested in implementing major reconfiguration mechanisms like an entire configuration backup, configuration reposition or service replication. In this respect, the decision modules may need complex artificial intelligence algorithms for evaluation of availability metrics but should also use simpler estimation techniques like the ones referred on [6].

Since the two types of modules will be deployed separately, the decision modules can be implemented as part of a global policy management network already deployed on the network services management domain. Also, the high-level language for definition of the policy management rules could be the same as the language used for configuration of monitoring modules.

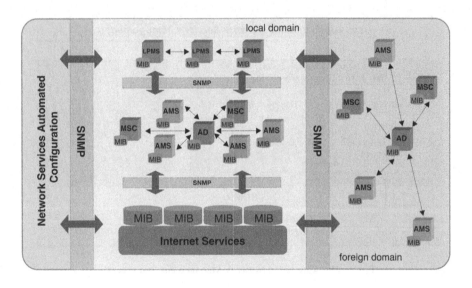

Fig. 2. Distributed Monitoring Architecture

3.2 Monitoring Module

The monitoring process is distributed through a set of monitoring servers, which are named Active Monitoring Servers (AMS) or Monitoring Server Candidates (MSC), according to whether the monitoring state is active or in standby mode respectively. Both types of monitoring modules should be deployed on several locations inside the monitoring domain and, if possible, on some strategic locations outside the monitoring domain, as showed on Figure 2. This distribution is mandatory since the reliability of the continuous measurement of the QoS metrics for a network service depends on the number and location of the AMSs. The group of MSCs is continuously and dynamically maintained through manual or automatic manipulation of configuration parameters, that is, the methodology to determine which servers become active for monitoring of a specific network service could be human based or automated through a decision module server. In the last case, two possible mechanisms, which complement each other, will be described in Section 4.

The monitoring sub-system should support the use of independent databases technologies for storage of the monitoring configuration parameters. It could also support the use of database systems external to the monitoring framework. This added flexibility could add resilience and speed up implementation and deployment times, since already existing distributed or local database systems could be used.

3.3 Automated and Distributed Network Services Configuration

The framework also should support a network services configuration sub-system. The aim of this part of the management system is to support network services configuration, re-configuration, backup or replication procedures taking into account the notified decisions made by AD modules on the monitoring sub-system or as a result of higher level user management applications. The internal design and specification of this configuration sub-system is out of the scope of this text.

3.4 Module's Interaction and Management Information Model

The definition of such management system requires the specification of management communication protocols and information models for management data representation. This includes management data structures or objects, their encoding rules and the protocol operations that must be implemented to ensure a correct and efficient interaction between all modules in the system and between these modules and external lower level monitoring probes or higher level user management applications. Shorter development/implementation periods and universal support should be obtained if well known and widely deployed management frameworks and protocols are used.

The management framework proposed in [8], which adopts the eXtensible Markup Language (XML) as an option for representation of management objects in network management, is composed by XML-based Manager, a XML/SNMP gateway for legacy SNMP (Simple Network Management Protocol) agents and a XML-based agent. The author indicates some useful features of XML for the transfer of a large amount of management data using HTTP, claming similar efficiency when compared to pure SNMP frameworks and considering resource utilization and processing time. Nevertheless, this information model and communication paradigm presents some drawbacks as reviewed on [7], mainly scalability problems introduced by the XML/SNMP gateway. The monitoring performance of SNMP-based solutions was also compared to Web Services-based frameworks by Aiko Prass *et al.* [9] and for many relevant performance aspects (like bandwidth usage, round trip delays, CPU time and memory requirements) the SNMP framework was considered a much better solution. It becomes obvious that the use of a protocol and an information model that was not originally specified for network management systems should be avoided as much as possible.

At this moment we have chosen the SNMP framework as the supporting deployment technology for the lower level management data representation and transfer between component modules, as depicted on Figure 1 and 2. The Internet Network Management Framework is the most dominant and widely available framework for network management.

For representation of higher level configuration parameters and monitoring metrics we can use the Structure of Management Information (SMI) [10] language, creating special configuration and decision rules databases. These can be implemented as special Management Information Bases (MIBs) [11]. Alternatively, this higher level management data could be represented by XML Schemas, which provide better features to represent data with a higher level of functionality. The tests realized in [12] showed that the XML encoding process can be faster than the ASN.1 Basic Encoding Rules (BER) encoding process if small data structures are used, although, BER decoding is always faster than XML decoding, independently of the data size.

4 Distributed Monitoring Architecture

Some existing distributed monitoring frameworks use the concept mobile agent as a means to distribute the monitoring process. Two types of agent mobility are characterize in Fuggetta *et al.* [4]: strong mobility, which has the ability to migrate both the code and the execution state of one network element; and weak mobility, which allows only code transfer across different network elements. The performance of both approaches is assessed in [5] and weak mobility presents a favorable performance factor, capable of minimizing the detection time for changes on the network state. But in our model, even the weak mobility technique could be problematic since we are after very short network services reconfiguration or replication periods. We use a mechanism that can be seen equivalent to a weak mobility technique but is much more efficient in terms of system response in case of relocation of a monitoring module, that is, in reality, implemented as a simple change in the sate of a standby module to an active module (and the replaced active module simple change of its state to inactive or to standby).

As depicted on Figure 2, we could consider two layers on our monitoring process: an higher layer composed by the monitoring servers and a lower layer composed by all the monitoring probes (the ones already implemented by standard monitoring agents on the network). The pseudo mobility mechanism, implemented through AMSs and MSCs, is only supported and explicitly effective on the higher layer. Its algorithm must guarantee that the network services availability metrics evaluated by the monitoring process are not disguised by an incorrect operation of one or a group of active monitoring modules.

The correct network location of the monitoring servers is of extreme importance when asserting the validity of its evaluated monitoring metrics. This requisite implies a complex algorithm to calculate the monitoring servers' location. Several existing techniques could help and some were studied. Jacobson's traceroute [17] was the first network tool that is widely used to determine the route taken by packets to diagnose network problems by building network maps. The inaccuracy of traceroute can occur when the path measurements go through a load balance router. They can distribute their traffic across multiple paths by using routing policies and circle, loops and diamonds [18] are common anomalies found when calculating the path from one source to multiple destinations. Paris traceroute [18] is an attempt to avoid the load balance problem. The reference works on [19] and [20] present an extension to Paris traceroute with the capability to construct a multipath map between a source and a destination by implementing an adaptive stochastic probing algorithm. Several probe methods were compared in different scenarios by Luckie *et al.* [21] and the method used by ICMP-Paris demonstrate the best performance. ICMP-Paris and UDP were also were compared with CAIDA's Archipelago (Ark) [22] dataset, which is the evolution of the Skitter infrastructure, where data is continuously sent by active probes to randomly select network destinations.

Taking into account the more accurate measurements of Paris traceroute and the concepts behind projects like Skitter and Archipelago, a methodology for distribution of the monitoring servers will be derived. This mechanism should be implemented on the AD modules and should calculate an efficient and resilient monitoring topology of AMSs and MSCs.

5   Conclusion

In this paper, we presented a distributed and automated management system with focus on the monitoring sub-system. The framework should be capable of monitoring all Internet services and make automated decisions based on the evaluated high level availability metrics. The system specification was based on other state-of-the-art network-services monitoring projects or technologies, that, in some cases, represent standard, well known and widely used mechanisms on the Internet, and, in other cases, represent new state-of-the-art approaches to network services monitoring. This article presents an ongoing research and development work that will enter the experimentation in the next months, so, we expect that experimental results could be presented in the near future.

References

1. Matthews, W., Cottrell, L.: The PingER project: active Internet performance monitoring for the HENP community. IEEE Communications Magazine 38, 130–136 (2000)
2. Fu, Y., Cherkasova, L., Tang, W., Vahdat, A.: EtE: Passive End-to-End Internet Service Performance Monitoring. In: USENIX Annual Technical Conference (2002)
3. Chen, T.M., Hu, L.: Internet performance monitoring. In: IEEE, pp. 1592–1603 (2002)
4. Fuggetta, A., Picco, G.P., Vigna, G.: Understanding code mobility. IEEE Trans. Soft. Engineering 24, 342–361 (1998)
5. Chen, T.M., Liu, S.S.: A model and evaluation of distributed network management approaches. IEEE J. on Selec. Areas in Comm. 20, 850–857 (2002)
6. Gunter, M., Braun, T.: Internet service monitoring with mobile agents. IEEE Network 16, 22–29 (2002)
7. Choi, M., Jong, J.: Performance Evaluation of XML-based Network Management. In: 16th IRTF-NMRG meeting (2004)

8. Choi, M., Hong, J., Ju, H.: JXML-Based Network Management for IP Networks. ETRI J. 25, 445–463 (2003)
9. Pras, A., Drevers, T., van de Meent, R., Quartel, D.: Comparing the Performance of SNMP and Web Services based Management. In: IEEE electronic Transactions on Network and Service Management, vol. 1 (2004)
10. Rose, M., McCloghrie, K.: Structure and Identification of Management Information for TCP/IP-based Internets. RFC 1155 (May 1990)
11. McCloghrie, K., Rose, M.: Management Information Base for Network Management of TCP/IP-based internets. RFC 1156 (May 1990)
12. Chadwick, D., Mundy, D.: Comparing the Performance of Abstract Syntax Notation One (ASN.1) vs eXtensible Markup Language (XML). In: TERENA Networking Conference (2003)
13. ITU-T Rec. X.701: Information Technology – Open Systems Interconnection. Systems Management Overview (1992)
14. Moore, B., Ellesson, E., Strassner, J., Westerinen, A.: Policy Core Information Model. RFC 3060 (February 2001)
15. Distributed Management Task Force (DMTF): Common Information Model (CIM) Standards, http://www.dmtf.org/standards/cim/
16. Boyle, J., Cohen, R., Herzog, S., Rajan, R., Sastry, A.: The Common Open Policy Service Protocol. In: Durham, D. (ed.) RFC 2748 (January 2000)
17. Jacobson, V.: Traceroute (1989)
18. Augustin, B., Cuvellier, X., Orgogozo, B., Viger, F., Friedman, T., Latapy, M., Magnien, C., Teixeira, R.: Avoiding traceroute anomalies with Paris traceroute. In: ACM SIGCOMM Internet Measurement Conference, pp. 153–158 (2006)
19. Augustin, B., Friedman, T., Teixeira, R.: Multipath tracing with Paris traceroute. In: Workshop on End-to-End Monitoring Techniques and Services, pp. 1–8 (2007)
20. Augustin, B., Friedman, T., Teixeira, R.: Measuring load-balanced paths in the Internet. In: 7th ACM SIGCOMM conference on Internet measurement, pp. 149–160 (2007)
21. Luckie, M., Hyun, Y., Huffaker, B.: Traceroute probe method and forward IP path inference. In: 8th ACM SIGCOMM conference on Internet measurement, pp. 311–324 (2008)
22. Archipelago measurement infrastructure, http://www.caida.org/projects/ark/
23. RIPE NCC DNS Monitoring Services, http://dnsmon.ripe.net/dns-servmon/about.html
24. Weimer, F.: Passive DNS replication. In: FIRST Conference on Computer Security Incident Handling (2005)
25. Newman, H.B., Legrand, I.C., Galvez, P., Cirstoiu, C.: MonALISA: A distribute monitoring service architecture. In: 2003 Conference for Computing in High-Enery and Nuclear Physics (2003)
26. Choi, Y., Hwang, I.: In-service QoS monitoring of real-time applications using SM MIB. Int. J. Net. Man 15, 31–42 (2005)

Network Partitioning and Self-sizing Methods for QoS Management with Autonomic Characteristics

Romildo Martins da Silva Bezerra[1] and Joberto Sérgio Barbosa Martins[2]

[1] Federal Institute of Education, Science and Technology of Bahia (IFBA)
Salvador – Bahia – Brazil
romildo@ifba.edu.br
[2] Salvador University (UNIFACS)
Salvador – Bahia – Brazil
joberto@unifacs.br

Abstract. The increasing complexity, heterogeneity and unpredictability of networks make the task of managing these systems highly complex. The autonomic computing paradigm brings innovative solutions to the network management area with new adaptable and "clever" solutions which should consider a heterogeneous functionality in a wide variety of fields. One of the challenges involved in proposing autonomic solutions for QoS management consists of dealing with the inherent complexity and proposing solutions with feasible execution time. In brief, the autonomic solution effective response time should be fast enough so that it could be applied before any new important network state change. In this paper, a framework that uses self-partitioning methods is considered as the basis for evaluating a solution to the problem of dynamic LSPs setup allocation in a general MPLS network topology focusing on keeping a low complexity solution and aiming at achieving the best possible problem solution identification response time.

Keywords: Network Management, Network Partitioning, Self-sizing Networks, Bandwidth Allocation, Autonomic Computing, Computational Complexity, Quality of Service, MPLS, LSPs.

1 Arguing on the Basic Characteristics of Autonomic Frameworks - An Introduction

The identification of an autonomic computing solution typically requires the definition of an autonomic framework which considers the autonomic paradigm main requirements such as adaptability, management of heterogeneous functionality and to promote intelligent interactions by using learning and reasoning techniques [1]. Besides that, one of the major challenges existing when proposing an autonomic solution is that they typically have to deal with highly complex networks and should have a feasible execution time. Indeed, the needed autonomic approach has to be computed in a feasible time in order to generate a "possibly

C.S. Hong et al. (Eds.): APNOMS 2009, LNCS 5787, pp. 151–160, 2009.

valid solutions" applicable to the network being managed. Thus, we have to look for a possible set of solutions and methods arranged in an adequate autonomic framework such that we try to reduce the inherent complexity and, at same time, we try to improve execution time performance in order to get a feasible execution time. As such, approaches applied to lower network complexity and to promote feasible execution time or improved performance are considered as "autonomic characteristics" and are recommended for an autonomic framework.

Network self-sizing capabilities have been explored in global and local approaches [2]. In brief, self-sizing consists in employing the ability to compute and allocate network resources using either a global or a local perspective. Discovering community structures or network clusters in huge and complex networks has been extensively studied by the research community [3] [4] [5] [6] . Network clustering algorithms have been proposed with great success in social networks, collaboration networks, gene networks, among others.. As a result, algorithms which discover communities in networks are able to partitioning the network and could be adjusted and/or adapted to take into account the autonomic system's complexity and response time requirements existing for autonomic network management.

In brief, the approach proposed by this paper consists of applying the self-sizing and partitioning methods considering the scope of an autonomic network management framework in order to achieve a less complex computational scenario resulting in a feasible execution time for the autonomic solution. The fact of being able to compute in an autonomic way in a complex network topology scenario is considered a basic characteristic that autonomic systems should be engaged in. The paper evaluates the proposed partitioning and self-sizing approaches by establishing a huge amount of LSPs in a MPLS network with a significant number of nodes and random topology.

In the next sections the autonomic network management framework is introduced. In section 3 we argue on the use of self-sizing and partitioning methods in autonomic frameworks. Section 4 evaluates a group of partitioning algorithms and proposes a new partitioning algorithm which considers the autonomic characteristics required. Section 5 evaluates the partitioning algorithm proposed considering the establishment of a set of LSPs in a random MPLS network with a significant number of nodes. Section 6 presents the final considerations and future works.

2   Autonomic Network Management Framework

The basic framework model adopted [7] to support dynamic resource allocation algorithms with autonomic capabilities is able to manage quality of service in computer networks using a Full Policy-based Management (FPBM) approach as shown in Figure 1. The model uses a traditional Policy-based Management strategy [8] to manage the policy lifecycle. To create new policies in an autonomic way according to the new state of the network it is added a knowledge layer, replacing the policy editor typically used in Policy-based Management systems.

By adopting this structure, the policies used by the execution plane are generated dynamically by the decision plane, replacing human intervention in this function. This new autonomic management model divides the autonomic network management for quality of service (QoS) in three planes: information, decision and execution planes. The model planes structure and interrelation are show in Figure 1.

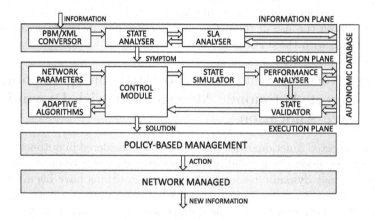

Fig. 1. Autonomic Management Network Model [7]

The self-management process runs in a cyclic way, i.e. the network continuously report its state, independently of the occurrence of problems, thus enabling self-optimization and supporting problems prevention (self-healing). The network state is analyzed (information plane), a solution is found (decision plane), and this solution is converted into a high-level policy (execution plane). The policies are translated into network technologies and mapped according with device capabilities. The policy distribution is done by a Policy Decision Point.

It follows a brief description for each plane.

- Information Plane - The information plane receives data which contains the network state and converts it to a standardized format in a high-level language (XML format) that can be treated by the framework. This conversion provides extensibility to the model since the creation of new converters makes it compatible with other sub-systems or applications. After that, the current network state (snapshot) is analyzed in order to verify its accordance with the Service Level Agreement (SLA) looking for the maintenance of pre-defined network parameters being managed.
- Decision Plane - The decision plane receives the network symptom and finds a solution that meets the specified SLA, if it exists. It is important to observe that the search for a solution, at first, considers an optimum solution which is the best solution for a set of solutions. Meanwhile, the overall algorithm complexity and time limitation in the searching process are considered. For example, an optimum solution to a problem that is no longer occurring is not the best strategy to be used.

- Execution Plane - The execution plane receives the diagnosis (solution) and generates a policy through the policies compiler. Before being implemented, the policy goes through a phase of syntax and semantics validation that does not compromise the autonomic information base.

In terms of autonomic network management model described, the decision plane is the focus. As indicated, the overall algorithm complexity and time limitation in the searching process for the best solution have to be considered. The following sections argue on the use of self-sizing and partitioning methods as a possible solution to deal with these issues.

3   Using Network Self-sizing and Partitioning Capabilities to Support Autonomic Management with Dynamic Resource Allocation

One of the expected functionalities that should be considered in autonomic management system design is its ability to support a dynamic resource allocation approach. Various dynamic resource allocation algorithms have been proposed and, in brief, they provide optimized results based on various criteria and computed based on near real time traffic data collected with various different measurements approaches. When considering an autonomic management framework, the dynamic resource allocation algorithm presents an additional requirement: it should scale to support large and complex network topology structures while keeping an "affordable" execution time. In our novel autonomic network management model, we propose an adaptable solution at the decision plane which promotes the utilization of dynamic resource allocation algorithms by applying simultaneously self-sizing and partitioning principles to the network node structure.

Networks with self-sizing capabilities have the ability to allocate resources like link capacity automatically and adaptively using online gathered data traffic information. Resource allocation decisions can be realized on self-sizing networks "globally" or "locally". A global self-sizing approach means a more centralized solution typically based on a snapshot of the network. A local self-sizing approach means that individual nodes should compute their new states based, typically, on locally acquired information.

In [2], it is argued that an important aspect of the self-sizing capability is the computation time required to find the optimal allocation. This computation time depends on the efficiency of the algorithm, and also on the number of possible solutions for the allocation problem. The number of possible allocations N can be represented as follows:

$$N = k^{lm} \tag{1}$$

where k, l and m are the routing alternatives, services supported and number of source-destination (SD) pairs respectively. As such, the inherent complexity and resulting computation time have a strong dependency on the network node structure and grouping or communities.

Partitioning is a second principle proposed to be used in our autonomic framework in order to better support its autonomic characteristics while making use of dynamic resource allocation algorithms. Partitioning corresponds to the identification of community structures in networks. According to [4], community structure is a property argued to be common in many networks corresponding to the division of network nodes into groups within which the network connections are dense, but between which they are sparse. Community structures in networks have been extensively studied and the concept is closely related to the idea of graph partitioning in graph theory [3] [4] [5] [6]. In terms of the autonomic network management model support for the adaptable use of dynamic resource allocation algorithms the objective is to develop a recursive algorithm that searches a maximum group of k communities with a maximum size of m nodes keeping connections among k groups as sparse as possible. In network terms this means that the partitioning algorithm should as much as possible to minimize network traffic among partitioned communities.

As such, the set of principles adopted for the autonomic framework model is to apply both partitioning and self-sizing methods in order to reduce the inherent complexity and to promote an improved overall response time in complex computer network structures.

4 A Recursive Partitioning Algorithm Based on Network Topology Density

A computer network is an undirected connected graph $G = (V, E)$ of sets such that $E \supseteq |V^2|$, thus, the elements of E are two-element subsets of V^1. In graph theory, the elements of V are the vertices (nodes or points) of graph G and the elements of E are its edges (lines).

In this paper will use the following notations and definitions:

- The order of a graph G is $|V|$ (vertex number). In this paper $n = |V|$.
- The vertex degree is the number of edges that connect to it.
- The density of a graph is the ratio between the number of edges and the number of possible edges.

In general, the task of finding an exact solution to a graph partition is considered a NP-complete problem, making difficult to solve it for huge network structures. Many areas such as social networks, web graph, ecosystems and others use graphs partitioning algorithms to find common properties or relationships between the vertices. The property that seems to be common to many networks is to define community structures. In other words, the main objective is to divide network vertices into sub-graphs that have higher density.

When looking for higher density sub-graphs, one obtains a set of vertex where the relationship between them is stronger, i.e., the number of edges that one is larger. In the context of computer networks, this represents a greater number

---

[1] Edges of type (e_i, e_i) will not be considered and (e_i, e_j) is equal to (e_j, e_i).

of paths between a vertices set. These vertices subsets of higher density are seen as sub-domains. The edges that do not belong to any sub-domain are links that allow inter-domain communication. As such, the algorithm originally used for finding communities will be effectively used in our context to identify the vertices for sub-domain creation. The algorithm will receive a graph G (network topology) and number n indicating the required number of sub-domains arrangements and then creates a new graph with the interconnection of communities, in effect, sub-domains (Figure 2).

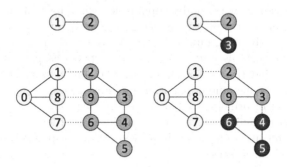

Fig. 2. Graph with two partitions and three partitions

The algorithms "walktrap"[2] [3], "fast greedy" [6], "eigenvector" [5] and "edge betweenness" [4] were initially evaluated to find communities (partitioning nodes) from large computer network domains. The algorithm evaluation was based on two metrics classified in terms of their relevant impact for dynamic resource allocation complexity and execution time algorithms support: minimum number of edges not belonging to the domain (higher density sub-domains) and execution time. The eigenvector algorithm was immediately discarded because it does not generated sub-graphs from certain pre-defined network sizes. The evaluation was then carried out with related graphs, not complete, with the minimum degree equal to two and randomly generated with n equal to 10, 50 and 100. The results are indicated in Table 1.

The "edge betweenness" algorithm showed the best behavior and was chosen as the algorithm to be used as the basic block towards the generation of sub-graphs with high average density, considering the three parameters analyzed (execution time, density and sub-domains size standard derivation). In brief its partitioning into sub-domains resulted in a maximum number of edges within the generated sub-graphs and the algorithm presented a satisfactory execution time. Moreover, the variation in the sub-graphs cardinality was the lowest.

As the next step, it is necessary to have a new algorithm that considers complexity and execution time in order to adequately support autonomic characteristics. A new "edge betweenness with dense sub-domain partitions" algorithm

---

[2] This algorithm uses random paths and their performance may not remain stable, because it is not deterministic.

Table 1. Algorithms evaluation results considering relevant metrics in order to find communities for huge computer networks. Best choices are in bold.

Parameters	execution time			density			standard deviation		
	seconds			edges not used			sub-domain size		
Algorithms — Nodes	10	50	100	10	50	100	50	100	250
edge betweenness	0.141	**0.172**	0.872	5	17	**30**	**2.05**	**6.83**	12.09
fast greedy	**0.125**	0.188	1.106	5	17	35	4.27	7.06	14.55
walktrap	0.126	0.181	**0.186**	5	19	33	3.90	11.98	**11.72**

enforcing a greater network density was then implemented in language "R" [9] following the steps illustrated in Algorithm 1.

```
1 input-graph(mynet, k);
   /* group vertices with edge.betweenness in k groups        */;
2 wtc ← edge.betweenness.community(mynet);
3 mynetcsize ← array(community.to.membership(mynet, wtc$merges,
   steps=find.steps(mynet,k))$csize);
4 mynetmembership ← array(community.to.membership(mynet, wtc$merges,
   steps=find.steps(mynet,k))$membership);
   /* create a list of vectors with subdomains membership     */
   subdomainslist ← vector("list", dim(mynetcsize));
5 subdomainslist ← create.subdomainslist(mynetcsize,mynetmembership);
   /* create subgraphs                                         */;
6 subdomain=list();
7 for i = 1 to dim(array(subdomainslist)) do
8    ⌊ subdomain[[i]] ← subgraph(mynet,array(subdomainslist[[i]]));
   /* subgraphs union                                          */;
9 subdomainsunion ← graph.empty(n=0, directed=FALSE);
10 subdomainsunion ← subgraphs.union(subdomain, subdomainslist);
   /* identify interdomains edges                              */;
11 sdiff ← graph.difference(mynet,subdomainsunion);
   /* create the new graph                                     */;
12 mynetup ← newgraph(mynet,subdomainsunion,sdiff);
```

Algorithm 1: The new "edge betweenness with dense sub-domain partitions" algorithm

The proposed "edge betweenness with dense sub-domain partitions" algorithm enforces a "network density" objective. It receives a graph (*mynet*) and a partitions number . It returns to the user a set of high density sub-graphs and creates a new graph (*mynetup*) indicating the relationship between the new set of sub-domains (Figure 2).The goal achieved by the derived algorithm is to aggregate the traffic on sub-graphs of higher density, reserving the edges not belonging to any sub-graphs only for communication between domains. Being so, the network is partitioned as the interconnection of a set of high-density subnets.

The "edge betweenness with dense sub-domain partitions" algorithm uses a score for an edge and measures the number of shortest paths through it. This procedure is repeated for all vertices for . The main idea of this new based community structure detection algorithm is that it is likely that edges connecting separate modules have high edge betweenness as all the shortest paths from one module to another must traverse through them.

5   Self-sizing with Local Control and Partitioning Algorithm Evaluation for LSPs Setup in a MPLS Network

We consider now the allocation of resources using a self-sizing approach at sub-domain level in order to evaluate the effectiveness and performance improvements created by the partitioning approach used. The sequence of actions taken to evaluate the partitioning algorithm used is as follows:

1. The "edge betweenness with dense sub-domain partitions" algorithm receives network data (topology, number of partitions).
2. The "edge betweenness with dense sub-domain partitions" algorithm generates the set of high-density sub-domains (partitioning).
3. A set of sub-graphs is created and interconnection edges are identified (domains interconnections).
4. An allocation resource algorithm is invoked and applied in order to establish a huge set of LSPs through the entire network.
5. The resource allocation algorithm allocates LSPs for intra-domain traffic.
6. The resource allocation algorithm allocates LSPs for inter-domain traffic.
7. The sequence of actions 5 and 6 is repeated whenever the chosen resource allocation algorithm must interact in order to setup LSPs interactively due to restrictions occurring in terms of bandwidth allocation or another allocation parameter constraint.

The resource allocation algorithm used for evaluating the partitioning algorithm was the "penalty-based LSP allocation algorithm" [10]. This algorithm, in brief, allocates LSPs according to their non allocation penalty and required bandwidth assigned by the network manager. In effect, for the specific purpose of evaluating the network partitioning into sub-domains any resource allocation algorithm could be used. Our option was to evaluate the partitioning effect with a dynamic allocation algorithm that will be used in the context of the autonomic network management framework.

The evaluation results are illustrated in Figure 3 considering the following evaluation scenarios. First, the setup of 3.000, 6.000 and 12.000 LSPs respectively. Second, the distribution intra-domains and inter-domains traffic among partitioned sub-domains follows the parameters 50%-50%, 80%-20% and 90%-10% respectively.In the first evaluation scenario (Figure 3a) a set of 3.000 LSPs was established considering the full interconnected network and the partitioned one with different configurations for inter-domain traffic. The execution time evaluated indicated an improvement in the overall execution time achieved with reductions of approximately 10,25% to 29,37% depending on the inter-domain traffic considered.

In the second evaluation scenario (Figure 3b) sets of 3.000, 6.000 and 12.000 LSPs were setup assuming an inter-domain distribution traffic of 50%-50%. The evaluation results show that the execution time was improved by approximately 11,07% (3.000 LSPs) to 31,68% (with a greater set of established LSPs) by using the partitioning algorithm.

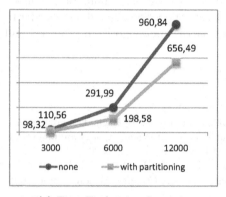
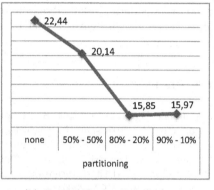

(a) First Evaluation Scenario   (b) Second Evaluation Scenario

Fig. 3. LSPs setup allocation execution time evaluation with network partitioning

6 Final Considerations and Future Work

The management of heterogeneous functionality, the need of adaptability and the application of learning and reasoning techniques to support intelligent inter-action are considered an essential requirement for achieving autonomic management in most frameworks.

This paper introduces a new partitioning algorithm applicable to the management of quality of service in autonomic computer networks and, by applying the self-sizing method, the evaluation results show that it promotes an overall improvement in the autonomic framework performance (approximately between 10% to 30%) in terms of a specific resource allocation algorithm. The scenario considered was the one where a dynamic allocation resource algorithm is used to set up a significant amount of LSPs in a huge MPLS network generated at random. In effect, dynamic resource allocation is a critical area since autonomic systems should derive on the fly new network states for, typically, highly complex topologies while keeping a feasible execution time to derive these solutions. In other words, the most basic result shown in this paper is that by applying partitioning and self-sizing is possible to improve the autonomic system performance.

Based on the results, authors argue that the partitioning of complex computer network topologies coupled with the capability of self-sizing methods could promote the employment of dynamic resource allocation algorithms to derive new solutions in complex networks and, as such, will pave the road to the creation of more clever and effective autonomic management systems.

In terms of the autonomic framework model proposed, the solution described effectively manages heterogeneous functionality, since the decision layer finds a solution by adopting a set of established methods and paradigms (partitioning and self-sizing) in a specific and focused way. Additional simulations will be carried on considering others dynamic resource allocation algorithms in order to evaluate dependencies with respect to the partitioning algorithm proposed and its performance.

References

1. Jennings, B., van der Meer, S., Balasubramaniam, S., Botvich, D., Foghlu, M., Donnelly, W., Strassner, J.: Towards autonomic management of communications networks. Communications Magazine, IEEE 45(10), 112–121 (2007)
2. Nalatwad, S., Devetsikiotis, M.: Self-sizing networks: local vs. global control. In: Proceedings of IEEE International Conference on Communications, June 2004, vol. 4, pp. 2163–2167 (2004)
3. Pons, P., Latapy, M.: Computing communities in large networks using random walks. Journal of Graph Algorithms and Applications 10(2), 191–218 (2006)
4. Newman, M.E.J., Girvan, M.: Finding and evaluating community structure in networks. Physical Review E 69 (2004)
5. Newman, M.E.J.: Finding community structure in networks using the eigenvectors of matrices. Physical Review E 74 (2006)
6. Clauset, A., Newman, M.E.J., Moore, C.: Finding community structure in very large networks. Physical Review E 70 (2004)
7. Bezerra, R.M.S., Martins, J.S.B.: A sense and react plane structured autonomic model suitable for quality of service (qos) management. In: Proceedings of 5th International IEEE Workshop on Management of Ubiquitous Communications and Services, vol. 1 (Maio 2008)
8. Bezerra, R.M.S., Martins, J.S.B.: Functional decoupling principle applied to network device and qos management. In: Proceedings of 7me Colloque Francophone of Gestion of Rseaux et of Services, June 2006, vol. 1, pp. 47–58 (2006)
9. R Development Core Team: R: A Language and Environment for Statistical Computing. R Foundation for Statistical Computing, Vienna, Austria (2009) ISBN 3-900051-07-0
10. Bezerra, R.M.S., Martins, J.S.B.: Penalty-based lsp allocation algorithm. Technical Report TR-01-AN2009, DMCC (2009)

A Scheme for Supporting Optimal Path in 6LoWPAN Based MANEMO Networks*

Jin Ho Kim and Choong Seon Hong**

Department of Computer Engineering, Kyung Hee University, Korea
{jinhowin,cshong}@khu.ac.kr

Abstract. In this paper, we focus on the scheme for the route optimization in 6LoWPAN based MANEMO environments. If 6LoWPAN mobile routers for supporting NEMO protocol are organized by nested NEMO, this nesting of 6LoWPAN mobile router makes the extremely un-optimal route between 6LoWPAN nodes and its correspondent node. This is so, because all of the packets should be forwarded through the bi-directional tunnel which is established between 6LoWPAN mobile router and its home agent. Therefore, the more the number of nested levels, the packet route will become increasingly complex and it causes more end-to-end packet delay. This problem can be handled efficiently by applying MANEMO, which provides integration of MANET and NEMO protocols. With the application of MANEMO in 6LoWPAN, we can avoid the nested NEMO since MANET consisting of 6LoWPAN mobile routers can directly communicate with each other. In this paper, we propose an interoperable architecture between MANEMO and 6LoWPAN to solve the un-optimal route problem of the nested NEMO. With proposed scheme, we can provide the route optimization to communicate between 6LoWPAN node and correspondent node. Furthermore, we can also reduce the tunneling overhead and the end-to-end packet delay.

Keywords: 6LoWPAN, NEMO, MANEMO.

1 Introduction

6LoWPAN (IPv6 over Low power WPAN) [1][2] is a simple low-cost communication protocol that allows wireless connectivity in applications with limited power. The 6LoWPAN protocol adopts the IPv6 protocol stack for seamless connectivity between IEEE 802.15.4 [3] based networks and the IPv6 based infrastructure. The 6LoWPAN protocol could be more suitable for smaller devices with lower energy consumption. Also, it enhances the scalability and mobility of sensor networks. The network mobility (NEMO) [4] protocol maintains the session continuity of moving networks, which are group of nodes that constitute a subnet of a mobile router (MR), even when the MR dynamically changes its point of attachment to the Internet. So, if NEMO is applied in

---

* This work was supported by the IT R&D program of MKE/IITA. [2008-F-016-02, CASFI]
** Corresponding author.

C.S. Hong et al. (Eds.): APNOMS 2009, LNCS 5787, pp. 161–170, 2009.
© Springer-Verlag Berlin Heidelberg 2009

the 6LoWPAN network to provide mobility for 6LoWPAN nodes that constitute within a subnet of the 6LoWPAN MR, even though each 6LoWPAN node does not equipped with the mobility function, it can maintain connectivity with the Internet through the 6LoWPAN MR as a network unit.

However, if the 6LoWPAN MRs for supporting NEMO protocol are organized by nested NEMO as shown in Fig. 1, this nesting of 6LoWPAN MR makes the extremely un-optimal route between 6LoWPAN nodes and its correspondent node (CN). This is so, because all of the packets should be forwarded through the bi-directional tunnel which is established between 6LoWPAN MR and its home agent (HA). Therefore, the more the number of nested levels, the packet route will become increasingly complex and it causes more end-to-end packet delay. This problem can be handled efficiently by applying MANEMO (MANET for NEMO) [5], which provides integration of MANET and NEMO protocols. With the application of MANEMO in 6LoWPAN, we can avoid the nested NEMO since MANET consisting of 6LoWPAN MRs can directly communicate with each other. However, there is no scheme for the route optimization in 6LoWPAN based MANEMO environments. Therefore, in this paper, we propose an interoperable architecture between MANEMO and 6LoWPAN to solve the un-optimal path problem of the nested NEMO. With proposed scheme, we can provide the optimal path to communicate between 6LoWPAN node and CN. Furthermore, we can also reduce the tunneling overhead and end-to-end packet delay.

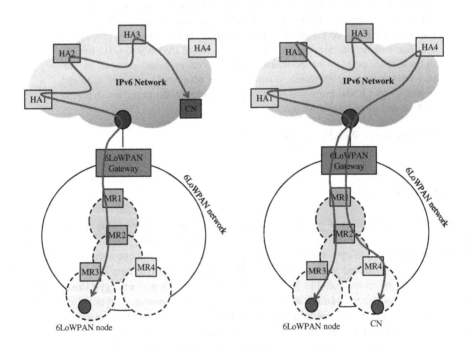

Fig. 1. Pinball routing problem in the nested NEMO

Our paper is organized as follows. In section 2, we briefly introduce the concept of 6LoWPAN and its packet format. In section 3, we describe a detailed scheme for supporting the optimal path in 6LoWPAN based MANEMO networks. Then we show performance analysis in section 4. Finally, we conclude in section 5.

2 Background

The challenge of 6LoWPAN is that IPv6 network and IEEE 802.15.4 network are totally different. The IPv6 network defines the maximum transmission unit (MTU) as 1,280 bytes, whereas the IEEE 802.15.4 packet size is 127 octets. Therefore, the adaptation layer is defined between the IP layer and the MAC layer to transport IPv6 packets over IEEE 802.15.4 links. The adaptation layer is responsible for fragmentation, reassembly, header compression, decompression, mesh routing, and addressing for packet delivery under mesh topology. The 6LoWPAN protocol supports the scheme to compress the IPv6 header from 40 bytes to 2 bytes. There are three types of 6LoWPAN dispatch header (1 byte) formats: addressing, mesh, and fragmentation [2]. All types of 6LoWPAN dispatch headers are orthogonal. The dispatch header indicates the information of the next header. For example, addressing dispatch indicates information of the IP or UDP header, and a mesh dispatch is used in mesh routing. A fragmentation dispatch indicates the information for fragmentation and reassembly of the packet. Actual headers are followed by its dispatch header. The header for a new function can be included by defining a new dispatch. A dispatch header has the following structure. If the 0 and 1 bits (the first 2 bits of the dispatch header) are 00, the next header is not a LoWPAN header. If they are 01, the next header is compressed or uncompressed IPv6, while 10 indicates a header for mesh routing and 11 indicates a header for fragmentation. If the dispatch pattern is 01000001, the following header is an uncompressed IPv6 header (40 bytes). If the pattern is 01000010, the following IPv6 header is fully compressed from 40 bytes to 2 bytes. When more than one LoWPAN header is used in the same packet, they should appear in the following order: mesh, fragmentation, and addressing header.

3 Proposed Mechanism

In this section, we explain detailed operation of a scheme for supporting the optimal path in 6LoWPAN based MANEMO networks. We define the 6LoWPAN packet format of router solicitation (RS) and router advertisement (RA) messages to discover a 6LoWPAN-MANEMO Gateway. Also, we define the 6LoWPAN format of 6LoWPAN prefix request (6LoPREQ) and 6LoWPAN prefix reply (6LoPREP) messages to discover the CN's prefix.

3.1 6LoWPAN-MANEMO Gateway Discovery

6LoWPAN networks are consisted of one 6LoWPAN-MANEMO Gateway and several 6LoWPAN based MANEMOs which support network mobility to 6LoWPAN nodes. Each 6LoWPAN based MANEMO has one 6LoWPAN MR and some 6LoWPAN nodes. The 6LoWPAN-MANEMO Gateway acts as a default gateway in the 6LoWPAN network.

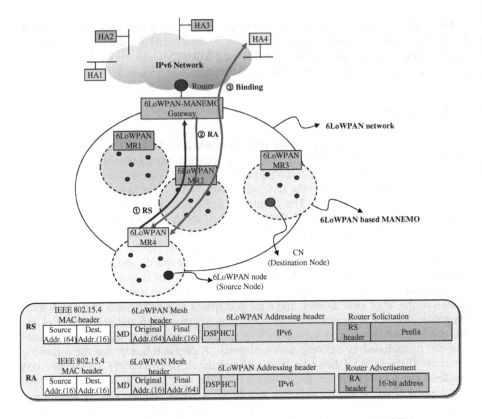

Fig. 2. 6LoWPAN-MANEMO Gateway discovery scenario and proposed RS/RA messages format

Fig. 2 shows the scenario when the 6LoWPAN MR joins a 6LoWPAN network using proposed RS and RA messages. When a 6LoWPAN MR4 moves to the new 6LoWPAN network area, it can detect its movement since the current PAN ID is different from the previous PAN ID by receiving the beacon message. If the 6LoWPAN MR4 is aware of the new 6LoWPAN network attachment, it sends a proposed unicast RS message as a 6LoWPAN packet format to the 6LoWPAN-MANEMO Gateway in order to join and notify attachment of the current 6LoWPAN network. The RS message should contain 6LoWPAN mesh header for the multi-hop routing. The original and final addresses in the mesh header of the 6LoWPAN RS message are set to the 6LoWPAN MR4's 64-bit MAC address and the 6LoWPAN-MANEMO Gateway's 16-bit address, respectively. We assume that all of 6LoWPAN-MANEMO Gateway's 16-bit address is 0x0001. So, the final address field of the 6LoWPAN mesh header can be set to 0x0001, which is fixed 16-bit address of the 6LoWPAN-MANEMO Gateway. The proposed RS message includes the 6LoWPAN MR4's prefix option, which is assigned to the ingress interface of the 6LoWPAN MR4. If the intermediate 6LoWPAN MR2 receives the RS message, then it just relays the RS message to the 6LoWPAN-MANEMO Gateway. The RS message is forwarded to the 6LoWPAN-MANEMO Gateway according to the final address of the mesh header. The 6LoWPAN-MANEMO Gateway obtains the 6LoWPAN MR4's information such as 64-bit MAC address, link-local address and 6LoWPAN prefix from the mesh header, the IP header and the RS option, respectively. The 6LoWPAN-MANEMO Gateway assigns a 16-bit address to the 6LoWPAN MR4 and it manages a list of all the 6LoWPAN MRs with 16-bit addresses which can be used for inside the 6LoWPAN network. Therefore, 6LoWPAN MRs do not require the 16-bit address collision avoidance mechanism. Upon receipt of the RS message, the 6LoWPAN-MANEMO Gateway sends a unicast proposed RA message to the 6LoWPAN MR4 to assign a 16-bit address which is only available within the 6LoWPAN network. The RA message contains the 6LoWPAN MR4's 16-bit address option which is assigned by the 6LoWPAN-MANEMO Gateway. The RA message should also contain 6LoWPAN mesh header for the multi-hop routing. The RA message will be delivered directly to the 6LoWPAN MR4 since both the source and destination of the RA message are unicast link-local addresses. The original and final addresses in the mesh header of the 6LoWPAN RA message are set to the 6LoWPAN-MANEMO Gateway's 16-bit address (0x0001) and the 6LoWPAN MR4's 64-bit MAC address, respectively. When the 6LoWPAN MR4 receives the RA message, its care-of address (CoA), which is the temporary IPv6 address at its current 6LoWPAN network attachment point, can be obtained by concatenating the prefix in the RA message, PAN ID in the beacon message and the assigned 16-bit address. The PAN ID and the 16-bit address are used as part of the IPv6 address. Finally, the 6LoWPAN MR4 registers binding between its home address and CoA with the HA.

3.2   6LoWPAN Mobile Network Prefix Discovery

Fig. 3 shows the scenario of 6LoWPAN prefix discovery of the Correspondent 6LoWPAN MR (CMR) using proposed 6LoPREQ and 6LoPREP messages. If the 6LoWPAN node in the 6LoWPAN MR4 wants to communicate with CN after the 6LoWPAN-MANEMO Gateway discovery has successfully finished, the 6LoWPAN MR4 should decide whether the CN is located in the current 6LoWPAN network or not. To accomplish this, the 6LoWPAN MR4 should perform mobile network prefix (MNP) discovery to find the location of the CN.

When the 6LoWPAN node sends the packet to the CN, the 6LoWPAN MR4 may receive this packet through its ingress interface. The 6LoWPAN MR4 discovers the destination prefix entry in the routing table to decide if the path to the CN is stored or not. If the route to the CN is not stored in the routing table, the 6LoWPAN MR4 sends the 6LoPREQ message to the 6LoWPAN-MANEMO Gateway including CN's prefix option to aware of where the CN is located. The 6LoWPAN-MANEMO Gateway can make a decision whether the CN's prefix is within the current 6LoWPAN network or external since it manages all of 6LoWPAN MRs and its MNP information by exchanging RS and RA messages. If the destination's prefix matches the CN's prefix in the mapping table, the 6LoWPAN-MANEMO Gateway replies 6LoPREP messages to the 6LoWPAN MR4. The 6LoPREP message includes a 16-bit address of CMR to which the CN belongs. On the other hand, there is no 6LoPREP header in the 6LoPREP message if the destination's prefix does not match. With MNP discovery, 6LoWPAN MR4 is able to get CMR's 16-bit address.

Upon receipt of the 6LoPREP message, the 6LoWPAN MR4 performs MANET routing protocol such as exchange RREQ and RREP messages to establish the optimal path between CMR and 6LoWPAN MR4 using the CMR's 16-bit address. With exchange 6LoPREQ and 6LoPREP messages, the optimal path is established between 6LoWPAN node and CN.

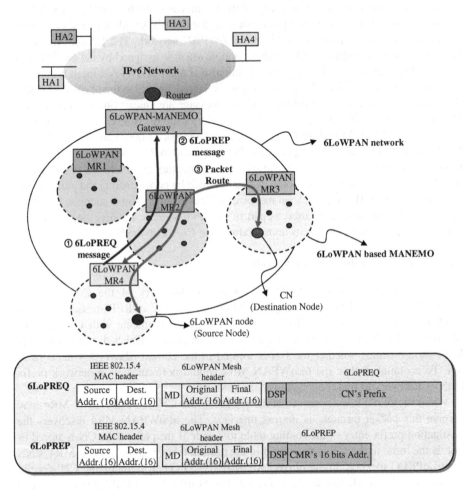

Fig. 3. 6LoWPAN prefix discovery scenario and proposed 6LoPREQ/6LoPREP messages format

Table 1 shows the all the necessary headers to send the 6LoPREQ message in detail. A mesh header is needed for the multi-hop routing between the 6LoWPAN MR and the 6LoWPAN-MANEMO Gateway. We define an another new dispatch header pattern, which is set to '00000001', for the 6LoPREQ message. The CN's prefix is included in the 6LoPREQ header to request matching the CMR's 16-bit address. Table 2 shows the all the necessary headers to reply the 6LoPREP message in detail. Also, we define a new dispatch header pattern, which is set to '00000010', for the 6LoPREP message. If the CN's prefix is within the current 6LoWPAN network, the CMR's 16-bit address is contained in the 6LoPREP header.

Table 1. 6LoPREQ message format and data

Header	Field	Data	Size
IEEE 802.15.4 MAC header	Source Address	6LoWPAN MR's 16-bit MAC address	2 bytes
	Destination Address	6LoWPAN Intermediate MR's 16-bit MAC address	2 bytes
6LoWPAN Mesh header	MD (Mesh Dispatch)	Original address flag=16-bit Final address flag=16-bit	1 byte
	Original Address	6LoWPAN MR's 16-bit address	2 bytes
	Final Address	6LoWPAN-MANEMO Gateway's 16-bit address	2 bytes
6LoPREQ header	DSP (Dispatch)	6LoWPAN Prefix Request (00000001)	1 byte
	6LoPREQ	CN's Prefix	8 bytes

Table 2. 6LoPREP message format and data

Header	Field	Data	Size
IEEE 802.15.4 MAC header	Source Address	6LoWPAN-MANEMO Gateway's 16-bit MAC address	2 bytes
	Destination Address	6LoWPAN Intermediate MR's 16-bit MAC address	2 bytes
6LoWPAN Mesh header	MD (Mesh Dispatch)	Original address flag=16-bit Final address flag=16-bit	1 byte
	Original Address	6LoWPAN-MANEMO Gateway's 16-bit address	2 bytes
	Final Address	6LoWPAN MR's 16-bit address	2 bytes
6LoPREP header	DSP (Dispatch)	6LoWPAN Prefix Reply (00000010)	1 byte
	6LoPREP	CMR's 16-bit address	2 bytes

Fig. 4 shows the proposed signaling flow for supporting the optimal path in 6LoWPAN based MANEMO networks. As mentioned above, the packet routing is optimized by MNP discovery and MANET routing protocol between 6LoWPAN MRs without bi-directional tunnel between 6LoWPAN MR and its HA.

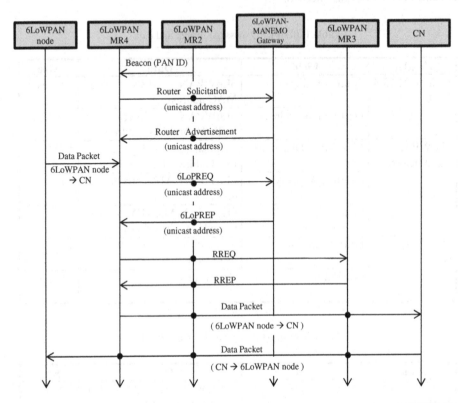

Fig. 4. Proposed signaling flow for supporting the optimal path in 6LoWPAN based MANEMO networks

4 Performance Analysis

To evaluate the performance of our proposed scheme, we perform simulation using NS-2 [6]. The IEEE 802.15.4 MAC protocol is employed as the Data Link layer. Table 3 shows the parameters for the simulation. We compare proposed 6LoWPAN based MANEMO and 6LoWPAN based NEMO schemes in terms of the tunnel overhead and end-to-end packet delay by increasing the number of 6LoWPAN MR in the 6LoWPAN network. The tunnel overhead and end-to-end packet delay are experienced by constant bit rate (CBR) traffic rate 10 packets per second, where each packet size is 50 bytes. In the simulation, we configure 6LoWPAN network with ten 6LoWPAN NEMOs deployed over an area of 100x100 square meters.

Table 3. Parameter for simulation

Parameter	Values
Area size	100m * 100m
Number of 6LoWPAN NEMO	10
Send packets	10 packets/sec
Data size	50 bytes
Transmission range	5m
Traffic type	CBR

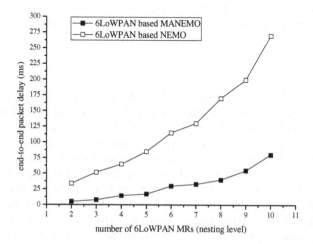

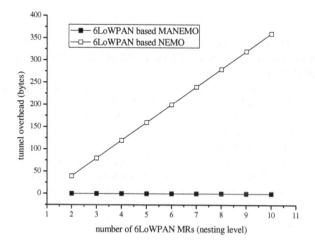

Fig. 5. End-to-end packet delay and tunnel overhead

Fig. 5 shows the result of the end-to-end packet delay between 6LoWPAN node and CN. In case of 6LoWPAN based NEMO scheme, if 6LoWPAN MRs are organized in nested NEMO, the path can become extremely un-optimal to communicate between 6LoWPAN node and its CN. Packets would travel through all HAs since all of packets should be forwarded through the bi-directional tunnel which is established between 6LoWPAN MR and its HA. The more the number of nested levels, the packet route will become increasingly complex and it causes more packets overhead such as pinball routing problem. However, in case of proposed 6LoWPAN based MANEMO scheme, which provides integration of MANET and NEMO protocol, 6LoWPAN MRs constitute ad-hoc network and they can directly connect to each other. Therefore, proposed 6LoWPAN based MANEMO can reduce the tunnel overhead and provide the optimized path to communicate between 6LoWPAN MR and CN. Also, it can minimize the end-to-end packet delay as shown in Fig. 5.

5 Conclusion

In this paper, we focus on the scheme for the route optimization in 6LoWPAN based MANEMO environments. With the application of MANEMO in 6LoWPAN, we can avoid the nested NEMO since MANET consisting of 6LoWPAN MRs can directly communicate with each other. Therefore, an interoperable architecture between MANEMO and 6LoWPAN can solve the un-optimal route problem of the nested NEMO. We define the 6LoWPAN packet format of RS and RA messages to discover a 6LoWPAN-MANEMO Gateway. Also, we define the 6LoWPAN format of 6LoPREQ and 6LoPREP messages to discover the CN's prefix. With proposed scheme, we can provide the optimal path to communicate between 6LoWPAN node and CN. Furthermore, we can also reduce the tunneling overhead and the end-to-end packet delay.

References

1. Kushalnagar, N., Montenegro, G., Schumacher, C.: IPv6 over Low-Power Wireless Personal Area Networks (6LoWPANs): Overview, Assumptions, Problem Statement, and Goals, IETF RFC 4919 (August 2007)
2. Montenegro, G., Kushalnagar, N., Hui, J., Culler, D.: Transmission of IPv6 Packets over IEEE 802.15.4 Networks, IETF RFC 4944 (September 2007)
3. IEEE Computer Society, IEEE Std. 802.15.4-2003 (October 2003)
4. Devarapalli, V., Wakikawa, R., Petrescu, A., Thubert, P.: Network Mobility (NEMO) Basic Support Protocol, IETF RFC 3963 (January 2005)
5. Wakikawa, R., Clausen, T., McCarthy, B., Petrescu, A.: MANEMO Topology and Addressing Architecture, IETF (January 2008) draft-wakikawa-manemoarch-01.txt
6. Network Simulator (ns) version 2, http://www.isi.edu/nsmam/ns

Bypassing Routing Holes in WSNs with a Predictive Geographic Greedy Forwarding

Minh Thiep Ha, Priyadharshini Sakthivel, and Hyunseung Choo

School of Information and Communication Engineering
Sungkyunkwan University, Korea
{thiepha,priya,choo}@skku.edu

Abstract. Applications in wireless sensor networks (WSNs) experience the routing hole problem. That is, the current node cannot forward to the destination, although it is the closest node, but not a neighbor of the destination. Accordingly, the packets from the source cannot be delivered to the destination. Jiang *et al.* recently proposed a SLGF approach to address this problem. However, SLGF still has long routing paths, since it uses the right-hand rule. In this paper, we describe Predictive Geographic greedy Forwarding, PGF. PGF uses information on the hole to build the virtual convex polygon, predict the routing path and choose the shorter path. PGF reduces the length and the number of hops of routing paths. Computer simulation shows our PGF scheme can reduce the average number of hops of routing paths by about 32% compared to Geographic greedy Forwarding, GF, and about 15% compared to SLGF.

Keywords: Routing hole problem; sensor networks.

1 Introduction

A typical wireless sensor network (WSN) is formed by a large number of distributed sensors together with an information collector [1]. These sensors are deployed to collect information and monitor situations in scenarios, such as forests, farmlands, harbors, and coal mines. They offer special benefits and versatility for wide applications in landslide prediction, object localization, and surveillance. However, due to the preservation of limited resources of sensor nodes, energy efficiency is a crucial objective of protocols designed in WSNs. Since routing is the most energy consumption operation of sensor nodes (hereinafter, referred to as nodes), it is very important to have a short routing path, *i.e.*, the number of nodes attending to forwarding operation is small. Greedy forwarding has been proven the most promising way for routing in WSNs. In greedy forwarding, each sensor forwards its packets to a 1-hop neighbor that is closer to the destination than it is. This method is repeated until the packet reaches the destination. However, geographic forwarding suffers from the so called *local minimum phenomenon*. In this, a node cannot forward its packet to its 1-hop neighbor, because all its 1-hop neighbors are further from the destination.

A BOUNDHOLE process [3] is used to construct the boundary of the hole, where the packets may be stuck, to mitigate the local minimum phenomenon.

C.S. Hong et al. (Eds.): APNOMS 2009, LNCS 5787, pp. 171–180, 2009.
© Springer-Verlag Berlin Heidelberg 2009

When a packet is stuck at a node, the routing process will start a perimeter routing phase, where the packet is routed by the downstream node, until it reaches a node that is closer to the destination than that stuck node. Then, the routing returns to the greedy forwarding. Recently, Jiang *et al.* [5] proposed a new information model in which nodes are marked as either safe or unsafe nodes. This safety information is used to bypass the hole. However, these approaches may experience a long detour path in the perimeter routing. In this paper, we propose PGF, a Predictive Geographic greedy Forwarding approach to reduce the number of forwarding nodes. The information of the boundary hole is used to build a virtual convex polygon, form the virtual routing paths, select the shorter one, and create the real routing path that includes neighbor nodes in the network. Therefore, PGF can achieve a short routing path, *i.e.*, the number of forwarding nodes is small.

We implement our proposed PGF scheme using a simulator built in C++ to evaluate performance. We also implement and compare our proposed scheme to Geographic greedy Forwarding [3], GF, which uses the BOUNDHOLE algorithm, and Safety Information based Limited Geographic greedy Forwarding [5], SLGF. Since these schemes address the routing hole problem using different approaches, we think they are good benchmarks for our proposal. Simulation shows PGF can reduce the average number of hops of routing paths about 32% compared to GF, and about 15% compared to SLGF. The remainder of the paper is organized as follows. In section 2, we briefly describe related work. Our proposed scheme PGF is presented in section 3. Section 4 provides the performance evaluation with simulation results. Finally, we conclude our work in the last section.

2   Related Work

Greedy-Face-Greedy (GFG) [2], Greedy Perimeter Stateless Routing (GPSR) [6], and Greedy-Other-Adaptive-Face Routing (GOAFR) [7] are currently the most popular methods to mitigate the local minimum issue. When the routing process becomes stuck at an intermediate node, it will start a perimeter routing phase. In this, the packet is routed by the "right-hand rule" counter-clockwise along a face of the planar graph that represents the same connectivity as the original network, until it reaches a node that is closer to the destination than the stuck node. Then, the routing returns to the greedy forwarding phase. In recent work [4], such a routing scheme is proved to guarantee delivery in any arbitrary planar graph. However, without sufficient shape information about the holes, such a routing may use a long detour path in the perimeter routing, compared to the shortest path to the destination. The work in [3] focused on the use of the hole area that contains the stuck nodes. The hole is detected at a stuck node, where the packet can get the local minimum in greedy forwarding routing [2,6]. A process called BOUNDHOLE is initiated to form a closed circle (also called the boundary of the hole or the boundary hole). The region enclosed by the boundary will be identified as the hole area (Fig. 1). For each node along the boundary, its successor node along the boundary in clockwise order (or counter

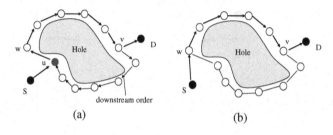

Fig. 1. (a) GF Routing. (b) SLGF Routing

clockwise order) is marked as the downstream node, *e.g.*, node w is downstream of node u, (or upstream node). Geographic greedy Forwarding, GF, is used to forward packets to the destination. In GF, when a package is stuck at a node in the boundary, *e.g.*, node u in Fig. 1a, it is forwarded to the downstream node of the current node, *e.g.*, node w in Fig. 1a, until it reaches a node that is closer to the destination than the stuck node, *e.g.*, node v in Fig. 1a. Then, the routing returns to the greedy forwarding phase.

Jiang *et al.* [5] recently proposed the Safety Information based Limited Geographic greedy Forwarding, SLGF. SLGF uses a new information model, in which nodes are labeled as safe or unsafe nodes in four quadrants. A node is labeled as an unsafe node in a quadrant if there is no safe neighbor in that quadrant. The unsafe information is spread until no new unsafe node is formed. SLGF tries to avoid forwarding packets to unsafe nodes, since using unsafe nodes will cause the local minimum. When a node has a packet, it tries to find a safe neighbor in the requesting zone, which is rectangle formed by its coordinate, and the coordinate of the destination to forward. If it cannot find a safe node in that area, it finds the first safe node in a counter clockwise direction. In the case where the destination is an unsafe node and is in the request zone of the current node, unsafe nodes are used to forward packets to the destination. However, both GF and SLGF approaches may experience a long detour path in the perimeter routing, since they find the forwarder in one direction only.

3   Predictive Geographic Greedy Forwarding Approach

We assume that each node knows its position, position of 1-hop neighbors, and position of the destination using either GPS or other location services. In this section, we describe the Predictive Geographic greedy Forwarding approach. It uses information about the hole to predict the short path to route between the source and destination nodes.

3.1   Overview

After checking if nodes are stuck nodes, the BOUNDHOLE [3] is run to build the boundary of the hole. The information of the boundary, *i.e.*, the position and

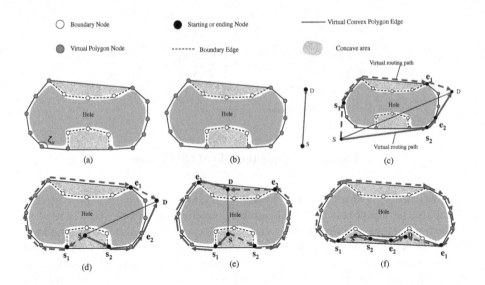

Fig. 2. (a) Boundary hole and virtual convex polygon; (b) Case 1; (c) Case 2; (d) Case 3; (e) Case 4; (f) Case 5

order of nodes in the boundary, is transferred to the source. When a source wants to transfer a packet to a destination, it first builds a virtual convex polygon of the hole. The source node determines its position and position of the destination compared to the virtual convex polygon to form two virtual routing paths in which adjacent nodes may be not forward to each other. By calculating the distance of two virtual paths, the source node chooses the shorter one to form the real routing path. The real routing path is formed using greedy forwarding between adjacent nodes that are not neighbors.

3.2   Forming the Virtual Convex Polygon

As mentioned in section II, after checking if nodes are stuck nodes, a BOUND-HOLE is run to build the boundary of the hole. When this process finishes, nodes know if they belong to a boundary. The node that gets the information of the hole, *i.e.*, the position of nodes in the boundary in the downstream order, will transfer this information to nodes in the boundary. In our proposal, the information about the hole is transferred to the source by flooding. The source node keeps this information to predict the path to the destination. Each time a source node wants to transfer to a destination, the source node builds a virtual convex polygon and determines its position, the position of the destination compared to the virtual convex polygon. The source node finds the nodes belonging to the virtual vertex polygon to build the virtual convex polygon. Node u calculates its *rotation angle* to check if it belongs to the virtual vertex polygon:

Rotation angle: Rotation angle of a boundary node u, ζ_u, is the angle spanned by a pair of adjacent neighbors of node u and is in the direction of the hole.

We can see that the rotation angle is formed by rotating the edge between u and its previous node in the boundary to the edge between u and its following node in the boundary counter clockwise. Node u belongs to the virtual vertex polygon if its rotation angle is less than 180 degrees, $i.e.$, $\zeta_u < 180^0$. This process is run until all nodes in the virtual convex polygon have a rotation angle less than 180 degrees. The virtual convex polygon is formed by connecting nodes satisfying the above condition. Nodes that are neighbors in the virtual convex polygon may not be neighbors in the hole boundary. Fig. 2a shows an example of the virtual convex polygon of a hole boundary. We also define the concave area. A concave area is the area that lies between the hole and the virtual convex polygon (Fig. 2). If the hole and the virtual convex polygon are the same, there is no concave area.

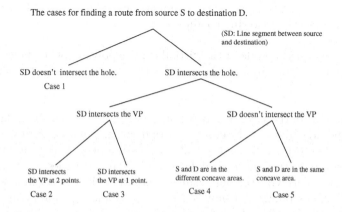

Fig. 3. Five cases for finding a route to the destination in PGF

3.3   Finding the Path to the Destination

After determining the shape of the virtual convex polygon, the source node finds the path to the destination based on its relative position and the position of the destination compared to the position of nodes in the virtual convex polygon. With different relative positions of the source S and destination D, we have a different way to find the path. The five cases are shown in Fig. 3. Each case is differentiated by the number of intersection points of the line segment SD with the hole boundary and the virtual convex polygon. These five cases are complementary. A different way to find the routing path is used for each different case. We will detail each case in the following section. The first case is the simplest case. In this case, the line segment SD does not intersect with the virtual convex polygon (Fig. 2b), we use greedy forwarding to transfer packets to the destination, because the hole does not affect the route between the source and destination. For the remaining cases, we can always find two paths to pass over the hole, $i.e.$, the path in the left side and the path in the right side of SD (Figs. 2c, 2d, 2e, or 2f), because the line segment SD intersects with the hole.

The routing path from the source to the destination will be the shortest path, if two paths are the shortest paths among paths in two sides and the chosen path is the shorter one between these two paths. We find the start and end nodes defined below to find these two paths. We have the routing path between the source and destination over the hole $S \Rightarrow u_1 \Rightarrow u_2 \Rightarrow ... \Rightarrow u_i \Rightarrow D$; u_1, u_2, ..., u_i are nodes in the boundary, then we name u_1 the start node and u_i the end node of this routing path. Each path has one start node and one end node. Therefore, we have two start nodes and two end nodes. Now, we will discuss how to find these nodes for each case. In the second case, the line segment SD intersects with the hole. SD also intersects with the virtual convex polygon at two points. That is, both S and D are outside the virtual convex polygon and the hole (Fig. 2c). The line segment SD divides the virtual convex polygon into two sides, *i.e.*, the left side and the right side. We only find the start and end nodes among nodes belonging to the boundary of the hole. A node u is a start node if it satisfies the following conditions:

1. Condition 1: u is a node of the virtual convex polygon of the hole.
2. Condition 2: The ray Su intersects with the virtual convex polygon at only u.

Since u is a node of the virtual convex polygon, *i.e.*, node in the boundary of the hole, and the ray Su intersects with the virtual convex polygon at only u, packets transferred from S to D through u will not be stuck at u. These conditions also make the routing paths more straightforward. The end nodes are found using the same conditions with the only change being that the ray Du intersects with the virtual convex polygon at only u. Nodes s_1, s_2 and e_1, e_2 are start and end nodes, respectively. After that, two virtual routing paths are formed as: $S \Rightarrow$ start node $\Rightarrow$ adjacent nodes in the virtual convex polygon $\Rightarrow$ end node $\Rightarrow D$.

From S, we transfer packets to the start node. Next, the adjacent node, *i.e.*, the neighbor node, of the start node is chosen as forwarder. If the start node is to the left side of the line segment SD, the chosen neighbor of the start node is the neighbor in the downstream direction. Conversely, if the start node is in the right side of the line segment SD, the chosen neighbor of the start node is the neighbor in the upstream direction. This selection is repeated until it meets the end node. Finally, we transfer from the end node to the destination (Fig. 2c). We calculate the total distance between nodes from S to D in two paths to choose a shorter path. The path that has the lesser distance will be selected as the routing path to transfer packets to the destination, *e.g.*, the solid red line in Fig. 2c. Since the distance between nodes in the above routing path, *e.g.*, from the source node to the start node s_2 in Fig. 2c, may be longer than the node radio range, we use greedy forwarding transfer to packets.

For the third case, the line segment SD intersects with the hole. SD also intersects with the virtual convex polygon, but at only one point. Since SD intersects with the virtual convex polygon at only one point, S or D must be inside the virtual convex polygon, *i.e.*, inside the concave area; the other is outside the virtual convex polygon, *e.g.*, node S is inside and node D is outside the virtual convex polygon in Fig. 2c. We also use the line segment SD to determine two sides and two routing paths. However, instead of dividing the virtual convex polygon, SD divides the boundary hole into two sides and categorizes nodes in the boundary hole into two sets, *i.e.*, the set in the left side and the set in the right side. The way to find the start nodes (or end nodes) for a node that is outside the virtual convex polygon is the same as for case 2, *e.g.*, node D in Fig. 2c finds two end nodes e_1 and e_2. For that node inside the concave area, we select from two sets of nodes, *i.e.*, the set of nodes in the left side and the set of nodes in the right side, two nodes that satisfy condition 2 as end nodes (or start nodes), *e.g.*, node S in Fig. 2c finds two start nodes s_1 and s_2. Next, two virtual routing paths are formed as: $S \Rightarrow$ start node $\Rightarrow$ adjacent nodes in the boundary hole $\Rightarrow$ adjacent nodes in the virtual convex polygon $\Rightarrow$ end node $\Rightarrow D$.

After getting to the start node from S, the adjacent node in the boundary hole, *i.e.*, the neighbor in the boundary hole, of the start node is chosen as the next forwarder. If the start node is in the left side (or right side) of the SD, the downstream neighbor (or the upstream neighbor) will be chosen as the next forwarder. This selection is repeated until finding a node in the virtual convex polygon, also in the boundary hole. From this point, only neighbors in the virtual convex polygon can be the next forwarder, *e.g.*, two red paths in Fig. 2c. After reaching the destination, we calculate the total distance of two paths and choose the shorter one for transferring packets, *e.g.*, the solid red line in Fig. 2c. As before, greedy forwarding is used to transfer packets between nodes that are not neighbors in the virtual routing path.

For the remaining cases, we use the same method as for case 3 to find start nodes, end node, s to form two virtual routing paths, and to select the shorter one for transferring packets.

3.4 Multiple Holes

For cases in which there are multiple holes between the source and the destination, the source node finds the routing path and transfers packets to the end node using the method as in the previous section. When the end node receives packets, it applies the same method as the source node to transfer packets to the destination. In Fig. 4, the packet from node S bypasses the first hole at node

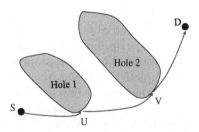

Fig. 4. Multiple holes

U, the end node in the routing path from S to D. U applies the same method as S and bypasses the second hole at node V. Finally, the packet arrives at the destination.

4   Performance Evaluation

In this section, we study the average-case performance of the proposed information model and routing algorithms, using a simulator built in C++. The performance metrics used in the evaluation are the number of hops and the length of the routing path.

In the simulations, nodes with a transmission radius of 20 meters are deployed to cover an area of $200m \times 200m$, under different deployment models. We evaluate approaches in two models. First, the nodes will be deployed uniformly. This is the ideal model (denoted by IA), in which the hole is only caused by sparse deployment. Usually, the size of a hole is very small. Second, we randomly set some forbidden areas inside the network area, where no nodes can be deployed. The forbidden areas, which may be irregular, are constructed to study the impact of larger holes on the proposed algorithms. Such a model is denoted by FA. We assume that the destination and the source are randomly selected, including both safe sources and unsafe sources. Before we test the routing performance for routing time, boundary information [3] is constructed for GF routing; safety information [5] is constructed for SLGF routing; and the virtual convex polygon is constructed for our PGF routing. Then, we test the networks when the number of nodes in the area is varied from 400 to 800 in increments of 50. For each case, 100 networks are randomly generated, and the average routing performance over all of these randomly sampled networks is reported. Fig. 5a shows the upper bound of the number of hops of the routing path. Fig. 5b shows the average number of hops of the routing path. As mentioned in section II, PGF can predict the routing path by forming the virtual routing paths, so the number of forwardings in PGF is small also.

Section II, when holes exist in the networks, GF routing may experience a long detour, because the packets are transferred to a node in the boundary hole and detour to downstream nodes before arriving at the destination. In the case of SLGF, because it uses the right-hand rule to select the next forwarder, although the right-side has the shorter path, the left-side path is still chosen. This causes a long detour. For PGF, the source node predicts the stuck nodes and chooses the shorter of the path on the left side and the path on the right side. As a result, PGF has a shorter path than GF and SLGF. As shown in Fig. 5a, PGF routing reduces the number of routing hops by about 15 percent compared to SLGF, and about 32 percent compared to GF. The above property still holds when more large holes occur under the FA model. In WSNs, the packet is forwarded hop-by-hop along the path. Reducing the number of hops can reduce end-to-end delay and furthermore support quick response to routing requests. Fig. 5c shows the corresponding average length of the entire routing path. Since PGF chooses the shorter path by comparing the total distance (length) of two paths, the length of the routing path in PGF is smaller than in GF and SLGF.

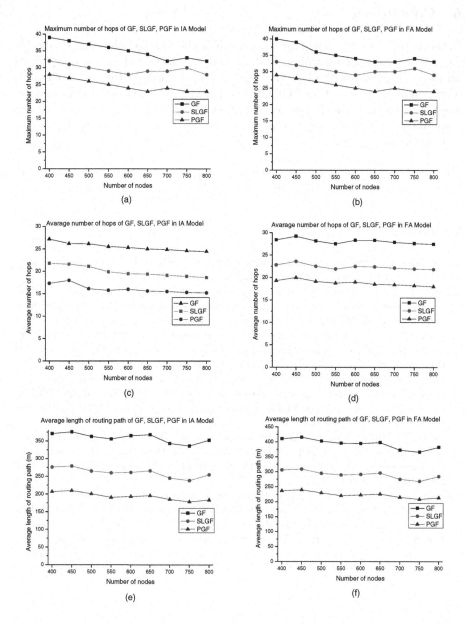

Fig. 5. (a) Maximum number of hops in IA model; (b) Maximum number of hops in FA model; (c) Average number of hops in IA model; (d) Average number of hops in FA model; (e) Average length of paths in IA model; (f) Average length of paths in FA model

5   Conclusion

In this paper, we have shown the way to bypass the routing hole using Predictive Geographic greedy Forwarding, PGF. A virtual convex polygon is built based on the information of the boundary hole to predict the routing path. Furthermore, simulations show that PGF efficiently reduces the number of hops and the length of routing paths. In future work, we will extend our work to increase the adaptability of the scheme so that it can be more efficient in networks where topology changes frequently.

Acknowledgments. This research was supported by MKE, Korea under ITRC IITA-2009-(C1090-0902-0046). Dr. Choo is the corresponding author.

References

1. Akyildiz, I.F., Su, W., Sankarasubramaniam, Y., Cayirci, E.: A survey on sensor networks. IEEE Communications Magazine 40, 102–114 (2002)
2. Bose, P., Morin, P., Stojmenovic, I.: Routing with guaranteed delivery in ad hoc wireless networks. In: Proc. The 3rd International Workshop on Discrete Algorithms and Methods for Mobile Computing and Communications, pp. 48–55 (1999)
3. Fang, Q., Gao, J., Guibas, L.: Locating and bypassing routing holes in sensor networks. In: Proc. The 23rd Annual Joint Conference of the IEEE Computer and Communications Societies (IEEE INFOCOM 2004), pp. 2458–2468 (2004)
4. Frey, H., Stojmenovic, I.: On delivery guarantees of face and comined greedy-face routing in ad hoc and sensor networks. In: Proc. The 12th Annual ACM/IEEE International Conference on Mobile Computing and Networking (ACM/IEEE MO-BICOM 2006), pp. 390–401 (2006)
5. Jiang, Z., Ma, J., Lou, W., Wu, J.: An Information Model for Geographic Greedy Forwarding in Wireless Ad-Hoc Sensor Networks. In: Proc. The 27th Annual Joint Conference of the IEEE Computer and Communications Societies (IEEE INFOCOM 2008), pp. 825–833 (2008)
6. Karp, B., Kung, H.: GPSR: Greedy perimeter stateless routing for wireless sensor networks. In: Proc. The 6th Annual ACM/IEEE International Conference on Mobile Computing and Networking (ACM/IEEE MOBICOM 2000), pp. 243–254 (2000)
7. Kuhn, F., Wattenhofer, R., Zollinger, A.: Worst-case optimal and average-case efficient geometric ad-hoc routing. In: Proc. The 4th ACM International Sysmposium on Mobile Ad Hoc Networking and Computing (ACM MobiHoc 2003) (2003)

Analysis of Time-Dependent Query Trends in P2P File Sharing Systems

Masato Doi, Shingo Ata, and Ikuo Oka

Graduate School of Engineering, Osaka City University
3–3–138 Sugimoto, Sumiyoshi-ku, Osaka 558–8585, Japan
{doi@n.,ata@,oka@}info.eng.osaka-cu.ac.jp

Abstract. In P2P file sharing systems, time dependent characteristics of query (query trends) for a file become much important to forecast the demand of the file in future. Prediction of future demand would be effective for the efficient use of the caching mechanism, however, the accurate prediction of query trend is difficult because patterns of query trends may differ significantly according to a nature of keyword used in the query. Identification of query pattern is one of important roles for accurate forecast of future query demand. In this paper, we propose a new method to classify measured query trends into some typical query patterns. We first measure query trend for each keyword in the most famous P2P file sharing system in Japan, and analyze the pattern of query trends by using clustering technique with Discrete Fourier Transform. We then apply our method to the measurement results and show that most of keywords can be categorized into one of four typical trend patterns.

1 Introduction

In recent years, Peer-to-Peer (P2P) file sharing system attracts attention and is used by many users. In P2P file sharing system, since each node works as the server and client, they can share and exchange information directly and equally. Therefore P2P file-sharing system is superior to client/server model in scalability and fault tolerance.

Now, there are many P2P file sharing systems (e.g., Napster, Gnutella, KaZaa, Cabos, Limewire, BitTorrent), and Winny and Share are the most popular file sharing systems in Japan.

In these P2P file sharing system, many of performance improvements such as search efficiency, load balancing, and anonymization is achieved by the use of caching mechanism. Since disk spaces for caching are limited, cache replacement algorithm is important for the efficient use of cached resource. As typical cache replacement algorithms are Least Recently Used (LRU), Least Frequently Used (LFU) and First In First Out (FIFO), and LRU is adopted in many file sharing system.

These cache replacement algorithms decide caches duration by referred time and the referred frequency, therefore caches of keywords that are accessed same

C.S. Hong et al. (Eds.): APNOMS 2009, LNCS 5787, pp. 181–190, 2009.

time or same frequency are processed as same condition regardless of the kind of contents.

However, cache replacement by checking the last accessed time is not always the best way to achieve the efficiency because the time dependent variance of query demand (we refer as *query trend* in this paper) may differ completely according to the type of keyword used in query. For example, a keyword related to an event is requested significantly around the time of the event, while in other periods there are few requests for the keyword. On the other hand, there are some requests at all time for an FAQ-like keyword. If we apply the LRU to event-related keywords, many of cache files has been created according to the recent requests of the keyword. However, after the time that the event has passed, the request for the keyword will decrease rapidly, and many of cached files will not be used which waste disk resources of peers. In order to solve this problem, it is necessary to introduce a new cache replacement algorithm which takes query trend of keyword into consideration.

Time-dependent trends of queries can be classified into some patterns. If we know which trend a keyword belongs to, we can predict how it will be requested. In this paper, we analyze time-dependent trends of queries in P2P file sharing system and classify trend patterns by clustering. Specifically, we collect keyword (filename) queries by crawling the P2P file sharing system. We then classify trend patterns by cluster analysis and analyze the nature of each cluster.

This paper is organized as follows: Section 2 introduces related works of popularity distribution and transition in some network applications. Section 3 describes the method of collecting search queries in P2P file sharing system which is most used in Japan now. In Section 4, we describe our proposed method of classifying query trends into typical patterns based on similarities of trends. We then present the experimental results of our analysis by applying measured query trends in section 5. Finally conclude our paper with future research topics in Section 6.

2   Related Work

In this section, we describe related works about popularity in web server and video-sharing site.

In [1], Padmanabhan et al investigated popularity distribution of contents and time-series transition of popularity based on data measured in web servers. As the result, [1] showed that the distribution of popularity of accessed contents roughly follows power-law as well as shown in other studies. [1] also analyzed time-series transition of popularity of top 100 contents and showed that 60-70% of contents still exists in top 100 contents after 5 days. However, [1] only considered popularity in web servers as a whole, and how changes popularity of individual content changes was not analyzed.

[2] studied User Generated Content (UGC) posted on Video-on-Demand (VoD) system such as YouTube. As a result of investigating popularity distribution of video, [2] showed popularity distribution almost follows power-law and

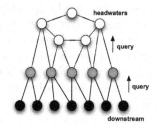

Fig. 1. Typical network topology of measured P2P file sharing system

fits best by power-law with an exponential cutoff. However, although [2] only analyzed the relation between days after video is posted and the number of accesses, [2] did not analyze the access patterns of individual content.

[3] analyzed transition of contents popularity in VoD system. In [3], patterns of transition of contents popularity are only two type (outbreak type, persistent type), however, they are insufficient to represent actual patterns of transition of contents popularity.

There are many studies about a cache replacement algorithm in P2P file-sharing system [4, 5, 6]. However, they did not take into consideration trend patterns.

As above, there are few studies about patterns of time-series transition of contents popularity, even if it exists, it performed only observation and speculation. Also there are no studies about trend patterns in P2P file sharing system. If we can classify patterns of query trends and identify which pattern a query trend for a keyword belongs to, we can predict demand of queries for the keyword in future. Predicting future requests enables designing new cache replacement algorithm that takes trend patterns into account for achieving more efficient cache utilization.

3   Measurement of Search Queries

In this section, we describe the method to measure search queries in P2P file sharing system which is most used in Japan. In Pure P2P network, since there is no management server, it is impossible to collect all search queries. We therefore consider the method of collecting queries as much as possible. In this paper, we develop a crawler program for collecting search queries.

3.1   Collecting Search Queries by Winny Crawler

In this paper, we focus on a P2P file sharing system which is mostly used in Japan. The topology is formed by kind of hierarchy according to line speeds of peers that are specified by users. Fig. 1 shows a typical network topology in the P2P system. Search queries are transmitted from peers with slow line to peers with high-speed line. Therefore, it has a nature that queries are gathered in peers with high-speed line. If we connect our peer by specifying higher link speed, our peer would be connected to the upper layer of the hierarchy and the peer can collect more keywords. Furthermore, since keyword of a query is encrypted by RC4, we develop a crawler which can decode encrypted queries.

Our crawler has node list and query list. Node list is the list of nodes available by monitoring advertisement messages from other nodes. When the crawler is needed to connect with another nodes, the crawler selects a node randomly from the node list. Query list handles information of search queries transmitted by other nodes. It includes identifier of query message, IP addresses of the node that generated the query and keyword exist. If the crawler receives same queries multiple times, the crawler does not discard queries with the same ID. At the analyzing phase, however we consider queries with the same ID to be single query. We describe the outline procedure of operation of procedure crawler as follows.

Outline of operation of Winny crawler

1. Create an initial node list which includes a set of nodes that the crawler firstly connects.
2. Connect to nodes listed in the node list by specifying that the node has high-speed link so that the node can be placed at the higher layer of the hierarchy.
3.   – If the crawler receives the search query message, the crawler adds the query message into the query list.
   – If the crawler receives the advertisement node from connected nodes, the crawler adds information of the node in advertisement message into the node list.
   – If the crawler receives the request of connection termination or unknown message from the connected node, the crawler terminates the connection to the node.

We corrected search queries by our crawler from April 1, 2009 to April 31, 2009. In the period, the crawler always connects about 500 nodes.

4   Method of Classifying Trend Patterns

In this section, we describe the method of classifying trend patterns of search queries. We define trend patterns as time-series of number of queries per day. We classify trend patterns of each keyword by cluster analysis.

4.1   Similarity of Time-Series Trends

To apply the clustering method, it is necessary to define distance between time-series transit of query. There are two methods which calculate in time domain in frequency domain respectively. In time domain calculation, distance between time-series transition is defined as Euclidean distance between vectors by considering time-series of transition as a vector.

However, Euclidean distance between vectors is not applicable directly because two time transitions are not completely synchronized, i.e., we have to slide the sequence of the trend to adjust the start of transitions to the same time.

In [7,8,9], DFT/DWT is used to calculate distance between time-series transition from similarity in frequency domain. However, if we use DWT the synchronization problem is still remained, and we therefore use the DFT for calculation of the similarity.

4.2 Similarity Derivation Based on Discrete Fourier Transform

DFT is Fourier Transform in discrete group and transformation from time domain to frequency domain. DFT of the time-series $x = [\mathbf{x_t}] = [x_0, x_1, \ldots, x_{n-1}]$ is represented by in the form of $f = [\mathbf{f_j}] = [f_0, f_1, \ldots, f_{n-1}]$, and is defined by

$$f_j = \sum_{k=0}^{n-1} x_k e^{-\frac{2\pi i}{n} jk} \quad j = 0, 1, \ldots, n - 1. \tag{1}$$

Note that i is the imaginary number ($i^2 = -1$) and DFT has symmetrical property like $f_j = f_{n-j}$. When the demention n of x_t is power-of-two, Fast Fourier Transform (FFT) can be applied to calculate in $O(nlogn)$.

Since our purpose is to categorize the *trend patterns*, absolute values on query trends (e.g., number of queries/hour) is not our focus. We therefore apply DFT to values normalized by maximum number of queries/hour appeared in the trend.

4.3 Distance between Time-Series Transition

After DFT/FFT, two time-series transitions p and q are transformed to $f(p)$ and $f(q)$ respectively. The distance between two time-series transitions can be considered as Euclidean distance of the coefficient sequence of $f(p)$ and $f(q)$ which is calculated from

$$Distance(p, q) = \sqrt{\sum_{j=1}^{\frac{2}{n}} (|f_j(p)| - |f_j(q)|)^2}. \tag{2}$$

4.4 Clustering Based on Similarity

We perform hierarchical clustering by distance obtained by Eqn. (2). We employ Ward's clustering method [10]. Ward's clustering method achieves to minimize the sum of squares of two arbitrary clusters. Distance between cluster t and cluster r is calculated by

$$S_{tr} = \frac{n_p + n_r}{n_t + n_r} S_{pr} + \frac{n_q + n_r}{n_t + n_r} S_{qr} - \frac{n_r}{n_t + n_r} S_{pq}, \tag{3}$$

t is the cluser obtained by combining p and q, and n_t, n_p, n_q, n_r are the element count in t, p, q, r respectively.

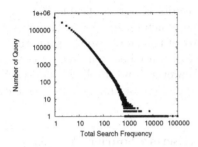
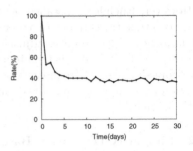

Fig. 2. Distribution of search frequency per keyword in Winny

Fig. 3. Change of keyword ranking

5   Measurement and Analytic Results

5.1   Statistical Characteristics of Measured Query Messages

We run our developed crawler for one month (from 2009/4/1 to 2009/4/31). After the measurement, we obtain 60,531,074 query messages having 3,899,980 keywords in total. The total number of unique nodes is 156,708. Fig. 2 shows total frequency of search distribution of each query in measurement period. In this figure, the horizontal axis represents the total search frequency, and the vertical axis represents the number of keyword corresponding the total search frequency. [11] presented that popularity distribution of contents follows power law. Distribution of search frequency per keyword in file sharing system also almost follows the nature of power law.

We also investigate how popularity changes in file-sharing system. We analyze how most frequent keywords vary by days. For this purpose, we first obtain top 100 most frequent keywords at the first day. We then obtain the ratio of keywords that remain top 100 keywords at the second day or more. Fig. 3 shows rate of remaining keywords in top 100 with progress of time. In this figure, the horizontal axis represents time (day), and vertical axis represents the rate of remaining keywords in top 100. We can observe that about 30% of keywords are still remained in top 100 after 30 days. Additionally, since about a half of keywords replaced rapidly (within 5 days).

5.2   Result of Classifying Trend Patterns Using Clustering

We perform clustering with 753 keywords with that have over 1,000 queries. We investigate transition of the number of queries every one hour. Then, in order to investigate rough transitions, we smooth transitions of the number of queries in 24 hours. We apply the moving average of the number of queries instead of the actual number of queries. The moving average can be obtained by

$$q_i = \frac{m_i + m_{i-1} + \ldots + m_{i-23}}{24}, \tag{4}$$

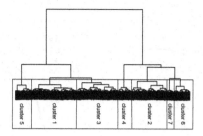

Fig. 4. Dendrogram of clustering

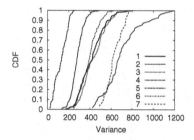

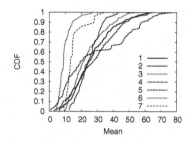

Fig. 5. Variance of the number of searches for each keyword

Fig. 6. Mean of the number of searches for each keyword

Where q_i is the i-th average of queries and m_i is the number of queries measured at time i. Additionally, we normalize transitions of the number of queries by the maximum number of queries.

We show dendrogram of clustering analysis in Fig. 4. We obtain 7 clusters in total shown in Fig. 4. In followings we analyze nature of each clusters.

Fig. 5 shows variance of the number of queries of each keyword by the cumulative distribution of each cluster. We can observe that the variance of cluster 5 is smaller than other clusters from Fig. 5. Therefore, the number of queries of keywords in cluster 5 keeps almost average value and it is found keywords in cluster 5 are searched constantly.

Fig. 6 shows mean of the number of queries of each keyword by the cumulative distribution of each cluster. This figure shows that mean for Cluster 6 is smaller than other clusters from fig. 6. In this paper, since we normalize transitions of the number of queries at maximum, small mean is caused by the keywords that are searched intensively in short time. Therefore, the smaller the mean is, the more the pattern of transition becomes sharpen. Hence, keywords in cluster 6 has the trend pattern that both increase and decrease of the number of queries are quite rapid.

Fig. 7 shows the duration from the time when the query rate exceeds 5% of the maximum and the time when the query rate reaches the maximum. The cumulative distributions of the duration are shown in these figures. Fig. 8 shows

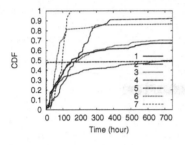

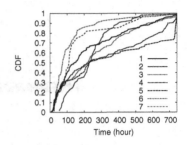

Fig. 7. Time concerning increasing to maximum, after exceeding 5% of maximum

Fig. 8. Time concerning decreasing to maximum to 5% of maximum

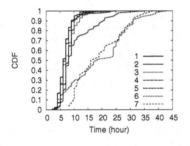

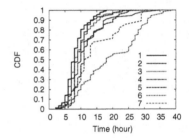

Fig. 9. Maximum continual increase time

Fig. 10. Maximum continual decrease time

the results of opposite side, i.e., the distribution of duration from the maximum to the 5% of the maximum. From these figures, Since Cluster 6 takes small value, Cluster 6 has the trend pattern that the number of searches increases rapidly and decreases rapidly. Additionally, Cluster 7 takes small value next to Cluster 6. Cluster 7 also have the trend pattern that the number of searches increases rapidly and decreases rapidly. On the other hand, Cluster 4 takes decent small mean and time, Cluster 4 has the trend pattern that the number of searches increases gradually and decreases gradually.

Fig. 9 and Fig. 10 show CDFs of the maximum duration time of increase of the number of queries for each keyword and that of decrease, respectively. From these figure, we can observe that the duration time of increase and decrease in Cluster 1, Cluster 2 and Cluster 3 have small values. Therefore, we consider that such clusters have the trend pattern that the number of queries varies drastically.

Thereby, it is found that trend patterns can be classified into four types. *Instant* and *temporal* have the nature that accesses concentrate on a specific period and *assurance* and *fickleness* have the nature that accesses happen constantly. Furthermore, in the trend pattern that accesses concentrate on a specific period, both time taken by increasing to maximum and decreasing 5% of maximum are 22 hours at the shortest. While the trend pattern that the

Table 1. Typical trend pattern

trend pattern	cluster number
instant	6, 7
temporal	4
assurance	5
fickleness	1, 2, 3

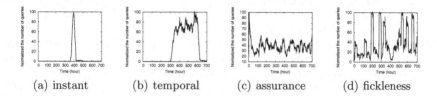

(a) instant   (b) temporal   (c) assurance   (d) fickleness

Fig. 11. Typical Trend Patterns

number of searches increases and decreases rapidly exits, the trend pattern that both time taken by increasing to maximum and time taken by decreasing to 5% of maximum take over 500 hours also exists. Additionally, in the trend pattern that the number of searches keeps constantly, while the trend pattern that has small variance of the number of searches and that is constantly searched exists, the trend pattern that has large variance of the number of searches and that the number of searches intensely changes also exists. Table. 1 shows aspect of each cluster and Fig. 11 shows typical trend patterns of each aspect.

6   Conclusion and Future Works

In this paper, we have considered search queries as index of popularity in P2P file sharing system and have classified trend patterns of search keywords from the numerical feature. First, we have collected search queries in P2P file sharing system and have collected data concerning time-series transitions of search keyword. We have then analyzed total number of queries distribution of each keyword. In the result, we have shown it has almost follows the nature of power law as reported in earlier studies. Then we have classified trend patterns by hierarchical clustering and have analyzed each cluster. In the result, we have shown trend patterns can be categorized into one of four typical types.

As future research topics, we need to analyze trend patterns of search keywords in other P2P file sharing systems. Moreover, for suggestion of new cache replacement algorithm, we need to consider the method to identify which trend a keyword belongs to and the optimal cache replacement algorithms for each pattern.

Acknowledgement

This work was partially supported by a Grant-in-Aid for Young Scientists (A) (No. 19680004) from the Ministry of Education, Culture, Sports, Science and Technology (MEXT) of Japan.

References

1. Padmanabhan, V.N., Qiu, L.: The content and access dynamics of a busy web site: Finding and implications. In: Proceedings of ACM SIGCOMM 2000, Stockholm, Sweden, August 2000, pp. 111–123 (2000)
2. Cha, M., Kwak, H., Rodriguez, P., Ahn, Y.-Y., Moon, S.: I tube, you tube, everybody tubes: Analyzing the world's largest user generated content video system. In: Proceedings of ACM IMC 2007, San Diego, CA, October 2007, pp. 1–14 (2007)
3. Mori, T., Asaka, T., Takahashi, T.: Analysis of popularity dynamics in youtube-like vod. In: IEICE Technical Report (NS2008-103), Japan, November 2008, vol. 108, pp. 87–92 (2008)
4. Cohen, E., Shenker, S.: Replication strategies in unstructured peer-to-peer networks. In: Proceedings of ACM SIGCOMM 2002, Pittsburgh, PA, August 2002, pp. 177–190 (2002)
5. Dunn, R.J., Gribble, S.D., Levy, H.M., Zahorjan, J.: The importance of history in a media delivery system. In: Proceedings of IPTPS 2007, Bellevue, WA (February 2007)
6. Obayashi, N., Asaka, T., Takahashi, T., Sakaki, J., Shinagawa, N.: Load balancing with content classification in P2P networks. IEICE Transactions on Communications J90-B, 720–733 (2007)
7. Wu, Y.-L., Agrawal, D., Abbadi, A.E.: A comparison of DFTand DWT based similarity search in time-series databases. In: Proceedings of ICKM 2000, New York, pp. 488–495 (2000)
8. Vlachos, M., Meek, C., Vagena, Z.: Identifying similarities, periodicities and bursts for online search queries. In: Proceedings of SIGMOD, June 2004, pp. 131–142 (2004)
9. Kontaki, M., Papadopoulos, A.: Efficient similarity search in streaming time seqences. In: Proceedings of SSDBM, Santorini Island, GR, June 2004, pp. 63–72 (2004)
10. Ward, J.H.: Hierarchical grouping to optimize an objective function. Journal of the American Statistical Association 58(301), 236–244 (1963)
11. Gummadi, K., Dunn, R., Saroiu, S., Gribble, S., Levy, H., Zahorjan, J.: Measurement, modeling, and analysis of a peer-to-peer file-sharing workload. In: Proceedings of SOSP 2003, Bolton Landing, NY (October 2003)

An Implementation of P2P System for Sharing Sensory Information*

Seung-Joon Seok[1], Nam-Gon Kim[1], Deokjai Choi[2], and Wang-Cheol Song[3,**]

[1] Dep. of Computer Eng., Kyungnam Univ., Kyungnam, Korea
{sjseok,atom}@net.kyungnam.ac.kr
[2] School of Electronics and Computer Eng., Chonnam Nat'l Univ., Gwangju, Korea
dchoi@chonnam.ac.kr
[3] Dep. of Computer Eng., Jeju Nat'l Univ., Jeju, Korea
philo@jejunu.ac.kr

Abstract. So far, numerous wireless sensor networks have been deployed indepedently and restrictly used for specific services as isolated islands. Recently it is considered to exploit the sensor networks already deployed by others or open its sensor networks for sharing the sensor data. In this paper, we introduce our empirical development of a P2P system for sensor information. We design a gateway node which includes a sink function for sensor nodes and plays a role of a peer node of the proposed P2P system. And we modify an open P2P framework of Pasty for adopting sensor, sensor network and sensor information. This P2P system is referred to as sensor P2P. To test the utility of our sensor P2P system, we deploy this system in KOREN (KOrea advanced Research and Education Network) and develop an application using which a user can search a sensor network with key words, find the position of the sensor network on the digital map and receives sensor data from the sensor network.

Keywords: Sensor Data Sharing, Sensor P2P, Pastry.

1 Introduction

In ubiquitous computing era, many wireless sensor networks (WSNs) has been developed and deployed in many places. The sensor data, however, is accessible to designated users, but there has been no good means to be open to wisher so far. Recently some people have tried to connect the local sensor networks distributed through the Internet, and make them easily-accessible.

There are two kinds of approaches on sharing sensory information up to date. The one is to use servers which store the sensor data received from sensor network. Sensor networks, which want to open their sensor data, send it to designated server(s) and users, which have need of the sensor data, ask the server for the data. The other is the

---

* This work was supported by the research and test program using KOREN/APII/TEIN of National Information Society Agency (NIA) in 2009.
** Corresponding author.

C.S. Hong et al. (Eds.): APNOMS 2009, LNCS 5787, pp. 191–200, 2009.

P2P technologies. The P2P system is used to share local digital contents, such as mp3 and video, text, among other users through the Internet. In this case, sensor (network) and user are both P2P peers, so user directly asks a specific sensor (network) for the wishful sensor data.

In this paper, we use the P2P technologies to share the sensor data through Internet. It is because the P2P system is flexible compared to the servers' approach in some points. The structure of P2P system is more scalable. Centralized servers can be a bottleneck point for dynamic and huge sensor data. P2P system may simply provide direct interaction between peers: sensors and users.

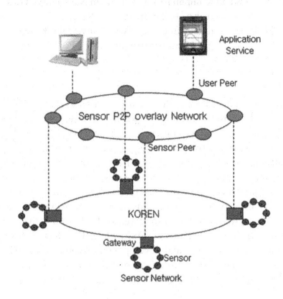

Fig. 1. The sensor P2P overlay network

In this paper, we introduce an empirical development of a P2P system using which user can search sensor network network with key words and directly ask the desired sensor data through Internet. Firstly, we design a gateway node which inlcudes a sink function for sensor nodes and plays a role of a peer node of the proposed P2P system. And we modify an open Pastry architecture for adopting sensor information. This P2P system is referred to as sensor P2P. Finally this paper introduces our experiment of the sensor P2P system in which we deloy the sensor P2P system in KOREN(KOrea advanced Research and Education Network) and test the P2P system using an application software deveploped to find sensor network and receive sensor information.

2 Background

As mentioned previous, there are two kinds of approaches on sharing sensory information up to date: server based approach and P2P based approach.

The server based approaches require data owners to register their data source to designated servers. SensorWeb [1] focuses on large scale sensor networks and sharing sensor data. SenseWeb requires all sensing data to register in a Web Server. GSI [2] proposed a software architecture in which servers process sensor data and databases store applications and sensor data. SensorMap [3] mainly focuses on spatial data and builds a portal that shows real-time sensor data on a map. Other SensorPlanet [4], SensorBase [5], and so on, exploits centralized data storage for sharing sensor data [6].

There has been recent works to connect the sensor networks through the P2P overlay network. TChord [7] proposes a Chord-based P2P protocol called Tiered Chord for sensor networks. [8] designs an overlay network built on Internet infrastructure as the large-scale sensor networks for military applications. Content-based query routing is implemented on top of IP routing by using Sun Microsystem's JXTA platform [9]. P2P UPnP [10] combines the Universal Plug and Play (UPnP) and JXTA to implement a uniform way to access and present different services from various sensor networks by installing a P2P Bridge at the remote WSNs. This bridging component exposes the WSNs to a P2P network and enables the UPnP Gateway to discover remote sensor nodes through the P2P substrate and to instantiate UPnP Proxies for them to ensure client connectivity. Also, Distributed Hash Tables (DHTs) are used to discover resources on the P2P substrate by querying for their descriptions.

3 Implementation of Sensor P2P Framework

In this paper, we propose a P2P architecture to share sensor data based on Pastry [6]. Pastry is one of the structured P2P frameworks which provide routing functionality in the overlay network using the DHT. We assume that peer nodes, called sensor peer, may have their sensors but the others, called user peer, may just want to join the P2P overlay network without local sensors. We consider the gateway node, which is located between sensor network and Internet, as the peer node of sensor P2P for simplification of our sensor P2P framework. The gateway node is composed of a sink module, peer module, communication protocols, and applications. The sink module gathers sensor data from sensor node, processes the data for sensor P2P services, and send the processed data to interior peer module. The peer module should join the sensor P2P overlay network as a peer node and play a role of the peer node of the sensor P2P. The gateway node typically connects to Internet and also has a sensor network interface, such as ZigBee, to communicate with sensor nodes. Figure 2 depicts the architecture of gateway node.

3.1 Implementation of Sensor Peer

As mentioned above, the gateway node, which is located between sensors and sensor users, is considered as sensor peer. The sensor peer node is located between sensor network and Internet and has the sensor P2P stack described in the next section. The sensor peer has one or more sensor nodes attached to the interior sink module using ZigBee communication. The sensor peer provides the sensor data collected from local sensor nodes to user peers which directly ask the data. Also it may want to share sensor data with other sensor peer.

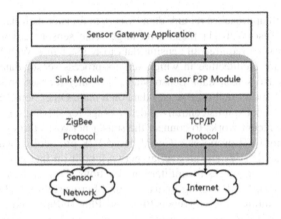

Fig. 2. The Architecture of gateway node (sensor peer)

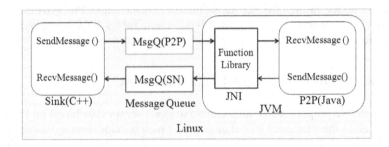

Fig. 3. The Interface between Sink and P2P modules

We develop the sensor peer in a small sized computer, such as UMPC, whose OS is Linux 2.6 and mainly use C++ programming language but the sensor P2P module based on FreePastry open source code to operate over JVM. So we should consider how to interface sink module and sensor P2P module within sensor peer node. Figure 3 shows our approach for the interface which makes two message queues for each data direction using POSIX message queue and defines two kinds of read and write functions for sink and P2P modules.

The sink module gathers raw sensor data from local sensor nodes through ZigBee interface and make the data into two kids of sensor P2P messages: real-time sensor data and past sensor data. The real-time sensor data is formed as a short size message format and the past sensor data also is formed as file type. Figure 4 and 5 shows these two types of sensor P2P messages. The real-time sensor message is composed of time, area, organization, sensor node information, the number of sensor, sensor type, sensing value, GPS information. The sink module sends the real-time sensor data to the local sensor P2P module whenever receives the request from the P2P module. Also, the P2P module may send the request message for the past sensory information to the sink module. The past sensor data file has two parts: file header and file body. The file header describes the information of sensor node.

Time (hour:min:sec)	Location 1	Location 2	Sensor Node Info.	Num of Sensor	Sensor Type	Value	...	GPS info

Fig. 4. Real-time sensor message format

```
Location(City/Province) : Gyeongnam
Organ(Office) : KyungnamUniv
Latitude : 35.1794102353440
Longitude : 128.5554919623680
Node info : EngineeringBuilding1
Sensor Type : humidity-temperature-illumination
Date : 2008-12-21
```

File header

```
22:21:34      humidity        25.74
22:21:37      humidity        25.78
22:21:37      temperature     16.73
22:21:37      illumination    169.92
22:21:37      humidity        25.78
22:21:41      humidity        25.81
22:21:41      temperature     16.71
22:21:41      illumination    170.29
22:21:41      humidity        25.81
22:21:44      humidity        25.85
22:21:44      temperature     16.73
22:21:44      illumination    169.56
```

File Body

(a) File contents

location	sensor node	sensor type	date (year-month-day)	start time (hour:min:sec)	store interval

(c) File name

Fig. 5. Past sensor data file format

3.2 Architecture of Sensor P2P Module

As mentioned above, this paper implements a P2P architecture to share sensor data based on Pastry which is a novel framework and provide open sources of FreePastry [11, 12]. Users and sensor nodes who want to share sensor data can join the sensor P2P overlay network as the peer node(user peer or sensor peer) and may consequently build an overly network with other peers. Also users who want sensor data can join the sensor P2P network as user peer. Figure 1 shows a conceptual overlay network among sensor networks. Each sensor peer may represent a gateway node connecting several sensors. When joining the Pastry overlay network, each peer become a node in a Pastry ring through the routing functionality of the Pastry. It could be started from the bootstrap node and done by the joining node's getting routing rows from the routing tables of visiting nodes.

Our purpose to use the structured P2P architecture in this paper is to make a flexible and scalable system for sharing the sensor data with no server. However, these days a lot of users want to use the web interface to do something in the Internet, so we have thought that a web server can be a proxy peer for non P2P devices. Of cause, the web server should install the P2P stack and have functions for the proxy service. So,

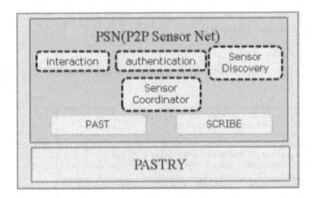

Fig. 6. Sensor P2P module(Stack)

the Pastry functions may be provided to the prompt users through the web interface. And, the prompt users don't need to install the P2P stack in their computer. This paper referred to this kind of web server as a web peer. The other reason to add the web peer is that the Pastry needs a bootstrap node. In the Pastry, the first joined node can be the bootstrap node and the other node can join the Pastry ring by sending the join request to the bootstrap node. So, the bootstrap node needs to become widely known and be more stable than usual Pastry peers.

All kind of peer node has the sensor P2P module (stack) composed of Pasty and several sub-modules shown in figure 6.

Interaction sub-module: The peer can send a control signal and get the reply back or give some data to its sensor and get the response through this sub-module. This sub-module is for interaction between sensors. For this sub-module we need further study about model for interaction between sensors.

Authentication sub-module: Users should be authenticated before getting the sensor data or doing any interactions to the other sensors. Current system just has a simple authentication sub-module by a pair of ID and password. The owner of sensors as a sensor peer stores peers' pairs of ID and password and permits access from only peers having correct ID and password.

Sensor Discovery sub-module: This sub-module provides the search function by keywords related to the sensors. The structured P2P uses the DHT to find the requested contents by using the content's exact name without servers such as DNS, so it supports only the exact matching. It is because as the DHT is based on the hash function, similar keywords produce entirely different results. However, users usually want to search using interesting a couple of keywords and get the list of the available contents. We have designed the sensor discovery sub-module for a search function in the Pastry. We assume that every sensor has one or more keywords and the owner of the sensor register the keywords when he or she joins the Pastry ring. A couple of points should be considered for design of the search function as follows: As the node ID and the object ID are hashed in the Pastry, every ID is independent each other. So, object ids hashed with keywords related to the sensor must be independent from an object ID hashed with sensor's name. But we should search the sensor by using one or more related keywords. Secondly, we should find where the owner of the sensor can register the keywords. This question arises because we don't want to lean on any server system.

Sensor Coordinator sub-module: When a sensor is shared among multiple users, if the status of the sensor is changed by a request from a user, other users may not understand the sensor data. This coordinator may permit the requesting peer proper operations according to the privilege level gotten after the authentication module. And when there is any change in the status of the sensor this module may notify the change to the related peers.

Real-time data sharing sub-module (Scribe) [13]: The real-time sensor data is monitored in this module based on the overlay multicast - Scribe built on top of Pastry. When a sensor peer wants to send its data, it makes a topicID as the multicast session identifier. Then any peers interested in the sensor can search the sensor's ID using the Sensor Discovery sub-module and join the multicast session for the sensor to get the real-time data.

Past data sharing sub-module (PAST) [14]: The already generated data (we call it past data) is stored in the P2P storage module - PAST built on top of Pastry. Any peers who want to get the past data can first search the sensor's id using the Sensor Discovery sub-module and get the data through PAST built on top of Pastry.

As mentioned above, the sink module makes a past sensor data file every time interval, such as each minute, where both processed sensor data and sensing time are stored. When a peer node is interested in the past data of a sensor, the peer can get the past sensor data file specified the sensor's information and stored time through PAST sub-module. The data has been stored in unit of one minute in our implementation.

3.3 Operation of Sensor P2P Framework

The FreePasty is an open source code to implement the Pastry in Java and we modify the Release 2.0_04 of FreePastry to develop sensor module. We implement a Topic List server to manage the Topic List by storing information about sensors on the Pastry ring in its database. So, each peer node can search a list of sensors that they want by contacting the Topic List server.

Figure 7 shows two operations of sensor P2P for real-time sensor data and past sensor data respectively. First sensor P2P framework's operation to share real-time sensor data is as follows:

① When a Sensor Peer joins the Pastry overlay network and wants to multicast its sensor data, firstly it registers its TopicID to the TopicListServer and is ready to send the data through Scribe module.

② A peer wants to monitor a couple of sensors and requests a list of live sensors to the TopicListServer.

③ The peer gets a list of sensors, and can choose a sensor. The peer joins the TopicID – multicast session ID for a chosen sensor.

④ The peer receives the multicasted sensor data through the Scribe module from the sensor peer.

⑤ When the sensor emitting the data wants to finish, it first sends the termination packet to the TopicListServer and finish the multicast.

As we described, the architecture to share the past sensor data is based on the PAST built on top of the Pastry. We also uses the TopicListServer to provide the search function for past sensor data. So, the TopicListServer keeps the sensor's history even after the sensor peer sends the termination message to finish the real-time data session. The sensor P2P framework's operation to share past sensor data is as follows:

① A Sensor Peer stores the generated sensor data every minute while multicasting it in real-time. When the Sensor Peer stores the data, it combines the sensor name and the time like "sensor1: 200810010800" and hash it to generate an object ID for storing it as a file in the PAST. Then, the past sensor data is stored using the object ID.

② A peer who interested in the past sensor data asks a list of sensor names to the TopicListServer, and selects one among replied sensor names.

③ The peer decides a period through its Graphic User Interface (GUI) and sends a query to the PAST. Then, the request can be sent to a node responsible for the object Id hashed with a name combined from sensor name and the starting time of the period like "sensor1: 200810010800".

④ The peer get the past sensor data from the object ID.

4   Deployment and Experiment of the Sensor P2P Framework

To test our sensor P2P framework to sharing sensor data, we have deployed 6 sensor networks in five different universities which are accessible to KOREN and one foreign university of Kyushu university. Each sensor network is composed of several sensor nodes, which have temperature, illumination, humidity sensors, and a gateway node connecting to both Internet and ZigBee network. Also we develop a test application using which a user can search a sensor network with key words, find the exact position of the sensor network on the digital map and receives real-time and past sensor data from the sensor network. Figure 7(a) shows the location of sensor node indicated on the map and real-time sensor value. Figure 7(b) shows the file contents of past sensor data.

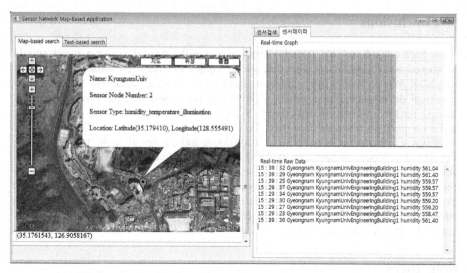

(a)   Finding the location of sensor node and receiving the real-time sensor data

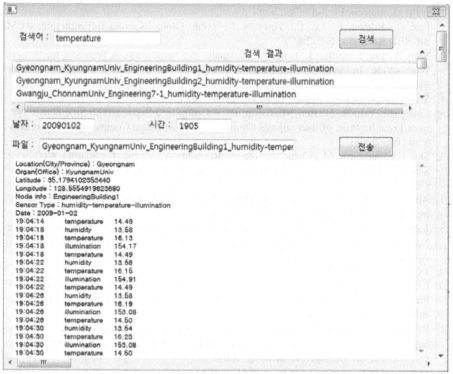

(b)   receiving the past sensor data file

Fig. 7. Test application results

5 Conclusions

It has been considered that already deployed and underutilized sensor network should be shared with other users and that some application requires sensor networks in which so many sensors are scattered world-wide. In this paper, we introduce our empirical development of a sensor P2P framework based on Pastry framework for sharing sensor information effectively. We defined a sensor peer for a gateway node and a uer peer for a sensor data user. To test the utility of our sensor P2P system, we deployed sensor networks including sensor nodes and a gateway node in some university and developed an application using which a user can search a sensor network with key words, find the position of the sensor network on the digital map and receives sensor data from the sensor network. We think our system could be deployed to enable users easily access sensor data, and could be flexible and scalable when integrating WSNs in the real world. In the future, study about interactive sensors may make us enhance our system more.

References

1. Kansal, A., Nath, S., Liu, J., Zhao, F.: SenseWeb: An Infrastructure for Shared Sensing. IEEE Multimedia 14, 8–13 (2007)
2. Fok, C., Roman, F., Lu, C.: Towards a Flexible Global Sensing Infrastructure. ACM SIGBED Review 4, 1–6 (2007)
3. Nath, S., Liu, J., Miller, J., Feng, Z., Santanche, A.: SensorMap: A Web Site for Sensors World-Wide. In: Proc. of International Conference on Embedded Networked Sensor Systems (2006)
4. SensorPlanet, http://www.sensorplanet.org/
5. Chang, K., Yau, N., Hansen, M., Estrin, D.: A Centralized Repository to Slog Sensor Network Data. In: Proc. of International Conference on Distributed Computing in Sensor Systems (June 2006)
6. Shu, L., Hauswirth, M., Cheng, L., Ma, J., Reynolds, V., Zhang, L.: Sharing Worldwide Sensor Network. In: Proc. of International Symposium on Applications and the Internet (2008)
7. Ali, M., Langendoen, K.: A Case for Peer-to-Peer Network Overlays in Sensor Networks. In: Proc. of WWSNA 2007 (April 2007)
8. Jiang, G., Chung, W., Cybenko, G.: Semantic Agent Technologies for Tactical Sensor Networks. In: Proc. of SPIE Conference on Unattended Ground Sensor Technologies and Applications (September 2003)
9. JXTA, https://jxta.dev.java.net/
10. Isomura, M., Riedel, T., Decker, C., Beigl, M., Horiuchi, H.: Sharing sensor networks. In: Proc. of 26th IEEE International Conference on Distributed Computing Systems Workshops (July 2006)
11. Rowstron, A., Druschel, P.: Pastry: Scalable, decentralized object location, and routing for large-scale peer-to-peer systems. In: Guerraoui, R. (ed.) Middleware 2001. LNCS, vol. 2218, p. 329. Springer, Heidelberg (2001)
12. http://freepastry.rice.edu/FreePastry/
13. Castro, M., Druschel, P., Kermarrec, A.-M., Rowstron, A.: SCRIBE: A large-scale and decentralized application-level multicast infrastructure. IEEE JSAC 20(8) (October 2002)
14. Druschel, P., Rowstron, A.: PAST: A large-scale, persistent peer-to-peer storage utility. In: Proc. of HotOS VIII (May 2001)

The Proposal of Service Delivery Platform Built on Distributed Data Driven Architecture

Jianfeng Zhang[1], Yuki Kishikawa[1], Kentaro Fujii[1], Yousuke Kouno[2],
Takayuki Nakamura[2], and Kazuhide Takahashi[1]

[1] Network Development Department
NTTDoCoMo, Inc.
3-5 Hikari-no-oka, Yokosuka-shi, Kanagawa, 239-8536, Japan
Tel.: +81-468-40-3773, Fax: +81-468-40-3784
{chou,kishikawa,fujiike,takahashikazu}@nttodocomo.co.jp
[2] Management System Development Department, DoCoMo Technology
3-5 Hikari-no-oka, Yokosuka-shi, Kanagawa, 239-8536, Japan
Tel.: +81-468-40-6470, Fax: +81-468-47-4649
{kouno,nakamurata}@docomo-tech.co.jp

Abstract. SDP (Service Delivery Platform) is a recommended system platform for NGN (Next Generation Network) that is expected to resolve two common system running problems: one is functional aspects such as high availability and providing effective maintenance methods, the other is cost aspects such as simplifying development method of a service. However, SDP is still on the way to be standardized, and its architecture has two problems: the congestion of service requests and flexibility of enabler, a service component of SDP. This paper explains how to build a SDP by adopting Distributed Data Driven Architecture and how our system resolves the problems by evaluating the prototype.

Keywords: NGN, SDP, Open Source Software, Distributed Computing, Service Orchestration.

1 Introduction

NIPPON TELEGRAPH AND TELEPHONE CORPORATION lunched commercial service of NGN (Next Generation Network) in March, 2008. The standardization of NGN is progressing rapidly by a standardization organization, for example, ITU-T (International Telecommunication Union Telecommunication Standardization Sector) [1]and ETSI TISPAN (European Telecommunications Standards Institute, Telecoms and Internet converged Services and Protocols for Advanced Network)[2]. The ubiquitous society is coming, which provides some new services by uniting dissimilar networks, such as a telephone and the Internet.

The system which provides various services by NGN needs to clear many technical conditions, such as the availability of a fixed-line telephone level, the openness of a function and advanced security. SDP (Service Delivery Platform) is advocated as a platform system by which these conditions can be fulfilled. As for SDP, regardless of

C.S. Hong et al. (Eds.): APNOMS 2009, LNCS 5787, pp. 201–210, 2009.
© Springer-Verlag Berlin Heidelberg 2009

domestic and overseas, the major companies like IT vendors already put various products on the market, and some companies have actually introduced SDP as a service platform.

However, advocated SDP is now under standardization. The architecture using SDP is insufficient of consideration about processing congestion in service orchestration, and consideration about flexibility in introducing or changing a service component which is called an enabler. In this paper, authors propose the construction method of SDP by using D3A (Distributed Data Driven Architecture) [3], and evaluate its effectiveness. The evaluation results show that our approach is useful to solve above mentioned problems of SDP.

2 Action Assignment and Solution of SDP

2.1 Action Assignment of SDP

We describe a definition and concept of SDP below. SDP is a service offer foundation for supporting efficient development and operation of services deployed on NGN, and cooperation of some work applications. Figure 1 shows the outline of SDP.

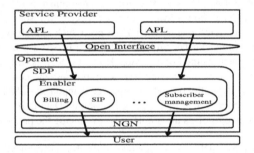

Fig. 1. The outline of SDP

Generally, SDP is requested to satisfy all the following concepts.

· Improvement in development efficiency
· TCO reduction
· High system availability
· Standardization correspondence
· Function openness
· Improvement in operation efficiency
· Advanced security

Due to space limitation, we can only focus on improvement of development efficiency which has close ties to this paper's concept. About the other details of SDP, please refer to literature [4] [5] [6] [7] [8] [9].

2.2 To Improve the Development Efficiency

There are various methods to improve development efficiency such as offering a development supportive tool. The simplest and the most effective method is introducing the concept of SOA (SOA: Service Oriented Architecture) that need not to develop but reuse existent functions to avoid redundant development. Therefore, it is recommended that SDP offers the reusable software called an enabler. An enabler is the component in which general-purpose services are coordinated per function. To improve the development efficiency, it is significant to develop a service along with the concept of SOA by combining and cooperating enablers and applications. It is unnecessary to develop general-purpose functions for each service, such as the billing processing; as a result, TCO will decrease.

2.3 Assignments and Realizable Plans to Service Orchestration

The detailed specification of SDP is still under standardization now, and there is no clear regulation, but the improvement in the development efficiency by the service orchestration and reduction of TCO mentioned above are especially important requirements to fulfill in the SDP concept. However, in the method of service orchestration proposed at present, the consideration about the following points is not enough.

There is BPEL (BPEL: Business Process Execution Language) as one choice which realizes service orchestration in SDP [10]. In BPEL, when orchestrating multiple services, a start-up sequence of service components are defined to the scenario called a workflow. The BPEL server is used for control of a workflow. It calls each component according to a workflow, and controls whole services. If a lot of service requests occur simultaneously in this system, the throughput of BPEL server will become a bottleneck.

2.4 Realization of the Distributed Processing Method

As the solution to the problem mentioned in section 2.3, we propose an architecture using the autonomous scenario of D3A (D3A: Distributed Data Driven Architecture). D3A is based on the concept of SOA, thereby it has high compatibility with SDP. In the system built as D3A, a whole function to realize the service is divided into each small function defined as LE (LE: Logical Element). As the minimum constituent unit of D3A, an LE is considered to be equivalent to an enabler of SDP.

Service orchestration of LE in D3A is realized in the scenario called LPD (LPD: Logical Path Data). As mentioned above, since BPEL is centralized workflow control, the control section becomes congest when many workflows occur simultaneously. As this measure, LE takes LPD around D3A system without preparing a control sever of LPD. By using the method that each LE receiving a LPD decides the next LE to invoke, the LPD control will be left to each LE which is processing. Therefore, there is no congestion problem at the control section. An LE operates on the common platform called LPDD (LPDD: Logical Path Data Driver). LE performs independence control of LPD through LPDD. All the LPDs used on a system are stored in LPDM (LPDM: Logical Path Data Manager), and LPDM returns an LPD according to the demand from LE. LPDM only manages LPDs and actual workflow control processing is not performed by LPDM.

An LPD is called by a user operation, a timer control of LE, or a trigger from other systems. LPD is described in XML script and has a logic section and a data section. Figure 2 shows the composition of LPD.

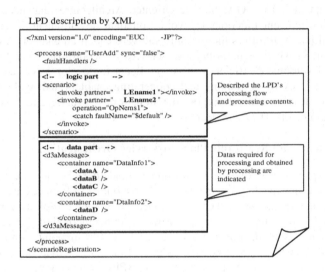

Fig. 2. The composition of LPD

Once an LPD is called, LPDD interprets the destination LE indicated in the logic section of the LPD and performs LPD transmission to the next LE for processing. An LE which has received an LPD executes a process defined in the logic section. If necessary, the LE will transpose own processing result data to the data section, and will transmit the LPD to the following destination LE. Even if a LPD is processed in each LE, the consistency of data can be guaranteed because the LPD itself has the data. There are synchronous mode and asynchronous mode to transmit an LPD. In synchronous mode, source LE executes the next process after waiting for the response of destination LE. In asynchronous mode, source LE promptly executes the next process without waiting for the response of destination LE [11]. Figure 3 shows the operation image of LPD. By using this method, we can solve the congestion centralized problem of BPEL.

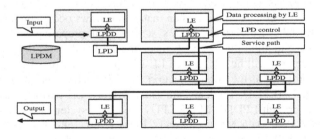

Fig. 3. The operation image of LPD

2.5 Realization of Enabler in D3A

In the method of service orchestration proposed at present, the consideration about the following points is also not enough.

When developing new services, the function which already exists as an enabler will be mounted by service orchestration. However, it is difficult to accept in case a new functional addition and change are required. Moreover, when a required enabler is not offered in introduced SDP, it is necessary to purchase another vendor's enabler, or to develop an enabler originally. In this case, a developer have to consider an interface between the enabler and SDP.

As a solution to this problem, we propose a method to construct an enabler using OSS (OSS: Open Source Software). In addition, we apply the architecture that absorbs the difference of the interface between SDP and OSS.

3   LE Container

3.1   Outline of LE Container

As a solution over the problem described in section 2.5, we propose using open source software as an enabler.

When using OSS as an enabler, an interface between OSS and SDP needs to be developed. When the number of OSS we want to use increases, the amount of development is not to be disregarded. This problem is also occurs when D3A is used as a framework of SDP. In order to reduce this burden, we have devised the structure which turns the OSS into LE at low cost. In this paper, we call it the LE container. Since OSS does not have an interface with D3A, the LE container absorbs the interface difference between OSS and LPDD. Specifically, LE container calls the arbitrary functions of OSS according to a configuration of the logic section of LPD passed from LPDD. If circumstances require, the processing result of OSS will be reset as the data section of LPD and be passed to LPDD. In addition, LE container performs optimal processing to the notice of LE starting/stopping and health check from LPDD. Figure 4 shows the mounting image of OSS using LE container.

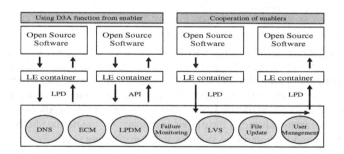

Fig. 4. The mounting image of OSS by LE container

We were concerned with the following problems in the case of substituting for opens source software as an enabler.

1 Since OSS is offered by various developers and used in various situations, there is no unity in a user interface.
2 In many cases, there are no operation management functions, such as failure management or redundant management.

By turning an OSS into an LE, the OSS can use a function required for the operation management which D3A offers as LPD or Java API via LE container, we can operate all the OSS on the common management-function system. We enumerate the operation management functions which D3A has in below. In addition, these functions are also mounted as each LE, respectively.

1. DNS
2. Element Configure Management Function (ECM)
3. Logical Path Data Information Management Function (LPDM)
4. Failure Monitoring Function
5. Redundant Management Function
6. Load Balancing
7. File Update Function
8. User Management Function

D3A can offer mostly all the functions required as SDP at high level such as TCO reduction, high system availability, and operating efficiency improvement by the functions described above. Therefore, the service offer foundation which has the pliability of open source software and the requirements needed for SDP can be built by combining D3A and LE container.

3.2 Mounting of LE Container

In this prototype, the LE container was implemented by using Java. Some LE container frameworks are constituted by the abstract class of Java, or configuration file of XML script. Although adjustments of configuration files are indispensable, development of inheritance class is arbitrary for each OSS. By adding such necessary minimum development, OSS can easily become LE. Detailed implementation method of LE container is shown below.

[Start-Stop processing of OSS]
To start or stop OSS, usually some shells get executed. Therefore, LE container has the function to execute a shell when it has received start or stop demands from LPDD, Thereby, starting/stopping remote functions of D3A are applicable to OSS only by writing the name of starting/stopping shells and the directories in which they are located to the configuration file of LE container. In case OSS needs to be started or stopped from a command line, a developer just needs to create the shells which execute these command lines and uses them as starting/stopping shells.

[Initialization processing at the time of LE starting or stopping]
There are some LEs that have to register especial DNS names at the time of starting. LE container performs this processing according to the contents of a common

configuration file. In case independent configurations are required for each OSS, a developer can make a unique configuration file separately so that LE container may read this file at the time of starting. As the example of a unique configuration file in need, a developer can implement a batch process performed at fixed time by defining the starting time of the process and the API of OSS to call in the file. When LE stops, LE container deregisters DNS names and releases the resource of the server on which it stays.

[Permanent Resident Processing Thread]

The LE container starts a permanent resident processing thread automatically after running of OSS. In this prototype, this permanent resident processing thread is implemented as an abstract class of Java. When process monitoring, periodical batch processing, etc. of OSS are required, what a developer has to do is to create the Java class which inherits this permanent resident processing thread class and to customize it. About the function to analyze the data division of LPD, the implementing forms of a data division differ for every open source software. Therefore, we develop in the inheritance class of LE container for each open source software of every using Java API for LPD management offered in D3A. Figure 5 shows the outline of LE container operation implemented in this time.

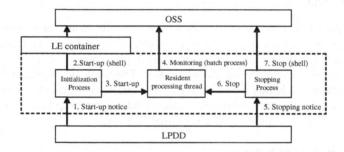

Fig. 5. The outline of LE container operation

4   Prototype of Communication Service

We aim to construct the system which provides communication services such as a VoIP telephone call, a chat, etc. and offer the service for maintenance work to our network.. We call this system used as this purpose "Communication Service." This time, we have evaluated our proposal system by using the prototype system construction for communication service by implementing the LE container mentioned above and turning OSS into LE.

4.1   Overview of Prototype System

This section describes the software construction and outline of prototype system. The prototype system adopts three OSSs called Asterisk® (IP-PBX)[12],IRC (chat)[13], and Aipo4 (groupware)[14]. These three OSSs are monitored by permanent resident processing threads.

Regarding Asterisk, the login information is acquired from the user profile database which D3A holds, the function which assigns each user the telephone number used by Asterisk is implemented. Hence, single sign-on using a D3A user is realized. Additionally, the GUIs for choosing a telephone number to send a call and for establishing a telephone conference are implemented. In order to set up these functions, we usually have to modify the configuration file of Asterisk directly. This time we have constructed the easy and flexible operation.

Regarding Aipo4, we have implemented the service orchestration which made it possible to browse a schedule list collectively. The LE container that manages Aipo4 extracts the operating schedule reserved by the existing system, and registers it in the database of Aipo4.

5 Evaluation

In this section we will evaluate this prototype system from the viewpoints of both cost and performance.

5.1 Cost Evaluation

The development cost for LEnized OSS was is about 100 lines per component. The development cost for the implementation of an individual GUI in Asterisk was several thousand lines, and the development cost for the implementation of service orchestration in Aipo4, was several thousand lines as well. The whole development cost of the prototype system, including development cost of the three OSSs, was less than ten thousand lines. In summary, these results show that we can keep lower implementation cost of SDP by using the architecture based on D3A.

5.2 Performance Evaluation

In order to verify that there is no problem with performance in our aiming communication service, we have conducted a load test on the LEnized OSS by using Asterisk. The following subsection describes the machine specification and test results.

[Machine specifications used for evaluation]
OS: Red Hat® Enterprise Linux® ES release 4
CPU: Intel® Xeon® 3.60GHz
MEMORY: 8GB

[Test method]
Asterisk LE is separately implemented in two servers, and environment as shown in Fig. 6 is built using the dial plan function of Asterisk. We have dialed the No.3900 (No.3800, No.3700, etc. unit interval/100) of Asterisk LE1 by using a soft phone, and verified the tone quality. The test composition is shown in Fig. 6.

Our desired value was the maximum simultaneous 20~40 telephone calls, which is a standard value of the small and medium business-oriented IP-PBX. However, the results (value $R \fallingdotseq 84.7$ etc.) have shown that the sound quality was excellent while processing 400 simultaneous telephone calls by server, and there was no instantaneous interruption and delay.

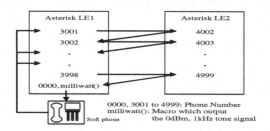

Fig. 6. The test composition of Asterisk

6 Conclusions and Future Work

In recent years, the SDP attracts attention as a service platform of the NGN. It is important for SDP to offer high failure resistance and an efficient operation function at low cost. Although there were advocates for certain architectures of enabler's service orchestration, it seems that there is not sufficient consideration about two problems: one is flexibility of an enabler, and the other is congestion of control processing server with service orchestration.

This paper has presented a novel approach to solve the problems mentioned above. We have proposed lower-cost SDP architecture based on the D3A, and evaluated its effectiveness. The design and implementation of communication service prototype, as described in this paper can be improved further.

We plan to implement the data warehouse to upgrade service quality, and the data processing software to offer a variety of IP multimedia services by using OSS. We also plan to implement and evaluate more value-added services. In the future, we will continue to enhance security and to supply communication service to mobile market.

References

1. ITU-T, http://www.itu.int/ITU-T/ngn/index.html
2. ETSI TISPAN, http://www.etsi.org/tispan/
3. Akiyama, K., Fujii, K., Kanauchi, M., Watanabe, T., Takahashi, K., Tanigawa, N.: The Proposal of the Distributed Data Driven Architecture to Realize IP. IEICE Trans. Commun (Japanese Edition) 102(706), 1–6 (2003)
4. IT pro, 31st, SDP,
   http://itpro.nikkeibp.co.jp/article/COLUMN/20071025/285489/?ST=ngn,
   http://itpro.nikkeibp.co.jp/article/COLUMN/20071025/285479/?ST=ngn

5. How to design the system for NGN,
   http://itpro.nikkeibp.co.jp/article/COLUMN/20061122/
   254604/?ST=ngn
6. Heard the IBM's NGN strategy,
   http://wbb.forum.impressrd.jp/feature/20071106/497/#zu02,
   http://wbb.forum.impressrd.jp/feature/20071113/501
7. Mitaku, I.: Trend of Servcie Delivery Platform(SDP). Business communication 44(10), 16–19 (2007)
8. BPEL4WS,
   http://www.ibm.com/developerworks/library/specification/
   ws-bpel/
9. Akiyama, T., Kon, T., Takahashi, K., Jinguji, M.: Economic technology of operation system -The Distributed Data Driven Architecture. NTT DoCoMo Technical Journal 13
10. Asterisk, http://www.asterisk.org/
11. IRC, http://www.ircnet.jp/servers.html
12. Aipo4, http://aipostyle.com/

LCGT: A Low-Cost Continuous Ground Truth Generation Method for Traffic Classification*

Xu Tian, Xiaohong Huang, and Qiong Sun

Research Institute of Networking Technology, Beijing University of Posts and
Telecommunications,
100876, Beijing, P.R. China
{tianx,sunq,huangxh}@buptnet.edu.cn

Abstract. Recently, with the progress of research on accurate traffic classification (TC), the major obstacle to achieving accurate TC is the lack of an efficient ground truth (GT) generation method. A firm GT is important for exploring the underlying characteristics of network traffic, building the traffic model, and verifying the classification result, etc. However, current existing GT generation methods can only be made manually or with additional high-cost DPI (deep packet inspection) devices. They are neither too complicated nor too expensive for research community. In response to this problem, we present LCGT, a low-cost continuous GT generation method for TC. Based on LCGT, we propose a novel updateable TC system, which can always reflect the features of up-to-date traffic. While we have found LCGT to be very useful in our own research, we seek to initiate a broader discussion to guide the refinement of the tools. LCGT is located on: http://code.google.com/p/traclassy

Keywords: ground truth, traffic classification, low-cost, automated.

1 Introduction

Accurate classification and categorization of network traffic according to application type is an important element for network trend analyses, dynamic access control, lawful interception, and intrusion detection. For example, network operators need to identify the requirements of different users from the underlying infrastructure and provision appropriately. Thus, real-time traffic classification is the core component of emerging QoS-enabled products [1] and real-time traffic monitor.

With the development of Peer-to-Peer (P2P) applications, the classical approach to traffic classification using well-known port numbers of the official IANA list [2], is far from accurate. To avoid drawbacks of port-based classification, payload-based techniques have been developed by many industry products. However, these techniques are impossible to deal with proprietary protocols or encrypted traffic. As a result, most

---

* China 973 Programme (No. 2009CB320505), Project 60811140347 supported by NSFC-KOSEF, Specialized Research Fund for the Doctoral Program of Higher Education(200800130014), Project 60772111 supported by National Natural Science Foundation of China.

C.S. Hong et al. (Eds.): APNOMS 2009, LNCS 5787, pp. 211–220, 2009.

current traffic classification techniques rely on distinguishing distinct behavior patterns among different applications when communication on a network. So many ML(machine learning) algorithms have applied on traffic classification, e.g. using supervised learning method like Decision Tree[3], NativeBayes[4][5], Bayesian Neural Network[6], and unsupervised learning method like K-means[7], EM[8], DBSCAN[8] algorithms.

For ML-based traffic classification techniques, there are three major steps in total: feature selection, traffic model built and traffic identity verification. Feature selection is to select the statistic characteristics of flows, based on which traffic models are built and incoming traffic is classified. Although different levels of features can be extracted, an absolute accurate ground truth is needed. Accordingly, traffic models are normally built with the selected features and specific ML algorithms. Therefore, the models built by ground truth should not only be quite accurate, but it should also keep up with current network condition. A static traffic model is not quite useful in reality and neither does an out-of-date ground truth. In a word, an accurate and up-to-date ground truth is obviously very important. Otherwise, there is no way to test whether the classification methods are efficient or not.

However, current research work on ground truth generation methods is far from enough. It is mostly because ground truth generation itself is also a traffic classification problem. Although it does not have strict real-time performance requirement, it must achieve approximate one hundred percent accuracy. As a result, researchers have to use out-of-date techniques to pre-classify the traffic.

Existing ground truth generation methods can only be made with high-cost DPI devices or manually. With regard to DPI-related generation method [9], devices are quite expensive so that only limited researchers can make use of them. Manual-based techniques rely on port-base traffic classification and exhaustive packet payload inspection, which is rather complicated. It may cost researcher several months to make a good ground truth dataset. Some other researchers [10] choose to use the public available traffic before 2004, when applications can be uniquely identified by port-based method. So far, the only publicly available ground truth [11], which was generated manually from locally captured traffic, was in 2004. Although the traffic models have been changed dramatically over the past five years, researchers in [3][4] [9] [12] still use this ground truth. As a result, the reliability of these traffic classification methods is questionable.

In this paper, we firstly propose a novel low-cost continuous ground truth generation (*LCGT*) method for traffic classification. With a carefully designed dynamic policy, it can automatically collect the users' application-related information to help servers build continuous ground truth for traffic classification. Based on our *LCGT* tool suite, we further propose an integrated increment-traffic classification system, which can always reflect the features of up-to-date traffic.

The remainder of this paper is organized as follows. In Section III, the architecture of LCGT is proposed. In Section IV, LCGT based traffic classification system is proposed. Experiment evaluation results are presented in section VI. Finally, Section VII concludes the paper with a summary.

2 LCGT Design

2.1 Overall Architecture

The motivation behind our LCGT is that although it might be rather difficult to identify the traffic application type from backbone network, the classification process can be much easier at the end host. Thus, the user can determine the application type together with its corresponding packet information. However, since the characteristics of end host traffic might be different with backbone point due to complicated network condition, we should still monitor the mirrored packet in the backbone point and select the ones that are produced by specific end hosts. With this guideline, we can design a continuous notifying policy between the users and servers at the backbone point and a firm ground truth can be generated accordingly.

Our LCGT for traffic classification consists of four parts, as shown in Fig. 1: 1) traffic flow monitoring agent (TFMA), 2) application-proof flow collector, 3) traffic capturer & Ground Truth filter, and 4) ground truth database server.

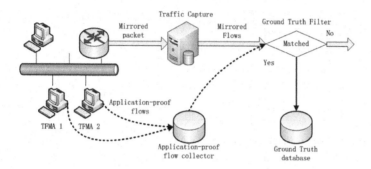

Fig. 1. The overall architecture of ground truth generation

1)   Traffic flow monitoring agent (TFMA)
TFMA is for collecting the packet trace for every running process in the OS (only for Windows XP/Vista currently). TFMA monitors the OS's process list and constantly collects the connection information of every process. It saves the log composed of lines of traffic summary, e.g. five tuples (source IP, source port, destination IP, destination port, L4type), and application name. Periodically, TFMA sends its saved log to the application-proof flow collector every Δt.
2)   Application-proof flows collector (AFC)
AFC is to collect application-proof traffic flow records from TFMAs . It then sends a list of the application-proof traffic flows to GT filter with a period of ΔT, which is greatly larger the Δt.
3)   Traffic capturer & Ground truth filter (GT filter)
The traffic capture is to capture continuous mirrored packets using libpcap, aggregate them into flows, and we calculate most possible real-time features that might be used in ML-based traffic classification. When finishing calculation, the traffic capture will send the flow information to GT filter. Then the GT filter can check the traffic

information provided by AFC to determine the specific application-proof flows. Finally, the filter sends all the flow information (5tuple and flow features) that appeared in the AFC to GT database.

4)   Ground Truth Database (GT database)

The GT database keeps a dynamic flow list as ground truth dataset by using a Dynamic Application-Proof Flows Selecting algorithm (DAPFS), which will be described in the following section 3.4. It should not only make a balance for traffic records between different applications, but also maintain a reasonable amount of instances for further traffic classification.

2.2   Workflow of LCGT

In our design, the ground truth can be automatically generated without any manual classification process. It also designed to be continuous in order to provide network operators an up-to-date ground truth. The time-ordered sequence chart of the overall ground truth generator is depicted in Fig. 2

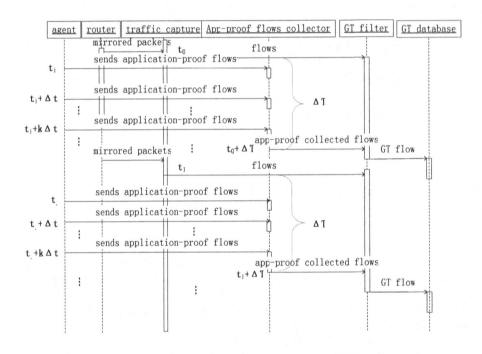

Fig. 2. Time ordered sequence chart of the overall ground truth generator

Here, t_0 is the start time of the first round generation for LCGT. The traffic capturer begins to capture the packets mirrored from the router, aggregate them into flows, and send them to the GT filter constantly. The ground truth is generated from a time window of ΔT, which is far greater than Δt, in order to make sure that there are enough flow records for most applications.

Since several running TFMAs have been deployed in the network and their behaviors are almost the same, we can take TFMAi for example to explain the mechanism.

As shown in Fig. 2, t_i is the time when the AFC receives the first application-proof flow records. Since TFMAi keeps sending application-proof flows record every Δt, we can conclude that $t_0 < t_i < t_0 + \Delta t$. During the period of $(t_0, t_0 + \Delta T]$, the AFC receives k records sent by TFMAi, where $t_i + k\Delta t \leq t_0 + \Delta T < t_i + (k+1)\Delta t$. AFC keeps a list for these k records The AFC then saves the records from all the TFMAs during $(t_0, t_0 + \Delta T]$, and send them to GT filter at $t_0 + \Delta T$. These records are kept in a active *app-list* with host flow information.

The GT filter saves each flow and statistic features in its database during $(t_0, t_0 + \Delta T]$. Once the filter receives application-proof flow records from the AFC, it begins to scan its own flow database, and check if the flow is in the *app-list*. If matched, the corresponding flow summary will be sent to the GT database, as a candidate of ground truth.

Finally, the GT database utilizes a selective algorithm to make ground truth dataset from the incoming candidate ground truth flows. When finished, the overall system will start a new round of ground truth generation from $t_1 \geq t_0 + \Delta T$. The only difference between the new round and first round generation is that the new one will utilize a refinement algorithm in the GT database, which can keep the GT database up-to-date and efficient. Both the selective algorithm and refinement algorithm used in the GT database is named *Dynamic Application-Proof Flows Selecting algorithm (DAPFS)*, which will be explained in the following section.

In a word, it is feasible to obtain continuous automated ground truth dataset by using the architecture mentioned above.

2.3 Dynamic Application-Proof Flows Selecting Algorithm (DAPFS)

In GT database, there are three factors that may have an effect on the quality of our continuous ground truth. The first factor is the number of instances in the ground truth. If the number is too few, the ground truth would not enough for ML-based traffic classification; while if the number is too large, it may cause over fitting problem and consume too much space. The second one is the imbalance of dataset. Different applications produce different number of traffic flows. For instance, there are only a little FTP flow in our experiment, therefore, it is impossible to collect enough flows within ΔT; On the other hand, since there are a large amount of HTTP and P2P flows in the Internet, we have to limit the amount of these flows in our ground truth. The third one is whether the ground truth is continuous and up-to-date. An outdated ground truth cannot stand for the current state of network traffic. Even worse, it may mislead our research result.

In order to deal with these three problems, we have developed a *DAPFS* in the GT database. Generally speaking, DAPFS will keep a maximum of M records for each application in memory within a certain period, say ΔT. Thus, the key point is the definition of M, which will have great impact on the overall performance. It is define as follows:

$$M=\max\{U,\min(V,Vmax)\} \tag{1}$$

where U is a user defined number, which balanced the performance and efficiency of ground truth for dynamic update and training a traffic classification engine; V is the number of records that each application can product during the latest ΔT seconds; V_{max} is the maximum number of records that allows every application products during the one ΔT seconds.

The explanation behind (1) is as follows. In DAPFS, each application maintains a FIFO for its candidates. The length of FIFO is no longer than M. Therefore, the index M can help us keep our GT database more efficient and flexible, and avoid over fitting problem meanwhile. Besides, the DAPFS can work for many rounds and accumulate rare flows, whose $V \leq U$, until the number of this application's flow is up to U. Fig. 3 depicts the algorithm in reality.

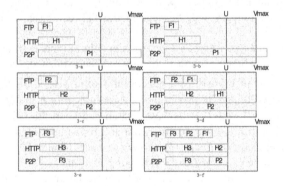

Fig. 3. Demonstration of how DAPFS helps ground truth generation. 3-a, 3-c,3-e are ground truth flow candidates produced during the first, second, and third ΔT; 3-b, 3-d, 3-e are flows summaries that finally saved in GT database as ground truth dataset.

As shown in Fig. 3, we make a reasonable assumption that the incoming number of ground truth candidates follows: FTP<<HTTP<P2P. In the first ΔT generation round, DAPFS keeps all the candidates (Fig. 3-a), whose number is smaller than V_{max}, as a ground truth (Fig. 3-b). When second round comes, FTP ground truth is still growing; the HTTP's summation of these two round is larger than U, therefore the DAPFS forget the oldest ($H2+H1-U$) flows; due to the abundance of P2P, the length of P2P's FIFO is limited to V_{max}. In the third round, the summation of FTP has almost reached U. HTTP and P2P is limited to U. As a result, after several round of generation, DAPFS will keep every application's record number in their corresponding FIFOs between U and V_{max}. In other words, the GT database will continuously get an up-to-date, dynamic, smart ground truth dataset since then.

3   LCGT Based Traffic Classification System (LCGT-TC)

Ground truth is the basis of traffic classification system. With our proposed LCGT, a dynamic incremental traffic classification system can be accomplished. In this section, we propose a LCGT based traffic classification system.

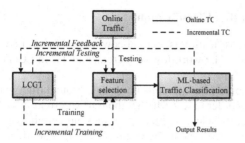

Fig. 4. LCGT based Traffic Classification System

Our LCGT-based TC system is depicted in Fig. 4, which includes online traffic classification and incremental traffic classification processes.

3.1 Online Traffic Classification

The process of online traffic classification is similar to traditional traffic classification methods, which includes training phase and testing phase.

(1) Testing phase

We can use LCGT as a training data, in which existing known applications, such as HTTP, P2P, EMAIL, ATTACK, etc., have been labeled accurately. We can choose a certain amount of training data in order to make the model better reflecting current traffic model.

Subsequently, feature selection block can then select some features, which can have high accuracy for traffic classification algorithm. In experiment result, we find that the features of the first twenty packets in one flow can have excellent overall performance. The features are listed in the following, which can be calculated separately in the forward and backward directions:

1)   inter-arrival time (minimum, maximum, mean and standard deviation)
2)   total bytes in IP packet (minimum, maximum, mean and standard deviation)
3)   bandwidth

These features are independent of flow length (thus allowing online calculation before an application has finished), which require no knowledge of packet contents (thus minimizing privacy intrusion or dependency on particular packet payload encoding).

Finally, a machine-learning algorithm can be chosen to construct an accuracy traffic model with incremental update ability. In LCGT-TC, we use ensemble algorithm, a famous data stream mining algorithm, to achieve online traffic classification and update ability.

(2) Testing phase

The testing phase in LCGT-TC is quite simple. It captures online real-time traffic, extracts the features used in pervious training phase, and classify it in the traffic model. Finally, it will pass the classification result, together with the traffic features, to the query system.

3.2 Incremental Traffic Classification

Incremental traffic classification is a brand new feature in our LCGT-TC compared to traditional TC systems. The workflow of Incremental TC in LCGT-TC is normally as follows:

First, a certain amount of continuous up-to-date ground truth will periodically goes into the TC system in order to test whether the traffic current model is still efficient or not. Specifically, feature selection block will pick up certain features used in online traffic classification for each incoming instance. Traffic classification block will classify the incoming instance based on current traffic model.

Second, if the accuracy of incremental test is below a certain threshold, it will feedback to LCGT indicating the traffic model in use is no longer fit for current traffic anymore and it will notify LCGT to incrementally update the traffic model.

Finally, once receives the message to update the traffic model, LCGT will then send up-to-date ground truth for incrementally update the model. In the same way, feature selection and traffic model will be updated correspondingly.

4 Experiment Evaluation

In this section, we present experiment evaluation result for LCGT-TC system, compared to best traditional algorithms, e.g., C45.[10] and BayesNet[11].

4.1 Data Preparation

In our experiment, ten TFMAs are deployed in our research lab. Specially, Δt is set to 30 second and ΔT is set to 30 minute. Then online traffic classification and incremental traffic classification are done automated. In this section, we use the incremental feedback result to evaluate the accuracy of traffic classification. For category C_i that contains $N(Ci)$ instances, we consider True Positive (TP) and Ture Negative (TN) as our metrics:

$$Rate(TP_{ci}) = 1 - N(FP_{ci}) \Big/ \sum_{j \neq i} N(C_i)$$

$$Rate(TN_{ci}) = N(TP_{ci}) / N(C_i) \tag{2}$$

Where $N(FPci)$ and $N(TPci)$ are the number of instances incorrectly and correctly classified as protocol Ci, respectively.

4.2 Accuracy Evaluation

To evaluate the accuracy, we compare our LCGT-TC with traditional best ML-based algorithm, e.g. C4.5 and BayesNet. Since traditional traffic classification algorithms do not have change detection ability, their models cannot update during the whole classification process. Here, we specify that Dj represents the j-th dataset of traffic in which there is no concept drift, and Cji means the i-th chunk of traffic dataset in Dj. Cji is only used in ensemble algorithm in our LCGT-TC system.

In each step of ensemble classification, there are five classifiers in memory to deal with the ever-changing traffic model. The classifier with worst accuracy is then crossed out and the other four better ones are kept. We can see from the result that the accuracy decreased dramatically when a change happens. Therefore, the classifiers selected in each phase are {C14, C15, C16, C17}, {C21, C24, C25, C28}, {C33, C35, C36, C37} and {C41, C42, C44, C47}. The whole process is depicted in Table1.

Table 1. Classifiers selected in every phases

i	C1i(%)	Selected classifiers after C2i
1	95.325	C14(79.35),C15(79.75),C16(79.00),C21(97.90)
2	95.775	C14(78.95),C15(79.39),C21(92.47),C22(97.80)
3	95.634	C15(79.56),C21(92.52),C22(97.84),C23(97.84)
4	95.812	C21(92.85),C22(93.08),C23(92.63),C24(97.85)
5	96.178	C21(92.44),C22(92.63),C24(92.81),C25(98.31)
6	96.019	C21(92.18),C24(93.08),C25(94.42),C26(97.93)
7	96.434	C21(90.72),C24(93.20),C25(94.81),C27(87.39)
8	95.507	C21(92.60,C24(92.77),C25(94.28),C28(97.88)
	C2i(%)	**Selected classifiers after C3i**
1	95.16	C21(80.98),C24(82.90),C25(83.12),C28(81.42)
2	95.80	C24(82.50),C25(82.47),C28(80.95),C31(89.71)
3	95.91	C24(82.70),C25(82.84),C31(90.25),C32(90.10)
4	95.70	C25(82.63),C31(90.50),C32(90.258),C33(91.16)
5	95.06	C31(90.28),C32(90.12),C33(90.86),C35(90.25)
6	95.28	C31(90.08),C32(89.96),C33(90.63),C352(90.79)
7	95.62	C31(90.38),C33(90.87),C35(90.97),C36(90.57)
8	95.99	C33(90.83),C35(90.79),C36(90.55),C37(91.29)
	C3i(%)	**Selected classifiers after C4i**
1	97.78	C33(85.16),C35(85.92),C37(86.81),C38(86.81)
2	98.12	C35(85.93),C37(87.00),C38(86.00),C41(96.83)
3	97.78	C37(86.83),C38(85.87),C41(96.34),C42(95.74)
4	97.37	C37(86.96),C41(96.23),C42(95.71),C43(95.25)
5	98.153	C41(96.47),C42(95.67),C43(95.25),C44(96.27)
6	98.50	C41(96.38),C42(95.63),C44(96.02),C45(93.54)
7	98.09	C41(96.48),C42(95.73),C44(96.06),C46(95.47)
8	97.69	C41(96.17),C42(95.35),C44(95.91),C47(95.187)

The overall accuracy of LCGT-TC is depicted in Table2. We can see that the average accuracy of LCGT-TC is beyond 95%, which is the best compared to C4.5 and BayesNet. With the help of LCGT, LCGT-TC can also illustrate the current traffic and keep former model at the same time, so it is more robust than C4.5 and BayesNet. Therefore, LCGT-TC can get good performance when dealing with real world traffic by using a continuous dynamic model.

Table 2. Algorithm Evaluation Result

Algorithm	accuracy	D1(%)	D2(%)	D3(%)	D4(%)
C4.5	TP	97.6	85.9	96.7	82.3
	TN	99.7	84.1	96.0	82.3
BayesNet	TP	91.1	87.0	83.7	88.5
	TN	98.5	81.6	86.8	80.6
LCGT-TC	TP	95.4	94.5	95.3	95.2
	TN	95.2	94.0	90.3	95.3

5 Conclusion

In this paper, we propose a novel low-cost continuous ground truth generation method. Based on *LCGT* tool suite, we further propose an integrated updateable traffic classification system, which can always reflect the features of up-to-date traffic. For future work, we plan to deploy more agents that collect the real-time ground truth dataset and attain the full potential of our LCGT and LCGT-TC.

References

1. Ubicom Inc., Solving Performance Problems with Interactive Applications in a Broadband Environment using StreamEngine Technology, August 14 (2007)
2. http://www.iana.org/assignments/port-numbers
3. Williams, N., Zander, S., Armitage, G.: A preliminary performanc comparison of five machine learning algorithms for practical IP traffic flow classification. Special Interest Group on Data Communication (SIGCOMM) Computer Communication Review 36(5), 5–16 (2006)
4. Andrew, W.M., Denis, Z.: Internet traffic classification using Bayesian analysis techniques. In: ACM International Conference on Measurement and Modeling of Computer Systems (SIGMETRICS), Banff, Alberta, Canada (June 2005)
5. Livadas, C., Walsh, R., Lapsley, D.: Using Machine Learning Techniques to Identify Botnet Traffic. IEEE, 967–974 (2006)
6. Auld, T., Moore, A.W., Gull, S.F., et al.: Bayesian Neural Networks for Internet Traffic Classification. IEEE Transactions on Neural Networks (1), 223–239 (2007)
7. Bernaille, L., Teixeira, R., Salamatian, K.: Early Application Classification. In: The 2nd ADETTI/ISCTE CoNEXT Conference, Lisboa, Portugal (December 2006)
8. McGreor, A., Hall, M., Lorier, P., et al.: Flow clustering Using Machine Learning Techniques. Passive&Active Measurement workshop 3015(4), 205–214 (2004)
9. Pietrzyk, M., Urvoy-Keller, G., Costeux, J.-L.: Revealing the Unknown ADSL Traffic Using Statistical Methods. In: Proc. 1st COST TMA Workshop, pp. 75–83 (2009)
10. Iliofotou, M., Kim, H., Pappu, P., Faloutsos, M., Mitzenmacher, M., Varghese, G.: Graph-based P2P Traffic Classification at the Internet Backbone. In: IEEE 12th Global Internet Symposium (in Conjunction with IEEE INFOCOM 2009) (April 2009)
11. Moore, A.W., Zuev, D., Crogan, M.: Discriminators for use in flow-based classification, Technical Report, RR-05-13, Department of Computer Science, Queen Mary, University of London (August 2005)
12. Canini, M., Li, W., Moore, A.W., Bolla, R.: GTVS: Boosting the Collection of Application Traffic Ground Truth. In: Proc. 1st COST TMA Workshop (2009)

Adaptive Coverage Adjustment for Femtocell Management in a Residential Scenario

Sam Yeoul Choi, Tae-Jin Lee, Min Young Chung, and Hyunseung Choo*

Department of Mobile System Engineering
Sungkyunkwan University, Korea
{id3102,choo}@skku.edu

Abstract. Femtocell is an emerging technology for expanding cell coverage and increasing data rate, and its users deploy in their own premises by themselves. Therefore, femto base stations cannot go through manual cell planning procedure as the mobile communication base stations do. Femto base station sets its parameters, such as transmit power or pilot power, considering the surrounding radio environment. Due to its automated cell planning and parameter setting, leakage of the power to the outside of a house can occur and lead to highly increased number of unwanted handover events of macrocell users, which will finally lead to higher call drop probability. In this paper, we propose a coverage self-optimizing scheme to decrease the call drop probability in a real residential scenario. Our proposed scheme prevents unwanted handovers and comprehensive simulation results show that it improves performance of network in terms of handover and call drop probability.

Keywords: Femtocell, Self-optimization, Auto-configuration, Self-organization.

1   Introduction

Conventional telecommunication systems did not take indoor coverage into serious consideration, therefore indoor call error was not an important issue. However, due to explosive growth of indoor voice and data usage, coverage increment has become a significant matter. Femtocell technology is a promising solution for such coverage extension. Femtocell base stations are a small base stations that users deploy in their houses or offices to increase indoor coverage and data rate. It enables standard mobile devices to connect to a mobile operator's network using residential DSL or cable broadband connections. It also provides low cost FMC (Fixed Mobile Convergence) services [1].

However, there still exist some problems to actualize such technology. Interference mitigation, frequency allocation, and self-organization are the problems that need to be solved. Minimizing interference with existing networks is required to maintain conventional network performance and maximally utilize femtocell abilities concurrently. Moreover, efficient frequency allocation mechanism needs to be developed to efficiently utilize frequency bands. Related issues are frequency planning/management, interference avoidance, and frequency hopping. Unlike conventional base stations that

---

* Corresponding author.

C.S. Hong et al. (Eds.): APNOMS 2009, LNCS 5787, pp. 221–230, 2009.

are planned and deployed by operators, femtocells are likely to be deployed by users in each of their premises. Therefore it is expected that a large number of femtocells are deployed in a distributed way without specific regulation, which makes it difficult to manage them. To efficiently utilize femtocell, it is required to set its own parameters and perform self-optimization during its operation. Such issues are recently being standardized by organizations such as 3GPP and IEEE [2]. In 3GPP, they use the terminology NodeB (NB) as the macro base station, and Home NodeB (HNB) as the femto base station. In this paper, we focus on self-optimization issue and a novel approach for coverage optimization is proposed.

Subsequent sections of this paper are organized as follows: Prior contributions in femtocell coverage adjustment are reviewed in Section 2. Section 3 describes our proposed scheme to effectively minimize the call drop probability. We evaluate the performance of our scheme compared with other schemes in Section 4. Finally, we give the advantages of our proposed scheme and concluding remarks in Section 5.

2   Related Works

In co-channel environment, since macrocell and femtocell use the same frequency band, frequent handover event might occur due to pilot power leakage of femtocell. The leaked power can significantly affect the macrocell User Equipments (UEs). To reduce such affection, controlling the power of femtocell is required. As related works of this paper, three different schemes for power adjustment, fixed power configuration, auto power configuration, and dynamic power control, are introduced. Furthermore, we introduce three access modes, open, closed, and hybrid access mode, widely used in femtocell operation.

2.1   Power Configuration Schemes

In [3], authors introduce three power configuration schemes. First one is fixed power configuration. Pilot power of femtocell is set to a constant value in this scheme. The power of all femtocells covered by a macrocell is fixed to meet a certain condition. It is set to -5 dBm in this paper as shown in Fig. 1(a). Fixed pilot power is not for usage in practical environment, it is only used for comparison with other schemes.

With auto power configuration scheme, the femtocell sets the pilot power according to the received power of the macrocell. The pilot power of a femtocell is set so that the received power from the attached macrocell is equal to that from the femtocell at target range r meter away from the femtocell. That is, the femtocell pilot power is adjusted to obtain a target cell radius of 10 meters if this is possible, limited by its maximum pilot transmit power. Path loss model was used to describe the method and the pilot power of the femtocell is set as follows:

$$P_{femto} = min(P_{macro} + G(\theta) - L_{macro}(d) + L_{femto}(r), P_{max}), \qquad (1)$$

where $38.5 + 20log_{10}(r)$ is used for $L_{femto}(r)$, which is the path loss model from the femtocell to the target radius r excluding any wall losses. P_{macro} is the pilot power of the macrocell sector where the femtocell is located. $G(\theta)$ is antenna gain value where

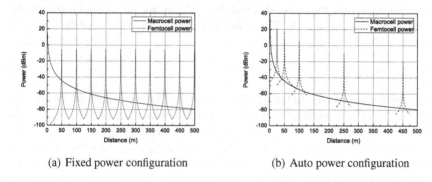

(a) Fixed power configuration   (b) Auto power configuration

Fig. 1. Simplified example of femtocell power configuration

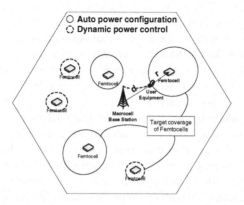

Fig. 2. Diagram of femtocell coverage within a macrocell area

θ is the angle of the received signal from the macrocell. $L_{macro}(d)$ is the path loss model for the macrocell signal, and $28 + 35log_{10}(d)$ is used also without any wall losses. P_{max} is the maximum pilot power set to 125 μW at the edge of the macrocell to the maximum value of 125 mW close to the macrocell. Fig. 1(b) shows a simplified example of femtocell power configuration using Eq. (1). Moreover, auto-configured power and coverage of femtocells using pathloss model as in Eq. (1) are depicted in Fig. 2. It describes the coverage of femtocells set within a macrocell area.

The authors propose a scheme named, dynamic power control, that adjusts the femtocell coverage determined by the usage of the femtocell UEs. When there are UEs that are in active mode (in a call or data communication), the pilot power is set to the same value as in the auto power configuration scheme. If all the UEs connected to femtocell are in idle mode, the pilot power of femtocell is reduced by 10 dB. Moreover, to prevent the femtocell UEs from being handed over to the macrocell, UE's idle mode cell reselection threshold is increased by 10 dB. So that the UEs will remain camped on the femtocell despite the reduced power. Fig. 1(b) shows the reduced power of the femtocell using this scheme. All UEs are in idle mode so the coverage of the femtocell is reduced by 10 dB.

2.2 Access Modes

There are three access modes of femtocell operation [4]. Open access mode allows anyone to get access to femtocells. Since there is no restriction of access, when the signal received from femtocell is stronger than the signal received from macrocell, anyone can get accessed to femtocell as long as there is bandwidth left to serve the anonymous user. When users who are not registered access a femtocell then the owners of the femtocell will be allocated with less bandwidth, so a way to financially or technically compensate the owners is required (discount for femtocell AP or provide faster data rate, for instance). In this paper, open access mode is applied to all femtocells. If a user passing nearby a femtocell is not registered and the femtocell is operating as closed access mode, handover event will not occur and the user will experience interference. The reason for choosing open access mode is that our purpose is to show performance improvement by proving less handover occurrence probability.

With closed access mode users are registered to femtocells in advance so that only permitted users can get service from registered femtocells. In this case when users not permitted to access a femtocell enter into coverage of the femtocell, handover is not allowed and significant interference will occur bringing deterioration in quality of communication. Instead users registered to femtocell are guaranteed with enough bandwidth. Two access modes are mixed to form hybrid access mode. For example, in a department store where many people exist, this method can be useful. Employees will be registered to a femtocell and as long as there remains accessible bandwidth anyone can also access to the femtocell with minimum bandwidth allocated to them.

3   Proposed Scheme: Adaptive Coverage Adjustment (ACA)

Power configuration method proposed in conventional research fixes the pilot power as initially set using Eq. (1) even if there is only one user actively communicating with the femtocell, thereby the coverage of the femtocell also remains fixed. Which implies that leakage of the pilot power to outer region is likely to happen when users in active mode exist inside the region. This leakage might lead to unwanted handover twice, from macrocell to the femtocell and back again in a very short period of time. Handover event in open access mode occurs when any UEs connected to macrocell approach close to a femtocell and the received signal from the femtocell is 4 dB stronger than that received from the macrocell for over 500 ms. When there exists strong signal leakage outside the house, users just walking outside the house will easily receive such strong signal from femtocells, and handover to the femtocell and back to macrocell again will happen. To prevent such situation in advance, we propose an adaptive coverage adjusting method even when active users exist in the femtocell. In ACA, even when active users are in the house, the femtocell operates self-optimization and adjusts the power and coverage to alleviate power leakage and detailed description of the algorithm is

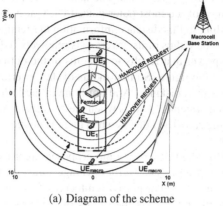

(a) Diagram of the scheme

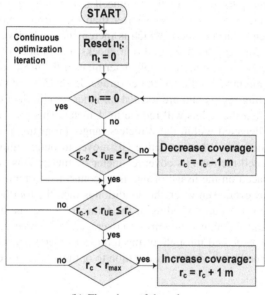

(b) Flowchart of the scheme

Fig. 3. Proposed scheme: Adaptive Coverage Adjustment

written below. With our proposed scheme, the coverage of femtocell is guaranteed to support enough quality of service and the probability of overall handover event will be reduced.

Fig. 3(a) depicts the scenario of our proposed scheme, adaptive coverage adjustment. It shows a situation where one UE is in active mode, namely UE_2 out of the three femtocell users within the household. UE_{mac} is a user equipment connected to the macrocell that approaches the femtocell. As the received signal at UE_{mac} from the femtocell becomes 4 dB stronger than that from the macrocell, it sends handover request message to the macro base station then it requests to the femto base station to perform handover procedure [5]. Receiving the handover request message, the femto-cell measures the distance of the most outer placed UE that is in active mode (in this scenario, UE_2). According to this distance information the femtocell reduces the pilot power iteratively so that the radius of its coverage is relatively reduced meter by meter, until it reaches certain power that results to a radius of the dotted circle, as indicated with the dotted arrow in the figure. So that radius of the coverage will become about 1 m longer than the measured distance. And until one of the following two conditions are met, the coverage is maintained. One is when the active user moves farther from the femtocell maintaining the call. The other is when there is no more handover request for certain period of time t.

Fig. 3(b) shows a flowchart describing the proposed algorithm more precisely by showing the conditions and actions taken for coverage adjustment. r_{UE} is the distance from the femtocell to the most outer placed femtocell UE. r_c is the radius of current coverage of the femtocell and r_{c-i} is the radius of the coverage reduced by i meters from r_c (i.e., the pilot power of the femtocell is reduced by certain level, until its radius of coverage becomes i meter shorter than the current radius). n_t is number of handover events during time t, which is incremented whenever a handover request message is received. The first condition checks the number of handover event occurrence. The second condition iteratively determines if the most outer placed UE is within 2 meters from current radius of the coverage. If all UEs are not placed within the 2 meters inspecting region and are located on inner region, then the femtocell reduces its power so that the radius will reduce by 1 meter. This process continues until an active UE is detected within the 2 meters range. From this point on, the third condition starts to operate. When an active UE moves to outer region of the current coverage, the femtocell recovers its coverage meter by meter. This is because in this paper, priority is placed on the usage of the users connected to femtocell, rather than macrocell users. However, even when the condition is met, the fourth condition determines whether or not the current radius of the femtocell coverage has exceeded the maximum radius, and maintains the current coverage. The whole process explained in this paragraph is executed throughout the femtocell operation, which is the self-optimization performed by the femto base station to optimize the power and the coverage of them[1].

4   Performance Evaluation

We consider a real residential scenario located in London with 130 femtocells deployed and develop a system-level simulator using C language to show performance improvement of the network when our proposed algorithm is applied.

4.1   Scenario Description

Fig. 4 shows the chosen region in London for our simulation [3, 6-8]. The map shows an area of 500m × 500m covered by a single macrocell sector in a dense residential

---

[1] Our proposed algorithm may also use signal to interference ratio (SINR) to adjust power of femtocells.

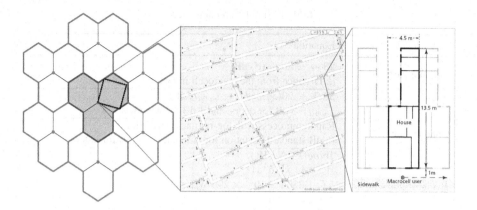

Fig. 4. Residential area for the simulation

area. Each square placed on the middle map denotes a femtocell deployed within a house. This area consists of many densely packed rows of compact houses which all have similar layouts as shown in form of floor plan in Fig. 4. These houses are located very close to footpath where macrocell users pass by. Moreover, they all have windows in the front room directly facing the footpath, which leads to a significant signal leakage since wall loss through a window is less than a wall. These conditions provide a worst case scenario of a co-channel femtocell deployment.

Femtocells are deployed in 10 percent of all houses in the area of our scenario. This is approximately equivalent to half of all the subscribers of a market-leading UK mobile operator deploying a femtocell in their homes [9]. This is about 130 femtocells deployed within this area. The simulation was performed with the femtocells located from 1 to 14 meters away from the footpath. Each house that uses femtocell is assumed to have at least one user at all times, and the number of users in a femtocell is limited to 4. Each of them has a 100 milli-Erlang voice traffic call model. It is also assumed that they follow an evening mobile television (TV) usage model derived from the mobile TV trial results [10]. TV sessions last on average 24 minutes, with a 16 percent busy hour mobile TV session attempt. The voice call model of a macrocell user forms an exponential distribution call duration with a mean of 100 seconds. They are modeled to pass this region starting from a random entry point and moving to a random exit point, at a speed of 1 meter per second. To simulate the handover probabilities, parameters are used as shown in Table 1. In our simulation we assumed that the pilot power is 1/10 of data transmission power for both macrocell and femtocell, which is common practice in UMTS. It is assumed that a handover is triggered when a pilot signal of a new cell is 4 dB stronger than that from the current cell for a time of 500 milliseconds (ms). When a handover event occurs, the time of the handover procedure is assumed to be 650 ms.

For all the power adjusting methods introduced in section 2 and 3, the probability of handover occurrence was measured using simulations. A total number of 100 iterations were performed per call with random distribution of the femtocells within a sector of the central macrocell.

Table 1. Simulation Parameters

Models	Equation
Outdoor path loss	$28 + 35log_{10}(d)$ dB, d: Distance from the macrocell in meters
Indoor path loss	$38.5 + 20log_{10}(d) + L_{walls}$ dB, L_{walls}: 3 dB for doors, and 1 dB for windows
Shadow fading	Random process with log-normal distribution (8 dB standard deviation for outdoor, and 4 dB for indoor).
Receiver noise power	$10log(kTNFW)$, W: 3.84×10^6 Hz kT: $1.3804 \times 10^{23} \times 290$ W/Hz NF_{femto}: 7 dB NF_{macro}: 4 dB
Antenna gain	$G(\theta)dB = G_{max} - min[12(\frac{\theta}{\beta})^2, G_s]$, β: 70/180, angle where gain pattern is 3 dB down from the peak G_s: 20 dB, sidelobe gain level in dB G_{max}: 16 dB, maximum gain level in dB

4.2 Results of Handover Impact

The result of the handover probability is shown in Fig. 5. The handover probability of macrocell user per call using 'fixed power configuration' scheme, shows an average of 65.45 percent. This result shows that it is not feasible to use this scheme in practical situation, due to very high handover probabilities. We assume that there is a 2 percent probability that a handover leads to a call drop. This means that each macrocell call has an approximately 2.61 percent chance of call dropping due to handovers resulting from the deployment of the femtocells. For 'auto power configuration', the handover probability per call is 22.21 percent. It has improved significantly compared to the previous scheme. Using the same assumption as in the previous scheme description, 0.89 percent call drop probability is obtained. Using 'dynamic power control', average handover probability per call is 8.44 percent. 0.34 percent of call drop probability is derived from the result. This scheme shows a significant improvement compared to the two previous methods. However, there still remains probability of call drop occurrence.

Our proposed scheme shows a result of 1.84 percent of handover probability, which results in 0.07 percent call drop probability, very close to zero. The reduced numerical value not only shows reduced number of call drops, but also the substantial decrease in signaling overhead required to support the handover procedures. The comparison of call drop results is shown in Table 2. With the results of the first three schemes, they show the necessity of auto-configuration abilities to provision the femtocells properly as described before. With our proposed scheme, an importance of self-optimization can be derived. By performing continuous optimization process, co-channel network with femtocells deployed together can be maintained and managed.

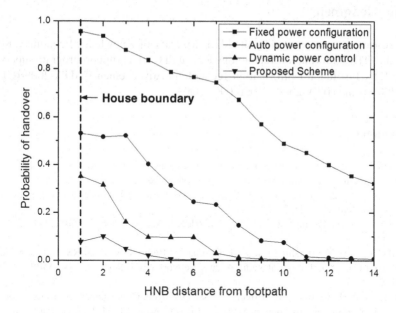

Fig. 5. Handover probability

Table 2. Call drop probabilities for our proposed scheme compared to three previous schemes

Scheme	Call drop probability
Fixed power configuration	2.61 percent
Auto power configuration	0.89 percent
Dynamic power control	0.34 percent
Proposed scheme	0.07 percent

5 Conclusion

Since femtocells are not manually deployed by operators, they require auto-configuration abilities for self provisioning of their power and cell size. Moreover, service providers are not capable of controlling all femtocells manually throughout their operation. Therefore self-optimization is required for femtocells to operate well with macrocells and other femtocells. We proposed a more efficient mechanism of controlling the pilot power of the femtocell to decrease macrocell user's handover probability due to leakage of femtocell power. Our proposed algorithm uses the distance of an active femtocell UE to control the pilot power. This leads to a significant reduction of the handover probability as well as the call drop probability compared with conventional power controlling schemes.

Acknowledgment

"This research was supported by the MKE(Ministry of Knowledge Economy), Korea, under the ITRC(Information Technology Research Center) support program supervised by the IITA(Institute for Information Technology Advancement)"(IITA-2009-(C1090-0902-0046)) and (IITA-2009-C1090-0902-0005).

References

1. Femto Forum, http://www.femtoforum.org/femto/
2. 3rd Generation Partnership Project, http://www.3gpp.org/
3. Claussen, H., Lester, H., Louis, G.S.: An Overview of the Femtocell Concepts. Bell Labs Technical Journal 13(1), 221–246 (2008)
4. 3rd Generation Partnership Project: HNB/HeNB Access Control. 3GPP TD S2-090733 (2009)
5. 3rd Generation Partnership Project: General packet Radio Service (GPRS) enhancements for Evolved Universal Terrestrial Radio Access Network (E-UTRAN) access. 3GPP TS 23.401 (2009)
6. Lester, T.W.H., Claussen, H.: Effects of User-Deployed, Co-channel Femtocells on the Call Drop Probability in a Residential Scenario. In: 18th Annual IEEE International Symposium on Personal, Indoor and Mobile Radio Communications, Athens, pp. 1–5 (2007)
7. Claussen, H.: Performance of Macro- and Co-channel Femtocells in a Hierarchical Cell Structure. In: 18th Annual IEEE International Symposium on Personal, Indoor and Mobile Radio Communications, Athens, pp. 1–5 (2007)
8. Claussen, H., Lester, T.W.H., Louis, G.S.: Self-Optimization of Coverage for Femtocell Deployments. In: Wireless Telecommunications Symposium, California, pp. 278–285 (2008)
9. United Kingdom, Office of Communications (Ofcom), The Communications Market 2006, Annual Communications Market Report (August 2006), http://www.ofcom.org.uk/research/cm/cm06/
10. Arqiva and O2: Oxford Mobile TV Trial Final Results, http://www.dvb-h.org/PDF/060626.Oxford-Final-Results.pdf
11. Hobby, J.D., Claussen, H.: Deployment Options for Femtocells and Their Impact on Existing Macrocellular Networks. Bell Labs Technical Journal 13(4), 145–160 (2009)

New Modeling for Traffic Engineering in FMC Environment

Takashi Satake

NTT Service Integration Laboratories
Nippon Telegraph and Telephone Corporation
9-11 Midori-cho 3-chome, Musashino-shi, Tokyo 180-8585, Japan
satake.takashi@lab.ntt.co.jp

Abstract. Unassumed teletraffic patterns, such as those caused by extra-high bit-rate mobile user traffic that is actively moving in a network, will become common on telecommunication networks in the 4G era. Unexpectedness and unevenness of their distribution will make it difficult to use "fundamental traffic," which is the conventional concept of reference traffic used in Japan, based on the E.500 ITU-T series, for network resource dimensioning. For 4G, new radio access technologies are being intensively developed, but the research on traffic engineering is still insufficient. In this paper, a new modeling and magnification factor for reference traffic to handle such cases is proposed. It is called "redundancy under general traffic in a closed network." Its calculation method approximated by using a normal distribution is described, and its accuracy on various network topologies is studied by Monte Carlo simulations.

Keywords: Traffic engineering, 4G, FMC.

1   Introduction

The facilities needed for desired communication quality in the Japanese telecommunication network are conventionally calculated using the fundamental traffic[1] based on the E.500 ITU-T standardization series [1]. The most important statistics determining the fundamental traffic are the following two:

- The traffic in the busiest hour of the day
- The average traffic in the 30 busiest hours of the year

In the old public switched telephone network (PSTN), the daily, weekly, and yearly variation patterns are considered to be almost identical. This is because telephone use in the old network was largely restricted to conversation, and statistical processing like that used to determine the fundamental traffic played a role of a filter that could remove noise and still be able to reflect trends. A good

---

[1] The fundamental traffic is defined in the telecommunications facilities rules in Japanese law.

C.S. Hong et al. (Eds.): APNOMS 2009, LNCS 5787, pp. 231–240, 2009.

balance between cost and performance could be obtained by using knowledge of the average traffic on the 30 busiest hours in a year, and knowing the busiest hour of the day reduced the amount of measurement work and statistical processing.

However, determining the busiest hour in the 4th generation (4G) mobile network is difficult because large and unevenly distributed traffic flows occur suddenly in that environment[2]. Furthermore, the measurement and analysis of the traffic statistics requires the use of complicated procedures such as packet analysis. In facilities calculation engineering, we use time series data provided relatively easily by traffic measurement functions such as the management information base (MIB) in network elements.

Network measurements that were conscious of traffic have been reported and analyzed in Refs. [2] and [3]. In addition, Ref. [3] has reported the remarkable recent growth of unusual traffic such as P2P traffic in the Internet in Japan and has pointed out the necessity of considering this traffic when designing future networks.

The introduction of the next-generation network (NGN) [4] will require not only a simple traffic-handling technique that ensures communication quality and the fairness of shared facility costs but also a common standard to be based on a clear rule that each carrier or service provider can use to calculate the amount of facilities needed for the traffic.

Previously, a reference traffic model for P2P traffic was proposed [5]. The focus was on the link traffic dispersion due to so-called heavy users being unevenly distributed in a network, and a network traffic model that reflects the uneven distribution of increasing traffic load was proposed. In this paper, following the model in Ref. [5], we present examples of reference traffic on the next-generation fixed and mobile convergence (FMC) networks.

2   Traffic Modeling and Metric Definition

2.1   Modeling Network Traffic

We model network traffic as follows.

- In a closed network with a definite topology linking m edge nodes, we fix the number of users N.
- There are only two kinds of users: heavy users n_h and light users n_l. That is, $N = n_l + n_h$[3].
- (Reflecting uneven distribution) Each user is distributed independently on each edge node pair in the network. .
- (Adjustment of unevenness #1) We assume the heavy user ratio $r_h = n_h/N$.
- (Adjustment of unevenness #2) We assume a traffic magnification T_h such that light user traffic: heavy user traffic = $1 : T_h$.

---

[2] It is pointed out in Ref. [3] that the time of the daily peak in Internet traffic shifts year by year.

[3] Modeling by two kinds of users is often seen in recent network performance evaluations, for example, in Refs. [6][7][8].

2.2 Definition of "Redundancy under General Traffic in a Closed Network"

We define the necessary bandwidth $B_L(r_h)$ of each link L in the closed network by using the following procedures.

1. Each user's traffic is generated on random edge node pairs.
2. The traffic distribution of each link is defined.
3. The 100α percentile value of the traffic distribution of the link is considered the necessary bandwidth B_L.

Here we normalize this value by the average traffic $T_L(r_h)$ of the link. That is,

$$R_L(r_h) \equiv \frac{B_L(r_h)}{T_L(r_h)} \tag{1}$$

and

$$\hat{R}_L \equiv \max_{r_h} R_L(r_h). \tag{2}$$

We call this $\hat{R}_L$ redundancy under general traffic in a closed network.

2.3 Analytical Derivation of Approximated $\hat{R}_L$

We derive the $\hat{R}_L$ defined in Eq. (2) from considerations of basic probability theory and statistics.

Traffic distribution in link L. From the viewpoint of a link in the network, the situation in which each user is distributed on random edge node pairs in a closed network is equivalent to when "the traffic of each user passes the link L with a probability p_L." Therefore, if we assume the number of users occupying link L to be a random variable x, x follows binominal distribution $\mathbf{B}(N, p_L)$ of the number of users N in the entire network and the success probability p_L (the probability of user appearance per link). That is, the probability distribution function $f(x)$ that expresses the probability $P(X = x)$ is

$$f(x) = {}_N C_x \cdot p_L^x \cdot (1 - p_L)^{N-x}. \tag{3}$$

Generally, if the number of samples (the number of trial times) is large, the binominal distribution is hard to calculate but can be approximated by a normal distribution[4]. Because the main consideration in this paper is carrier networks,

---

[4] When we assume $q \equiv 1 - p$, the binominal distribution $\mathbf{B}(n, p)$ can be approximated by the normal distribution $\mathbf{N}(np, npq)$ if at least one of the following conditions is satisfied in practice:

- $\min(np, nq) > 10$.
- $0.1 < p < 0.9$ and $npq > 5$.
- $npq > 25$.

N can be supposed to be large enough. Hereafter we assume $f(x)$ with the normal distribution $\mathbf{N}(Np_L, Np_L(1 - p_L))$.

Decision of link L bandwidth. The number of light users per link obeys the distribution of $\mathbf{B}(n_l, p_L) \approx \mathbf{N}(n_l p_L, n_l p_L(1 - p_L))$, and the number of heavy users per link obeys the distribution of $\mathbf{B}(n_h, p_L) \approx \mathbf{N}(n_h p_L, n_h p_L(1 - p_L))$. In addition, from the network modeling described in subsection 2.1, we know that the heavy user traffic is T_h and the light user traffic is 1. Because we can suppose that the choice between the edge nodes is independent for each user, we know from the additivity of normal distributions that the traffic distribution per link can be considered to be

$$\mathbf{N}\left((n_l + T_h n_h)p_L, (n_l + T_h^2 n_h)p_L(1 - p_L)\right). \tag{4}$$

In this paper, we simplify the analysis by using the Williams approximation [9][5] for the cumulative distribution function of normal distribution. As for $\mathbf{N}(\mu, \sigma^2)$,

$$P(-x \le X \le x) = \sqrt{1 - \exp(-\frac{2}{\pi}(\frac{x - \mu}{\sigma})^2)}. \tag{5}$$

Now we assume x_α with the value of x giving a 100α percentile value $(0.5 < \alpha < 1)$, and to simplify the expression of the numerical formula,

$$A(\alpha) \equiv \sqrt{-\frac{\pi}{2}\log(1 - (2\alpha - 1)^2)}. \tag{6}$$

Transforming Eq. (5) and substituting (6) , we thus get

$$x_\alpha = \mu + \sigma A(\alpha). \tag{7}$$

We can decide the bandwidth of the link on the basis of this x_α. From the assumption in subsection 2.1, $N = n_h + n_l, r_h = n_h/N$. Substituting Eq. (4) into (7), we get

$$B_L(r_h) = N(1 + (T_h - 1)r_h)p_L$$
$$+ \sqrt{N(1 + (T_h^2 - 1)r_h)p_L(1 - p_L)}A(\alpha). \tag{8}$$

Derivation of $\hat{R}_L$. We know from Eq. (4) that the average traffic $T_L(r_h)$ in a link L with the heavy user ratio r_h is

$$T_L(r_h) = (n_l + T_h n_h)p_L. \tag{9}$$

Substituting Eqs. (8) and (9) into (1), we obtain

$$R_L(r_h) =$$
$$1 + \frac{A(\alpha)}{\sqrt{N}} \frac{\sqrt{1 + (T_h^2 - 1)r_h}}{1 + (T_h - 1)r_h} \sqrt{\frac{1 - p_L}{p_L}}. \tag{10}$$

---

[5] The absolute error of the Williams approximation is $|\varepsilon(x)| \, 3.2 < \times 10^{-3}$[9].

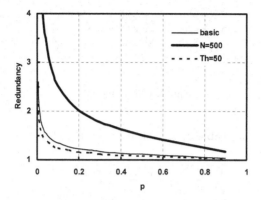

Fig. 1. Redundancy $\hat{R}_L$ vs. p_L curves calculated using Eq. (12)

We assume r_h such that $R_L(r_h)$ takes the maximal (the maximum) with $\hat{r_h}$:

$$\hat{r_h} = \frac{1}{1 + T_h}. \tag{11}$$

Then, when we substitute Eq. (11) into (10), we get

$$\hat{R}_L = R_L(\frac{1}{1 + T_h})$$
$$= 1 + \frac{A(\alpha)}{\sqrt{N}} \frac{1 + T_h}{2\sqrt{T_h}} \sqrt{\frac{1 - p_L}{p_L}}. \tag{12}$$

2.4   Interpretation of Eqs. (11) and (12)

In Fig. 1, we show $\hat{R}_L$ vs. p_L curves calculated using Eq. (12) and the parameters $N = 10000$, $T_h = 100$, and $\alpha = 0.99$ as the basic condition. We also show the corresponding curves calculated when changing one parameter from the basic condition ($T_h = 50$ or $N = 500$). We can see that $\hat{R}_L$ decreases when p_L increases. From this, we confirm that p_L is a metric of the traffic collection effect in a closed network. That is, it is an index of the economy of scale. We also see that $\hat{R}_L$ decreases if N increases, and from this we can understand that N represents a grain of the user traffic in a closed network in which each particle becomes relatively small if N increases. On the other hand, $\hat{R}_L$ increases if T_h increases, and from this we can understand that T_h expresses a difference of the 2-dimensional call class and also represents a grain of the user traffic in which a relatively large particle is included in the network if T_h increases. $\hat{R}_L$ also increases if α increases[6].

Finally, we see from Eq. (11) that $\hat{r_h}$ is a parameter depending on only T_h, and when $T_h \gg 1$, $\hat{r_h}$ is approximately equal to the reciprocal of T_h.

---

[6] From the analysis of Eq. (6), $\frac{dA(\alpha)}{d\alpha} > 0$. Hence from Eq. (12), $\frac{\partial \hat{R}_L}{\partial \alpha} > 0 \, (0.5 < \alpha < 1)$.

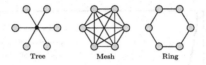

Fig. 2. 6Node models

Table 1. $\hat{R}_L$ calculated by Eq. (12)

Topology	Basic	$N = 500$	$T_h = 50$
6Tree	1.2543	2.1374	1.1816
6Mesh	1.6125	3.7391	1.4374
6Ring	1.2275	2.0173	1.1624

3   Examples

3.1   Simple Topology

We exemplified numerical confirmation for the approximate calculation that we used when deriving $\hat{R}_L$ in subsection 2.3 and considered the properties of $\hat{R}_L$ through Monte Carlo simulations of various numerical examples in typical network topologies by using the Crystal Ball®2000 [10]. For comparison, we unified threshold α of the bandwidth of 0.99 in the following calculations.

First, $\hat{R}_L$ was calculated by Eq. (12) under the condition with the parameters used in subsection 2.4 and simple 6-node network topologies (Fig. 2); tree[7], mesh, and ring topologies were assumed. p_L in each topology of 6 nodes is given by the basic graph theory as follows.

$$p_{Tree}(6) = \frac{1}{6} \approx 0.1667$$

$$p_{Mesh}(6) = \frac{1}{6(6-1)} \approx 0.0333$$

$$p_{Ring}(6) = \frac{6+2}{8(6-1)} = 0.2$$

The results are shown in Table 1. Next, Monte Carlo simulations were carried out with the same parameters. The simulation results of $R_L(r_h)$ as a function of heavy user ratio r_h are shown in Fig. 3. That the maximal to show $\hat{R}_L$ in each case agreed with the calculated value in Table 1 was confirmed by observation.

---

[7] We inserted an intermediate node purely for relaying traffic in which user traffic does not occur in the 6-edge node tree network.

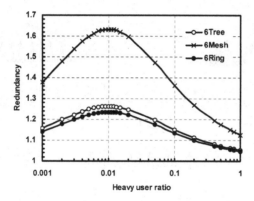

Fig. 3. Monte Carlo simulation results showing $R_L(r_h)$ as a function of r_h (the basic condition; 6 node)

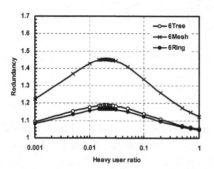

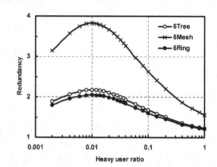

Fig. 4. $R_L(r_h)$ in 6Node models with $T_h = 50$

Fig. 5. $R_L(r_h)$ in 6Node models with $N = 500$

Regarding $\hat{r_h}$. From Eq. (11), $\hat{r_h}$ in the basic condition is given by

$$\hat{r_h} \approx 0.0099.$$

$\hat{r_h}$ of all topologies obeyed this calculated value in the simulation results (Fig. 3). Moreover, it was observed that $\hat{r_h}$ shifted approximately twice when only T_h from the basic parameters was reduced to $1/2$ (Fig. 4). From this, it was confirmed that $\hat{r_h}$ depends on only T_h being equivalent with Eq. (11).

Influence of N. We show the case where only N is reduced from the basic condition in Fig. 5. Increase of $\hat{R_L}$ was observed in all topologies. From this, N expresses a kind of traffic grain as mentioned in subsection 2.4, and it was confirmed that $\hat{R_L}$ is influenced by N. At the same time, it was confirmed that the change of N does not affect $\hat{r_h}$ being equivalent with Eq. (11).

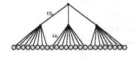

Fig. 6. Double tree a (24DTa)

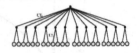

Fig. 7. Double tree b (24DTb)

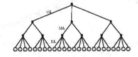

Fig. 8. Triple tree (24TT)

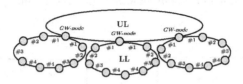

Fig. 9. Double ring (24DR)

3.2   Complex Topology

We show examples of cases of topologies with two or more layers: double tree a (24DTa) (Fig. (6)), double tree b (24DTb) (Fig. 7), triple tree (24TT) (Fig. 8), and double ring (24DR) (Fig. 9) linking between 24 edge nodes[8].

Inspection of $\hat{R}_L$. We assumed the basic condition with $N = 10000$, $T_h = 100$ as above. Then, we calculated p_L considering the network topologies. The detailed derivation is in Ref. [5]. With these parameters, $\hat{R}_L$ in the case of complex topologies was calculated from Eq. (12), as shown in Tables 2 and 3.

Simulation results are shown in Figs. 10 and 11. 24-UL1 represents UL of 24DTa and 24TT, and 24-LL represents LL of 24DTa, 24DTb, and 24TT. (Furthermore, 24-UL2 represents UL of 24DTb and 24TT-ML represents ML of 24TT.) In the case of complex topologies, it was confirmed by observation that the maximal value to show $\hat{R}_L$ agreed with each calculated value.

Table 2. $\hat{R}_L$ calculated by Eq. (12) in 24Node models

$\hat{R}_L$	24DTa	24DTb	24TT	24DR
Upper layer	1.2070	1.3865	1.2070	1.1557
Middle layer	-	-	1.2763	-
Lower layer	1.5455	1.5455	1.5455	See Table 3.

---

[8] We also inserted intermediate nodes purely for relaying user traffic.

Table 3. $\hat{R}_L$ calculated by Eq. (12) in LL of 24DR

$\hat{R}_L$	LL#1	LL#2	LL#3	LL#4
Lower layer	1.2891	1.3319	1.3963	1.5116

By the above, we exemplified that the redundancy under general traffic in a closed network, $\hat{R}_L$, and $\hat{r}_h$ can be calculated by Eqs. (11) and (12), respectively, if the parameters N, T_h, p_L, and α are fixed.

3.3 Consideration

Through the examples on simple and complex topologies, it was confirmed that $\hat{R}_L$ does not contradict the effects of an "economy of scale" from a qualitative view.

When we compared $\hat{R}_L$ under the same parameters on the same scale in the simple topologies (the number of edge nodes m is identical), $\hat{R}_L$ was the smallest in the case of the ring topology — the economy of scale was large — and was second smallest in the case of the tree topology. In the case of the mesh topology, $\hat{R}_L$ was the largest (Figs. 3–5).

In complex topologies, traffic stays below the middle layer and the number of substantial users decreases in the upper layer. Thus, the higher the layer in which the traffic is accumulated and the fewer the links in that layer, the smaller the $\hat{R}_L$ — the economy of scale was large (Fig. 10). In the case of the lower layer of a ring topology, the smaller the $\hat{R}_L$ of a link, the nearer that link is to the gateway node (connected to the upper layer) in which the traffic accumulates from the lower layer (Fig. 11).

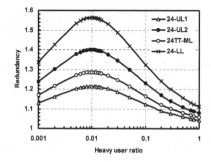

Fig. 10. $R_L(r_h)$ in 24DTa, 24DTb and 24TT

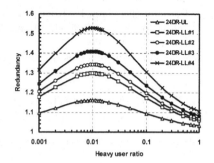

Fig. 11. $R_L(r_h)$ in 24DR

4   Conclusion

We focused on dispersion of traffic distribution caused by unevenly distributed mobile heavy users in a closed network and modeled the network traffic that reflected the environment where burst load and uneven distribution increase. We also defined a new magnification factor for reference traffic for network resource dimensioning that is independent of the old fundamental traffic measurement obtained using the busy hour of traffic.

Acknowledgements

The author would like to thank Mr. Yoshiyuki Chiba and Mr. Tsukasa Okamoto for their contributions to the early part of the study, and Dr. Haruhisa Hasegawa and Mr. Hideaki Yoshino for their many helpful suggestions.

References

1. ITU-T (former CCITT) Recommendation E.500(rev.1) (1992)
2. Sen, S., Wang, J.: Analyzing Peer-To-Peer Traffic Across Large Networks. IEEE/ACM Trans. Netw. 12(2), 219–232 (2004)
3. Cho, K., Fukuda, K., Esaki, H., Kato, A.: The Impact and Implications of the Growth in Residential User-to-User Traffic. In: Proc. of ACM SIGCOMM 2006, September 2006, pp. 207–218 (2006)
4. Morita, N., Imanaka, H.: Introduction to the Functional Architecture of NGN. IEICE Trans. Commun. E90-B(5), 1022–1031 (2007)
5. Satake, T., Chiba, Y., Okamoto, T.: Reference Traffic for Network Resource Dimensioning under Unexpected and Unevenly Distributed Traffic Load. IEICE Trans. Commun (in Japanese) J91-B(5), 734–745 (2008)
6. Piccolo, F., Neglia, G., Bianchi, G.: The Effect of Heterogeneous Link Capacities in BitTorrent-like File Sharing Systems. In: Proc. of 1st International Workshop on Hot Topics in Peer-to-Peer Systems, Hot-P2P 2004 (2004)
7. Clevenot, F., Nain, P., Ross, K.: Multiclass P2P Networks: Static Resource Allocation for Service Differentiation and Bandwidth Diversity. Performance Evaluation 62, 32–49 (2005)
8. Liao, W., Papadopoulos, F., Psounis, K.: Performance Analysis of BitTorrent-like Systems with Heterogeneous Users. Performance Evaluation 64, 876–891 (2007)
9. Williams, J.D.: An Approximation to the Probability Integral. Ann. Math. Statist. 17, 363–365 (1946)
10. Oracle Inc.: Crystal Ball ®, http://www.crystalball.com/

Experience in Developing a Prototype for WiBro Service Quality Management System

Dae-Woo Kim[1], Hyun-Min Lim[1], Jae-Hyoung Yoo[1], and Sang-Ha Kim[2]

[1] Network R&D Laboratory, KoreaTelecom, 463-1 Jeonmin-dong, Yuseong-gu,
Daejeon, 305-811, South Korea
{daewoo,hmlim,styoo}@kt.com
[2] Department of Computer Science, ChungNam National University, 220 Gung-dong,
Yuseong-gu, Daejeon 305-764, South Korea
shkim@cnu.ac.kr

Abstract. This paper describes development of KQIs (Key Quality Indicators) and a prototype of the KQIs monitoring system for WiBro (Wireless Broadband Internet) service provided by Korea Telecom. In this paper, we analyze WiBro service model based on user's perspective. Particularly, we essentially deal with the Web browsing service experienced by users as an internet service by WiBro Service. In addition, we focus on data bearer service with a wireless feature since it is the foundation of the WiBro Service. In developing the KQIs and KPIs, we use a top-down analysis method with user's perspective and a bottom-up analysis method to get the data sources to be measured. We categorize them to Accessibility, Retainability, Quality, Latency, Supportability, and Diagnosis as combined KQIs with user's perspective. Through this service model analysis, we identify the KQIs and KPIs based on the user to manage the WiBro service quality. And we analyze the gap between the developed quality indicators and the measurement of them in KT and implement the prototype of monitoring system for them.

Keywords: Wireless Broadband Internet Service, Key Quality Indicators, Key Performance Indicators, Service Model Analysis, Service Quality Management.

1 Introduction

As wireless markets mature and competition intensifies, an increasingly common expectation is a guaranteed quality of service, especially, in the high end of the market, for which mobile communications is rapidly becoming a vital business tool. In this competitive environment, Customer Experience Management is becoming a fundamental requirement for mobile network operators and service providers. Wireless service providers and network operators have to be able to monitor and manage service quality in order to bridge the gap between customers' perception of quality of service and the traditional methods for managing network performance. Also, with the growth of mobile services, it has become very important for an operator to measure the QoS and QoE of its network accurately and improve it further in the most effective and cost-efficient way to achieve customer loyalty and maintain competitive

C.S. Hong et al. (Eds.): APNOMS 2009, LNCS 5787, pp. 241–250, 2009.

edge. Therefore, we need service quality indicators to monitor and manage the service quality based on the customer's experience and a monitoring system for them.

In this paper, we deal with the service model analysis for WiBro Service to develop KQIs (Key Quality Indicators) and KPIs (Key Performance Indicators) to monitor and measure the service quality. In particular, we focus on the Web browsing service and data bearer service in the WiBro Service, because these are basic internet service and function with a wireless feature. For reference, WiBro service (Wireless Broadband Internet Service) is a service provided by KT (Korea Telecom) that is a wireless broadband internet technology developed by the Korean telecom industry. WiBro is the Korean service name for the IEEE 802.16e (mobile WiMAX) international standard [1]. To analyze the service model, we used the KQI development method from GB 923[2, 3]. And we identified and presented the KQIs and the KPIs for Web browsing service and data bearer in the WiBro Service through the service model analysis. Through these KQIs and KPIs, the operator could identify the root cause of why the service application performance is not met, or relate QoE measurements to other metrics collected in the network. And we reviewed if the KQIs and the KPIs could be measured in the network as gap analysis between the developed quality indicators and the measurement of them, and described the experience of developing the KQIs for WiBro Service and a prototype for monitoring them.

The rest of this paper is organized as follows: Section 2 gives a brief overview of the WiBro Service. Section 3 presents KQI development for Web browsing and data bearer service. Section 4 describes the categorization of KQIs. Section 5 presents experience of implementation of WiBro Service Quality Management System. Finally, we conclude our work and outline some future issues in Section 6.

2 Overview of WiBro Service

Based on All-IP, KT's WiBro service uses 2.3 GHz frequency, OFDM (Orthogonal Frequency Division Multiplex) technology and UICC (Universal Integrated Chip Card) authentication. The maximum downlink speed is 18.4 Mbps, and the average downlink speed per subscriber is 3 Mbps. The maximum uplink speed is 4 Mbps, and the average uplink speed per subscriber is 1.2 Mbps. KT's WiBro service via broadband is now available, beginning in Seoul and the surrounding metropolitan area. Building on its broadband success, KT has now launched the 'KT WiBro' service via broadband. As this service can be used anytime and anywhere within the service region, even while in motion, the key service types include Web surfing type, Platform type, Communication type, Life type, Fun type [4]. Generally, most of users of WiBro Service use the Web surfing type service. Therefore, in this paper, we basically focus on the Web surfing type service among the key services of WiBro service.

3 KQIs Development for WiBro Service

To develop KQIs and KPIs for Web browsing service as a basic service of WiBro Service, at first, we use a top-down method to analyze user's perspective for it and then use a bottom-up method to get the available data source as KPIs. According to the top-down method described in GB923 [2,3], we analyze the service scenario, the service delivery architecture, and the user service transaction flow on the service delivery architecture. From the analyzed service transaction flow, we will identify the Quality Indicator candidates. As the data bearer is being covered regardless of service type in service layer, we will handle the Web Browsing Service Usage Scenarios and User Tasks. In addition, we will concentrate on laptop users, since they constitute the majority of KT WiBro customers. Through this procedure, we developed the KQIs for the Web browsing service and data bearer of the WiBro service.

3.1 Analysis of Web Browsing Service Transaction Flow

When a request is sent through an action of the user, a set of transactions are applied to KT's WiBro service delivery architecture. In this set of transactions, in order to identify the customer-based metrics that capture the customer's perception of quality of service, customer service usage scenarios need to be mapped into the transaction flow under the WiBro service delivery architecture.

Thus, we need to study a customer service usage scenario step–by-step based on the passage of time, which is described below:

- A customer connects to the WiBro network, and a connection is successfully established.
- The customer accesses the WiBro web portal located in the KT server via his device, and the initial page is displayed.
- The customer selects news content from the initial web page, and news content provided by a content provider is displayed.
- The customer selects other news content while he or she moves around, and the content provided by a content provider is displayed.
- The customer does not use his or her device for a while, and returns home.
- The customer selects sports news content, and the content provided by a content provider is displayed.

Each of the actions from the individual timelines can be broken down to a discrete customer action, and it is important to use these to identify the KQIs in the WiBro service delivery chain. A high-level transaction flow diagram associated with the customer actions has been analyzed, and is shown in Figure 1.

From the transaction flow for the user's actions over WiBro service, KQIs have been identified in the matrix as follows : Network Entry Success/Failure Rate, Content Access Success/Failure Rate, Security Content Access (DRM (Digital Rights Management) related) Success/Failure Rate, Handover Success/Failure Rate, Content Download Success/Failure Rate, Idle Mode Exit Success/Failure rate.

Aside from the success/failure rate of key transactions, the key request and response timings have been identified as well such as Network Entry Latency, Content Download Time, Handover Latency, Idle Mode Entry Latency, Paging Latency (network re-entry), Idle Mode Exit Latency (network re-entry).

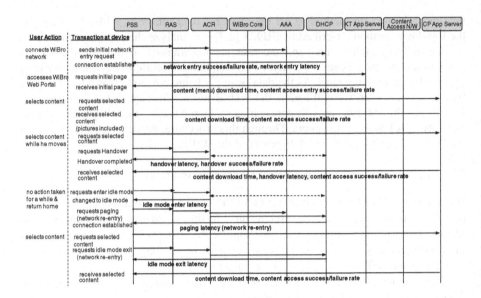

Fig. 1. Overall Transaction Flow Diagram for User Actions in the Web Browsing Service

The transactions described in Figure 1 are broken down to discrete system actions such as Network Entry Transaction Flow [1,5], Handover Transaction Flow [1,5], Web Browsing Transaction Flow [7]. In addition, these are analyzed by being broken down to detail discrete system actions. Through this analysis, the related KQIs have been identified like the KQIs related to the success/failure rate from user's transaction flow analysis and the KQIs related to the request and response timing. The detail KQIs are being handled in Section 4.

Apart from network engaged service transaction, from the point of view of service provider, we take billing component into consideration. There are the packet charging transaction flow between AAA and PPS (Pre-Paid Server) and the content charging transaction flow between back-end (service) platform and PPS for billable content charging. Also, in order to identify RF-based metrics, it is worthwhile to look at Air Interface Overview [8] and ITU-R 8F/1079-E [9]. Through these documents, which are focused on throughput and RF power (signal power, interference). As such, we can identify the KQI and KPI candidates by the analysis of transaction flows and the wireless features.

3.2 Analysis of Data Source

In order to monitor the identified KQIs, we need to analyze the data sources such as network element management systems, network management systems, operations support systems, and etc. From these systems, we can get the performance data related to the quality of service. The related data sources in KT include WiBro NMS, IP Premium NMS, WiBro EMS, IPAS, WKMS, NeOSS SA, and Service Server Management System. The provided information is shown in Table 1. The systems as data sources would be used when a SQM (Service Quality Management) system is made for WiBro service.

Table 1. Data Sources

Data Source	Provided Information
WiBro NMS	The performance data of the repeaters and L2 switches
WiBro EMS	The performance data for wireless access section including RAS, ACR
IP Premium NMS	The performance data from network elements of the core network.
IPAS	The quality of service of the core network
WKMS	The information of user terminal with UICC
NeOSS SA	The problem handing history of the customer
ICIS	The information of service provision for the customer
Service Server Management System	The performance data from the service servers related to application service.

4 Categorization of the KQIs

End-users tend to think of quality in very non-technical terms. Typical user concerns could be voiced like the customer perception in Figure 2. These types of non-technical concerns have been translated into a set of measurable quality factors by telecommunications standards bodies. Therefore, the KQI and KPI candidates can be mapped to the service quality categories perceived by customer, which are based on ITU-T E.800 [10] and they are tailored in accordance with WiBro characteristics.

After the categorization of the Quality Indicators, we can create a combined KQI that is representative of the related KQIs, as shown in Figure 2. Some KQIs could be classified as KPIs if they can be measured in the network element. As a result, the

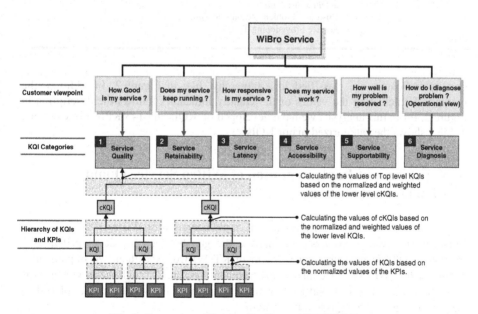

Fig. 2. Categories of Service Quality and KQI/KPI Tree

Table 2. Service Quality Categories and KQI candidates

Category	KQI candidates
Service Quality	• RF(Radio Frequency) Quality of RAS/Repeater • RAS/ Repeater Throughput • Quality of Transport Network • DSA(Dynamic Service Addition)/DSC(Dynamic Service Change)/DSD(Dynamic Service Deletion) Failure Rate • Quality of PSS
Service Retainability	• Abnormal Session Termination Rate • Inter Sector/RAS/ACR HHO(Hard Hand Over) Failure Rate • Inter Sector FA(Frequency Assignment) / RAS FA / ACR FA HHO Failure Rate • Location Update Failure Rate
Service Accessibility	• Network Entry Failure Rate • Content Access Failure Rate • PKM(Privacy Key Management) Failure Rate • Registration Failure Rate • DHCP Success Rate • PPS(Pre-Paid Server) Accessibility
Service Diagnosis	• Wireless Network Elements • Access Network Elements • PE(Provider Edge) Network Elements • WiBro Core Network Elements • Service Servers
Service Latency	• Web Browsing Latency • Handover Latency • Idle Mode Exit Latency • Network Entry Latency
Service Supportability	• Time from receipt of fixed-antenna-installation request to complete fixed-antenna-installation • Customer Problem Resolution Time • Service Provisioning Time • Billing Accuracy

Quality Indicators are classified in a hierarchy, like in Figure 2. In addition, the method of calculating the values of the Quality Indicators follows the description in Figure 2. From this analysis, we have developed 68 KQIs and 143 KPIs. For example, Table 2 shows the categorized major KQIs.

5   Experience of Implementation of WiBro Service Quality Management System

In this section, we describe experience in implementation of WiBro Service Quality Management System. At first, we analyze the gap between the developed quality indicators and the measurement of them in KT WiBro. And then, we implement a prototype for WiBro service quality management system. Through these, we check if the system with the developed KQIs is useful to manage the service quality of WiBro and discuss about the KQIs and the system.

5.1   Analysis of Gap between the Developed Quality Indicators and the Measurement of Them

In order to analyze the gaps between the developed quality indicators and the measurable data in KT WiBro network, it is worth looking at the service delivery chain where we can measure them. In addition, to capture the user perceived experience, the service quality is able to be divided into two levels. One is the "individual level" – the quality experienced by an individual customer, and the other is the "aggregated level" – the aggregation of the quality experienced by all customers or customer regional grouping [3]. Basically, the perceived quality is the value judgment actually made by a customer based on a combination of expectations and experience. Hence if we can capture service quality data of the individual level, then it is possible to distinguish between expected quality and actual experienced quality, so it would be the better way to use the individual level data as much as we can.

Basically, each level has its own benefits and drawbacks. The drawbacks of the individual level KPIs are mostly around cost, implementation efforts and usually what a lot of operators end up doing is a mixture of the two to get an overall adequate customer experience view.

At the moment, the individual level KPIs with regard to network connection and handover are not available at proactive data source of KT. It is because ACR does not generate the individual level log. However KT are trying to make it possible by requesting the ACR vendor, so the individual level KPIs with regard to network connection and handover may be available at WiBro EMS or other data source in future.

The analyzed gaps are:

- Visibility on metrics related to the use of the protocol between RAS and ACR supplied by the vendor.
- Metrics related to the individual level network connection establishment information for Accessibility and Retainability to the AAA and Application Service Servers like WiBro Portal and WiBro specific applications.
- Metrics related to the individual level handover information for Accessibility and Retainability on WiBro EMS.
- Metrics which allow the measurement and quality assurance of the content being provided.
- Validation of the content which requires the appropriate measurement criteria which can report on its accuracy.
- Metrics at the WiBro device end to provide insight into overall Web browsing latency, Handover latency, and Handover success rate close to the end user so that we can measure that network performance against the customer perceived service quality.

As the result of the gap analysis, the 21% of the KPIs on individual level and the 66% of the KPIs on aggregation level could be measured in KT.

5.2   Implementation of a Prototype for Service Quality Management System

To implement the prototype of the WiBro SQMS(Service Quality Management System), we use the data of the aggregation level KPIs from the related systems as data sources in sub-section 3.2. Because the implemented system was a prototype, we did not connect the related systems to SQMS in direct. Instead of it, we built the data base with the data like being connected with the related systems. The requirements of WiBro SQMS's functions referred to eTOM(Enhanced Telecom Operations Map) model[11,12] from TM Forum. The detail function requirements for WiBro SQMS are shown on Table 3.

Table 3. The detail functions for WiBro SQMS

Functions	Detail functions
Monitoring Service Quality	• Collecting the data from probes and the related systems • Managing rules of data filters for events and alarms • Managing the period of interworking with the related systems
Analyzing Service Quality	• Managing the relationship of service components • Translating the collected data to the KPIs and KQIs according to the formula of KPIs and KQIs • Control of collecting the service quality data
Improving Service Quality	• Modifying the formula of KPIs and KQIs in real time • Performing the planned activity for restoring service quality
Reporting Service Quality Performance	• Providing the tree view for the dependency of the service • Providing the customized GUI for users • Delivering the information of the service quality degradation to another processes
Creating Service Performance Degrada-tion Report	• Monitoring the status of service performance degradation (that is, monitoring if the service quality values cross the thresholds) • Creating the report for service performance degradation • Removing the duplicated data by rules
Tracking & Managing Service Quality Per-formance Resolution	• Retrieving the detail events for service configuration components • Managing and retrieving the history for the service quality degradation
Closing Service Per-formance Degradation Report	• Retrieving the progress about the cause analysis of service quality degradation, restoration and improvement on service quality
Managing Service Quality Modeling	• Managing the properties for the service components and data sources • Managing the dependency of the service components • Editing, adding and removing the KPIs and KQIs • Managing the service quality model and Retrieving the detail components • Converting the raw quality data to KPIs and KQIs

In this system, the user can choose and register the KQIs, and KPIs which are inter-ested in to the system. The view of the system is shown in Figure 3. The description of the numbers in Figure 3 is as follows: the number 1 is information of customer, the number 2 is tree of the KQIs and KPIs, the number 3 is current and previous status of the KQI or KPI, the number 4 shows the architecture of service and the components, the number 5 indicates the number of threshold violations of the quality indicator related to the component, the number 6 shows the report list for the service quality

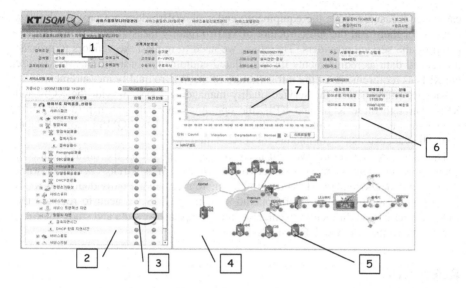

Fig. 3. View of WiBro Service Quality Monitoring

degradation, the number 7 shows the value variation of a KQI or a KPI on the time axis. With the SQMS in Figure 3, we could monitor the KQIs and KPIs for WiBro service quality and trace the cause components when the service quality is degraded.

But because the KQIs and KPIs are from the aggregation level data, we could not trace the customer's service quality correctly. In the case of service quality degradation happened by a lowering of the related KQIs' values, the operators check all of the related KQIs with his experience to find the root cause. Therefore, we need the rules to find the root cause with KQIs and KPIs. Also, we cannot find and trace the customer's service quality because we cannot get the individual level data from the data source. The data cannot be provided from the related systems in real time. In addition, the related systems such as EMS, NMSs, and so on are difficult to provide the data to the SQMSin real-time because they need time to collect the data and have burden to provide the data in real-time. As above, the SQMS with aggregation level data can monitor the service quality over all. But without the individual level data, SQMS cannot deal with the customers' complaints about the service quality.

6 Conclusion

In this paper, we developed KQIs and KPIs for WiBro service quality management by analyzing the service scenario, the service delivery architecture, the user's action transaction flow, the detail transaction flows such as network entry, handover, web browsing, and the charging transaction flows on service provider, in addition, the wireless features as top-down approach on user's perspective. And then, we analyzed the data sources to check if the KQI and KPI candidates are able to be measured in practice as the bottom-up approach. But the most of individual level KQIs and KPIs cannot be measured since the equipment vendor of WiBro does not provide the data sources for them and the infrastructure of WiBro quality measurement is not be prepared by the service providers. Therefore, we need to request the vendors to provide the related data as requirements for the related equipments. In addition, we will use the KPIs as requirements of the infrastructure of the WiBro quality measurement.

And we implemented the service quality management system with the KQIs and KPIs. Through these KQIs and KPIs, the operator can easily get grip of the entire service quality situation. But without the individual level data, SQMS is not sufficient to handle the customer's complaints for service quality.

Therefore, in near future, we will need to study to construct the infrastructure of quality measurement for respective network sections such as wireless access network section and backbone network section to be able to measure the individual level KQIs and KPIs. In addition, through the operation of the SQMS we need to regulate the KQIs and KPIs' weight factors to calculate an appropriate formula and study the rules to trace the root cause with KQIs and KPIs.

References

1. IEEE Standard 802.16e, IEEE Standard for Local and metropolitan area networks Part 16: Air Interface for Fixed and Mobile Broadband Wireless Access Systems-Amendment2: Physical and Medium Access Control Layers for Combined Fixed and Mobile Operation in Licensed Bands and Corrigendum 1 (Feburary 2006)
2. TM Forum, Wireless Service Measurements Handbook, GB923 version 3.0 (March 2004)
3. TM Forum, SLA Management Handbook, Volume 2 Concepts and Principles, GB917-2 version 2.5 (July 2005)
4. KT WiBro, http://www.ktmobile.co.kr
5. TTA TTAS.KO-06.0082/R1, Specifications for 2.3GHz band Portable Internet Serviec-Physical & Medium Access Control Layer (December 2005)
6. Hong, D.: 2.3GHz Portable Internet (WiBro) for Wireless Broadband Access, ITU-APT Regional Seminar (2004)
7. IETF RFC 2126, Hypertext Transfer Protocol – HTTP/1.1
8. Li, K.-H.: IEEE 802.16e-2005 Air Interface Overview, WiMAX Solutions Division, Intel Mobility Group (June 5, 2006)
9. TU-R 8F/1079-E, WiMAX Forum, Additional technical details supporting IP-OFDMA as an IMT-2000 terrestrial radio interface (December 2006)
10. ITU-T E.800, Telephone Network and ISDN Quality of Service, Network Management and Traffic Engineering – Terms and Definitions Related to Quality of Service and Network Performance Including Dependability (August 1994)
11. TM Forum, GB921 D, Enhanced Telecom Operations Map(eTOM) The Business Process Framework For The Information and Communications Services Industry Addendum D: Process Decompositions and Descriptions, Release 6.0 (November 2005)
12. TM Forum, GB921 D, Enhanced Telecom Operations Map(eTOM) The Business Process Framework For The Information and Communications Services Industry Addendum D: Process Decompositions and Descriptions, Release 6.0 (November 2005)

BDIM–Based Optimal Design of Videoconferencing Service Infrastructure in Multi–SLA Environments

Hao Shi, Zhiqiang Zhan, and Dan Wang

State Key Laboratory of Networking and Switching Technology, Beijing University of
Posts and Telecommunications, Beijing 100876, China
Haure.Shi@gmail.com, zqzhan@bupt.edu.cn, Dannier.Wong@gmail.com

Abstract. For a videoconferencing service provider, the infrastructure
design in multi-SLA (Service Level Agreement) environments plays a
pivotal role in the maximization of business profits. Traditional design
methods usually concern costs exclusively but overlook financial losses.
In this work, an optimal design methodology is proposed using Business-
Driven IT Management (BDIM) concepts, whereby numbers of servers,
routers and link bandwidth can be determined to minimize the sum of infrastructure costs and business losses. This method captures the linkage
between infrastructure and business impacts by modeling availability,
multi-SLA loads and transmission delay. A numerical example is discussed to demonstrate the value of the approach. A further conclusion is
that the optimal design can be used for dynamic service provisioning.

1 Introduction

The videoconferencing service is one of the most common IT services. In the past,
IT service providers aimed at reducing Total Cost of Ownership (TCO) only
when designing infrastructure according to Service Level Agreements (SLAs)
[1]. In effect, the ultimate expenses IT service providers paid were far beyond
original intention, due to they neglected the business losses incurred by SLA violations which may be very huge [2]. Therefore, IT service providers should also
quantify business losses, which can be guided by Business-Driven IT Management (BDIM) concepts. BDIM has been defined as *"the application of a set of
models, practices, techniques and tools to map and to quantitatively evaluate in-
terdependencies between business performance and IT solutions—and using the
quantified evaluation—to improve the IT solutions' quality of service and related
business results"* [3,4,5,6,7]. In this work, we will develop infrastructure cost and
business loss models to find the optimal infrastructure design based on BDIM.

The remainder of this paper is organized as follows: Sect. 2 formally clears
the problem to be solved; Sect. 3 proposes the optimal design after analyzing
infrastructure characteristics, multi-SLA loads and transmission delay; Sect. 4
considers an application of the method through a numerical example; conclusions
are provided in Sect. 5.

C.S. Hong et al. (Eds.): APNOMS 2009, LNCS 5787, pp. 251–260, 2009.

2   Optimal Design Problem Description

This section gives the scenario of videoconferencing IT infrastructure, then formally describes the optimal design problem.

In this context, we define the videoconferencing service (VCS) infrastructure as IT resources, or entities, in a connection among each other. Moreover, multi-SLA environments are the situations in which an IT service provider delivers VCS to multiple customers with a single set of IT infrastructure. Typically, the VCS infrastructure exhibits a star network architecture with each client branch being a child node and the IT service provider being a central node. Figure 1 illustrates a functional, static viewpoint of a single node in this star schema.

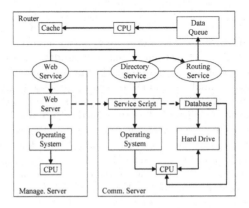

Fig. 1. A Single Node Scenario

The management server serves as a web server to provision web services, while the communication server plays the role of a directory server. The router is responsible for storing and forwarding loads within threshold time. In this paper, we only consider servers, routers and communication links, due to the fact these items cost much more than video cameras and other trivial entities.

From the view of BDIM, the IT infrastructure design confronts a compromise problem. On the one hand, the infrastructure availability and performance will improve as its redundancy arises; but TCO will increase as well. On the other hand, TCO will decrease when the redundancy drops; however, the availability and performance will also deteriorate, which will further induce greater probability of SLA violation incidents and thus more losses. Therefore, the optimal infrastructure design is the one that can achieve the goal of minimizing the sum of TCO and business losses, which can be formally posed as Tab. 1.

In this table, the delay D_{e2e} is the end-to-end video transmission delay, and X denotes the number of SLAs. The load-balanced machines enable the infrastructure to process input loads while the standby machines provide the required availability. In following sections, we will investigate $C\left(\Delta T\right)$ and $L\left(\Delta T\right)$ in multi-SLA conditions.

Table 1. Formal Optimal Problem Description

Find:	N_{ms} : the total number of Manage. servers M_{ms} : the number of load-balanced Manage. servers N_{cs} : the total number of Comm. servers M_{cs} : the number of load-balanced Comm. servers N_{r} : the total number of routers M_{r} : the number of load-balanced routers B : the bandwidth of communication links
By Minimizing:	$C(\Delta T) + L(\Delta T)$ the total financial impacts on the business over ΔT time
Subjecting to:	$N_{\mathrm{ms}} \geq M_{\mathrm{ms}}, \ N_{\mathrm{cs}} \geq M_{\mathrm{cs}}, \ N_{\mathrm{r}} \geq M_{\mathrm{r}}, \ B > 0,$ $D_{e2e} \leq$ threshold and constraints of X SLAs

3   Problem Formalization

The VCS infrastructure costs equal the aggregation of each individual part expense, while business losses amount to the monetary penalty caused by SLA violations which can be either of: 1) infrastructure is unavailable; 2) transmission delay exceeds the threshold. The availability has to do with IT entity dependencies, and the transmission delay correlates closely with both video loads and infrastructure. Thus, let us examine availability, multi-SLA load characteristics and end-to-end transmission delay in sequence.

3.1   The VCS Availability Model

At present, it is extremely hard to propose an universal strategy to model dependencies of any IT infrastructure [8,9]. Based on VCS features, we choose E-R diagram to depict IT entities and their dependencies, as shown in Fig. 2.

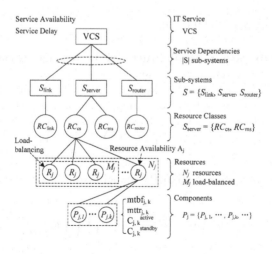

Fig. 2. The Dependencies of VCS Entities

From the figure, VCS depends on three subsystems—server (S_s), router (S_r), and link subsystems (S_l). Hence, VCS is available if and only if all three subsystems are ready to use. According to multiplication principle, the VCS availability is equivalent to three subsystem availability's product: $A_s A_r A_l$. In actual practice, Operating Level Agreements (OLAs) between operators and providers specify the communication link availability A_l. Now let us focus on A_s and A_r.

The Server Subsystem Availability (A_s). The server subsystem includes two types of resource classes: RC_{cs} and RC_{ms}. Typically, RC_j is a cluster made up of N_j identical, independent servers, M_j of which are in load-balanced status while the rest ($N_j - M_j$) are running in standby mode. Each individual resource R_j in RC_j is composed by a component set $P_j = \{P_{j,1}, \cdots, P_{j,k}, \cdots\}$. All the components must be operational for the resource to be operational. For instance, a single RC_{cs} may be constituted by hardware, operating system and communication software components. The composition of resource RC_{ms} resembles RC_{cs}, which is omitted in Fig. 2.

Now we proceed to evaluate A_s based on the method proposed in [6]. In the light of combination theory and standard availability theory, the availability of servers in the central node yields: $A = \prod_{j \in RC} A_j$, where A_j is RC_j availability. A resource class is available when M_j out of N_j are available. So we have $A_j = \sum_{k=M_j}^{N_j} \left[\binom{N_j}{k} \left(A_j^R \right)^k \left(1 - A_j^R \right)^{N_j-k} \right]$, where A_j^R represents R_j availability and can be obtained from Mean-Time-To-Repair (MTTR) and Mean-Time-Between-Failures (MTBF) values, which concludes: $A_j^R = \prod_{k \in P_j} \left[\frac{mtbf_{j,k}}{mtbf_{j,k} + mttr_{j,k}} \right]$. For simplicity, we assume without loss of integrity that standby machines are needed only in the central node. Finally, the server subsystem reliability is:

$$A_s = \prod_{j \in RC} \left[\left(\sum_{k=M_j}^{N_j} \left[\binom{N_j}{k} \left(A_j^R \right)^k \left(1 - A_j^R \right)^{N_j-k} \right] \right) \prod_{i=1}^{M} \left(\prod_{k \in P_j} \frac{mtbf_{j,k}}{mtbf_{j,k} + mttr_{j,k}} \right) \right] \quad (1)$$

where M is the number of child nodes.

The Router Subsystem Availability (A_r). A single resource class RC_r forms the router subsystem. For an element $R \in RC_r$, it is also a component set $P = \{P_1, \cdots, P_k, \cdots\}$. Suppose there are N_j routers, and M_j out of them operate in load-balanced mode. It is required that at least M_j must be operational simultaneously for the router subsystem to be available. Thus we conclude the router availability of the central node: $\sum_{k=M_j}^{N_j} \left[\binom{N_j}{k} (A_R)^k (1 - A_R)^{N_j-k} \right]$, where A_R is a single router availability which equals $\prod_{k \in P} \frac{mtbf_k}{mtbf_k + mttr_k}$. As a result, the router subsystem availability model is obtained:

$$A_r = \left[\sum_{k=M_j}^{N_j} \left[\binom{N_j}{k} (A_R)^k (1 - A_R)^{N_j-k} \right] \right] \cdot \prod_{i=1}^{M} \left(\prod_{k \in P} \frac{mtbf_k}{mtbf_k + mttr_k} \right) \quad (2)$$

3.2   Characterizing Multi–SLA Loads

This part makes clear statistical characteristics of multi-SLA loads, providing the basis of next subsection.

Multi-SLA loads can be evaluated from three aspects. Firstly, all the X SLAs are not necessarily, exactly the same. The more branches an organization has, the more traffic loads it creates. Secondly, clients usually don't hold video meetings concurrently due mainly to meeting time preferences and time zone differences. Finally, differentiated videoconferencing instances last diverse duration. Based on these features, we can use Tab. 2 to specify multi-SLA loads.

Table 2. Example Multi-SLA Loads

Duration	SLA 1	SLA 2	$\cdots$	SLA X
08:00 − 10:00	23%	18%	$\cdots$	29%
10:00 − 12:00	42%	31%	$\cdots$	12%
14:00 − 16:00	26%	30%	$\cdots$	15%
16:00 − 18:00	9%	21%	$\cdots$	44%
Availability	97.2%	98.2%	$\cdots$	98.5%
Delay (ms)	98	95	$\cdots$	90
ϕ_a (\$/h)	1680	1830	$\cdots$	1980
ϕ_d (\$/h)	1420	1560	$\cdots$	1750
Branches	10	11	$\cdots$	7

The "Duration" represents time frames when clients have meetings, and the "SLA" values corresponding to time slots stand for the possibility that meetings fall into them. The next four rows indicate Service Level Objectives, while the last one describes numbers of branches. The data in this table can be calculated from conference memoranda or service log files using data mining techniques.

Assume t denotes a certain frame. There may be 0 up to X instance(s) running in t with distinct probabilities. Here we use $P_{k,t}$ to describe the probability the kth client holds video conferences in frame t, i.e., the SLA values corresponding to the duration in Tab. 2. Therefore, the probability of no instance is $\prod_{k=1}^{X}(1 - P_{k,t})$, one instance of $\sum_{k=1}^{X}\left[\prod_{j=1}^{k-1}(1 - P_{j,t})\right]P_{k,t}\left[\prod_{j=k+1}^{X}(1 - P_{j,t})\right]$, and so on. Equation (3) summarizes the general case.

$$P_t^j = \sum_{i=1}^{|C|}\left[\prod_{m\in C_i}P_{m,t}\cdot\prod_{n\notin C_i}(1 - P_{n,t})\right], \quad C = \left\{\binom{X}{j}\right\} \qquad (3)$$

The set C consists of all possible j-out-of-X combinations (C_i). The symbol P_t^j expresses the probability j VCS instances are running in time frame t, and $j \in [0, X]$. Assume there are I_t instances during t. The mathematical expectation of I_t can thus be expressed as: $E(I_t) = \sum_{j=0}^{X}j\cdot P_t^j$. Finally, the quantity of target loads is the least integer larger than or equal to the maximum $E(I_t)$:

$$I = \text{CEIL}\left[\max_{t\in T}\{E(I_t)\}\right] \qquad (4)$$

where T represents the set including all time periods.

3.3   The VCS End-to-End Transmission Delay Model

Numerous studies have concluded for natural hearing VCS delay should be approximately 100 ms [10]. A great many factors contribute to the delay, such as video capture, compression, transmission, recovery, etc. Here we focus on two principal elements—the time spent in routers and propagation [11,12]. Figure 3 presents the details of the delay spent in a single-hop router.

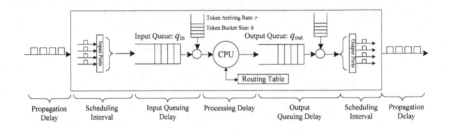

Fig. 3. The Composition of Delay Spent in a Single–hop Universal Router

The propagation delays in Fig. 3 equals the largest data packet length divided by the link bandwidth. The scheduling interval is the time spent by a rate-based scheduler to serve an incoming packet. It corresponds to the transmission time of a packet served with the reserved rate R according to the Fluid Flow Model [13]. The Simple Token Bucket (STB) filter [14] is used for traffic shaping. A bucket holds up to b tokens. For each sent data unit, one token is consumed while new tokens are generated into the bucket with filling rate r. When the bucket is empty, arriving packets are simply discarded. Consequently, the total end-to-end transmission delay in a network consisting of K hops under multi-SLA environments can be given by following model:

$$D_{e2e} = \max_{1 \leq j \leq K} \left[\frac{b}{R} + I \cdot \sum_{j=1}^{K} \left(\frac{L_j}{R} + \frac{L_j}{\gamma_j} + P_{r,j} \right) \right] \tag{5}$$

where j represents the jth hop in total K hops. The variable I consists with the target load in (4). The symbol L_j indicates the packet length of the IP data flows including Ethernet headers, and γ_j stores the jth link bandwidth. In addition, $P_{r,j}$ is the propagation delay.

3.4   The Infrastructure Cost and Business Loss Models

This subsection aims to set up infrastructure cost and business loss models on the foundation of the infrastructure availability and delay models given in Sect. 3.1 and Sect. 3.3.

The Infrastructure Cost Model. Resources in standby mode usually cost lower than those in load-balanced, such as equipment depreciation, license cost, electricity fee, etc. For the server subsystem, the total costs can be obtained by adding the cost of every component $P_{j,k}$ in each resource class. Let $C_{j,k}^{act}$ and $C_{j,k}^{sta}$ denote the $P_{j,k}$ cost under load-balanced and standby mode respectively, then the costs consumed by servers in period ΔT is: $C_s(\Delta T) = \Delta T \cdot \sum_{j=1}^{|RC|} \left(\sum_{l=1}^{M_j} \sum_{k=1}^{|P_j|} C_{j,k}^{act} + \sum_{l=1}^{N_j-M_j} \sum_{k=1}^{|P_j|} C_{j,k}^{sta} \right)$, where j runs over all resource classes, l runs over resources and k runs over all components.

For the subsystem S_r, several identical resources constitute a resource class RC_r. Thus we have $C_r(\Delta T) = \Delta T \cdot \left(\sum_{l=1}^{M_j} \sum_{k=1}^{|P_j|} C_{j,k}'^{act} + \sum_{l=1}^{N_j-M_j} \sum_{k=1}^{|P_j|} C_{j,k}'^{sta} \right)$, where $C_{j,k}'^{sta}$ and $C_{j,k}'^{act}$ individually represent the cost of a single router in standby and load-balanced mode.

Generally speaking, the communication link cost is a function of an independent variable—bandwidth B. Accordingly, the overall infrastructure cost consumed in period ΔT, i.e., $C(\Delta T)$ in Tab. 1, is finally acquired:

$$C(\Delta T) = C_s(\Delta T) + C_r(\Delta T) + C_l(B, \Delta T) \tag{6}$$

The Business Loss Model. This part gives the business loss model from the view of BDIM.

First, SLA penalty will take effect when the infrastructure fails to work. Suppose the penalty rate is ϕ_a in this case. The business loss caused by unavailability is $L_A = \Delta T \cdot \phi_a \cdot (1 - A_s A_r A_l)$. Second, penalty will also be activated if the transmission delay exceeds the threshold, though the service may be available. Under this circumstance, let penalty rate be ϕ_d, then we have loss: $L_D = \Delta T \cdot \phi_d \cdot (A_s A_r A_l) \cdot P_{d>100}$, where d is the D_{e2e} in (5), and $P_{d>100}$ is the probability the delay goes beyond 100 ms. The thresholds are specified in SLAs and may be other values rather than 100 ms. Thus the probability P_d can be noted as $\{1, d > 100; 0, d \leq 100\}$. Hence, the entire loss the IT service provider suffers is:

$$L(\Delta T) = \Delta T \cdot [\phi_a (1 - A_s A_r A_l) + \phi_d (A_s A_r A_l) P_d] \tag{7}$$

So far, we have induced the infrastructure cost model (6) and the business loss model (7), which means the connection between IT infrastructure and business values has been established. In conclusion, the optimal infrastructure design problem stated in Tab. 1 is finally solved:

$$\text{Optimal Design} = \min [C(\Delta T) + L(\Delta T)] \tag{8}$$

4   A Numerical Example

The purpose of this section is to evaluate the proposed method through a complete example. We use Tab. 2 and Tab. 3 together to set required parameters, which are typical for recent technology. In Tab. 3, tuples (a, b, c) represent the values for a single management server, communication server, and router.

Table 3. Parameters for Example Scenario

Parameters	Values
ΔT	1 month
$C_{j,k}^{act}$ ($/month)	hw = $(1400, 1400, 1550)$ os = $(270, 250, 320)$ as = $(180, 170, 0)$
$C_{j,k}^{sta}$ ($/month)	hw = $(1200, 1200, 1300)$ os = $(190, 180, 290)$ as = $(160, 150, 0)$
C_l [$/(month · MB/s)]	1200 (MB/s is the bandwidth unit)
$(A_{ms}, A_{cs}, A_r, A_l)$	$(99.81\%, 98.1\%, 99, 89\%, 99.9\%)$ [6] (these values are calculated from appropriate MTBF and MTTR values)

Let us first design the infrastructure in a conventional pattern without business considerations, using a tuple $(N_{ms}, N_{cs}, N_r, M_{ms}, M_{cs}, M_r, B)$ to represent a detailed design. The meaning of each variable has been elaborated in Tab. 1.

The cheapest infrastructure design here must be $(31, 31, 31, 31, 31, 31, 1)$. Obviously, this design cannot meet customer requirements owning to the intolerant delay—156 ms. In order to process video loads in time and guarantee no server, router or communication link are saturated, the design must be $(N_{ms}, N_{cs}, N_r, M_{ms}, M_{cs}, M_r, B)=(32, 32, 32, 32, 32, 32, 3)$. This configuration has availability of 85.2% which is still inadequate. The designer may add one more standby machine in each resource class, generating a design $(33, 33, 33, 32, 32, 32, 3)$, monthly cost of \$185550, delay of 97 ms and availability of 95.1%, where the designer may rest according to the traditional infrastructure design method. Nevertheless, this design is not optimal as demonstrated below.

The fault of the conventional infrastructure design method lies in that business losses are not taken into account. Table 4 compares six infrastructure designs using the BDIM-based design methodology proposed above.

Each line in Tab. 4 represents a special design. The second column gives the design details. The third and fourth columns show the total cost of ownership and business loss (in \$/Month) separately, and the fifth column sums them. The last column indicates how much the business losses are by adopting that

Table 4. Infrastructure Designs Using BDIM

No.	Infrastructure Design	TCO ($/Month)	Loss ($/Month)	TCO + Loss ($/Month)	Cost of Choosing Wrong ($/Month)
1	$(32, 32, 32, 32, 32, 32, 1)$	178,480	4,203,340	4,381,820	3,861,712
2	$(33, 33, 33, 32, 32, 32, 3)$	185,550	4,105,590	4,291,140	3,771,032
3	$(33, 33, 33, 32, 32, 32, 4)$	186,750	3,980,452	4,167,202	3,647,094
4	$(34, 34, 34, 33, 33, 33, 4)$	192,290	327,818	520,108	0 (Optimal)
5	$(35, 35, 35, 34, 34, 34, 5)$	199,030	322,480	521,510	1,402
6	$(36, 36, 36, 35, 35, 35, 6)$	205,770	318,470	524,240	3,132

particular design compared to the optimal one (No. 4) which has availability of 98.4%, delay of 92 ms. The optimal design has the lowest overall cost plus loss. This table apparently demonstrates traditional designs can cost millions of dollars loss per month compared to the BDIM-based optimal design.

Finally, we show how sensitive the optimal design is to variations in video-conferencing loads. Consider Fig. 4 which shows the total cost plus business loss as loads increase from 4 to 12.

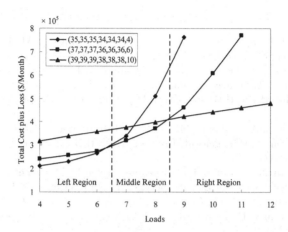

Fig. 4. Loads' Impacts on Total Cost plus Loss

Three broken lines are shown in the figure. The load values are divided into three regions: the left region is from 4 to 6, in which the first design (diamond tag) is optimal; the middle region (7 to 8) has a best design of (37, 37, 37, 36, 36, 36, 6) noted by square tag; the right region (9 to 12) has an optimal design of (39, 39, 39, 38, 38, 38, 10) (triangle tag). Judging from this observation, the optimal design always follows the bottom-most line in any region. Therefore, the business loss model described in this paper can be used for dynamic service provisioning since it seizes appropriate flex points.

5   Conclusions

This paper has proposed a methodology that can be used to design VCS IT in-frastructure based on BDIM in multi-SLA situations. Four kinds of technological metrics are integrated: availability, delay, multi-SLA characteristics and business impacts through a business loss model. The proposed method can find the op-timal videoconferencing infrastructure design by fulfilling the minimization of TCO plus the financial losses on account of infrastructure deficiencies.

Much more work can be undertaken to improve the results. First, a better business loss model including scheduling strategies may be used to approximate reality more faithfully. Second, the star topology should be developed to more generalized structures to reflect other application scenarios. Eventually, server resources may be multiplexed to provision various services, such as disk storage, e-commerce, etc, which will increase IT service providers' risk resistance capacity.

Acknowledgments. This work was supported by the Funds for Creative Research Groups of China (60821001) and 863 Project (2008AA01Z201).

References

1. Ameller, D., Franch, X.: Service Level Agreement Monitor (SALMon). In: Proc. – Int. Conf. Compos.–Based Softw. Syst., ICCBSS (2008)
2. Beal, A., Mossé, D.: From E–business Strategy to IT Resource Management: A Strategy-centric Approach to Timely Scheduling Web Requests in B2C Environments. In: 3rd IEEE/IFIP Int. Workshop Bus.-Driven IT Manage., pp. 89–97 (2008)
3. Sauvé, J., Moura, A., Sampaio, M., et al.: An Introductory Overview and Survey of Business–driven IT Management. In: First IEEE/IFIP Int. Workshop Bus.-Driven IT Manage. (2006)
4. Moura, A., Sauvé, J., Bartolini, C.: Business–driven IT Management – Upping the Ante of IT: Exploring the Linkage between IT and Business to Improve both IT and Business Results. IEEE Commun. Mag. 46(10), 148–153 (2008)
5. Sauvé, J., Marques, F., Moura, A., et al: SLA design from a business perspective. In: Schönwälder, J., Serrat, J. (eds.) DSOM 2005. LNCS, vol. 3775, pp. 72–83. Springer, Heidelberg (2005)
6. Sauvé, J., Marques, F., Moura, A., et al: Optimal Design of E–commerce Site Infrastructure from a Business Perspective. In: Proc. Annu. Hawaii Int. Conf. Syst. Sci., vol. 8 (2006)
7. Moura, A., Sauvé, J., Bartolini, C.: Research Challengers of Business–drvien IT Management. In: Second IEEE/IFIP Int. Workshop on Bus.-Driven IT Manage. (2007)
8. Scheibenberger, K., Pansa, I.: Modelling Dependencies of IT Infrastructure Elements. In: 3rd IEEE/IFIP Int. Workshop on Bus.-Driven IT Manage., pp. 112–113 (2008)
9. Ralha, C.G., Gostinski, R.: A Methodological Framework for Business–IT Alignment. In: 3rd IEEE/IFIP Int. Workshop Bus.-Driven IT Manage., pp. 1–10 (2008)
10. Karlsson, G.: Asynchronous Transfer of Video. IEEE Commun. Mag. 34(8), 118–126 (1996)
11. Ke, C.H., Chilamkurti, N.: A New Framework for MPEG Video Delivery over Heterogeneous Networks. Comput. Commun. 31(11), 2656–2668 (2008)
12. Zhang, X., Li, C., Li, X.: Hybrid Multicast Scheme for Many–to–Many Videoconferencing. In: Proc. – Int. Conf. Netw. Comput. Adv. Inf. Manage., NCM, vol. 2 (2008)
13. Zhang, H.: Service Disciplines for Guaranteed Performance Service in Packet–switching Networks. Proceedings of the IEEE 83(10), 1374–1396 (1995)
14. Tanenbaum, A.S.: Computer Networks, 4th edn. Tsinghua University Press (2005)

Contention Window Size Control for QoS Support in Multi-hop Wireless Ad Hoc Networks

Pham Thanh Giang and Kenji Nakagawa

Department of Electrical Engineering, Nagaoka University of Technology,
Kamitomioka 1603–1, Nagaoka-Shi, Niigata, 940–2188 Japan
{giang@kashiwa,nakagawa@}.nagaokaut.ac.jp

Abstract. IEEE 802.11 MAC protocol for medium access control in wireless *Local Area Networks* (LANs) is the *de facto* standard for wireless ad hoc networks, but it does not necessarily satisfy QoS requirement by users such as fairness among flows or throughput of flows. In this paper, we propose a new scheme aiming to solve per-flow fairness problem and achieves good throughput performance in IEEE 802.11 multi-hop ad hoc network.

We apply a cross-layer scheme, which works on the MAC and link layers. In the link layer, we examine each flow and choose suitable contention window for each output packet from the flow. The aim of the mechanism is achieving per-flow fairness. In our proposed scheme, the MAC layer of each station tries to evaluate network condition by itself then adjust channel access frequency to improve network utilization and ensure per-node fairness. Performance of our proposed scheme is examined on various multi-hop network topologies by using Network Simulator (NS-2).

Keywords: cross-layer, per-flow/per-node fairness, bandwidth utilization, multi-hop wireless, IEEE 802.11.

1 Introduction

In multi-hop ad hoc network, throughput and fairness performances are very important. However, IEEE 802.11 standard [1] does not provide good throughput and fairness for stations in some asymmetry topologies [2,3]. When stations cooperate to forward packets from other stations through the ad hoc network, each station has to transmit both the direct flow, which is generated by the station, and forwarding flows, which are generated by its neighboring stations. The station also shares the channel capacity with its neighboring stations. The contentions in the MAC and link layers affect the performance of the network. Due to the MAC layer contention, the allocated bandwidth for each station cannot ensure per-flow fairness. In the link layer, the direct flow and forwarding flows contend for the buffer space. Obviously, the direct flow gets more advantage than forwarding flows [4,5,6].

C.S. Hong et al. (Eds.): APNOMS 2009, LNCS 5787, pp. 261–272, 2009.

In this paper, we propose a new scheme in order to achieve good fairness in both the MAC and link layers. We also aim to ensure good throughput performance and help disadvantaged flows get better throughput. A modification of the MAC layer will let each station estimate network condition and determine fair bandwidth allocation for them. By tuning back-off time, they can get suitable allocated bandwidth. In the link layer, we use *Round Robin* (RR) queue for each flow, and monitor output packet to ensure per-flow fairness. We use the *Network Simulator* (NS) [7] to evaluate our proposed scheme in some unfairness situations relying on some asymmetry topologies.

The rest of the paper is organized as follows: Section 2 reviews some related works. Section 3 describes our proposed scheme. Section 4 evaluates our proposed scheme by comparing with original FIFO scheduling in IEEE 802.11 standard [1], Shagdar's method [5] and PCRQ scheduling [8]. Finally, Section 5 concludes the paper.

2   Related Work

The fairness performance at the MAC layer has been an active research field in the past several years. The protocols MACA [9] and its extension MACAW [10] use the four-way RTS/CTS/Data/ACK handshake signals to reduce collisions caused by hidden terminals in the network. The protocol MACAW has been standardized in the IEEE 802.11 [1] as *Distributed Coordination Function* (DCF). However, the RTS/CTS scheme in DCF does not solve all unfairness in bandwidth in case of asymmetric links. Several schemes for improving the fairness of MAC protocol have been proposed in the literature [11,12,13]. Moreover, Li et al. [14] investigated *Extended Inter-Frame Spacing* (EIFS) problem, i.e., the fixed EIFS value leads to unfair bandwidth allocation for each stations. They proposed flexible EIFS values based on a measurement of the length of *Sensing Range* (SR) frame. However, the length of SR frame cannot be always recognized because of the spatial reuse of the bandwidth. Moreover, those research mainly consider the MAC layer fairness problem, they do not consider per-flow fairness problem.

Jangeun et al. [4] pointed out the weak point of FIFO scheduling in multi-hop networks and then proposed various queuing schemes. Each scheme has offered different degree of fairness. However, their research is based on the ideal MAC layer fairness assumption that cannot be satisfied. Thus, their schemes do not give good performance in the real networks. Shagdar et al. [5] and Izumikawa et al. [6] also focused on the contention of direct and forwarding flows, and proposed scheduling algorithms by using RR queue. The DCF mechanism are modified in [5] to achieve the bandwidth utilization by sending all the packets at the head of RR queues continuously without delay by back-off algorithm. However, their solutions have the same problem as [4] due to the unsatisfactory assumption that the MAC layer gives fair bandwidth allocation. PCRQ scheduling [8] also uses RR queue with three algorithms to control input/output packets and the turn of RR queue. It achieves good result in per-flow fairness. In PCRQ scheduling, the MAC layer fairness is also improved indirectly because PCRQ scheduling gives a delay for a packet from heavy flows in order to help disadvantaged flows get a chance in using channel bandwidth. However, it slightly degrades the total throughput performance.

3   Proposed Scheme

The reason of unfairness problem in multi-hop ad hoc network is due to both the MAC and link layers contention. In this section, we propose a scheme to improve per-flow fairness in wireless ad hoc networks. In the MAC layer, we make a minor modification to the original DCF channel access mechanism to help each station can achieve suitable allocated bandwidth. In the link layer, we use RR queue with a monitoring algorithm to ensure fair between flows.

3.1   MAC Layer Modification

In IEEE 802.11, the DCF algorithm behaves unfair for throughput and allocated bandwidth for asymmetric topology. To avoid this, our solution tries to obtain distributed fair allocated bandwidth by controlling back-off time. By decreasing back-off time, the channel access frequency of a station is increased and also the given bandwidth for a station will be increased and vice versa.

MAC layer modification lets each station determine the current allocated bandwidth and the fair allocated bandwidth for themselves. If a station finds out its given bandwidth less than its fair allocated bandwidth, it will increase its channel access frequency in order to achieve better allocated bandwidth. To evaluate the current utilization bandwidth of a station, the station measures time, which uses for transferring packets during an evaluating time.

$$Current\_utilization = \frac{Transferring\_time}{Evaluating\_time} \tag{1}$$

To determine a fair allocated bandwidth for a station, the station observes the channel and counts the number of flows in its carrier sensing range. The flows in the transmission range are identified with MAC and IP addresses of both source and destination. The flows, which are out of its transmission range but in its carrier sensing range, are considered as one flow. The station also examines the number of sending flows, which it is generating and forwarding. The ratio of the number of sending flows to the number of total flows in its carrier sensing range should be proportion of fair share bandwidth.

$$Fair\_utilization = \frac{Sending\_flows}{Total\_flows} \tag{2}$$

When a station finds out that the current utilization in (1) is smaller than the fair utilization in (2), it means that is not fair for the station in using bandwidth.

The station will try to increase the channel access frequency by turning its back-off time as the following change of contention window size.

$$CW_{new} = \frac{Current\_utilization}{Fair\_utilization} CW \qquad (3)$$

3.2   Link Layer Modification

In the link layer, the direct flow's queue tends to occupy completely the buffer space and to seize almost bandwidth from forwarding flows. Therefore, the proposed scheme monitors output packets from RR queues to ensure them using bandwidth fairly. A heavy offered load flow is required sending some packets with longer back-off time, and then more bandwidth is left for receiving packets from its neighboring stations. Thus, throughputs of forwarding flows are improved. A packet at the head of the queue for flow i is marked when it is dequeued, at the following probability;

$$P_{i\_marked} = \begin{cases} 0, & \text{if } qlen_i \leq ave \\ \gamma \dfrac{qlen_i - ave}{(n-1)ave}, & \text{if } qlen_i > ave \end{cases} \qquad (4)$$

where γ is an output weight constant. Packets from a heavy offered load flow may be marked with probability in range 0 to γ. In case one queue is full while other queue is empty, packets may be marked with probability γ. If the queue length is smaller or equal to the average queue length, all packets will be dequeued without being marked. When packet is marked, a cross-layer signal is sent to the MAC layer to require send the packet with longer back-off time:

$$CW_{new} = \kappa * CW \qquad (5)$$

where κ is a delay weight constant.

4   Performance Evaluation

We now evaluate the performance of our proposed scheme by comparing with the original FIFO scheduling in IEEE 801.11 standard [1], Shagdar's method [5] and PCRQ scheduling [8] on various topologies of multi-hop wireless ad hoc networks. We use *Network Simulator* (NS-2) [7] for evaluation. The simulation parameters are shown in Table 1. In PCRQ scheduling, we use the same parameters as in [8]: the input weight constant $\alpha = 2.0$, the hold weight constant $\beta = 0.3$, the output weight constant $\gamma = 0.3$ and delay time $\delta = 1[ms]$. In the proposed scheme, we set the output weight constant $\gamma = 0.3$ and the delay weight constant $\kappa = 2$.

The fairness and throughput performance metrics are evaluated.

- Fairness index: We use the fairness index, which is defined by R. Jain [15] as follows:

$$\text{Fairness Index} = \frac{(\sum_{i=1}^{n} x_i)^2}{n \cdot \sum_{i=1}^{n} x_i^2} \qquad (6)$$

Table 1. Parameters in the simulation

Channel data rate	2[Mbps]
Antenna type	Omni direction
Radio Propagation	Two-ray ground
Transmission range	250[m]
Carrier Sensing range	550[m]
MAC protocol	IEEE 802.11b (RTS/CTS is enable)
Connection type	UDP/CBR
Buffer size	100[packet]
Packet size	1[KB]
Simulation time	100[s]

where n is the number of flows, x_i is the throughput of flow i. The result ranges from $1/n$ to one. In the best case, i.e., throughput of all flows are equal, the fairness index achieves one. In the worst case, the network is totally unfair, i.e, one flow gets all capacity while other flows get nothing, the fairness index is $1/n$. In this paper, the fairness index is evaluated based on goodput at the destination station.

– Total throughput: The average of total goodput of all flows in the network.

4.1 Scenario-1

Scenario-1 includes a chain of three stations with two flows. The coordinates of stations are shown in Fig. 1. This topology is also known as *Large-EIFS* problem [14], which is described in Fig. 2. In this scenario, stations S1 and S2 are in one transmission range. Stations S1 and R are also in another transmission range. Stations S2 and R are out of transmission range but in carrier-sensing range. At the last state of four-way handshaking process from sender S1 to receiver R, R sends an ACK frame in reply to a data frame from S1, then S2 detects the ACK frame, but cannot decode it. Thus, S2 must wait an EIFS before accessing the channel, while S1 waits a DIFS, which is much smaller than the EIFS. Li et al. [14] has proved that allocated bandwidth for S1 is four times greater than S2 because of the *Large-EIFS* problem.

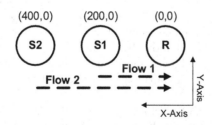

Fig. 1. Scenario-2: The *Large-EIFS* problem

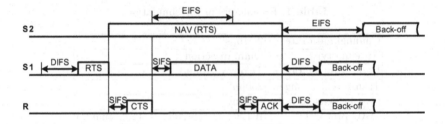

Fig. 2. Unfairness in bandwidth due to *Large-EIFS* problem

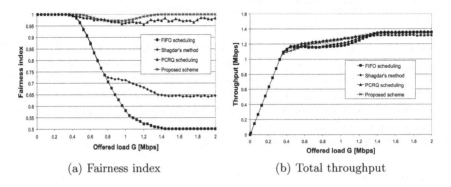

(a) Fairness index   (b) Total throughput

Fig. 3. Simulation results in Scenario-1

We examine network performances in this scenario by letting the stations S1 and S2 generate traffic at the same offered load G to R. The performance metrics are evaluated versus offered load G.

Fairness indices are shown in Fig. 3(a). When the offered load is small, all scheduling methods get perfect fairness index. When the offered load becomes larger, because a common queue is used in FIFO scheduling, the direct flow gradually occupies completely the buffer space then the fairness index becomes very bad. In Shagdar's method [5], even different queue is used for each flow, but the throughput of the forwarding flow is limited by the allocated bandwidth of S2 which is too much smaller due to *Large-EIFS* problem [14]. Thus, the fairness index in Shagdar's method is not good. In PCRQ scheduling [8], input and output packets to RR queues is monitored, the per-flow throughput becomes fairer and also the allocated bandwidth at the MAC layer is improved indirectly. Thus, PCRQ scheduling achieves good fairness index. In the proposed scheme, both the MAC and link layers fairness are considered, we also achieve very good fairness index.

The total throughputs of all flows are shown in Fig. 3(b). When the offered load is small, throughputs in all method are similar. When the offered load becomes greater, PCRQ scheduling uses bandwidth slightly less efficiently than the others do. Because PCRQ scheduling makes a delay for sending to give chance for receiving in order to achieve the MAC layer fairness indirectly. In the proposed scheme, S2 increase its channel access frequency to get the MAC layer fairness. It also improves network utilization and the proposed scheme achieves better throughput than PCRQ scheduling.

4.2   Scenario-2

Scenario-2 is *three-pairs* scenario. The coordinates of stations are shown in Fig. 5. In this scenario, stations S1-S2 and S2-S3 are out of transmission range but in carrier sensing range. Stations S1-S3 are out of carrier sensing range and the two external pairs S1-R1 and S3-R3 are independent from each other. Thus, two external pairs contend bandwidth with only the central pair S2-R2 while the central pair contends with both external pairs. In this topology, the central pair cannot access the medium in saturated state.

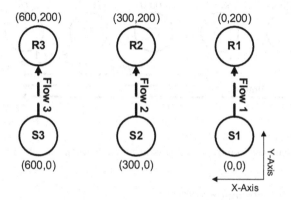

Fig. 4. Scenario-2: three pairs scenario

We examine fairness problem in this topology by letting the stations S1, S2 and S3 generate traffic at the same offered load G to R1, R2 and R3, respectively. The performance metrics are evaluated versus offered load G.

Fairness indices are shown in Fig. 5(a). When the offered load is small, all scheduling methods get perfect fairness index. When the offered load becomes larger, FIFO scheduling does not help the central pair to access the medium. However, both PCRQ scheduling and Shagdar's method also do not have better results than FIFO scheduling. The reasons are explained as follows. PCRQ scheduling only works in the link layer, it does not have information of flows out of transmission range. Therefore, it cannot improve the MAC layer fairness. Shagdar's method works in both the MAC and link layers. It modifies back-off algorithm in order to achieve per-flow fairness and network utilization. However, Shagdar's method cannot ensure the MAC layer fairness and it fails to achieve per-flow fairness in this topology. In the proposed scheme, the central pair finds out that it gets less bandwidth than its fair allocated bandwidth. It will improve

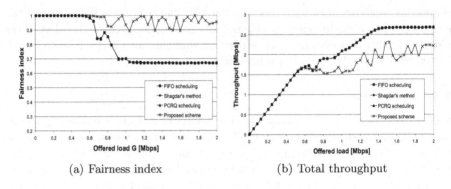

(a) Fairness index   (b) Total throughput

Fig. 5. Simulation results in Scenario-2

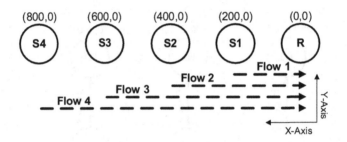

Fig. 6. Scenario-3: a five-station chain with four flows

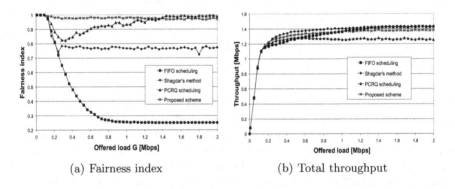

(a) Fairness index   (b) Total throughput

Fig. 7. Simulation results in Scenario-3

the channel access frequency by reducing its back-off time. Thus, the proposed scheme can achieve the good MAC layer fairness and per-flow fairness.

The total throughputs of all flows are shown in Fig. 5(b). When the offered load is small, throughputs in all method are similar. When the offered load

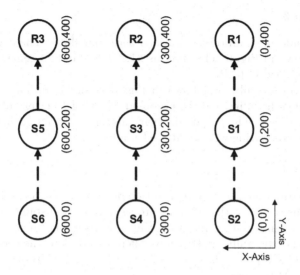

Fig. 8. Scenario-4: Grid scenario

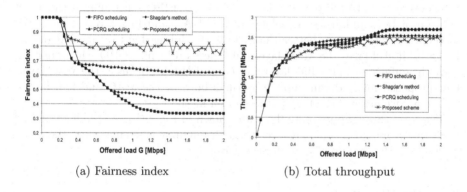

(a) Fairness index   (b) Total throughput

Fig. 9. Simulation results in Scenario-4

becomes greater, the proposed scheme achieves total throughput smaller than the other methods as the trade-off between fairness and throughput. In the other methods, the central pair cannot access the channel bandwidth and two external pairs can use maximum channel bandwidth. The total throughput can be twice of maximum channel bandwidth. In the proposed scheme, the MAC layer fairness is ensured, the central pair can achieve a half of maximum channel bandwidth. It leads two external pairs also can use a half of maximum channel bandwidth. Thus, the total throughput is only one and a half of maximum channel bandwidth. It is obviously smaller than in the other methods.

4.3   Scenario-3

Scenario-3 includes a chain of five stations with four flows. The coordinates of stations are shown in Fig. 6. The stations S1, S2, S3, and S4 generate traffic at the same offered load G to R.

The performance results are shown in Fig. 7(a) and 7(b). The results show that both PCRQ scheduling and the proposed scheme get better fairness performance than the others. The proposed scheme also have advantage in throughput performance by comparing with PCRQ scheduling.

4.4   Scenario-4

Scenario-4 is a grid scenario with high station density and high traffic density as in Fig. 8. In this scenario, the columns are separated by distance greater than the transmission range but smaller than the carrier sensing range. The stations in a column generate traffic at the same offered load G to the receiver in the same column. This topology faces both *Large-EIFS* and *Three-pairs* unfairness problems at the MAC layer.

The fairness between six flows are shown in Fig. 9(a). At the high offered load, each method achieves different fairness level due to contention between the forwarding flow and direct flow at the link layer, *Large-EIFS* and *Three-pairs* problems at the MAC layer. The fairness index of FIFO scheduling is the worst because of both unfairness problems in the MAC and link layers. Shargdar's method achieves fairness better than FIFO scheduling due to using RR queue. However, it still faces the MAC layer unfairness problem. PCRQ scheduling can solve the unfairness in the link layer and *Large-EIFS* problems at the MAC layer. Our proposed method can improve fairness both at the MAC and link layers. Therefore, we achieve the best fairness in all methods.

The total throughputs of all flows are shown in Fig. 9(b). At the high offered load, the proposed scheme achieves total throughput is smaller than the other methods as the same reason with Scenario-2.

5   Conclusion

Our proposed scheme is divided in two parts. One works on the MAC layer. We slightly modified back-off algorithm in order to achieve fair allocated bandwidth for each station. The other works on the link layer. We controlled RR queue in order to help each flow use bandwidth fairly. The cross-layer signal is sent from the link layer to the MAC layer to required the MAC layer sends the packet from the heavy offered load flow with longer back off time. The parameter γ can be chosen any value from zero to one and the parameter κ can be chosen any value greater than one. The same reason with PCRQ scheduling, the large value of these parameters may degrade bandwidth utilization. In our simulations, we chose γ equal to 0.3 and κ equal to 2.

In the simulation results, we evaluate our method by comparing with some other methods. The results showed that our method achieved good performances in *Large-EIFS* problem, *three-pairs* problem scenarios and also complex topologies as *long-station chain* and *grid* scenarios. In these topologies, FIFO scheduling has only one queue, and then it cannot solve the link layer contention. Shagdar's method uses RR queue, but also works ineffectively because the allocated bandwidths at the MAC layer are not suitable for forwarding and direct flows in the link layer. PCRQ scheduling can achieve good per-flow fairness. However, it fails to keep good per-flow fairness of flows, which do not come to the link layer. The proposed scheme works in both the MAC and link layers for aiming to improve both the MAC and link layer fairness. Therefore, we have good results in per-flow fairness even in the MAC layer unfairness topologies. Moreover, the proposed scheme gets better throughput performance than PCRQ scheduling compared to link layer topologies as *Large-EIFS* and *long-station chain* topologies. Because PCRQ scheduling tries to get the MAC layer fairness indirectly by giving a delay of heavy offered load flow to give a chance for forwarding flows. It affects to throughput performance. The proposed scheme adjusts back-off time to let disadvantaged station can get better channel bandwidth. Therefore, throughput performance in the proposed scheme also improves. In the MAC layer unfairness topologies, fairness between stations is improved by our method then total throughput is reduced as the trade-off between fairness and throughput.

References

1. Department, I.S.: IEEE 802.11 wireless lan medium access control (MAC) and physical layer (PHY) specifications. ANSI/IEEE Standard 802.11 (1999)
2. Xu, S., Saadawi, T.: Does the ieee 802.11 mac protocol work well in multihop wireless ad hoc networks? Communications Magazine, IEEE 39(6), 130–137 (2001)
3. He, J., Pung, H.K.: One/zero fairness problem of MAC protocols in multi-hop ad hoc networks and its solution. In: Proceedings of the International Conference on Wireless Networks (2003), pp. 479–485 (2003)
4. Jangeun, J., Sichitiu, M.: Fairness and qos in multihop wireless networks. IEEE Vehicular Technology Conference 5, 2936–2940 (2003)
5. Shagdar, O., Nakagawa, K., Zhang, B.: Achieving per-flow fairness in wireless ad hoc networks. Elec. Comm. in Japan, Part 1 89(8), 37–49 (2006)
6. Izumikawa, H., Sugiyama, K., Matsumoto, S.: Scheduling algorithm for fairness improvement among subscribers in multihop wireless networks. Elec. Comm. in Japan, Part 1 90(4), 11–22 (2007)
7. The Network Simulator: ns-2, http://www.isi.edu/nsnam/ns/
8. Giang, P.T., Nakagawa, K.: Scheduling algorithms for performance and fairness over 802.11 multi-hop wireless ad hoc networks. In: IEICE Tech. Rep. NS2008-21, pp. 19–24 (June 2008)
9. Karn, P.: Maca: A new channel access method for packet radio. In: ARRL/CRRL Amateur Radio 9th computer Networking Conference, pp. 134–140 (1990)
10. Bharghavan, V., Demers, A., Shenker, S., Zhang, L.: MACAW: a media access protocol for wireless lan's. In: Proceedings of the conference on Communications architectures, protocols and applications, pp. 212–225. ACM, New York (1994)
11. Nandagopal, T., Kim, T.E., Gao, X., Bharghavan, V.: Achieving mac layer fairness in wireless packet networks. In: Proceedings of the 6th annual international conference on Mobile computing and networking, pp. 87–98. ACM, New York (2000)

12. Bensaou, B., Wang, Y., Ko, C.C.: Fair medium access in 802.11 based wireless ad-hoc networks. In: Proceedings of the 1st ACM international symposium on Mobile ad hoc networking & computing, pp. 99–106. IEEE Press, Los Alamitos (2000)
13. Jiang, L.B., Liew, S.C.: Improving throughput and fairness by reducing exposed and hidden nodes in 802.11 networks. IEEE Transactions on Mobile Computing 7(1), 34–49 (2008)
14. Li, Z., Nandi, S., Gupta, A.K.: Ecs: An enhanced carrier sensing mechanism for wireless ad-hoc networks. Computer Communication 28(17), 1970–1984 (2005)
15. Jain, R., Chiu, D.M., Hawe, W.: A quantitative measure of fairness and discrimination for resource allocation in shared conputer systems. Technical Report TR-301, DEC Research Report (1984)

Tackling the Delay-Cost and Time-Cost Trade-Offs in Computation of Node-Protected Multicast Tree Pairs

Visa Holopainen and Raimo Kantola

TKK Helsinki University of Technology, Department of Communications and
Networking, P.O. Box 3000, 02015 TKK, Finland
{visa.holopainen,raimo.kantola}@tkk.fi

Abstract. While there are many algorithms for finding a node-protected
pair of multicast trees in a graph, all of the existing algorithms suffer from
two flaws: Firstly, existing algorithms either have a long running time or
produce high-cost tree pairs (Time-Cost Trade-Off). Secondly, existing
algorithms aim to produce a low-cost tree pair, but do not take delay
constraints into account (Delay-Cost Trade-Off). We tackle these trade-
offs by introducing an iterative algorithm, which finds a low-delay tree
pair very quickly and then iteratively seeks lower-cost pairs. We present
a sequential version of our algorithm; as well as an architecture for implementing it in a computing cluster, along with the corresponding master-
and slave-algorithms.

It turns out that the success percentage of our algorithm drops as the
average node degree decreases, and as the *share of multicast destination
nodes increases*. Fortunately, in these cases performance of the existing
algorithms improves, which means that our algorithm complements the
existing algorithms. Two additional benefits or our algorithm are: (1) it
can be used in multi-homing scenarios with almost no modifications, and
(2) it can be used in both directed and undirected graphs.

Keywords: Protection, Multipath routing, Heuristics, Cluster computing, Redundant/Protected/Disjoint/Colored Multicast/Steiner trees.

1 Introduction

The context of this work is a Guaranteed Services Network - for example a
Metro Ethernet. In this network a management system, located at Network
Operations Center, computes point-to-point tunnels and (static) multicast trees
based on customer orders. As the term "guaranteed" implies, fast connectivity
restoration is of utmost importance. Hence the management system needs to
compute protection tunnels and trees before network component failures occur.

A *Node-Protected Multicast Tree Pair* (NP-MTP) effectively protects multicast sessions against single-link and single-node failures. Unfortunately, even
without delay constraints, the problem of finding the lowest-cost NP-MTP is

C.S. Hong et al. (Eds.): APNOMS 2009, LNCS 5787, pp. 273–282, 2009.

$\mathcal{NP}$-hard, since the underlying *Steiner Tree Problem* is $\mathcal{NP}$-hard [3]. These reasons motivate us to develop a probabilistic heuristic algorithm for finding NP-MTPs.

The most important aspects of our algorithm are the following: In most realistic graphs our algorithm produces a low-delay NP-MTP very quickly while, on the other hand, it can be used for iteratively finding lower-cost NP-MTPs until the lowest-cost NP-MTP is found. Our algorithm is easy to parallelize, which makes is practical. Furthermore, with almost no modifications, our algorithm could be used in multi-homing scenarios, where the source and destination nodes of the two trees are different[1]. Also, our algorithm works equally well in directed and in undirected graphs.

The rest of this paper is organized as follows. Subsection 1.1 gives preliminaries. Section 2 summarizes the most relevant related work. Section 3 first describes a sequential version of our algorithm accompanied with an example, and then an architecture and the corresponding master- and slave-algorithms for implementing the algorithm in a computing cluster. Section 4 presents performance evaluation regarding the sequential version of our algorithm, and finally Section 5 concludes the work.

1.1   Preliminaries

Let $G = (V, E)$ be a graph, where V is the set of nodes (vertices), and E the set of directed links (edges). Correspondingly, let $|V|$ be the number of nodes, and $|E|$ the number of links. Let s be the multicast source node, $D \subseteq V - s$ the corresponding set of multicast destination nodes, and $|D|$ the number of destinations.

We define NP-MTP as follows: two trees T^{red} and T^{blue} form a NP-MTP if after any single failure of $v \in V - s$ (or $e \in E$) s can still connect each non-failed $d \in D$ via at least one of the two trees.

In this work, cost of tree T ($T.cost$) refers to the amount of bandwidth that needs to be reserved from E to accommodate T in G. In other words, without loss of generality, $T.cost$ is the number of links T spans.[2] In this work, delay of tree T ($T.delay$) refers either to the maximum or to the mean over the sums of link delays from s to each $d \in D$ via T.

2   Related Work

We borrow the concept of parallelizing an optimization problem from [1]. However, the specific problem addressed in this work is completely different.

In [2] the authors present a heuristic for finding Multicast Graphs (MGs) - that is - acyclic graphs rooted at s that provide at least two node-disjoint paths

---

[1] However, in this case we would probably need some signaling with the customer network, which is not addressed in this work.

[2] We could also use any (random) numbers as link costs in our algorithm. However, the selected cost metric is sufficient for determining our algorithm's performance.

from s to each $d \in D$. Their algorithm finds a lowest-cost node-disjoint path pair
from s to each $d \in D$ and combines the pairs to form the MG. The resulting
MG is close to optimal in cost-sense. An improvement in [3] also takes delay
constraints into account. For the special case $D = V - s$ a minimum-cost MG
can be computed in polynomial time [8]. However, MGs can be used for $1 + 1$
protection only, while NP-MTPs can be used for $1 : 1$ protection. Hence G could
accommodate less pre-emptible "extra" traffic (e.g. best-effort Internet-traffic)
if we would use MGs instead of NP-MTPs.

For the case $D = V - s$ [9] presents an algorithm for computing a NP-MTP
given a MG. However, there seems to be no easy way to make this algorithm
usable if $D \neq V - s$. This is a possible topic of future research.

The problem of finding *colored trees* (NP-MTPs for which $D = V - s$) is well
covered by previous work. The first algorithm addressing this problem is the one
in [4]. Building on that, improved heuristics are presented in [5] and [6]. While the
latter algorithms are close to optimal if $D = V - s$, they are suboptimal if this is
not the case. Also, these algorithms have cubic complexity. While algorithms [5]
and [6] have cubic complexity, Algorithm 4 in [7] finds low-cost NP-MTPs for
the case $D = V - s$ in linear time (however, it uses a different cost metric than
this paper).

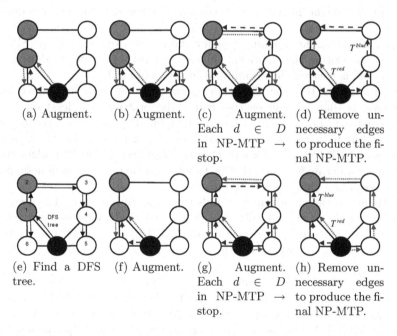

(a) Augment.   (b) Augment.   (c)   Augment.   (d) Remove un-
                                Each $d \in D$ necessary edges
                                in NP-MTP $\rightarrow$ to produce the fi-
                                stop.   nal NP-MTP.

(e) Find a DFS   (f) Augment.   (g)   Augment.   (h) Remove un-
tree.   Each $d \in D$ necessary edges
                                in NP-MTP $\rightarrow$ to produce the fi-
                                stop.   nal NP-MTP.

Fig. 1. Constructing a NP-MTP using Algorithm 7 of [6] (a-d), and Algorithm 4 of [7]
(e-h). Steps c-d and g-h modify the algorithms to the problem addressed in this paper,
that is, to find a NP-MTP from s to D, where D can be any subset of $V - s$.

Figure 1 presents the logics of Algorithm 7 in [6] (subfigures a-d) and Algorithm 4 in [7] (subfigures e-h) modified to the problem addressed in this paper. There the black node is s and gray nodes constitute D. The algorithm illustrated in subfigures a-d finds loops (starting and ending to existing T^{red}) which minimize the cost of new T^{red} at each step, and uses the loops to update the NP-MTP (dashed and dotted lines); while the one in subfigures e-h utilizes Depth First Search (DFS) tree and back-edges in loop forming.

3 Our Algorithm

First we describe a sequential version of our algorithm. For initialization the sequential algorithm computes a Breadth First Search (BFS) tree from s to $V - s$. Let us denote this tree by T^{init}. The following describes one iteration of the algorithm.

First the algorithm modifies G by assigning a random (integer) weight for each link in E from interval $[1, \max(1, i/dist)]$, where i denotes the iteration count, and $dist$ denotes distance of the link's end node from s in T^{init}. This means that during early iterations, weights are adjusted very little. Also, weights of links that are far from source are adjusted less on average than those close to source. From here on the link weight assignment created at this phase is denoted by W_i^1.

By using this weight assignment policy and the Shortest Path First (SPF, i.e. Dijkstra's) algorithm the algorithm creates trees with short delays during early iterations. This idea will be exemplified in subsection 3.1.

After producing W_i^1, the algorithm removes one outgoing link of s from E. The reason is that for a given tree pair to be protected, at least one of the links originating from s must not be shared among the two trees. The algorithm selects the link to be removed in a round-robin manner. After this, the algorithm computes an SPF tree T^1 from s to D in the modified G. The SPF algorithm is modified to prefer nodes in D, that is, to visit a node in D before a node not in D, if the resulting path costs would be equal.

At this point, the algorithm first adds the removed link to E. The algorithm re-modifies G by setting all link weights in E to random values form interval $[1, \max(1, i/dist)]$, and by removing each link of T^1 from E. The algorithm also increases incoming link weights of the internal nodes of T^1 among E (in $O(|E|)$-time). This is done because while computing the second tree, we want to avoid the internal nodes of T^1 to some extent. The amount by which the incoming link weights are increased depends on the number of found NP-MTPs, as well as on whether a link's start and/or end node is in D. If the algorithm has not yet found NP-MTPs, the weights are increased much. This helps the algorithm to find NP-MTPs with a higher probability. However, on average the NP-MTPs have a higher sum cost. If the algorithm has found many NP-MTPs, the weights are increased only slightly. This causes the algorithm to find a NP-MTP rarely, but when it does, sum cost of the NP-MTP is usually low. Also, weights of links having start or end node in D are increased less. From here on the second weight

Algorithm 1. Sequential version of our algorithm for a single multicast tree request; topology changes during computation are *not* addressed

---

Input: Graph G, source node s, destination nodes D, max. runtime t_{max}, delay constraint d_{max}.

Output: Two NP-MTPs (T^{red}, T^{blue}) and (T_c^{red}, T_c^{blue}).

* Let (T^{red}, T^{blue}) be the minimum-cost NP-MTP, and (T_c^{red}, T_c^{blue}) the delay-constrained NP-MTP (NULL at this point);
* Let c_{min} be the minimum sum cost of T^{red} and T^{blue}, and c'_{min} the minimum sum cost of T_c^{red} and T_c^{blue} (infinity at this point);
* Let $N_{NP\text{-}MTP}$ be the number of found NP-MTPs (zero at this point);
* Let i be the iteration count (zero at this point);
* Compute a BFS tree T^{init} from s to $V - s$;

while $(t_{current} - t_{start}) < t_{max}$ **do**

    $i \leftarrow i + 1$;

    // ----------------------- Generate W_i^1 --------------------

1.   * Set weight of each $e \in E$ (not destined to s) to a random integer from interval $[1, \max(1, i/dist)]$, where $dist$ is the distance of end node of e from s in T^{init};

    * Remove one link which originates from s from E in a round-robin manner;

    * Compute SPF tree T^1 from s to D in G (prefer nodes in D);

    * Add the removed link to E;

    // ---------------------- Generate W_i^2 ---------------------

    * Set link weights using procedure **1**;

    for *each link $e \in T^1$* **do**

        * Remove e from E;

        * Let v be the end node of e;

        if *v is not a leaf node* **then**

            for *each neighbor u of v that is neither father or son of v in T^1* **do**

                * Increase weight of link $(u \rightarrow v)$ by a factor inversely proportional to $N_{NP\text{-}MTP}$ (increase less if u and/or v is in D);

            end

        end

    end

    * Compute SPF tree T^2 from s to D in G (prefer nodes in D);

    * Add each removed link to E;

    if *each $d \in D$ is in T^2* && *T^1 and T^2 are node-protected* **then**

        $N_{NP\text{-}MTP} \leftarrow N_{NP\text{-}MTP} + 1$;

    end

    if *each $d \in D$ is in T^2* && *T^1 and T^2 are node-protected* && $(T^1.cost + T^2.cost) < c_{min}$ **then**

        $T^{red} \leftarrow T^1$; $T^{blue} \leftarrow T^2$; $c_{min} \leftarrow T^1.cost + T^2.cost$;

    end

    if *each $d \in D$ is in T^2* && *T^1 and T^2 are node-protected* && $(T^1.cost + T^2.cost) < c'_{min}$ && $(T^1.delay + T^2.delay) < d_{max}$ **then**

        $T_c^{red} \leftarrow T^1$; $T_c^{blue} \leftarrow T^2$; $c'_{min} \leftarrow T^1.cost + T^2.cost$;

    end

end

return (T^{red}, T^{blue}), and (T_c^{red}, T_c^{blue});

assignment is denoted by W_i^2. At this point the algorithm computes a second SPF tree T^2 from s to D in the re-modified G.

Now that the algorithm has the two trees T^1 and T^2, it needs to check whether they are node-protected or not. This can be done in $O(|V|)$-time by first setting temporary trees T_{temp}^{red} and T_{temp}^{blue} to hold only s, and then backtracking from each $d \in D$ one at a time to the existing T_{temp}^{red} and T_{temp}^{blue} via T^1 and T^2, each time checking whether any node on the backtracked paths (excluding d and s) was present in both paths, and if not, updating T_{temp}^{red} and T_{temp}^{blue} to hold the nodes on the backtracked paths.

If T^1 and T^2 were node-protected, the algorithm computes their sum cost. If it is lower than the lowest found so far, the algorithm updates best NP-MTP to point to the newly found tree pair, that is, the algorithm sets $T^{red} = T^1$ and $T^{blue} = T^2$. Next the algorithm checks whether the tree pair meets the delay constraint. If it does, and its sum cost is smaller than the smallest found delay-constrained pair cost, the minimum-cost delay constrained tree pair is updated ($T_c^{red} = T^1$ and $T_c^{blue} = T^2$).

Pseudocode of the sequential version of our iterative algorithm is presented in Algorithm 1. Time-complexity of one iteration is determined by the complexity of Dijkstra's algorithm, so time-complexity of Algorithm 1 is $O(I(|E|+|V|log|V|))$, where I is the number of executed iterations.

3.1 Example of Our Algorithm's Logic

Figure 2 gives a very simple example of our algorithm's logic. In each subfigure, a solid line represents a link, a solid arrow with a small arrowhead a link whose other direction (the direction against the arrow) has been removed before computing T^1 (but not before T^2), dotted arrows represent T^1 and dashed arrows T^2. Numbers next to links represent their weights (weight of a link is 1 if not shown).

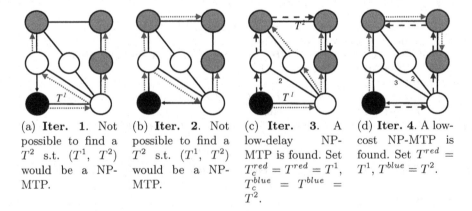

(a) **Iter. 1.** Not possible to find a T^2 s.t. (T^1, T^2) would be a NP-MTP.

(b) **Iter. 2.** Not possible to find a T^2 s.t. (T^1, T^2) would be a NP-MTP.

(c) **Iter. 3.** A low-delay NP-MTP is found. Set $T_c^{red} = T^{red} = T^1$, $T_c^{blue} = T^{blue} = T^2$.

(d) **Iter. 4.** A low-cost NP-MTP is found. Set $T^{red} = T^1$, $T^{blue} = T^2$.

Fig. 2. Constructing NP-MTPs using our algorithm. Each link's delay is 1, $d_{max} = 3$.

Algorithm 2. Clustered version of Algorithm 1 - master; simultaneous tree requests and topology changes during computation *are* addressed.

---

* Let **H** be an associative array which maps a pair (source, destinations) to a (graph, red_tree, blue_tree, cost, start_time, max_time, iteration_count, nrof_found_NP-MTPs) 8-tuple;

while *1* **do**
   // ---------------------- Fetch from DB_R_C ----------------
   for *each NP-MTP comput. requests in* **DB_R_C do**
2. $(s, D, t_{max}) \leftarrow$ Fetch the NP-MTP comput. request from **DB_R_C**;
      $G \leftarrow$ Fetch graph from **DB_G**;
      $\mathbf{H}[(s, D)] = (G, \text{NULL, NULL}, \infty, \text{current sys. time}, t_{max}, 0, 0)$;
   end
   // ---------------------- Send to DB_R --------------------
   for *each item* $((s, D) \rightarrow h) \in$ **H do**
      $t_r = h.\text{max\_time} + h.\text{start\_time} - t_{current}$;
3.   * Send 5-tuple $(t_r, h.\text{iter}, h.\text{nrof\_found\_NP-MTPs}, s, D)$ to **DB_R**;
      $h.\text{iter} \leftarrow h.\text{iter} + 1$;
   end
   // ---------------------- Fetch from DB_S ------------------
   for *each solution 5-tuple sol in* **DB_S do**
6.   * Let h point to $\mathbf{H}[(sol.s, sol.D)]$;
      $h.\text{nrof\_found\_NP-MTPs} \leftarrow h.\text{nrof\_found\_NP-MTPs} + 1$;
      if $sol.cost < h.cost$ || $h.graph \,! = sol.G$ **then**
         $h.\text{red\_tree} \leftarrow sol.T^1$; $h.\text{blue\_tree} \leftarrow sol.T^2$;
         $h.graph \leftarrow sol.G$; $h.cost \leftarrow sol.cost$;
      end
      * Remove *sol* from **DB_S**;
   end
   // ---------------------- Send to DB_S_C --------------------
   for *each item* $((s, D) \rightarrow h) \in$ **H do**
      if $(t_{current} - h.\text{start\_time}) > h.\text{max\_time}$ **then**
7.   * Send $(s, D, h.\text{red\_tree}, h.\text{blue\_tree})$ 4-tuple to **DB_S_C**;
         * Remove $((s, D) \rightarrow h)$ from **H**;
      end
   end
end

It is easy to see that with two first iterations (subfigures (a) and (b)) the algorithm does not find a NP-MTP. However, with the weight assignment of subfigure (c) (W_i^1 and W_i^2 are assumed to be same), a NP-MTP is found. There cost of T^1 is 5 and maximum delay of T^1 is 3, which satisfies delay constraint $d_{max} = 3$ (here delay of T^2 is ignored). Again, with the weight assignment of subfigure (d) a NP-MTP is found. Now cost of T^1 is 4 and maximum delay of T^1 is 4. This example highlights the delay-cost trade-off.

Algorithm 3. Clustered version of Algorithm 1 - slave.

	while *1* **do**
4.1.	* Fetch a ($time_{remaining}$, i, $N_{NP\text{-}MTP}$, s, D) 5-tuple from **DB_R** (probability of fetching a given 5-tuple depends inversely on its $time_{remaining}$);
4.2.	* Fetch G from **DB_G**;
	* Compute T^1 and T^2 as presented in Algorithm 1;
	if *each $d \in D$ is in T^2 && T^1 and T^2 are node-protected* **then**
5.	| * Send solution (G, s, D, ($T^1.cost + T^2.cost$), T^1, T^2) to **DB_S**;
	end
	end

3.2   A Cluster Computing Implementation of Our Algorithm

Figure 3 presents an architecture of a cluster-based implementation of our algorithm. There the dotted arrows correspond to writing to a database and dashed arrows correspond to reading from a database. Databases increase maintainability and fault-tolerance, since slave (or backup master) computers can come and go without affecting other components. The "NP-MTP computation request" is based on a customer order. It includes (1) a multicast source node, (2) a list of multicast destinations, and (3) the maximum time in seconds that can be used for computing the NP-MTP.

Algorithms 2 and 3 show a parallel version of Algorithm 1. To save space, delay-constrained tree computation is omitted from these algorithms. The databases (**DB_x**) in Figure 3 are the same as in those algorithms. Also, the numbers next to the arrows in Figure 3 correspond to the numbers in the left margins of Algorithms 2 and 3.

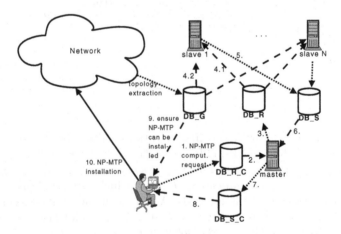

Fig. 3. Architecture of a cluster-based implementation of Algorithm 1

4 Performance Evaluation

Figure 4 presents performance evaluation results regarding the sequential version of our algorithm (denoted by A) and two benchmark algorithms (denoted by B and C). Lower bound of the optimal solution is denoted by D. For delay it was obtained by running SPF algorithm for each tested (s, D) pair and computing the mean of maximal delays; and for cost by assuming that all nodes in D can be connected by a loop starting from and ending to s without using any nodes in $V - D$, in which case the optimal sum cost of trees is simply $2|D|$. We generated ten vertex-redundant Waxman-graphs using the method of [3] with $|V| = 500$ (average degree 7.98). For each graph we tested 40 (s, D) pairs using each algorithm, so that $(|D|+1)/|V| = 0.25$, $(|D|+1)/|V| = 0.5$, $(|D|+1)/|V| = 0.75$, or $(|D|+1)/|V| = 1$.

The maximum delay of T^{blue} as produced by algorithm A was about 20% larger on average than that of T^{red}, while algorithms B and C produced same delays

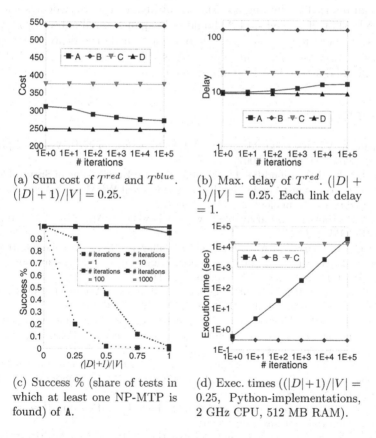

(a) Sum cost of T^{red} and T^{blue}. $(|D|+1)/|V| = 0.25$.

(b) Max. delay of T^{red}. $(|D|+1)/|V| = 0.25$. Each link delay $= 1$.

(c) Success % (share of tests in which at least one NP-MTP is found) of A.

(d) Exec. times $((|D|+1)/|V| = 0.25$, Python-implementations, 2 GHz CPU, 512 MB RAM).

Fig. 4. Performance of our algorithm (A), benchmark B (see Fig. 1 (e-h)), and benchmark C (see Fig. 1 (a-d)). Lower bound of the optimal solution is denoted by D.

for T^{red} and T^{blue}. It can be seen that while algorithm B is fast, it produces large costs and delays; whereas algorithm C is slow. By definition, algorithms B and C find one NP-MTP in each test, while (empirically) success percentage of algorithm A drops as a function of $(|D|+1)/|V|$ if iteration count is less than 1000. We also noticed that success percentage of of algorithm A drops as the average node degree is decreased. However, due to space limitations the corresponding graph is omitted.

5   Conclusions

We presented an iterative probabilistic algorithm which tackles the delay-cost and time-cost trade-offs in computation of Node-Protected Multicast Tree Pairs (NP-MTPs). We presented a sequential version of the algorithm, as well as a parallel architecture along with the corresponding master- and slave-algorithms. Performance evaluation showed that our algorithm is effective. While Algorithm 4 in [7] can be used for finding NP-MTPs quickly, it yield large delays and costs for the NP-MTP if the NP-MTP should only span a small subset of network's nodes. Also, since it uses DFS tree, it does not tackle the delay-cost trade-off intrinsically. However, since this algorithm is fast and always finds a NP-MTP, it could be used as the first step of our algorithm.

References

1. Holopainen, V., Ilvesmäki, M.: A Parallel Architecture for IGP Weights Optimization. In: Proc. APNOMS (2007)
2. Singhal, N.K., Sahasrabuddhe, L.H., Mukherjee, B.: Provisioning of Survivable Multicast Sessions Against Single Link Failures in Optical WDM Mesh Networks. Journal of Lightwave Technology 21, 2587–2594 (2003)
3. Irava, V.S., Hauser, C.: Survivable Low-Cost Low-Delay Multicast Trees. In: Proc. GLOBECOM (2005)
4. Medard, M., Barry, R.A., Finn, S.G., Gallager, R.G.: Redundant trees for preplanned recovery in arbitrary vertex-redundant of edge-redundant graphs. IEEE/ACM Transactions on Networking 7, 641–652 (1999)
5. Balasubramanian, R., Ramasubramanian, S.: Minimizing Average Path Cost in Colored Trees for Disjoint Multipath Routing. In: Proc. ICCCN (2006)
6. Xue, G., Chen, L., Thulasiraman, K.: Quality-of-service and quality-of-protection issues in preplanned recovery schemes using redundant trees. IEEE Journal on Selected Areas in Communications 21, 1332–1345 (2003)
7. Zhang, W., Xue, G., Tang, J., Thulasiraman, K.: Faster algorithms for construction of recovery trees enhancing QoP and QoS. IEEE/ACM Transactions on Networking 16, 642–655 (2008)
8. Frank, A.: Rooted k-connections in digraphs, Tech. Rep., Egervary Research Group on Combinatorial Optimization (2006)
9. Bejerano, Y., Koppol, P.: Optimal Construction of Redundant Multicast Trees in Directed Graphs. In: Proc. INFOCOM (2009)

An Algorithm for the Measure Station Selection and Measure Assignment in the Active IP Network Measurement*

Yongguo Zeng, Lu Cheng, Ting Huang, and Zhiqing Li

State Key laboratory of Networking and Switching Technology,
Beijing University of Posts and Telecommunications, Beijing, P.R. China, 100876
ygzeng2010@gmail.com

Abstract. It is critical to obtain accurate network information for network management and the improvement of network service quality. Network measurement is an efficient method to obtain such information. Now the network measurement models and algorithms have become a research hot spot; and the core problem is how to cut down the cost of device deployment and reduce the effects caused by measurement. In this paper, we discuss the problem of measure station selection and measure assignment in active IP network, and propose a greedy algorithm to select stations and a method based on link usage cost to assign measures. Simulation results show that our method is efficient in station selection and measure assignment.

Keyword: network measurement; greedy algorithm; station selection; measure assignment.

1 Introduction

With the enlargement of network scale and the emergence of various new network services, network status information is critical for network configuration optimization and network management, and it is also important for network service and security [1]. Now many researchers and government organizations have focused on network measurement, and the main research field contains network measurement technology, measurement models, and measurement information collection frames [2]. According to the mode of measurement, network measurement is generally separated into two broad categories: passive measurement and active measurement. Passive measurement does not involve sending probe packets into the network, so it will not increase the network loads or affect the performance of the network. Active measurement must send probe packets to target nodes to measure end-to-end performance statistics such as latency, bandwidth, package loss rate and some other network performance parameters. Active measurement can also be used to test the network connectivity and

---

* This work was supported in part by the 973 project of China (2007CB310703), Fok Ying Tong Education Foundation (111069).

C.S. Hong et al. (Eds.): APNOMS 2009, LNCS 5787, pp. 283–290, 2009.

infer fault routers or links. In this paper, we discuss the problems of measure station selection and measure assignment.

There is much research on measure station selection and measure assignment currently. Literature [3] discusses the station deployment of multicast routing. Approaches presented in [4], [5] are based on shortest path routing tree, using simple greedy algorithm to find minimal station set and measure assignment. The problem of fault-tolerant and dynamic adaptive measure station placement and measure assignment is proposed in [6], [7], but it is impossible to solve it in sparse link network. Literature [6], [8] introduce a new concept Arity, and the literature [6] proves that the optimal set of measure station is the set of high Arity nodes, so to select high Arity nodes can improve the efficiency of station selection. A Solution based on single measure station is proposed in [9], [10], and such a method will cause network traffic converging and overloading, even seriously affect the network's performance. Literature [11], [12] discuss measure station selection under the constraint of maximum distance and how to strike a distribution of measure stations when the number of measure stations is given. In essence, both of the two are an extension of the greedy algorithm to adapt to the conditions attached.

Our study will make foundation on the shortest path routing tree, and it will reduce the number of measure stations by using greedy algorithm for the not covered link. In order to reduce the impact to network when measurement is taken, we introduce the rate of link bandwidth usage to evaluate the link usage cost, and then use the greedy algorithm to select the measure set with lowest cost. According to the available bandwidth, we can update measure set dynamically.

2 Problem Definition

2.1 Definition of Measure Station Selection Problem

The deployment of measure stations can be mapped to a general edge covering problem in graph theory, defined as follows.

Definition 1. Given an network $G(V, E)$, where V denotes the set of nodes, E represents the edges between two nodes. It is assumed that under the current routing protocol, the links set measured by node $u \in V$ recorded as $T_u \in E$, then the problem of measure station deployment could be transformed to seek minimal nodes set $S \subset V$ which makes $\bigcup_{u \in S} T_u = E$.

In order to seek minimal measure station set, the links measured by each node must be known first. For the autonomous system using the OSPF (Open Shortest Path First) routing protocol, measurable link set of one node is considered to be the link set contained in the routing tree in [5] and others. In this paper, we also define the link set contained in the routing tree as the measurable link set of one node. The problem to seek minimal measure stations under the condition of knowing measurable link set of each node is NP hard, and this has been proved in [5], [6].

2.2 Definition of Measure Assignment Problem

When the number and the locations of measure stations are confirmed, it is necessary to optimize the measure assignment, which means, for every link, to determine which station is responsible for measuring. The link could be measured by multiple nodes requires an optimal selection solution to reduce the impact to the network when measuring.

Definition 2. Given a network $G(V, E)$, the measure station set $S \subset V$ and the link set to be measured $M \subset E$ are provided, the link can be measured by station $u \in S$ is T_u. If the set $D = \{p(u, v) \mid u \in S, v \in V\}$ ($p(u, v)$ is a path in the routing tree T_u) is a measure assignment, then for any $\forall e(v, w) \in M$, there must be $\exists u \in S$, which makes $e \in T_u$, where $p(u, v) \in D$ and $p(u, w) \in D$ (path $p(u, w)$ is longer than path $p(u, v)$ by only one edge $e(v, w)$).

Through sending measures to path (u, v) and (u, w) , we can get the performance information about the link $e(v, w)$. Active measurement brings about additional network traffic which will affect the behavior of the network. Measure assignment is usually optimized according to different aims, such as minimal router number the probe packets past, minimum number of probe packets, and shortest detect path. All solutions of this problem are NP-hard [5], but using greedy algorithm can obtain an approximate solution.

According to the characters of active measurement, our ambition is to reduce the network traffic load caused by active measurement such that the performance of the network will not decrease. This is also conducive to the accuracy of the measurement. To this end, we introduce the rate of available link bandwidth r to evaluate the link usage cost, $r = A / W$, where A is the link bandwidth available, W is the total bandwidth, and the cost of a probe past one link is $C = 1 - r$. In this method, we take many factors into account, such as the link bandwidth constraints, the routing ability, queuing delay and so on. Using the rate of available bandwidth could be more comprehensive to assess the link situation, because the absolute available bandwidth in the link may have low available bandwidth rate and in such a case this link may have high queuing delay and packet loss rate. Sending measures through low rate of available bandwidth will cause link load increasing, which may cause more packets' loss, network delay, jitter and so on. That means the cost of measurement is relatively higher. But through the links with high rate of available bandwidth, the original data on the network traffic will be affected much less, and the impacts on the measurement affected by external factors will be much smaller. So seeking the optimal measure assignment solution can be defined as follows.

Definition 3. In the network $G(V, E)$, for the link $e \in E$, the usage cost is $C(e)$, and the measure cost is denoted as $CM(e)$. The measure station set $S \subset V$ and the link to be measured set $M \subset E$ have been provided. The problem to seek optimal measure assignment solution is to find a measure set $D = \{p(u, v) \mid u \in S, v \in V\}$ ($p(u, v)$ is a path in the routing tree T_u) which satisfies definition 2, and makes the value of $\sum\limits_{e \in M} CM(e) = \sum\limits_{p(u,v) \in D} C(u,v)$

$(C(u,v) = \sum\limits_{e \in p(u,v)} Ce)$ minimum.

The problem of seeking minimum cost measure assignment could be reduced to the problem in [5], and the solutions of these problems are NP hard [5]. The next section will discuss how to strike a greedy algorithm for the approximate solution.

3 Approximate Algorithm of Station Selection and Measure Assignment

The measure station selection and measure assignment are NP-hard, so there is still no optimal solution in polynomial time complexity. Currently, solutions for this problem basically adopt the idea of linear programming or greedy algorithm to obtain an approximate optimal solution.

3.1 The Approximate Solution to the Measure Station Selection

It has been proved that the best collection of measure stations is a collection of high-Arity nodes in [6], [8]. In [5], a greedy method is proposed which obtains the approximate solution through the nodes with high degree, and this method is similar with the one based on high-Arity nodes. Algorithm in this paper also takes the greedy strategy to get a solution. This greedy strategy does not rely on maximum degree nodes, but to select the nodes that will measure the unmeasured links as many as possible. According to the definition of the problem of measure station selection in section 2, approximation algorithm is as follows.

Algorithm 1. Measure Station Selection

```
Input: Network  G(V,E), the routing tree T_u of each node u ;

Output: Node set  S ⊂ V , and ∪_{u∈S} T_u = E .
MeasureStationSelection()
    S ← Ø, Q ← Ø, R ← E ; /* S is the set of selected station,
Q is the set of links can be measured by stations in S ,
R is the links set that can't be measured by stations in
S . */
    Select the max degree node u ,  S ← u ,  Q ← T_u ,  R = R - T_u ;
    /* First choose the largest degree node will reduce
the height of the routing tree, which make the cost of
measurement much lower relatively*/
    While ( Q ≠ E )
    {
        for ( ∀u ∈ V - S )
```

```
           select the max  T_u ∩ R  ( u ∈ V − S )
       then S = S ∪ u ,   Q = Q ∪ T_u ,   R = R − T_u
    }
```

Because all nodes cover same nodes in the autonomous system at begin, the algo-rithm selects the largest degree node first, by doing this the height of the routing tree and the cost of measurement can be reduced. According to the greedy algorithm based on uncovered links, we can use stations as few as possible to measure the whole net-work links.

3.2 An Approximate Solution to Measure Assignment Based on the Cost of Link Usage Rate

When the measure station set has been provided, the optimal measure assignment solution is just to find a measure set that makes the minimal measure cost. For one link $(v, w) \in E$ to be measured, from station $u \in S$, we should send two probe packets $p(u, v)$ and $p(u, w)$. So the measure cost for link (v, w) is $CM(v, w) = C(u, v) + C(u, w)$. If the probe $p(u, v)$ or $p(u, w)$ has been sent in other measure behavior, then it can be reused, so the measure cost of link (v, w) can be reduced to $C(u, w)$ or $C(u, v)$.

To get the answer, first of all, we should obtain the rate of available bandwidth based on past experience or the outcome of the last measurement, and calculate the usage cost value of each link, denoted as $C(e) = 1 - r_e (e \in E)$. Then for each link $(v, w) \in M$ to be measured, set the initial measure value $C(u', v) + C(u', w)$, and the measure station $u' \in S$, which makes $C(u', v) + C(u', w) \le C(u, v) + C(u, w)(\forall u \in S)$, $(v, w) \in T_{u'}$. In the solving process, choose the link with minimal measure cost each time, mark and assign the measure station to it. Because the probes which have been sent are reusable, we should update the measure value for each unassigned link after one link is selected to reduce the cost value of the probe which can be reused. The whole process of approxi-mation algorithm is as follows.

Algorithm 2. Measure assignment

```
Input:  Network  G(V,E) ,  the  routing  tree  T_u  of  measure
        station u ∈ S ,  the  link  set  to  be  measured  M ⊂ E ,  the
        usage  cost  value  C(e)  of  each  link ;
Output: Measure assignment D = { p(u,v) | u ∈ S, v ∈ V} ,  which make
        the    Σ     C(u,v) get  a  minimal  value;
            p(u,v)∈D
MeasureAssignment()
```

```
for ( ∀(v,w) ∈ M )
  Compute initial measure value  CM(v,w) = C(u',v) + C(u',w) of
link   (v,w) ,   the   measure   station   u' ∈ S   satisfy
C(u',v) + C(u',w) ≤ C(u,v) + C(u,w)(∀u ∈ S) ,  and  (v,w) ∈ T_u' ;
End for;
While (M ≠ ∅)
    Select the minimum  CM(v*,w*) ;
    M ← M - (v*,w*) ,   D = D ∪ p(u',v*) ∪ p(u',w*) /* remove the link
selected and assign measure program for it*/
    for ( ∀(v,w) ∈ M && (v,w) ∈ T_u' )
    if( v ∈ {v*,w*} || w ∈ {v*,w*} ) (for example v == w* )
      then compute  CM(v,w) = Min(CM(v,w), C(u',w)) ;/* from
    station u' to measure the link (v,w) , just need to send
probe p(u',w) , the probe  p(u',v) can be reused */
```

Due to the dynamic change of routers and network traffic, the link usage cost will be also changed dynamically. Therefore, the measure assignment can't be always an optimal assignment program. In practice, the measure solution can be assigned dynamically by central control node according to the latest routing trees and link usage cost. At first, the initial link usage cost may be scored based on past experience or link distance, and then we can calculate the cost by the available bandwidth data measured last time.

4 Emulation and Analysis

In the experiment, we use nem [13] as the network generator, and adopt waxman [14] network model to generate network topology. Waxman is the most widely used topology generate model in network research. By setting the three parameters n (denotes the number of nodes in network), α (control density of edge in the generate topology), β (control the average degree of nodes), we can get different network topology. The distance and bandwidth utilization of each edge is set randomly, and this does not affect the experiment.

For the problem of measure station selection, we simulate the random station selection method and the simple maximum degree node priority algorithm introduced in [5] to compare with the algorithm discussed in this paper. The number of measure station selected by the three algorithms are denoted as N_r, N_s and N_o, and the result of the experiment is shown in Tab. 1. Similarly, on the issue of measure assignment, we also simulated three algorithms which are random assignment algorithm, simple assignment algorithm introduced in [5] and the greedy algorithm based

Table 1. Station Number Contrast

N	α	β	Avg Degree	Max Degree	N_r	N_s	N_o
400	0.3	0.02	2.83	9	6	10	5
400	0.3	0.04	3.935	14	16	45	8
400	0.3	0.08	5.025	22	60	43	13
400	0.6	0.02	3.595	12	12	20	7
400	0.6	0.04	4.98	20	27	24	13
400	0.6	0.08	6.13	32	78	48	18

Table 2. Measure Cost Contrast

N	α	β	Station Number	C_r	C_s	C_o	Station Number	C_r	C_s	C_o
400	0.3	0.02	5	2391	1518	873	10	2075	890	475
400	0.3	0.04	8	2228	1115	612	45	2053	269	137
400	0.3	0.08	13	2127	914	519	43	1726	180	117
400	0.6	0.02	7	2300	1349	772	20	1995	579	308
400	0.6	0.04	13	2267	1056	567	24	1816	485	266
400	0.6	0.08	18	2028	788	469	48	1715	121	73

on link usage cost proposed in this paper. The assignment simulation is based on the stations selected by algorithm proposed in [5] and our algorithm. The simulation data are shown in Tab. 2.

The data of experiment demonstrate that, our method can dramatically reduce the number of measure stations. When the measure station set is determined, the measure assignment obtained by our algorithm gets a much lower cost than other method. The low measure cost means the measure activity has tiny impacts on the network. At the same time, the measure activity will also get an accurate result. On the other hand, from Tab. 1, we can see random selection method may be better than the maximum degree node priority method. In the Tab. 2, for the same network topology, the measure cost is much lower when the number of measure station grows bigger. Even in the different network, the network with more edges has a lower measure cost due to more measure stations, so how to reduce the number of measure stations and obtain the best measure assignment at the same time is an extremely meaningful problem.

5 Conclusion

In this paper, we discuss the problem of measure station selection and measure assignment in active IP network. We propose a greedy algorithm to minimize the number of measure stations. In addition, for the problem of measure assignment, we introduce the condition of link usage cost. According to the rate of available bandwidth on the link, we select the measure solution which has the lowest total cost, and dynamically update the assignment solution according to the real network situation. The experiment shows that our method effectively reduces the number of measure stations and the cost of the measure activity. Optimization of the measure assignment depends on the selection of measure stations, and fewer measure stations may lead to a high measure cost. So how to trade off between the number of measure stations and the cost of measurement is very difficult in research. And it is definitely worth further studying.

Reference

1. Zhang, H., Fang, B.: Internet measurement and analysis. Journal of Software 14(1), 110–116 (2003)
2. Cai, Z., Liu, F., et al.: Network measurement model and its optimization algorithm. Journal of Software 19(2), 419–431 (2008)
3. Adler, M., Bu, T., Sitaraman, B., Towsley, D.: Tree layout for Internet network characterizations in multicast networks. In: Proc. of the NGC 2001, pp. 189–204. Springer, Heidelberg (2001)
4. Breitbart, Y., Dragan, F., Gobjuka, H.: Effective network monitoring. In: Proc. of the IEEE ICCCN 2004, pp. 394–399. IEEE Communication Society, Los Alamitos (2004)
5. Bejerano, Y., Rastogi, R.: Robust monitoring of link delays and faults in IP networks. IEEE/ACM Transactions On Networking 14(5) (2006)
6. Kumar, R., Kaur, J.: Efficient beacon placement for network tomography. In: Proc. of the ACM SIGCOMM IMC, pp. 181–186. ACM Press, New York (2004)
7. Nguyen, H., Thiran, P.: Active measurement for multiple link failures diagnosis in IP networks, pp. 185–194. Springer, Heidelberg (2004)
8. Horton, J., Lopez-Ortiz, A.: On the number of distributed measurement points for network tomography. In: Proc. of the ACM SIGCOMM IMC, pp. 204–209. ACM Press, New York (2003)
9. Breitbart, Y., Chan, C.Y., Garofalakis, M., Rastogi, R., Silberschatz, A.: Efficiently monitoring bandwidth and latency in IP networks. In: Proc. of the IEEE INFOCOM 2001, pp. 933–942. IEEE Communication Society, Los Alamitos (2001)
10. Walz, J., Levine, B.: A hierarchical multicast monitoring scheme. In: Proc. of the NGC on Networked Group Communication, pp. 105–116. ACM Press, New York (2000)
11. Jamin, S., Jin, C., Jin, Y., Raz, D., Shavitt, Y., Zhang, L.: On the placement of Internet instrumentation. In: Proc. of the IEEE INFOCOM 2000, pp. 295–304. IEEE Communication Society, Los Alamitos (2000)
12. Bartal, Y.: Probabilistic approximation of metric space and its algorithmic applications. In: Proc. of the 37th Annual IEEE Symp. on Foundations of Computer Science, pp. 184–193. IEEE Communication Society, Los Alamitos (1996)
13. Magoni, D.: Network topology analysis and internet modelling with Nem. International Journal of Computers and Applications 27(4), 252–259 (2005)
14. Waxman, B.M.: Routing of Multipoint Connections. IEEE Journal on Selected Areas in Communications 6(9), 1617–1622 (1988)

Volume Traffic Anomaly Detection Using Hierarchical Clustering

Choonho Son, Seok-Hyung Cho, and Jae-Hyoung Yoo

Network R&D Laboratory
463-1, Jeonmin-dong, Yuseong-gu,
Daejeon, Korea
{choonho,sjho,styoo}@kt.com

Abstract. In a large backbone network, it is important to detect shape traffic fluctuation for servicing robust network. However, there are too many interfaces to monitor the characteristics of traffic. First we collect volume traffic of boundary link. From the volume traffic, we make groups which have similar traffic patterns by hierarchical clustering algorithm. This result shows that most of traffic has similar patterns, but some traffic which is far from centroid has an anomaly traffic pattern. This paper gives a hint for network operators that which traffic has to be checked out.

Keywords: Anomaly Detection, Volume Traffic, Hierarchical Clustering, Network Management.

1 Introduction

In a large backbone network, a traffic management is the most important function to provide robust network services, because the traffic patterns directly reflect the users' behaviors. The traffic of backbone link is an aggregation of users' traffic. It means that malicious traffic such as DoS attack can degrade the performance of others.

Traffic anomaly detection is required to provide fairness of traffic sharing. The accurate way to detect anomaly is decoding packets. However it has some problems such as trespass on person's privacy. Moreover it consumes much resources of router. The next solution is analyzing flow of traffic. A flow is a 4-tuple element, source IP, destination IP, source port and destination port. Each flow represents a continuous communication between client and server. Therefore, flow analysis is an alternative method of anomaly detection. Firewall and IDS (Intrusion Detection System) usually adopt flow analysis algorithm to detect anomaly hosts. Packet inspection and flow analysis are almost impossible to detect anomaly in a large backbone network, because a backbone network consists of thousands of interfaces and monitoring every interfaces gives huge resource consumption in a router.

The reasonable solution for anomaly detection in a large backbone network is analyzing volume traffic of links. This is a trade-off between accuracy and speed. The speediness compared by accuracy is more important, because network operator can do

C.S. Hong et al. (Eds.): APNOMS 2009, LNCS 5787, pp. 291–300, 2009.
© Springer-Verlag Berlin Heidelberg 2009

temporary post-analysis such as traffic mirroring or deep packet inspection after detecting volume traffic anomaly.

In this paper, we detect anomaly from volume traffic by clustering algorithm. First we correct all traffic of boundary links in a backbone network, and then classify traffic by hierarchical clustering. These clusters can be divided into normal group and anomaly one. The characteristics of anomaly group are analyzed after classification.

The paper is organized as follows. Section II discusses related work. Section III describes the traffic classification method by hierarchical clustering algorithm. Section IV presents anomaly detection algorithms. Section V shows the experimental result in KT(Korea Telecom) and the reason of anomaly detection. Lastly, Section VI concludes this paper and discusses future work.

2 Related Work

The research on network traffic is one of the most important areas from PSTN (Public Switched Telephone Network) to Internet. The measurement of traffic [1], [2], [3], [4],[5] is basic research topic to detect the characteristics of network. After measurement studies, many valuable facts were introduced such as characteristics of P2P, efficient traffic engineering algorithm and anomaly detection [6],[7],[8],[9]. The scale of traffic analysis is various from host traffic patterns to access network. There are also researches on ISP network and worldwide network. In addition, the focuses on researches are also different. Some people analyze the traffic pattern of applications such as Skype or IPTV. Another researcher collects flow information from routers. Volume traffic analysis is rarely untouched area, because it is hard to collect large scale volume data.

Wide area traffic patterns such as traffic volume, packet size, and traffic composition in terms of protocols and application is analyzed in MCI backbone network [1]. The traffic patterns based on protocols and application is various [2]. In addition, it shows that traffic has a geographical locality. The characteristics of TCP flows such as round-trip times, out-of-sequence packet rates, and packet delay are analyzed in Sprint network [4]. The patterns of residential or non-residential traffic on Japanese ISP are well summarized [5].

Anomaly detection is extracting important information from traffic measurement data. Traffic pattern is considered as wave signal on a system. Traffic anomaly is analyzed by signal processing like noise detection in a signal process [6]. Traffic analysis from the point of statistics is researched, which uses Principal Component Analysis of volume traffic [7]. The observation of backbone link shows the variety of IP and TCP features, and gives header anomalies of packets [8]. Anomaly can be detected by flow analysis using PCA algorithm [9].

To analyze the characteristics of traffic, the measurement points are quite different. The most common measurement point is boundary link between two ISPs which has large volume traffic [1],[4]. The traffic patterns of domestic link and international link is compared [1]. Multiple measurement points such as domestic link, international link, customers' lines, and leased lines give various traffic patterns based on measurement point [5]. However, collecting multiple data-set is quite difficult work since many companies avoid providing data-set. This means that most of works for analyzing backbone traffic are based on sampled links [8],[10],[11],[12].

Our approach has following contributions:

- We analyze all traffic of boundary links in Tier 1 ISP.
- We give long-term traffic pattern of customer link.
- We introduce hierarchical clustering for grouping traffic patterns.

3 Volume Traffic Clustering

The traffic of boundary link in backbone network has similar traffic patterns, because it is an aggregated link which includes thousands of subscribers. In addition, many ISPs prefer star or hierarchical topology for manageability. This means that boundary links have similar capacities and speed. Monitoring boundary link in backbone network is the best point for analyzing traffic patterns. By clustering boundary links, normal and abnormal groups have different characteristics.

3.1 Hierarchical Clustering

Hierarchical clustering [13],[14],[15] groups data over a variety of scales by creating a cluster tree or *dendrogram*. The tree is not a single set of clusters, but rather a multilevel hierarchy, where clusters at one level are joined as clusters at the next level. This allows you to decide the level or scale of clustering that is most appropriate for your application.

To perform hierarchical cluster analysis on a data set, the basic process is this:

1. Start by assigning each item to a cluster, so that if you have N items, you now have N clusters, each containing just one item. Let the distances (similarities) between the clusters the same as the distance (similarities) between the items they contain.
2. Find the closest (most similar) pair of clusters and merge them into a single cluster, so that now you have one cluster less.
3. Compute distances (similarities) between the new cluster and each of the old clusters.
4. Repeat step 2 and 3 until all items are clustered into a single cluster of size N.

Step 3 can be done in different ways, which is what distinguishes single-linkage from complete linkage and average linkage clustering. In single-linkage clustering (also called the connectedness or minimum method), we consider the distance between one cluster and another cluster to be equal to the shortest distance from any member of one cluster to any member of the other cluster. If the data consist of similarities, we consider the similarity between one cluster and another cluster to be equal to the greatest similarity from any member of one cluster to any member of the other cluster. In complete-linkage clustering (also called the diameter or maximum method), we consider the distance between one cluster and another cluster to be equal to the greatest distance from any member of one cluster to any member of the other cluster. In average-linkage clustering, we consider the distance between one cluster and another cluster to be equal to the average distance from any member of one cluster to any member of the other cluster.

3.2 Volume Traffic Monitoring

The boundary link in Figure 1 is connection between backbone and access network. The backbone network consists of routing protocol, but access network consists of switching area such as Ethernet or FTTH. The capacity of boundary link is usually 1G, 2.5G, or 10G. We collect interface counter values of boundary router or service edge router via SNMP. The MIB values of interface traffic are ifInOctets and ifOutOctets [16] which are total number of octets received or transmitted from interface, including framing characters. We collect every five minutes and calculate the transmission rate or bps.

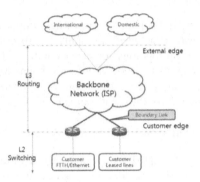

Fig. 1. Boundary link at ISP for data collection

The customers consist of two kinds as residential customer or non-residential customer. The residential customer is home subscriber which uses FTTH or Metro Ethernet. On the other hands, non-residential customer is based on leased line including most of companies and governments. The number of boundary link in residential customer is about 590, and the number of boundary link in non-residential customer is about 160.

3.3 Hierarchical Clustering of Volume Traffic

The first step for hierarchical clustering is calculating similarities called as *correlation coefficient* [17] which indicates the strength of a linear relationship between two random variables. In this case, random variable is time-series volume traffic of interface like Figure 2.

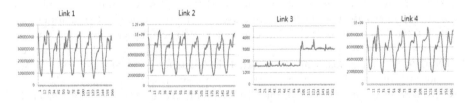

Fig. 2. Volume traffic of two boundary links among one week. X axis is time, and Y axis is transmission rate (Bit per second)

The correlation coefficient $\rho_{X,Y}$ between two volume traffic X and Y with expected values μX and μ_Y and standard deviation σ_X and σ_Y is defined as:

$$\rho_{X,Y} = \frac{\text{cov}(X,Y)}{\sigma_X \sigma_Y} = \frac{E((X - \mu_X)(Y - \mu_Y))}{\sigma_X \sigma_Y}, \tag{1}$$

where E is expected value operator and cov means covariance. Table 1 represents correlation coefficient of four volume traffic. Once the proximity between volume traffic has been computed, complete linkage algorithm groups each objects into clusters. The hierarchical, binary cluster tree created by the linkage function is most easily understood when viewed graphically like Figure 3. This graph clearly shows that Link 1 and Link 2 have highly similar pattern, and Link 3 is the most abnormal traffic among them.

Table 1. Proximity matrix (Correlation Coefficient) of four volume traffic

	Link 1	Link 2	Link 3	Link 4
Link 1	1	0.957	-0.211	0.952
Link 2	0.957	1	-0.174	0.948
Link 3	-0.211	-0.174	1	-0.183
Link 4	0.952	0.948	-0.183	1

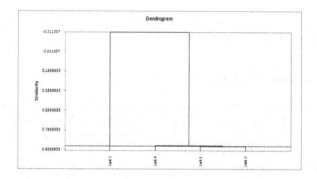

Fig. 3. Dendrogram, hierarchical binary tree created by the complete linkage function

4 Anomaly Detection

The anomaly detection algorithm divides the clusters into normal cluster and anomalous one. The rationale behind this approach is the assumption that normal and anomalous traffic form different clusters in the space. There are two analysis processes. The first one is detecting volume anomaly in whole traffic. This method assumes that there is huge or very small volume traffic in a data set. This analysis gives network operator the warning signal, for example, there is traffic concentration (high volume traffic) or leakage (low volume traffic). The second one is detecting anomalous traffic pattern compared by others. This method assumes that most of boundary links have similar traffic patterns.

4.1 Volume Anomaly Detection

The capacities of boundary links are usually 1 Gigabit Ethernet, 2.5 Gigabit Packet over SONET, or 10 Gigabit Ethernet. Our intuition is that the volume traffic of each capacity will have similar traffic pattern. In this case, there will be three clusters, which represent 1GE, 2.5 PoS, and 10GE. However, there is abnormal traffic like new attached interface or DoS attacked link. This traffic will have different position like Figure 4.

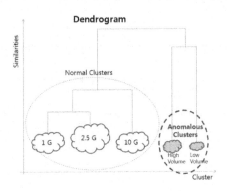

Fig. 4. Dendrogram, normal clusters and anomalous clusters, anomalous clusters consist of high volume cluster and low volume cluster

4.2 Pattern Anomaly Detection

The purpose of pattern anomaly detection is discovering abnormal traffic patterns above boundary links. Backbone traffic generally has periodicity like daily and week-ly circulation. To detect pattern anomaly, the volume traffic has to be standardized before calculating hierarchical clustering.

We use *zscore (Z)* function which returns a centered, scaled version of traffic X. It has mean zero and standard deviation one unless X is constant. This is usually used to put data on a same scale before further analysis.

$$Z = (X - mean(X)) \,/\, std(X) \qquad (2)$$

where Z is zscore and std means standard deviation. After zscore function is applied, next step is same with volume anomaly detection.

5 Experimental Result

We analyzed traffic logs from February 1 to February 7 in 2009. The volume traffic has 1-hour resolution. The number of boundary link from customer is 590 which include every residential customer who uses FTTH or Metro Ethernet.

5.1 Analysis of Volume Anomaly

First experiment analyzed uplinks of boundary traffic. Figure 5 shows dendrogram, which consists of 63 clusters. We manually divide normal clusters and anomalous clusters. Cluster 51, 26 and 33 are included in anomalous clusters and Cluster 1 is the largest normal cluster. Figure 6 shows traffic patterns of Cluster 51 which constitutes anomalous cluster and sample traffic patterns of Cluster 1 which is normal cluster.

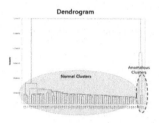

Fig. 5. Dendrogram of volume anomaly detection, there are 60 normal clusters and 3 anomalous clusters

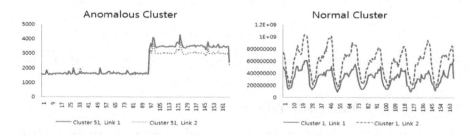

Fig. 6. Sample traffic pattern of anomalous cluster (left), and normal one (right)

5.2 Analysis of Pattern Anomaly

The second analysis is detecting anomaly patterns among boundary links. Figure 7 shows the dendrogram of pattern anomaly detection which data are standardized by zscore function. The hierarchical clustering gives four clusters. Clusters 1 and 2 are normal clusters and Cluster 3 and 4 are anomalous clusters. There are 7 links which are included in anomalous clusters.

Figure 8 shows the difference among two anomalous clusters. Cluster 3 and 4 has relatively constant traffic pattern than the pattern of normal clusters. Cluster 3 has one traffic jump among monitored period, but Cluster 3 does not have traffic jump. The reason why seven links belong to anomalous cluster is that these links are new installed interfaces and there is no subscriber at that time. The small traffic of anomalous links may be routing packets and SNMP monitoring traffic, because routing tables is consistently updated and network management system collects routers' status.

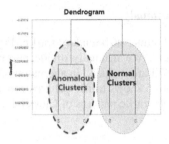

Fig. 7. Dendrogram of pattern anomaly detection, there are 2 normal clusters and 2 anomalous clusters

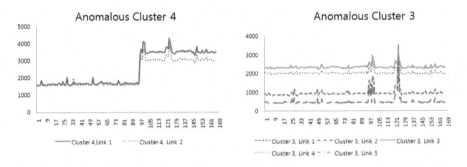

Fig. 8. Sample traffic patterns of anomalous clusters 4 and 5, X axis is hour and Y axis is transmission rate (bit per second)

5.3 Analysis of Normal Traffic Patterns

There are two normal clusters in Figure 7. We analyze the traffic patterns of them. The Cluster 1 consists of 569 links which are distributed in whole country. However Cluster 2 consists of 14 links which have locality. The place of links is Seoul, the capital of Korea. In addition, Cluster 1 and 2 have slightly different traffic patterns like Figure 9, even though the subscribers of link are residential internet users.

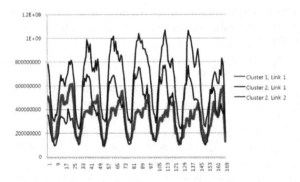

Fig. 9. Sample traffic patterns of normal clusters 1 and 2, X axis is hour and Y axis is transmission rate (bit per second)

The thick graph of Figure 9 is the traffic pattern of Cluster 1, the others are that of Cluster 2. The pattern of Cluster 1 usually has a peak transmission rate at evening, but the patterns of Cluster 2 have a peak transmission rate at noon.

6 Conclusion

In this paper, we introduce anomaly detection approach that applies Hierarchical Clustering algorithm from large volume traffic in backbone network. After hierarchical clustering of volume traffic, we divide every cluster into normal cluster and anomalous cluster. The traffic of anomalous cluster represents low traffic volume among boundary links. This is the detection of volume anomaly. The second work is detecting anomaly patterns after standardizing the volume traffic. This shows that anomalous traffic has relatively constant traffic patterns compared by normal traffic. From the analysis of anomaly pattern, we can obtain some different traffic patterns among normal clusters. Most of traffic belongs to one cluster in normal cluster, which is distributed in whole country. However some traffic has geographic locality and high transmission rate at noon, even though all monitored traffic is residential internet traffic.

This work is a static analysis from monitored traffic. We have a plan to develop real-time anomaly detection using hierarchical clustering. In this work, we choose hourly data with one week for analysis. It is necessary to decide which time interval is best for detecting anomaly. Finally, we are going to analyze whole traffic which includes access link, domestic and international link with other ISPs.

References

1. Thompson, K., Miller, G., Willder, R.: Wide area Internet traffic patterns and characteristics. IEEE Network 11, 10–23 (1997)
2. McCreary, S., Claffy, K.: Trends in wide area ip traffic patterns-a view from ames internet exchange. Technical report, CAIDA (2000)
3. Brownlee, N., Claffy, K.: Internet measurement. IEEE Internet Computing 08, 30–33 (2004)
4. Fraleigh, C., Moon, S., Lyles, B., Cotton, C., Khan, M., Moll, D., Rockell, R., Seely, T., Diot, C.: Packet-level traffic measurements from the sprint ip backbone. IEEE Network 17(6), 6–16 (2003)
5. Fukuda, K., Cho, K., Esaki, H.: The Impact of Residential Broadband Traffic on Japanese ISP Backbones. ACM SIGCOMM Computer Communications Review 35(1), 15–21 (2005)
6. Barford, P., Kline, J., Plonka, D., Ron, A.: A signal analysis of network traffic anomalies. In: ACM Internet Measurement Workshop, pp. 71–82 (2002)
7. Lakhina, A., Crovella, M., Diot, C.: Diagnosing Network-Wide Traffic Anomalies. In: ACM SIGCOMM, pp. 219–230 (2004)
8. John, W., Tafvelin, S.: Analysis of Internet Backbone Traffic and header Anomalies observed. In: 7th ACM SIGCOMM conference on Internet Measurement Conference (2007)
9. Lakhina, A., Crovella, M., Diot, C.: Mining Anomalies Using Traffic Feature Distributions. ACM SIGCOMM Computer Communications Review 35(4), 217–218 (2005)

10. Kim, H., Fomenkov, M., Barman, D., Faloutsos, M., Lee, K.Y.: Internet Traffic Classification Demystified: Myths, Caveats, and the Best Practices. In: ACM CoNEXT (2008)
11. Karagiannis, T., Papagiannaki, K., Faloutsos, M.: BLINC: Multilevel Traffic Classification in the Bark. In: ACM SIGCOMM (2005)
12. Karagiannis, T., Broido, A., Faloutsos, M.: Transport Layer Identification of P2P Traffic. In: 4th ACM SIGCOMM conference on Internet Measurement Conference (2004)
13. Jonhson, S.C.: Hierarchical Clustering Schemes. Psychometrika, 241–254 (1967)
14. Clustering, http://en.wikipedia.org/wiki/Cluster_analysis
15. Clustering,
    http://home.dei.polimi.it/matteucc/Clustering/tutorial_html/
    hierarchical.html
16. RFC 1213, http://www.ietf.org/rfc/rfc1213.txt
17. Correlation, http://en.wikipedia.org/wiki/Correlation

Memory-Efficient IP Filtering for Countering DDoS Attacks[*]

Seung Yeob Nam[1] and Taijin Lee[2]

[1] Dept. of Information and Communication Engineering, Yeungnam University,
214-1, Dae-Dong, Gyeongsan-si, Gyeongbook, 712-749, Korea
[2] KISA, IT Infrastructure Protection Division,
7th floor West IT Venture Tower 78 Garakbondong, Songpagu, Seoul, Korea

Abstract. We propose a two-stage Distributed Denial of Service (DDoS) defense system, which can protect a given subnet by serving existing flows and new flows with a different priority based on IP history information. Denial of Service (DoS) usually occurs when the resource of a network node or link is limited and the demand of the users for that resource exceeds the capacity. The objective of the proposed defense system is to provide continued service to existing flows even in the presence of DDoS attacks, and we attempt to achieve this goal by discriminating existing flows from new flows. The proposed scheme can protect existing connections effectively with a reduced memory size by reducing the monitored IP address set through sampling in the first stage and using Bloom filters. We evaluate the performance of the proposed scheme through simulation.

1 Introduction

A DoS attack is a malicious attempt to disrupt the service provided by networks or servers. These days the power of a DoS attack is amplified by incorporating over thousands of zombie machines through botnets [1] and mounting a distributed denial-of-service (DDoS) attack. Although a lot of defense mechanisms have been proposed to counter DDoS attacks [2], it still remains as a difficult problem especially because the attack traffic tends to mimic normal traffic recently.

If a small number of machines are participating in a DoS attack to a selected server, the IP addresses of those attack machines might be detected using the approaches of [3,4] without managing per-flow states. When a very large number of machines are participating in a DDoS attack, it may be possible to detect malicious IP addresses based on the observation that the IP addresses in a flash crowd have usually been seen by the given web site before, while very few IP addresses in DDoS attacks are seen by the web site before [5]. Peng et al. [6,7] studied how to detect DDoS attacks by monitoring the increase of new IP

---

[*] This work was supported by the IT R&D program of MKE/KEIT. [2009-S-038-01, The Development of Anti-DDoS Technology].

C.S. Hong et al. (Eds.): APNOMS 2009, LNCS 5787, pp. 301–310, 2009.

addresses based on the above observation. However, Peng at al. [6,7] did not consider the memory requirement and per-flow information management can be a very challenging task at a high speed link or for a high profile website.

We focus on protection of a given subnet in this paper and we attempt to avoid per-flow state management by monitoring only the hosts that are accessed by many external IP addresses. Since DDoS victims are accessed by many machines, the victim can be easily detected even though we monitor smaller set of packets through sampling. Thus, a smaller number of IP addresses are managed with a reduced memory size in the proposed scheme.

2   Defense Mechanism

According to Jung et al. [5], when the number of clients increases during a DDoS attack, most of them are from new IP addresses that have not been seen before. Based on this observation Peng et al.[6,7] used IP history database, which is referred to as the List of Existing IP Addresses (LEA) in this paper, to discriminate DDoS attack flows from normal flows. We follow a similar approach. However, their scheme requires per-flow state management, which may not be feasible at edge routers because of computational complexity or excessive memory requirement. Thus, the proposed scheme reduces the processing and memory requirement burden on the edge routers by adjusting the amount of monitored traffic by sampling.

The proposed DDoS defense mechanism consists of two stages. In the first stage of the proposed defense mechanism, we select internal IP addresses with many external connections by sampling in order to reduce the memory size for LEA. In the second stage, the LEA is constructed only for the IP addresses selected in the first stage. After constructing LEA, if the number of external IP addresses accessing a specific internal IP address exceeds some threshold, then it is regarded as an indication of DDoS attack, and existing connections are served with a higher priority than new connection attempts based on the information in the constructed LEA. More detailed operation in each stage is described below.

2.1   First Stage of the Proposed Defense Mechanism

In the first stage, the source and destination IP address pairs, (SrcIP, DstIP), of arriving packets are sampled. We sample the (SrcIP, DstIP) pairs with the sampling probability of p_s. Let h_p be a uniform random hash function that maps (SrcIP, DstIP) pairs to $[0, 1)$. When s and d are the source and destination IP addresses of an arriving packet, if $h_p(s, d) < p_s$, then the IP address pair (s, d) is sampled.

In the first stage of the proposed defense mechanism, we manage one connection status table T_1 and one Bloom filter B_1 as shown in Fig. 1. The hash table T_1 counts the number of distinct IP addresses accessing a specific destination IP in a given time interval I_1. Since we intend to protect internal nodes from DDoS attacks currently, we consider only internal IP addresses for destination

IP addresses in T_1. $COUNT1(d)$ counts the number of IP addresses accessing a sampled destination address d. In order to check whether a sampled source IP address has been counted already or not, each sampled source IP address is registered in the Bloom filter B_1 which is M_1-bit long and has k hash functions as shown in Fig. 1(b). We use Bloom filters [8] in order to reduce the memory size and B_1 is shared among different destination IP addresses in order to raise efficiency of the limited memory space. In case of collision in Bloom filter B_1, $COUNT1(d)$, i.e. the number of IP addresses accessing d might be underestimated. The size of the Bloom filter B_1, i.e. M_1, should be large enough to keep the collision probability at a negligible level.

If the value of $COUNT1(d)$ exceeds a pre-specified threshold N_{th}^1, then the IP address d is considered as a potential victim and LEA is constructed for d in the second stage until the onset of an intensive DDoS attack. $COUNT1(d)$ counts only sampled source addresses accessing d. Thus, we need to note that the total number of IP address that actually accessed d is N_{th}^1/p_s on average when $COUNT1(d) = N_{th}^1$.

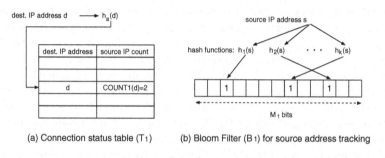

(a) Connection status table (T_1)   (b) Bloom Filter (B_1) for source address tracking

Fig. 1. The structure of tables managed in stage 1

Before considering the second stage, we investigate collisions in the hash table T_1. When a first packet from the source address s to d arrives at the router with the proposed DDoS defense system, the connection information for the destination address d is managed in the row $h_a(d)$ of the hash table T_1, where h_a is a uniform random hash function. When a destination address d_1 is mapped to the row r by the hash function h_a, if that row is occupied by a different address d_2, this kind of contention is called a collision in the hash table T_1. Since this kind of collision can occur with a non-negligible probability and can affect the performance of the defense system significantly, the entry management policy is very important in case of collision.

We manage the entries of the hash table T_1 based on the following idea. The IP addresses connected by many IP addresses need to be retained in the hash table because they might be the victim of a DDoS attack. Thus, we manage contention in hash table T_1 as follows. When a new IP address d_n arrives at the row occupied by other IP address d_o, if $COUNT1(d_o) > 1$, then we retain the entry for the old IP address d_o. If $COUNT1(d_o) = 1$, the new IP address replaces the entry for the old IP address.

2.2   Second Stage of the Proposed Defense Mechanism

If an IP address d is selected in the first stage because of the value of $COUNT1(d)$ larger than N_{th}^1, then there is a possibility that the IP address d has become a target of a DDoS attack. But, this expectation cannot be ascertained at this level because the first threshold N_{th}^1 is usually selected too low to be used for DDoS detection. The first threshold N_{th}^1 is only used to filter out the destination IP addresses accessed by small number of IP addresses, which are not likely a victim of a DDoS attack. Thus, in the second stage, we make a list of IP addresses which will be allowed to access the internal IP addresses selected in the first stage, i.e. LEA, in the presence of DDoS attacks.

For the destination IP addresses not selected in the first stage, no traffic regulation will be applied since the number of accessing IP addresses is small. For the destination IP addresses selected in the first stage, the following traffic control is enforced to support continued service to existing flows in the presence of DDoS attack. If the number of IP addresses accessing an internal IP address exceeds the second threshold N_{th}^2, which is larger than N_{th}^1/p_s, that internal IP address is considered to be under a DDoS attack and the packets from the IP addresses in LEA are served with a higher priority than the packets from other IP addresses. The packets from new IP addresses might be served with a lower priority or be dropped in case of severe congestion. Thus, the edge router equipped with the proposed defense system is performing *admission control* implicitly by limiting the number of external IP addresses connecting with an internal IP address based on LEA information. If the number of IP addresses accessing a specific internal IP address decreases below N_{th}^2, then new IP addresses can be registered in LEA until the number of connecting external IP addresses reaches N_{th}^2 again. If the number of external IP addresses decreases below a third threshold $N_{th}^3 (= N_{th}^1/p_s)$, then the corresponding internal IP address is considered not to be under DDoS attack and LEA for that IP address is not managed any more.

Hereafter, we investigate the management of LEA in more detail. In the second stage, we manage one connection status table T_2 and one Bloom filter B_2 as shown in Fig. 2. The hash table T_2 tracks the number of external IP addresses currently connected with each internal IP address selected in the first stage. $COUNT2(d)$ counts the number of IP addresses currently accessing the internal IP address d. One-bit field D is initialized to zero and set to one when a DDoS attack is detected for the corresponding internal address. $COUNT_p(d)$ retains old values of $COUNT2(d)$, and the detailed usage is explained below when the update of $COUNT2(d)$ is discussed in more detail. In order to reduce the memory size, the external IP addresses accessing a specific internal IP address are registered and managed in the Bloom filter B_2. To reduce the memory size further, B_2 is shared among all the internal IP addresses which were selected in the first stage, and thus, the Bloom filter B_2 manages LEA for the selected internal IP addresses. The major difference between B_1 and B_2 is that for B_2 a separate timer is allocated for each bit as shown in Fig. 2(b). $TIMER(i)$ represents the timer allocated for the i-th bit in the M_2-bit Bloom filter B_2. The field of $A(i)$ tracks the packet arrival from the external IP addresses which are using i-th bit in B_2 in a given time interval. The value of $A(i)$ is used for the update of $COUNT2(d)$, which will be described below. In our scheme, if there has been no packet between an IP address pair over some pre-specified time interval, then the connection between the IP address pair is considered to be disconnected. Thus, the timer in the Bloom filter B_2 is used to monitor the connectivity between internal and external IP address pairs.

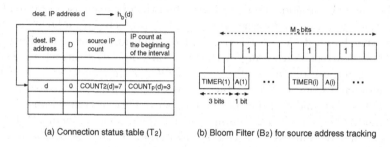

(a) Connection status table (T2)   (b) Bloom Filter (B2) for source address tracking

Fig. 2. The structure of tables managed in stage 2

More detailed operation of the timers is as follows. k_2 denotes the number of hash functions used in the Bloom filter B_2, and h_i^b denotes the i-th hash function used in B_2. When the internal IP address d is selected in the first stage and registered in T_2, if a first packet from a new source address s' to d is monitored by the router, then the value of $COUNT2(d)$ increases by one, the k_2 bit positions corresponding to $h_i^b(s')$ $(i = 1, 2, \ldots, k_2)$ of B_2 is set to one, and the values of the timers, $TIMER(h_i^b(s'))$ $(i = 1, 2, \ldots, k_2)$, are set to an integer R. The value of each timer is decreased by one periodically at the interval of I_T. Thus, when at least one timer for a given external IP address decreases down to zero after $(R-1) \times I_T$ seconds, the bit value corresponding to the expired timer is changed into zero and s' is considered to be disconnected from d. If a new packet from s' to d arrives before the timers expire, then the timers corresponding to the hash values of s' is set to R again. The detailed values of R and I_T need to be determined considering the packet interarrival pattern of normal flows.

Crovella et al. [9] has shown that the World Wide Web (WWW) traffic exhibits self-similarity and the silent times in the on-off pattern of Web traffic may be heavy-tailed, primarily due to the influence of user "think time." According to the measurement results of [10] http packet interarrival times still have a long-tailed distribution, but the ratio of packet interarrival times which is over 2 seconds is rather low, i.e. less than 3%. Thus, in this paper, if there is no packet exchange between a node pair over 300 seconds, the connection between the node pair will be regarded as being disconnected. Then, $(R-1) \times I_T$ should be no less than 300 seconds in the Bloom filter B_2 because an external IP address should be retained in B_2 at least 300 seconds. In order to reduce the memory size of the timer, we allocate three bits for R and R is set to 7. If we set I_T to 50 seconds, then $(R-1) \times I_T$ becomes 300 seconds. These values of R and I_T are the default parameters in our scheme.

In the proposed scheme, it is important to maintain the value of $COUNT2(d)$ consistent with the set of IP addresses remaining in the Bloom filter B_2 even after the timer expiration event. Let us assume that B_2 accurately manages the set of external IP addresses currently accessing an internal IP address d. If $COUNT2(d)$ is overestimating the number of external IP addresses currently connected with d because the timer expiration events are not reflected yet, then the proposed defense system will work conservatively. Although the utilization of the link or internal node might be lower than the optimal case because of the conservative response, the internal node d will be protected from DDoS attacks. On the other hand, if $COUNT2(d)$ is underestimating the number of external IP addresses currently accessing d, then the ongoing DDoS attack might not be detected because of the small value of $COUNT2(d)$, and the node d may not be protected from a DDoS attack in this case. Thus, we need to prevent $COUNT2(d)$ from underestimating the number of current connections.

Fig. 3 shows how $COUNT2(d)$ is managed in the proposed scheme. Whenever a new external IP address s'' accesses an internal IP address d, the value of $COUNT2(d)$ is increased by one. At the same time, s'' is also registered in the hash table B_2, the timers corresponding to the hash values of s'' is set to R, and the values of A's corresponding to the same hash values are set to one. The values of all A's are cleared to zero at the interval of $(R-1) \times I_T$. The A-clearing epoch corresponds to the thick broken line in Fig. 3. Thus, the A value tracks whether a new packet transmission is made from existing or new external IP addresses in each time interval of duration $(R-1) \times I_T$.

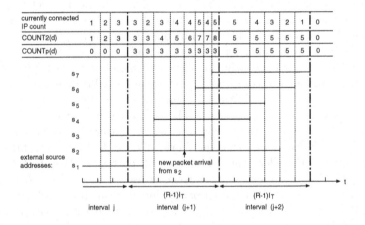

Fig. 3. The update of $COUNT2(d)$ and $COUNT_p(d)$ over time

In order to prevent $COUNT2(d)$ from underestimating the number of IP addresses currently connected with d, the value of $COUNT2(d)$ is increased promptly when a new external IP address accessing d is detected, but the expired connections in the current interval are reflected only at the end of the current interval in the following way. We need to note that when 10 new external IP addresses are detected in the j-th interval, if every address sends only one packet, then the timers for all those addresses will expire in the $(j+1)$-st interval, and the connection between them and d is considered to be disconnected. At the beginning of $(j+1)$-st interval, $COUNT_p(d)$ is set to the number of IP addresses connected with d at the end of the j-th interval, and maintained throughout that interval. The IP addresses counted in $COUNT_p(d)$ is deducted from $COUNT2(d)$ at the beginning of the $(j+2)$-st interval because the timers corresponding to them must have expired in the $(j+1)$-st interval if there was no other packet from those IP addresses. The value of $COUNT_p(d)$ is set to the newly calculated $COUNT2(d)$ at the beginning of each interval. If an IP address, which arrived before $(j+1)$-st interval, sends a new packet to d in the $(j+1)$-st interval, then $COUNT2(d)$ is increased promptly on packet arrival since the lifetime of that connection is extended by at least $(R-1) \times I_T$. This case corresponds to s_2 in Fig. 3. If we manage the value of $COUNT2(d)$ in this way, i.e. by increasing the value promptly and performing decrement later, we can maintain $COUNT2(d)$ not lower than the real IP count as shown in Fig. 3.

2.3   Algorithm

Finally, we summarize the proposed DDoS detection and control mechanism in Fig. 4. If $COUNT2(d) > N_{th}^2$, then existing flows registered in LEA, i.e. B_2, is served with a higher priority than new flows. When a new packet arrives from a malicious node's IP address s_m, if all the bit values corresponding to the hash values of s_m is already set to one by other existing IP addresses in B_2, then this is called a collision in B_2, and this bad flow will be served with a higher priority. Thus, the size of the Bloom filter B_2 needs to be large enough to keep the false negative probability due to the collision in B_2 sufficiently low.

Let R_{T_2} denote the number of rows in the hash table T_2 in Fig. 2. Then, the Bloom filter B_2 can accommodate only up to $R_{T_2} \times N_{th}^2$ external IP addresses because no more addresses are allowed if the value of $COUNT2$ is larger than N_{th}^2 for each internal IP address in T_2. k_2 is the number of hash functions used for B_2. After inserting n' IP addresses into a bit vector of size M_2, the probability that a particular bit is still zero is $(1 - 1/M_2)^{k_2 n'}$. The collision in Bloom filters occurs when a new external IP address observes one in every bit position indicated by k_2 hash functions. The collision probability of the arriving source IP addresses after the $(R_{T_2} N_{th}^2)$-th registered address is then

$$(1 - (1 - 1/M_2)^{k_2 R_{T_2} N_{th}^2})^{k_2} \approx (1 - e^{-k_2 R_{T_2} N_{th}^2 / M_2})^{k_2}. \qquad (1)$$

The right-hand size of (1) is minimized when $k_2 = \ln 2 \times M_2/(R_{T_2} N_{th}^2)$ [8]. It then becomes

$$(1/2)^{k_2} = (0.6185)^{M_2/(R_{T_2} N_{th}^2)}. \qquad (2)$$

According to (2), k_2 should be equal to or larger than 7 in order to keep the collision probability less than 1%. Thus, k_2 is set to 7 in this paper. In this case,

M_2 can be determined in terms of R_{T_2} and N_{th}^2 as

$$M_2 = \lceil k_2 R_{T_2} N_{th}^2 / \ln 2 \rceil \simeq \lceil 10.1 \times R_{T_2} N_{th}^2 \rceil. \qquad (3)$$

The size of M_1 can be determined in a similar way.

```
D(d) is initialized to zero.
if(packet arrival from an external IP s to an internal IP d)
    if(d is registered in T₂){   /* Then, d is in stage 2. */
        if( COUNT2(d) > N²th){
            /* Then, the traffic to d is scheduled with a strict priority policy. */
            if( s is in LBA, i.e. B₂)    the packet is sent to a higher priority queue;
            else                         the packet is sent to a lower priority queue;
        }
        else if( D(d) = 1 and COUNT2(d) < N³th){   /* The DDoS attack is over.*/
            delete d from T₂; D(d)=0; /* The entries in B₂ will soon expire.*/
        }
        else{   /* The LBA construction phase */
            if( s is registered in B₂){
                set the corresponding timers to R;
                if( there is any A(i) with the value of zero in the corresponding position)
                    set the values of A(i)'s to 1;   increase COUNT2(d) by one;
            }
            else{
                register s in B₂;
                set the corresponding timers to R;
                set the values of A(i)'s to 1;
                increase COUNT2(d) by one;
            }
            if( COUNT2(d) > N²th)   D(d)=1;
        }
    }
    else{ /* d is not in T₂. Then, d is in stage 1, and we sample (s,d) pairs */
        if( hp(s,d) < ps){   /* (s,d) pair is sampled.*/
            if( d is not in T₁){
                register d in T₁;
                COUNT1(d)=1;
                register s in B₁;
            }
            else{
                if( s is not in B₁){
                    increase COUNT1(d) by one;
                    register s in B₁;
                }
            }
            If( COUNT1(d) > N¹th)
                d is registered in T₂; d is cleared from T₁;
        }
    }
}
```

Fig. 4. Summary of the proposed DDoS detection and control algorithm

3   Numerical Results

We evaluate the performance of the proposed scheme through OPNET simulation. We consider how well existing traffic can be protected in the presence of DDoS attack, and measure it by the ratio successfully delivered traffic to the sent traffic, i.e. throughput. Fig. 5(a) shows the network topology for simulation. The proposed DDoS defense system is deployed at the edge router $R1$. $Z2$ clients are accessing U servers, and $Z1$ attackers are mounting attack on them later. $Z3$ external users are connected with V internal users before the DDoS attack. In order to consider the case of bandwidth congestion attack, we let all the traffic from the external network pass a single link between $R1$ and $R2$, whose link rate is 100 Mbps. In order to investigate the scheme when the attack traffic pattern is not discriminated from normal traffic pattern, we model both attack and normal traffic using on-off traffic, which has Pareto distributions for on and off times [9]. The probability density function is given as $p(x) = \alpha k^{\alpha} x^{-\alpha-1}, x \geq k$. The parameters are selected as $\alpha_{ON} = 1.2$, $\alpha_{OFF} = 1.5$, $k_{ON} = 0.167$, and $k_{OFF} = 10$. Then, the average on and off durations are 1 and 30 seconds, respectively. During on periods, we send 500 Byte packets at the constant rate of 1.55 Mbps. Then, the average rate of one flow is 50 Kbps. In our scenario, the number of attackers $Z1$ is set to 4000, and thus, the average rate of the aggregate attack traffic is 200 Mbps, much higher than the link rate between $R1$ and $R2$. The number of victim servers U is 10, the number of internal users V is 100, the number of normal clients accessing the victim servers $Z2$ is 200, and the number of external users accessing the internal users $Z3$ is set to 200.

Fig. 5(b) compares the throughput provided by the proposed system with the throughput obtained by FIFO queue with no defense mechanism. The DDoS attack is offered from 900 seconds, and we observe that much higher throughput is maintained by the proposed mechanism. We also observe that the proposed scheme effectively protects the users who are not the targets of the DDoS attack, but are in the same subnet as the DDoS victim nodes. If we let R_{T_1} and R_{T_2} denote the number of rows in T_1 and T_2, then the total required memory size is $R_{T_1}(48 + 10.1 N_{th}^1) + R_{T_2}(65 + 10.1 N_{th}^2)$. In the considered scenario, we set the parameters as $R_{T_1} = 50$, $R_{T_2} = 20$, $N_{th}^1 = 3$, and $N_{th}^2 = 50$. In this case, the required memory size is less than 2KB. The optimal parameter selection problem will be investigated further in the future.

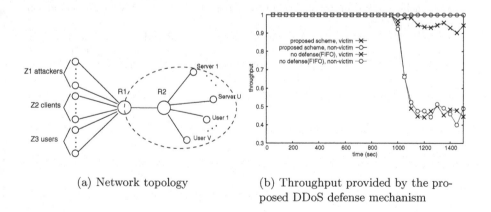

(a) Network topology

(b) Throughput provided by the proposed DDoS defense mechanism

Fig. 5. Simulation Result

4 Conclusions

We proposed a two-stage DDoS defense mechanism which protects a given subnet by serving existing flows and new flows with a different priority based on IP history information with a reduced memory size. The memory reduction is achieved by reducing the monitored IP address set through sampling in the first stage and using Bloom filters in both stages. The simulation result shows that high throughput can be maintained for existing flows in the presence of DDoS attack with a small memory size.

References

1. Dagon, D., Gu, G., Lee, C.P., Lee, W.: A Taxonomy of Botnet Structures. In: Proc. of Annual Computer Security Applications Conference (ACSAC) (December 2007)
2. Peng, T., Leckie, C., Ramamohanarao, K.: Survey of Network-Based Defense Mechanisms Countering the DoS and DDoS Problems. ACM Computing Surveys 39(1) (April 2007)
3. Estan, C., Varghese, G.: New Directions in Traffic Measurement and Accounting. In: Proc. of ACM SIGCOMM (August 2002)
4. Kompella, R.R., Singh, S., Varghese, G.: On Scalable Attack Detection in the Network. In: Proc. of ACM Internet Measurement Conference (IMC) (October 2004)
5. Jung, J., Krishnamurthy, B., Rabinovich, M.: Flash Crowds and Denial of Service Attacks: Characterization and Implication for CDNs and Web Sites. In: Proc. of World Wide Web (WWW) Conference (May 2002)
6. Peng, T., Leckie, C., Ramamohanarao, K.: Protecting from Distributed Denial of Service Attack Using History-based IP Filtering. In: Proc. of IEEE ICC, May 2003, pp. 482–486 (2003)
7. Peng, T., Leckie, C., Ramamohanarao, K.: Proactively Detecting Distributed Denial of Service Attacks Using Source IP Address Monitoring. In: Proc. of Networking Conference, May 2004, pp. 771–782 (2004)
8. Fan, L., Cao, P., Almeida, J., Broder, A.Z.: Summary cache: a scalable wide-area web cache sharing protocol, Technical Report 1361, Univ. of Wisconsin-Madison (February 1998)
9. Crovella, M.E., Bestavros, A.: Self-similarity in world wide web traffic: evidence and possible causes. IEEE/ACM Trans. Networking 5(6), 835–846 (1997)
10. Sun, Z., He, D., Liang, L., Cruickshank, H.: Internet QoS and traffic modelling. IEEE Proceedings 151(5), 248–255 (2004)

Framework Design and Performance Analysis on Pairwise Key Establishment

Younho Lee[1] and Yongsu Park[2]

[1] Dept. of Information and Communication Engineering, Yeungnam Univ., Korea
[2] College of Information and Communications, Hanyang Univ., Korea
yhlee@ynu.ac.kr, yongsu@hanyang.ac.kr

Abstract. Pairwise key establishment provides an effective way to build secure communication links among sensor nodes using cryptographic techniques. Up till now, researchers have devised numerous schemes that employ diverse cryptographic or combinatoric methods in order to provide high security, high connectivity and low storage overheads on the sensors. In this paper, we present a new framework on pairwise key establishment. We show that it can encompass most of the major previous schemes. Furthermore, we analyze the performance of the previous schemes using the proposed framework. When 100% connectivity is provided, under the same storage overhead on the sensor nodes, Blundo scheme and Blum scheme provide the highest security against intelligent adversaries while YG-L scheme is the best against random adversaries. When connectivity is under 100%, for random adversaries, location-aware schemes provide better security than non-location based schemes whereas for intelligent adversaries, all location based schemes' security is less than that of other schemes.

Keywords: Wireless sensor networks, network security, secure communications.

1 Introduction

While wireless devices are omnipresent in our lives, wireless sensor networks (WSN) are increasingly popular to deal with many monitoring problems, such as real-time traffic monitoring, wildlife tracking, bridge safety monitoring, etc. Sensor networks consist of a large amount of sensor nodes that are typically resource-constrained devices with wireless communication and data processing capabilities. These nodes collaborate with each other to build communication networks and to accomplish their task.

Since sensor nodes can be deployed in hostile environments, security concern is very important in WSN. Many security breaches or attacking methods have been found, such as routing attacks, communication eavesdropping/modification, impersonation, DoS (Denial-of-Service) attacks, etc. Moreover, resource constraints and wireless environments add difficulties for securing sensor networks.

In security services, authentication and secure communication are crucial and pairwise key establishment is one of the most popular and convenient techniques

C.S. Hong et al. (Eds.): APNOMS 2009, LNCS 5787, pp. 311–320, 2009.
© Springer-Verlag Berlin Heidelberg 2009

to achieve them. In this method, initially a key pool which contains a large amount of symmetric keys is generated. Then, each sensor node picks a set of keys from the key pool. After all the nodes are deployed, every two neighboring nodes try to find a common key with each other pair-wisely. By using the common key, they can establish a secure channel. If a sensor node wants to communicate with non-authenticated nodes, neighboring nodes that were already authenticated act as intermediates for authentication and key exchange.

Up till now, numerous pairwise key establishment schemes have been devised [9,5,4,6,12,10,8,7,11]. To increase performance they use diverse cryptographic or combinatoric techniques such as multivariate functions, linear algebra, finite field theory, and so on.

In this paper, we present a new framework on pairwise key establishment which can encompass most of the major previous schemes. Furthermore, we analyze the performance of the previous schemes using the proposed framework. When 100% connectivity is provided, under the same memory constraint on the sensor nodes, Blundo scheme and Blum scheme provide the highest security against intelligent adversaries while YG-L scheme is the best against random adversaries. When connectivity is under 100%, for random adversaries, location-aware schemes provide better security than non-location based schemes whereas for intelligent adversaries, all location based schemes' security is less than that of other schemes.

Our scheme can be viewed as an approach to increase trustworthiness as well as to enhance security of the systems to be protected. The rest of this paper is organized as follows. In Section 2 we describe previous work. Section 3 deals with preliminary information for describing our framework. In Section 4 we build a new framework and show that it compasses most of the major schemes. Then, In Section 5 we analyze the previous schemes by using the proposed framework and finally we offer some conclusions in Section 6.

2   Related Work

The first pairwise key establishment scheme is Eschenauer-Gligor scheme (a. k. a. EG scheme) [6]. This scheme consists of 3 phases. In the first phase, the scheme generates a key pool that contains s symmetric keys. Each node randomly selects w keys from the key pool. In the second phase, nodes are deployed and each pair of neighboring nodes tries to find a common key. If they find it, this key is used to secure the communication link between them. When the second phase is finished, a graph can be constructed, where vertices represent sensor nodes and an edge exists iff two nodes already established the common key. In the final phase, assume that a node i wants to establish another non-neighboring node j. If there is a path from i to j, intermediate nodes can act as intermediates to exchange common keys between i and j.

Blundo et al.'s method [2] can be used in pairwise key establishment as follows. [2] defines a special bivariate t-degree polynomial function $f(x,y) = \sum_{i,j=0}^{t} a_{ij}x^i y^j$ over finite field F_q, which has the property that $f(x,y) = f(y,x)$.

Each sensor i has his/her polynomial share, $f(i, y)$, which requires $(t + 1) \log_2 q$ storage overhead. Then, for any two sensor nodes i and j, they can calculate a common key $f(i, j) = f(j, i)$. This scheme is secure until up to t node capturing. If an adversary captures at last $t + 1$ nodes, he can compute $f(x, y)$ for all x and y.

Liu and Ning proposed an efficient key predistribution method [9] using the bivariate function of [2], which can be a combination of EG scheme and Blundo method. Instead of containing s symmetric keys in the key pool of EG scheme, in Liu and Ning's method (which will be called LN scheme hereafter) a key pool contains s polynomial functions. Then, each sensor randomly selects w functions. After deployment neighboring nodes try to find a common function. If they find it, they use Blundo method to share the common key.

DDHV [5] uses Blom scheme [1] that relies on the linear algebraic technique on finite fields. Even though the technique is completely different from [2], characteristics of Blom scheme is very similar to that of Blundo method. Moreover, characteristics of DDHV is very similar to that of LN scheme.

Recently many pairwise key establishment schemes use deployment information of sensor nodes to increase security and connectivity and to minimize storage overhead on sensors (we will call them *location-based* schemes, hereafter). DDHV-L [4] and LN-L scheme [10] can be considered as enhancements of DDHV and LN scheme, respectively. To the best of our knowledge, YG-L scheme [12] provides 100% connectivity among sensor nodes.

3   Preliminaries

In this section we briefly explain important performance criteria for pairwise key establishment and classify adversaries.

3.1   Performance Criteria for Pairwise Key Establishment

In this subsection we identify performance criteria for pairwise key establishment and explain their importance, as follows.

- **Space overhead for the sensor nodes:** In pairwise key establishment schemes, each node should store symmetric keys or key generation information to share keys with its neighboring nodes. When there are n nodes in WSN, we can consider a naive pairwise key establishment algorithm where each node stores $n - 1$ keys for all other nodes, which requires a large amount of space overheads. Generally, in large-scaled WSN there are numerous nodes whereas each sensor nodes have small memory (e.g., Berkeley Mica Motes have Atmega 128L micro-controller with 128KB flash memory [3]).
- **Connectivity:** When there is a severe memory constraint to store keys, sensor nodes may not find a common symmetric key with their neighbors, which results in failure of secure communication. Hence, connectivity is one of important aspects in pairwise key establishment.

- **Security:** As mentioned in Section 1, there are various attacks in WSN. In the first type of the attack, adversaries may acquire a certain node and extract keys to eavesdrop the secure communication or alter the communicated message. In the second, they can consume the battery of the nodes or block the network traffic. There are many other types of attacks. However, among various attacks, we focus on the first type of the attack, which is the most serious in pairwise key establishment.
- **Computation overhead:** Some pairwise key establishment schemes require that nodes should perform heavy computation such as $GF(q)$ operations whereas some others are very computationally efficient. In this paper we do not consider computation overhead to simplify analysis.
- **Communication overhead:** In WSN, each node consumes significant energy to transmit data via wireless channels. However, we cannot compare the communication overhead of the previous schemes since they are not fully optimized. We also omit this aspect for ease of analysis.

3.2   Classification of Adversaries in Pairwise Key Establishment

In this subsection we classify adversaries in pairwise key establishment as follows: random adversary, passive intelligent adversary, and active adversary.

- **Random adversary:** Random adversary is a naive attacker, who randomly select some nodes and extract key generation information from them. Then, he tries to eavesdrops the communication between other unattacked nodes.
- **Passive intelligent adversary:** We assume that he already has knowledge on the algorithms or protocols. Using this knowledge, he can selectively capture the nodes and then he tries to eavesdrops the communication between other unattacked nodes. However, we assume that he is not interested in forging nodes, modifying the communication messages, etc.
- **Active adversary:** In addition to the capability of passive intelligent adversary, he can use any method to interrupt/corrupt WSN, e.g., forging some nodes, jamming the communications, trying DoS attack, modifying the communication message, etc.

As in most previous work, we assume that there are only random adversaries and passive intelligent adversaries, which simplifies analysis.

4   Design of Framework on Pairwise Key Establishment

In this section we describe pairwise key establishment framework which encompasses most of the major previous schemes. In Section 4.1, we describe the new framework and in Section 4.2, we show that major previous schemes can be expressed via the proposed framework. In Section 5, we will analyze the previous schemes in terms of memory requirement, security and connectivity by using the proposed framework.

4.1   Overview of the Proposed Framework

Assume that there are n sensor nodes and 1 base station. The framework consists of 3 phases: key predistribution, shared-key discovery, and path-key establishment.

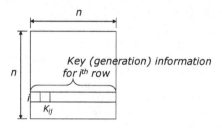

Fig. 1. Key matrix for threshold key space TS_k

- **Step 1. key predistribution:** In this step, a key pool S is generated where S consists of multiple key spaces TS_k $(1 \leq k \leq s)$, which we will call *threshold key spaces* hereafter. A threshold key space TS_k has the following properties.
  1. TS_k contains n^2 symmetric keys K_{ij} $(1 \leq i, j \leq n)$, which can be expresses as a *key matrix* (See Fig. 1). Note that among n^2 keys some of them can be identical.
  2. $K_{ji} = K_{ij}$, which implies that the key matrix is symmetric.
  3. each sensor node i selects ith row in the matrix and receives either n key values or certain information to compute them (both of which will be called *key generation information* hereafter).
  4. TS_k has a security parameter t_k. If an attacker captures some nodes and acquires $t_k + 1$ key generation information, then he can successfully reconstruct the matrix that is equal to n^2 keys. If that case occurs, we will call that TS_k is *revealed*.
  5. The storage overhead for ith key generation information is $\Theta(t_k)$.

  Note that in the key space $S = \{TS_k | (1 \leq k \leq s)\}$, each TS_k's security parameter t_k may not be the same with each other.

  After S is generated, each node i $(1 \leq i \leq n)$ selects w TS_k in S and store i'th key generation information for all selected TS_k. Hence, the space overhead for each node is $w \frac{\sum_{k=1}^{s} \Theta(t_k)}{s}$.

- **Step 2. Shared-key discovery:** After all the nodes are deployed in the field, neighboring two nodes (i, j) try to find some common TS_k with each other. If there exist at least z common $TS_{k_1}, ..., TS_{k_z}$, they can computes z common key $K_{ij}^x = K_{ji}^x$ for TS_{k_x} $(1 \leq x \leq k)$.

  Finally, node i, j compute a common key K using K_{ij}^x $(1 \leq x \leq k)$, e.g., $K = \oplus_{x=1}^{z} K_{ij}^x$. By using K, they can establish a secure communication channel.

- **Step 3. Path-key establishment:** When Step 2 is finished, status of secure channel establishment can be expressed as a graph where vertices represent sensor nodes and an edge exists iff two nodes already established the common key K.

  Assume that a node i wants to establish another non-neighboring node j. If there is a path from i to j, intermediate nodes can act as intermediates to exchange common keys between i and j, e.g., see Fig. 2.

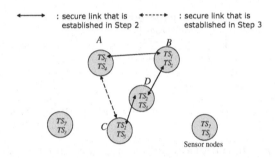

Fig. 2. An example for Step 2 and Step 3 in the framework

4.2   Relation between the Previous Schemes and the Proposed Framework

In this section we estimate feasibility of the proposed framework by showing that the major previous schemes can be expressed by the framework. We deal with the following previous algorithms: EG scheme [6], q-composite [3], Blom scheme [1], Blundo schme [2], DDHV [5], LN [9], DDHV-L [4], IOMY-L [8], and LN-L [10] (addition to these, the proposed framework can encompass other not-mentioned schemes such as [7,11]). The following table shows the summary of parameter values when those schemes are expressed via the framework.

- **EG scheme [6]:** Assume that in the framework, TS_k $(1 \leq k \leq s)$ has security parameter $t_k = 0$. This implies that when $t_k + 1 = 1$ nodes are captured, the matrix is revealed, i.e., TS_k matrix consists of all the same value K_k. Hence, key space $S = \{TS_k | (1 \leq k \leq s)\} = \{K_k | (1 \leq k \leq s)\}$. Each sensor node selects w TS_k from S. In Step 2, when the nodes are deployed, neighboring nodes find their common key and if there exists at least one common key, secure communication is established by using it, i.e., $z = 1$. Step 3 of EG scheme is identical to that of the framework.

- **q-composite [3]:** This scheme is identical to EG scheme except for the Step 2, as follows. After deployment, if two neighboring nodes have at least q common keys, they generate a common session key and establish secure channel. Hence, In the framework, $t_k = 0$ and $z = q$.

- **Blum scheme [2], Blundo scheme [1]:** Key space S consists of a single threshold key space, i.e., $S = TS_1$ and $t_1 = t$. Each node i has the ith

Table 1. Summary of parameter values when schemes are expressed via the framework

Scheme	num. of TS_k $(= s)$	num. of selected per node $(= w)$	security parameter $(=t_k)$	key composition parameter $(= z)$	deployment method
EG	s	w	0	1	random
q-composite	s	w	0	q	random
Blom	1	1	t	1	random
Blundo	1	1	t	1	random
DDHV	s	w	t	1	random
LN	s	w	t	1	random
IOMY-L	s	w	0	1	location-based
DDHV-L	s	5	t	1	location-based
LN-L	s	$1{\sim}2$	t	1	location-based

key generation information in the TS_1. After deployment, every neighboring nodes can calculate the common key $K_{ij}=K_{ji}$ by using their key generation information.

- **DDHV [5], LN [9] scheme:** These schemes are identically represented in our framework. In S, there are s TS_k, i.e., $S = \{TS_k|(1 \le k \le s)\}$. Security parameter t_k for TS_k is t for all k. Each sensor node selects w TS_k from S and stores their key generation information. After deployment, every neighboring node finds a common TS_k with each other. If exists, they can calculate the common key $K_{ij}=K_{ji}$ by using their key generation information in the common TS_k.

- **IOMY-L scheme [8]:** This scheme can be considered as a variant of EG scheme. Like EG scheme, key space $S = \{TS_k|(1 \le k \le s)\} = \{K_k|(1 \le k \le s)\}$ where $t_k = 0$. Unlike EG scheme that assigns random w TS_k for each sensor, IOMY-L scheme uses location information, i.e., each sensor i is assumed to be located according to the probability density function $f_i(x,y)$ where x, y represent indexes of 2-dimensional cell C_{xy}. Then, each TS_k corresponds to one of 2-dimensional cell C_{xy} and then every sensor has w TS_k according to $f_i(x,y)$.

- **DDHV-L [4]:** DDHV-L is an enhancement of DDHV with location information of sensors. As in DDHV, $S = \{TS_k|(1 \le k \le s)\}$ where $t_k = t$ for all k. Sensors are grouped into G_{ij} according to their expected deployment location cell C_{ij}. Then, from S subset S_{ij} corresponding C_{ij} is defined such that each S_{ij} has some TS_k and S_{ij} shares a few TS_k with its neighboring subset $S_{(i\pm1)(j\pm1)}$. Then, each sensor corresponding to G_{ij} randomly chooses w TS_k in S_{ij}. Remaining part is identical to DDHV-L.

- **LN-L scheme [10]:** LN-L scheme is an improvement of LN scheme with location information of sensors. As in LN scheme, $S = \{TS_k|(1 \le k \le s)\}$ where $t_k = t$ for all k. Sensors are grouped into G_{ij} according to their expected deployment location cell C_{ij}. Then, from S subset S_{ij} is defined such that each S_{ij} has a distinct TS_k. Then, each sensor corresponding to

C_{ij} chooses 5 TS_k from S_{ij}, $S_{(i\pm1)j}$, $S_{i(j\pm1)}$. Remaining part is identical to LN-L scheme.

- **YG-L [12]:** This scheme provides 100% connectivity by sharing at least 1 TS_k is shared between any neighboring nodes. As in LN scheme/DDHV, $S = \{TS_k | (1 \leq k \leq s)\}$ where $t_k = t$ for all k. Sensors are grouped into G_{ij} according to their expectedly located cell C_{ij}. S is partitioned into S_1 and S_2 where there are totally N cells C_{ij} and $|S_1| = N$. Then, each sensor corresponding to C_{ij} chooses one TS_k from S_1 and $1\sim2$ TS_k from S_2 such that any two nodes corresponding to neighboring cells have at least a common $TS_k \in S_2$.

5   Analysis

In this section, we analyze the previous schemes, by using the proposed framework, with respect to connectivity, security, and storage overhead for sensors, as follows.

5.1   Maximal Security for 100% Connectivity and Fixed Storage Overhead

To the best of our knowledge, there are 3 major schemes to provide 100% connectivity: Blundo scheme, YG-L scheme, and Blum scheme. We analyze security of them as follows.

Passive intelligent adversary: First we consider how to provide maximal security for passive intelligent adversaries under the condition of 100% connectivity and fixed storage overhead for sensor nodes. In the framework, the average storage overhead for each node is $w(\frac{\sum_{k=1}^{s} \Theta(t_k)}{s})$. Since we assume that connectivity is 100%, in the framework every pair of neighboring nodes should have at least 1 common TS_k.

If the average storage overhead for each node, $(=w(\frac{\sum_{k=1}^{s} \Theta(t_k)}{s}))$, is fixed, t_k has the maximum value for $w = 1$. Since the framework is safe until at most t_k nodes are captured for TS_k, the schemes for $w = 1$ have the strongest security and such schemes are Blundo scheme and Blum scheme.

Notice that all the location-based schemes such as YG-L scheme should have $w > 1$ by the following reason. Suppose that $w = 1$ in the framework. Since all the neighboring nodes should have at least one common TS_k, all the node have the single common TS_k, which is not a location-based scheme any more. Hence all location-based schemes should $w > 1$. I.e., security of location-based schemes is lower than that of Blundo/Blum schemes.

Random adversary: We consider security of the framework against random adversaries. If the average storage overhead for each node, $w(\frac{\sum_{k=1}^{s} \Theta(t_k)}{s})$, is fixed, t_k has the maximum value for $w = 1$. Since the framework is safe until at most

t_k nodes are captured for TS_k, the schemes for $w = 1$ have the strongest security and such schemes are Blundo scheme and Blum scheme for maximal t_k nodes capturing.

Consider the case of capturing more than t_k nodes. The passive intelligent adversary can select the nodes having the same TS_k and TS_k is revealed. However, random adversary simply select the node randomly. In this case, the probability of one TS_k is revealed is $1 - \sum_{i=0}^{t_k} x!/(x - i)!/i!(w/s)^i(1 - w/s)^{(x-i)}$ [9]. By using this equation, probability of compromised links in YG-L scheme and that in Blundo/Blum schemes can be compared (we omit the analysis for lack of space). As analyzed in [10,4,12], location-based schemes including YG-L scheme show better security than Blom/Blundo schemes, for large node capturing attack.

5.2   Acquiring Maximal Security and High Connectivity Given Fixed Storage Overhead

Recall that the storage overhead for ith key generation information is $\Theta(t_k) = c(t_k + 1)$. By using this, we consider how to acquire maximal security in the framework as follows.

Random adversary: We consider security of the framework against random adversaries. Assume that each TS_k has a security parameter t_k and the probability of choosing this for each node is p_k ($0 \le p_k \le 1$). Then, the average storage overhead for each node is $\sum_{k=1}^{s} p_k c(t_k + 1)$. If TS_k is revealed, the adversary can calculate $n^2 p_k$ keys.

Hence, if the adversary randomly attacks k nodes, the expected value of the number of the captured key is $\sum_{k \; s.t. \; kp_k > t_k}^{s} n^2 p_k$. Hence, given c, s, t_k, we should find p_k to minimize $\sum_{k=1}^{s} p_k c(t_k + 1)$ and $\sum_{k \; s.t. \; kp_k > t_k}^{s} p_k n^2$. These two values are smaller if p_k is smaller. However, p_k approaches to 0 connectivity becomes lower. Hence, location-aware schemes are better than other schemes since they can assign common TS_k for each neighbor while keeping p_k small.

Passive intelligent adversary: Recall that if the average storage overhead for each node, $w(\frac{\sum_{k=1}^{s} \Theta(t_k)}{s})$, is fixed, t_k has the maximum value for $w = 1$. Since the framework is safe until at most t_k nodes are captured for TS_k, the schemes for $w = 1$ have the strongest security and such schemes are Blundo scheme and Blum scheme.

Consider the case of capturing more than t_k nodes. If a passive intelligent adversary can select the nodes having the same TS_k and TS_k is revealed. Recall that in the all location-based schemes each neighbor should share TS_k with very high probability. Hence, by using this information he can selectively choose the nodes in the same cell or neighboring cell, which results in revealing of TS_k. Therefore, security of location-based schemes is lower than that of non-location based schemes.

6 Conclusion

In this paper, we presented a novel framework on pairwise key establishment. We showed that it can encompass most of major previous schemes which rely on diverse cryptographic or combinatoric methods. Moreover, we analyzed the performance of the previous schemes using the proposed framework. When 100% connectivity is provided, under the same memory constraint on the sensor nodes, Blundo scheme and Blum scheme provide the highest security against intelligent adversaries while YG-L scheme is the best against random adversaries. When connectivity is under 100%, for random adversaries, location-aware schemes provide better security than non-location based schemes whereas for intelligent adversaries, all location based schemes' security is less than that of other schemes.

References

1. Blom, R.: An Optimal Class of Symmetric Key Generation Systems. In: CRYPTO 1984 (1985)
2. Blundo, C., De Santis, A., Herzberg, A., Kutten, S., Vaccaro, U., Yung, M.: Perfectly-secure key distribution for dynamic conferences. In: Brickell, E.F. (ed.) CRYPTO 1992. LNCS, vol. 740, pp. 471–486. Springer, Heidelberg (1993)
3. Chan, H., Perrig, A., Song, D.: Random key predistribution schemes for sensor networks. In: Proceedings of IEEE Symposium on Security and Privacy (2003)
4. Du, W., Deng, J., Han, Y.S., Varshney, P.K.: A Key Predistribution Scheme for Sensor Networks Using Deployment Knowledge. IEEE Trans. on Dependable and Secure Computing 3(1) (2006)
5. Du, W., Deng, J., Han, Y.S., Varshney, P.K.: A Pairwise Key Predistribution Scheme for Wireless Sensor Networks. In: Proceedings of the 1st ACM workshop on Security of ad hoc and sensor networks (2003)
6. Eschenauer, L., Gligor, V.D.: A Key-Management Scheme for Distributed Sensor Networks. In: ACM CCS 2002, pp. 41–47 (2002)
7. Huang, D., Mehta, M., Medhi, D., Harn, L.: Location-aware Key Management Scheme for Wireless Sensor Networks. In: IPSN 2005 (2005)
8. Ito, T., Ohta, H., Matsuda, N., Yoneda, T.: A Key Pre-Distribution Scheme for Secure Sensor Networks Using Probability Density Function of Node Deployment. In: Proceedings of the 3rd ACM workshop on Security of ad hoc and sensor networks (2005)
9. Liu, D., Ning, P.: Establishment Pairwise keys in distributed sensor networks. In: ACM CCS 2003 (2003)
10. Liu, D., Ning, P.: Location-based Pairwise key establishments for static sensor networks. In: Proceedings of the 1st ACM workshop on Security of ad hoc and sensor networks (2003)
11. Mohaisen, A., Nyang, D.-H.: Hierarchical Grid-Based Pairwise Key Predistribution Scheme for Wireless Sensor Networks. In: Römer, K., Karl, H., Mattern, F. (eds.) EWSN 2006. LNCS, vol. 3868, pp. 83–98. Springer, Heidelberg (2006)
12. Yu, Z., Guan, Y.: A Key Pre-Distribution Scheme Using Deployment Knowledge for Wireless Sensor Networks. In: Römer, K., Karl, H., Mattern, F. (eds.) EWSN 2006. LNCS, vol. 3868, pp. 83–98. Springer, Heidelberg (2006)

MIH-Assisted PFMIPv6 Predictive Handover with Selective Channel Scanning, Fast Re-association and Efficient Tunneling Management

Igor Kim, Young Chul Jung, and Young-Tak Kim*

Dept. of Information and Communication Engineering, Graduate School, Yeungnam University. 214-1, Dae-Dong, Gyeongsan-si, Gyeongsangbuk-do, 712-749, Korea
ikim@ynu.ac.kr, callaguy@yumail.ac.kr, ytkim@yu.ac.kr

Abstract. In this paper, we investigate the major limitations of PFMIPv6, and explaine detailed lower layer procedures, which may significantly affect the handover performance. We propose a solution to reduce the overall channel scanning time by using IEEE 802.21 Media Independent Handover (MIH) services. MIH service is also used for efficient handover triggering and event notifications. In order to minimize the network attachment delay, we propose a fast re-association scheme for Mobile Node (MN). An efficient management of transient Binding Cache Entry (BCE) allows to eliminate out-of-order packet delivery and reduces the end-to-end delay at handover. Through the performance evaluations on OPNET network simulator, we validate that the proposed scheme provides better performance compared to PFMIPv6 and PMIPv6.

Keywords: IEEE 802.11, handover, PMIPv6, PFMIPv6, QoS, IEEE 802.21 MIH.

1 Introduction

Since most people want to continuously receive time-critical real-time multimedia services on portable devices attached to various wireless access networks, seamless secure mobility with guaranteed QoS provisioning across multiple wireless access networks is believed to be an essential feature in the next generation Internet. The QoS-guaranteed seamless multimedia service provisioning usually requires guaranteed bandwidth, limited jitter (e.g., less than 50 ms), and limited packet loss (e.g., less than 10^{-3}) during the handover.

In order to provide seamless mobile services, various mobile IP protocols, such as Mobile IPv4 (MIPv4) [1], MIPv6 [2], Fast Mobile IPv6 (FMIPv6) [3], Hierarchical Mobile IPv6 (HMIPv6) [4] and Proxy Mobile IPv6 (PMIPv6) [5], have been proposed in IETF. Especially, PMIPv6 provides the merits of network-initiated mobility that does not possess MIPv6 mobile node functionality.

---

* Corresponding author.

C.S. Hong et al. (Eds.): APNOMS 2009, LNCS 5787, pp. 321–330, 2009.
© Springer-Verlag Berlin Heidelberg 2009

A proxy agent in the network performs the mobility management signaling on behalf of the mobile node. PMIPv6 transparently provides mobility for mobile nodes within a PMIPv6 domain, without requesting any modification in the mobile node. Recently, a fast handover mechanism for PMIPv6 has been proposed as PFMIPv6 [6] that improves the performance during handover, such as attachment to a new network and signaling between mobility agents. PFMIPv6 provides operational procedures of proactive handover and reactive handover procedures. In PFMIPv6, however, the major time consuming components, such as channel scanning for the new access network (n-AN) and re-association to the new mobile access gateway (NMAG), are not specified in detail, and are defined as out-of-scope.

In this paper, we propose MIH (Media Independent Handover)-assisted PMIPv6-based fast handovers. The channel scanning time at mobile node is reduced by providing the related information for the active channels in neighbor APs through MIH. The detailed function of MIH [7] for collections of available wireless network resources, support of handover decision making and optimized tunneling & buffering are explained. This paper also proposes a proactive authentication in advance to minimize the handover time. We explain the detailed procedure of proactive authentication using the network initiated handover procedure between PMAG (previous MAG) and NMAG. The performance of the proposed MIH-assisted PFMIPv6 predictive handover has been analyzed using OPNET simulation.

The rest of this paper is organized as follows. In section 2, related work on PFMIPv6 is briefly explained. In section 3, the PFMIPv6 predictive handover with reduced channel scanning time, pre-authentication and transient Binding Cache Entry (BCE) management is explained in detail. In section 4, the performances of the enhanced PFMIPv6 handover are analyzed, based on OPNET simulation results. Especially, the packet delivery performances in uplink and downlink are evaluated individually. Finally, section 5 concludes this paper.

2   Background and Related Work

2.1   Basic Operations in Fast Handovers for PMIPv6 (PFMIPv6)

Recently, IETF MIPSHOP working group has been developing a protocol that attempts to reduce the handover latency of PMIPv6 [6]. The protocol called fast handovers for PMIPv6 (PFMIPv6) describes the necessary extensions to FMIPv6 for operations in PMIPv6 environment. PFMIPv6 defines two modes of operation: predictive and reactive fast handovers. In predictive fast handover, a bi-directional tunnel is established between PMAG and NMAG prior to the MN's attachment to the NAR. In reactive mode, this tunnel is established after MN is attached to the new access router (NAR).

Fig.1 depicts the predictive fast handover for PMIPv6. After MN completes the scanning phase and makes a decision to which AP it will perform handover, it sends a report to the previous access network (P-AN) containing MN ID and the new AP ID. The P-AN informs the PMAG about the MN's handover to the

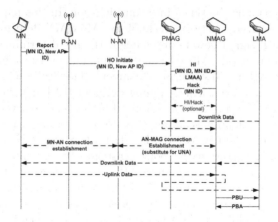

Fig. 1. Predictive fast handover for PMIPv6

new AN. PMAG sends a handover initiate (HI) message to NMAG transferring MN related context. NMAG replies with handover acknowledgement (HAck) message, and establishes a bi-directional tunnel between PMAG and NMAG. NMAG may also request PMAG to buffer the packets destined to the MN by setting U flag in HI message. The packet forwarding is restarted after HI/HAck message exchange with F flag set. Since MN doesn't participate in mobility signaling and cannot send unsolicited neighbor advertisement (UNA) to NMAG, a connection establishment notification may be regarded as a substitute for UNA. Just after the MN's attachment to the N-AN, the uplink and downlink packet flows go through the PMAG. NMAG may send proxy binding update (PBU) message to LMA, which updates MN's cache entry. After the PBU/PBA negotiation is completed, the uplink/downlink packets may be delivered through the NMAG directly.

The PFMIPv6 protocol operation may be regarded as one of the way to reduce the overall handover latency and packet loss. However, in order to minimize the overall handover latency lower layer procedures, such as channel scanning and re-association must be considered. In addition, in PFMIPv6 packet tunneling management is not optimal and may cause increase in end-to-end delay and packet out-of-order delivery.

2.2   Analysis of Limitations in PFMIPv6 Handover

The major goal of PFMIPv6 protocol is minimization of handover delay and packet loss during MN's handover. In addition, PFMIPv6 describes a context transfer related issues, so that the NMAG may acquire MN's profile not from MN or LMA, but from PMAG directly. PFMIPv6 defines two fast handover techniques: predictive and reactive. In this paper, we consider only predictive handover case and assume that MN initiates handover prior to its attachment to the new network. In order to provide a seamless handover, the access technology specific interactions must be taken into consideration. In this paper, we assume IEEE 802.11 [8] as access network technology. After MN detects that the link condition level (e.g. signal power) drops below the pre-determined threshold, it should start scanning mode. MN executes scanning on every channel and makes decision to which AP it will perform attachment. The decision is usually based on the received signal level. The scanning delay in IEEE 802.11 network may take quite long time varying from hundreds of millisecond to several seconds [9].

After MN finishes channel scanning it sends a report message to the access point with its ID and the new AP ID. In 802.11, however, MNs don't send this information directly to the access point. Instead of it, an Inter-Access Point Protocol (IAPP) is used, where the new AP sends notifications to the previous AP during the MN's re-association. This approach may not be applicable for predictive handover, since the handover process starts after MN's attachment to the new network. In addition, the HO Initiate message, which is used to inform PMAG about the MN's handover, is not defined in the specification, and is defined as out-of-scope.

PMAG has to send Handover Initiate (HI) message to NMAG containing the MN's profile information. However, it is not clear how PMAG may know NMAG's IP address, since it only receives a new AP ID and doesn't know to which MAG it belongs.

PMAG has to buffer packets destined to the MN if requested by NMAG by setting U flag in HI message. It is not defined by which criteria NMAG should request those buffered packets from PMAG. Furthermore, after handover phase is completed, MN starts sending data to NMAG, which in turn tunnels the packets to PMAG, which then tunnels those packets to LMA, which in turn delivers the packets further. This kind of routing may increase the end-to-end delay, especially if the distance between two MAGs is long. Nevertheless, NMAG may update binding by sending PBU message to LMA. The path switch in this case may induce an out-of-order packet delivery at the destination node. In addition, it is not defined what event should trigger this binding update.

Another possible problem in MAG is a neighbor queue overflow at the NMAG. A neighbor queue size is usually not large. Normally, routers consider a neighbor queue size of 3 packets [10]. This problem is caused by the new network attachment delay. The primary component of the attachment delay may be AAA authentication. The packet coming from PMAG may be delivered to NMAG, but MN may not complete the authentication phase, which may lead to buffer overflow.

3   PFMIPv6 Predictive Handover with Selective Channel Scanning, Pre-authentication, and Efficient Tunneling Management

3.1   MIH-Assisted PFMIPv6 Handover

We propose an MIH-assisted PFMIPv6 predictive handover as shown in Fig.2. When a MN initially attaches to an access network, MIH function (MIHF) of the

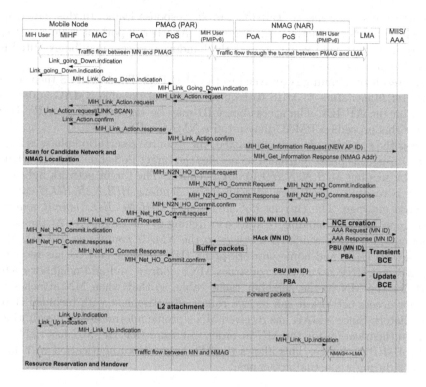

Fig. 2. Proposed MIH-assisted PFMIPv6 predictive fast handover

MN performs registration to MAG. After registration, MAG may receive event notifications from the MN. In addition, MIHF of MAG may retrieve information about the neighbor access networks to which MN may perform handover from Media Independent Information Server (MIIS). This information includes channel numbers on which neighbor access points are configured. When the link conditions deteriorate and the signal levels drops below the pre-defined threshold, a *Link_Going_Down* event is generated and delivered to the MIH Point of Service (PoS). MIH PoS sends the *MIH_Link_Action* request message commanding to start channel scanning. This command contains the channel numbers on which the MN has to perform scanning, so that the overall channel scanning time may be significantly reduced. In addition to this, any other scheme that allows data transmission during interleaving intervals of scanning may be used in order to minimize delay, such as described in [9]. There are two possible ways to execute scanning, i.e. active and passive mode. In case of passive scanning, the station "listens" every channel for the beacon frames transmitted by the neighbor APs. The scanning delay in this mode is dependent on the frequency of beacon generation. On the other hand, the frequent beacon generation may cause significant overhead. In case of active scanning, the station sends a probe request frames finding available AP. In this case the scanning time depends on the time the station has to wait for a probe response frame. In this paper, we assume that MN uses active scanning mode. Another important issue is setting the scanning threshold value, so that the handover process may be executed in proactive manner.

The result of scanning is returned to the MIH PoS. The MIH PoS then sends *MIH_Get_Information* to MIIS to request the list of possible NMAG addresses based on the AP IDs. The access point ID may be regarded as the AP's BSS ID. By this procedure PMAG may acquire NMAG's IP address.

After the scanning phase is completed, the resource reservation phase is initiated. During this phase, current PoS notifies the target PoS about the MN's handover by sending *MIH_N2N_HO_Commit* request. The target network indicates that it is ready to accept new attachment by sending *MIH_N2N_HO_Commint* response message. The current PoS now sends *MIH_Net_HO_Commit* request message to MN commanding to start a switching to the target AP. At the same moment, the PMAG sends HI message to the NMAG containing MN ID, MN interface ID (IID) and home network prefix (HNP). Since NMAG has received MN IID and HNP it can now construct MN's address and add neighbor cache entry for the MN to the neighbor cache table, so that when MN attaches to the new access network it can receive router advertisement with included HNP immediately. Since NMAG has received MN's profile information it may initiate AAA authentication for the MN in order to ensure that the MN has rights to access PMIPv6 service. This allows the MN to avoid authentication with new network during it's actual attachment.

3.2   Enhanced Management of Tunneling and Transient Binding Cache Entry (BCE)

NMAG also sends a PBU message to LMA requesting the LMA to create a transient binding cache entry (BCE). At any time instance there should be only one BCE that can forward packets in uplink and downlink direction and one that can allow forwarding only in uplink direction as it is defined in [11]. The newly created transient BCE allows only uplink packet transmission, while the old BCE allows bi-directinal transmission through the PMAG. When the MN is ready to perform switching between access networks it notifies current PoS with *MIH_Net_HO_Commit* response and undergoes L2 handover. The buffered packet from PMAG may now be delivered to NMAG, and the PMAG sends PBU message to LMA activating the previously created transient BCE, so that the packets may now be forwarded directly to/from NMAG. In PFMIPv6 specification there should be additional HI/HAck message exchange between PMAG and NMAG requesting buffered packets. However, it is not defined based on which criteria and when NMAG has to request the buffered packets from PMAG.

When the mobile node performs access network attachment, a *Link_Up* indication message is delivered to the new PoS, which indicates that the MN is ready to accept packet. From now all the packets can be forwarded through the NMAG-to-LMA tunnel directly without involving PMAG. There is also no need to update a binding cache table at LMA, since this step was already done by PMAG.

4   Performance Evaluation and Analysis

4.1   Simulation Model

We implemented PMIPv6, PFMIPv6 and the proposed MIH-assisted PFMIPv6 predictive handover scheme in OPNET [12] network simulator. The simulation topology is depicted in Fig.3. There is a single LMA and two MAGs within one PMIPv6 domain. MN performs horizontal handover from PMAG to NMAG while moving from BSS1 to BSS2. The BSS radius was set to 100 meters. IEEE 802.11b with DSSS was used as access network technology. A bi-directional 64 kbit/s PCM VoIP (RTP/UDP) session was used as application between MN and CN. In this paper, we assume that the MN moves at the pedestrian speed (10km/h). The performance of the proposed scheme was compared with PFMIPv6 and PMIPv6 in terms of mouth-to-ear delay, jitter, packet loss and Mean Opinion Score (MOS) value in uplink and downlink directions separately.

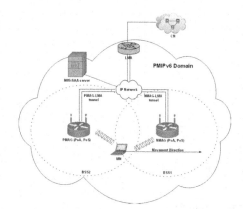

Handover Component (msec)	PMIPv6	PFMIPv6	Proposed PFMIPv6
Scanning time, T_{scan}	350	350	35
Open authentication, $T_{open-auth}$	0.6	0.6	0.6
Re-assiciation, T_{re-ass}	0.6	0.6	0.6
AAA authentication, T_{AAA}	200	200	N/A
HI transmission, T_{HI}	N/A	21	21
Hack transmission, T_{HAck}	N/A	21	21
PBU transmission, T_{PBU}	25	N/A	25
PBA transmission, T_{PBA}	25	N/A	25
RS transmission, T_{RS}	50	N/A	N/A
RA transmission, T_{RA}	30	N/A	N/A

Fig. 3. Simulation topology   **Fig. 4.** Handover delay components

4.2   Uplink Handover Performance Analysis

The mouth-to-ear delay performance depicted in Fig.5(a) shows that the proposed PFMIPv6 can perform better than PFMIPv6 and PMIPv6. This is due to reduced channel scanning time and pre-authentication in the proposed PFMIPv6. Additionally, the packets after handover process are delivered directly to NMAG, eliminating the unnecessary routing to PMAG. Mouth-to-ear delays for PFMIPv6 and PMIPv6 are much higher than in the proposed PFMIPv6 because the packets at the MN's buffer have to wait much longer time before transmission. In case of PFMIPv6, the packets may be delivered to NMAG after attachment immediately, but in case of PMIPv6 a PBU/PBA exchange with LMA and RS/RA exchange with NMAG have to be done. Nearly at 24.7 sec. of simulation time the delay is fluctuating in PFMIPv6. This is because packets arrive out-of-order after the path switching. This also can be seen from jitter

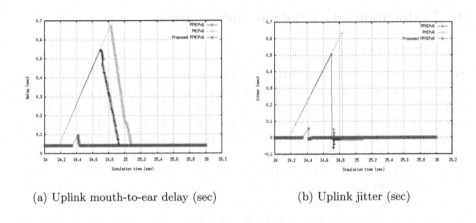

(a) Uplink mouth-to-ear delay (sec)   (b) Uplink jitter (sec)

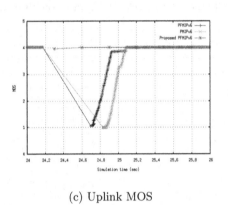

(c) Uplink MOS

Fig. 5. Uplink performance

performance depicted in Fig.5(b). The negative jitter value indicates that the
packets were delivered out-of-order. There was no out-of-order packets in uplink
direction in the proposed PFMIPv6 and PMIPv6 since in both cases the uplink
packets were delivered directly through the NMAG-to-LMA tunnel. There was
no packet loss in uplink direction. Once the handover process was started, all
the packets coming from the upper layer were buffered at the MAC queue of the
MN. Fig.4 shows the handover delay components.

4.3   Downlink Handover Performance Analysis

The downlink performance in depicted in Fig.6. The mouth-to-ear delay in the
proposed scheme was less compared to PFMIPv6 and PMIPv6. Performance
of the delay also indicates that there was out-of-order packet delivery in case
of the proposed scheme and PFMIPv6. This was due to the data transmission
path switch. Jitter performance also indicates out-of-order packet delivery and

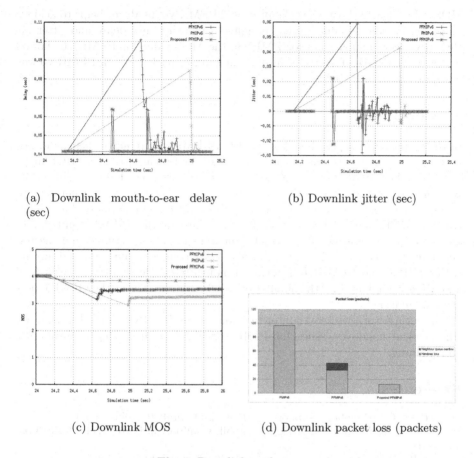

(a) Downlink mouth-to-ear delay (sec)

(b) Downlink jitter (sec)

(c) Downlink MOS

(d) Downlink packet loss (packets)

Fig. 6. Downlink performance

is slightly higher than 20 ms for the proposed PFMIPv6. There is a slight delay fluctuation at 25 sec. of simulation time in PMIPv6. This is due to the neighbor discovery procedure at NMAG. NMAG when receiving the first packet destined to the MN starts the neighbor discovery process. Once the MN is found, all packets may be delivered to it. The mouth-to-ear delay in PFMIPv6 is even higher than in PMIPv6. This is because the data packets during handover are forwarded through the PMAG. In case of PMIPv6, the packets are forwarded to PMAG until LMA receives PBU message sent by NMAG. The amount of packet loss, however, is much higher in PMIPv6 case compared to other schemes. The packet loss was caused by long scanning delay and access network attachment delay. In case of PFMIPv6, there was a neighbor queue overflow, which was set to four packets. The packets from PMAG were delivered to NMAG and buffered in the neighbor queue awaiting MN's attachment to the new network. In the proposed PFMIPv6 handover scheme only 13 packets were lost in downlink direction compared to 97 and 43 for PMIPv6 and PFMIPv6, respectively. The

MOS value clearly shows that the proposed PFMIPv6 handover helps to achieve much better voice quality. Even though the mouth-to-ear delay and jitter for PFMIPv6 were higher than for PMIPv6, the MOS value for PFMIPv6 showed better performance. This is because the amount of packet loss in PFMIPv6 was much less than in PMIPv6.

5   Conclusion

In this paper, we proposed an MIH-assisted PFMIPv6 predictive fast handover with reduced channel scanning time, fast re-association and transient BCE management. The proposed scheme is based on MIH services, which is providing information to reduce the overall channel scanning time. By using pre-authentication of the MN, the network attachment delay was reduced. Furthermore, handling a transient BCE allows delivery of data packets through the NMAG in uplink and downlink direction after MN's attachment immediately, i.e eliminating the unnecessary data transmission through PMAG. We compared our proposed scheme with PMIPv6 and PFMIPv6 in terms of mouth-to-ear delay, jitter, packet loss, and MOS value based on the simulation results obtained from the OPNET network simulator.

The future work will cover inter-technology fast PMIPv6 vertical handovers. The vertical handover issues for PMIPv6 will be studied deeply considering multihoming.

References

1. Perkins, C.: IP mobility support for IPv4. IETF Draft (October 2008)
2. Johnson, D., Perkins, C., Arkko, J.: Mobility support in IPv6. IETF Draft (June 2004)
3. Koodly, R.: Fast handovers for mobile IPv6. IETF Draft (July 2005)
4. Soliman, H., Flarion, Castelluccia, C., Malki, K.E., Bellier, L.: Hierarchical mobile ipv6 mobility management (HMIPv6). IETF Draft (August 2005)
5. Gundavelli, S., Leung, K., Devarapalli, V., Chowdhury, K., Patil, B.: Proxy mobile IPv6. IETF Draft (August 2008)
6. Yokota, H., Chowdhury, K., Koodli, R., Patil, B., Xia, F.: Fast handovers for proxy mobile IPv6. IETF Draft (March 2009)
7. Media Independent Handover Services: IEEE P802.21/D10.0 draft standard for local and metropolitan area networks (April 2008)
8. Wireless LAN Medium Access Control (MAC) and Physical Layer (PHY) Specifications: IEEE 802.11 standard for information technology (June 2007)
9. Wu, H., Tan, K., Zhang, Y., Zhang, Q.: Proactive scan: Fast handoff with smart triggers for 802.11 wireless lan. In: INFOCOM, Anchorage, Alaska, USA, May 6–12, pp. 749–757 (2007)
10. Liebsch, M., Le, L.: Inter-technology handover for proxy MIPv6. IETF Draft (February 2008)
11. Liebsch, M., Muhanna, A., Blume, O.: Transient binding for proxy mobile IPv6. IETF Draft (March 2009)
12. OPNET Technologies Web Site, http://www.opnet.com

Forecasting WiMAX System Earnings:
A Case Study on Mass Rapid Transit System

Cheng Hsuan Cho and Jen-yi Pan

621 Chia-Yi, Taiwan, Republic of China
Department of Communications Engineering,
National Chung Cheng University
kusojuo@gmail.com, jypan@comm.ccu.edu.tw

Abstract. The WiMAX network has been proposed as an alternative technology to provide wireless network service. This paper focuses on WiMAX system forecasting in Mass Rapid Transit (MRT) environments and describes numerous models, including a user model, application model, economic model, subscription model, and evaluation model. We simulated a WiMAX network using these models to forecast the profitability of a broadband access system in an MRT environment.

Keywords: Deployment, WiMAX, QoS, Economic, Simulation and Modeling.

1 Introduction

It is important to evaluate the overall performance of a WiMAX (worldwide interoperability for microwave access) network (based on the IEEE802.16 standard and the WiMAX forum network architecture). As compared to other telecommunication networks, a WiMAX network provides more services and applications, and it has a higher data rate. Standards for broadband wireless access (BWA) are being developed within IEEE Project 802, working group 16, also referred to as 802.16 d/e/m/j. To promote 802.16-compliant technologies, the WiMAX Forum was founded. According to the specifications proposed by the WiMAX forum, 802.16 technologies are attractive for a wide variety of applications including high-speed Internet access, WiFi hotspot backhaul, cellular backhaul, public safety services, and private networks.

In recent years, many studies have focused on WiMAX systems and their deployment. In addition, comparisons of WiMAX with different technologies such as 3G and LTE have been reported. [1][2][3] have focused on fixed WiMAX arrangements for urban, suburban, and rural environments. [1] addresses the deployment considerations for a wireless metropolitan area network based on the IEEE 802.16-2004 Air Interface Standards. [2] discusses a deployment in which some factors related to the PHY layer are considered, and the wireless link budget is calculated. [3] considers the physical performance and path loss model in a fixed WiMAX deployment. However, while all of these consider a fixed BWA scenario, they do not consider and discuss mobility issues in relation to the WiMAX standard, which is important for mobile users. Therefore, in this study, we focus on WiMAX system forecasting in a Mass

C.S. Hong et al. (Eds.): APNOMS 2009, LNCS 5787, pp. 331–344, 2009.

Rapid Transit (MRT) environment. We examine user/network aspects and present an evaluation of the system performance in an urban environment.

However, offering a service for mobile users while roaming is a complex issue. A WiMAX operator needs to satisfy requirements such as return on investment, system performance, reliability, and so on. The operator needs to consider and balance many factors to provide optimum service. The business case and system design need to address the following questions:

In which markets should we offer end-user WiMAX service?
How should we price the service?
How do we guarantee an appropriate level of quality?
How much will the annual capital and operational expenses be?

In this paper, we propose a framework called WiMAX Deployment for Mass Rapid Transit (MRT) System (WiDeMS). It is the result of heuristic research on designing a method to demonstrate the mobility issues in WiDeMS. It describes numerous models that provide solutions to the abovementioned questions and requirements. The remainder of this paper consists of four parts. Section 2 introduces the modeling technique and research steps of the WiDeMS framework. Section 3 describes the simulation scenario of WiDeMS. The simulation results are presented in Section 4. Finally, the conclusions are presented in Section 5.

2 Deployment Frameworks and Model Description

First, we discuss the planning and deployment of WiDeMS. There are some aspects that need to be clarified. For example, we investigate the operation and execution in an MRT system. The research involved in WiDeMS planning is divided into four steps, as shown in Fig. 1:

(1): Modeling stage: A design methodology and model for deploying a network, along with pricing. This stage includes the user, application, economic, and subscription models.

(2): WiMAX deployment stage: Clarifying the deployment scenario for the WiMAX deployment. The actual operating environment should be considered and investigated since it may affect the deployment strategy. For example, buildings, terrain, or tunnels in the service coverage area may affect the strategy.

(3): Network simulation stage: In this stage, all the factors related to WiDeMS are determined. The input argument, scenario description, physical layer definition, MAC layer definition, and global definition should be completed in this step.

(4): Evaluation stage: The evaluation stage analyzes some simulation results and predicts the performance using the evaluation model. In addition, if the simulation results are not as good as expected, stages (1), (2), or (3) should be revisited to rationalize the assumptions.

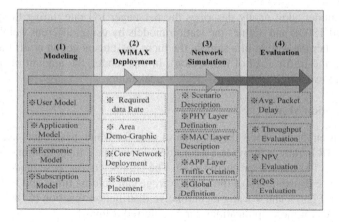

Fig. 1. Research steps involved in development of WiDeMS

The first stage describes and illustrates numerous models and basic assumptions. For our WiDeMS requirements, the station model, user model, traffic model, and economic model are well defined in this stage. In the modeling stage, a suitable mathematical or statistical technique that can be used in the simulation scenario is determined.

2.1 User Model

This model assumes the number of passengers at each station at different times. For example, if we have ten stations, the model should map each station to its correlated passengers. The model will also have its distribution (i.e., peak-rates from 1100–1300 and 1700–1900, and off-peak rates at other times). Fig. 2 shows a 3-D model for this

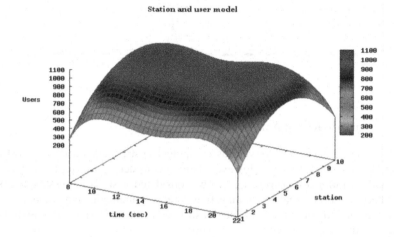

Fig. 2. User model with time deviation and stations

table. The model additionally describes the passenger I/O rate during the peak rates (at 1200 and 1800), and defines the station models by using different weights from 1 to 5. The higher the weight of a station, the more passengers get on or off trains at this station (i.e., the traffic load of the station). In this model, the amount of user traffic varies from low to high from the off-peak to peak hours.

2.2 Application Model

The application model shows the duration of time for which passengers use the applications on MRT-trains. Our research considered two types of applications: voice over IP (VoIP) and file transfer protocol (FTP). The parameters for these models are listed in Table 1. The distributions of the VoIP average call holding times and FTP file sizes are shown in Figs. 3 and 4 using a cumulative distribution function (CDF). Modeling methodologies for more applications can be referred from [4].

Table 1. Parameters of VoIP and FTP traffic model

Component	Factors
VoIP avg. call holding time	Exponential Distribution
VoIP frame length	20 ms
VoIP Frame size	128 bytes
FTP File Size	Truncated lognormal

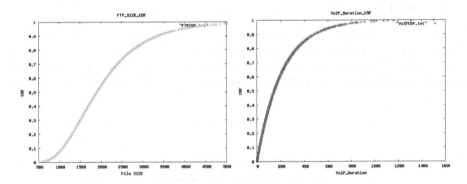

Fig. 3 and Fig. 4. FTP duration model and VoIP duration model

2.3 Economic Model and Assumptions

The economic model considers various equipment costs, market-related investment parameters, and other factors. However, before considering the profit, the annual capital and operational expenses need to be considered first. The investment cost is the first consideration for telecommunication/wireless communication operators.

First, we need to determine cost factors such as the capital expenditure and operation expenditure.

A.   CAPEX

The capital expenditure (CAPEX) includes the costs of the end-user device, WiMAX equipment (e.g., base station (BS)/subscriber station (SS)), backhaul link equipment, and server equipment. All the costs involved in the setup of the wireless environment are included in the CAPEX. For example, the spectrum license cost is included in the CAPEX. Many WiMAX operators own different licensed bands (e.g., 2.5G/3.5G), but some operate in un-licensed bands. In licensed bands, operators need to consider the size of the bands that need to be covered and their costs in the cases of continuous or non-continuous bands. The first WiMAX-certified products operated in the licensed 3.5-GHz frequency band, followed by systems in the 2.5-GHz licensed band and the 5.8-GHz license-exempt band. It is important thing to consider all these costs in the CAPEX.

B.   OPEX

The operational expenditure (OPEX) is the on-going cost of running a business. The OPEX of companies includes the salaries of workers and facility expenses such as rent and utilities.

In a WiMAX network structure, the OPEX includes costs associated with the operation and maintenance of the WiMAX and backhaul equipment. It also includes accounting expenses, advertising, office expenses, supplies, attorney fees, insurance, and property management. For network operation and maintenance, a detailed costing model should be used. We list some of the important equipment and operational expenses in Table 2. This includes factors related to deployment costs, while some investment factors were referred from [1], [2], [5], and [8] :

Table 2. Costs for MRT system

Investment factor	Deployment cost
WiMAX BS	$15000 per BS
WiMAX BS installation cost	$2500 per BS
License fee	$15000
BS site rental	$3000 per BS per year
Backhaul link equipment	$40000
SS installation cost	$500
SS site rental in MRT stations	$10000 per SS per year
Administration cost	10% of total investment
Maintenance cost	10% of total investment

As a second step, the WiMAX operator needs to calculate the total revenue/income. In this step, we determined our pricing model by using a two-part tariff strategy for a WiDeMS subscriber:

$$P_{(i)} = \sum_{u=1}^{N(i)} [B + (A * T_{u,(i)})] \tag{1}$$

$$N(i) = MSR(i) * S, \quad 0 : msr(i) : 1 \tag{2}$$

The formula (1) illustrates the profit in month i. The number of users (denoted by U) is a key factor in our framework. The B is basic service fee and A is denoted application service fee. $T_{u,(i)}$ f is the total spent time for a user in the i month. Also, we formulate the total number of subscribers in equation (2). $MSR_{(i)}$ is Monthly Subscription Rate (MSR) of the i month. The total passenger is represented by S. In this equation, the number of subscriber (we denote by N) is a key factor in our framework, the number of subscribers (overbooking rate) will change based on the service quality. If the quality (denoted by Q) is not good enough, some users will cancel their subscriptions. However, the monthly subscription rate (MSR) will increase constantly. The MSR can be calculated based on the most recent month's quality:

$$MSR(i) = MSR(i-1) + INC - BQ(i-1) \qquad (3)$$

The formula (3) illustrates the MSR in month i which is depending on last month. INC is a constant that indicates the subscription increase rate of the entering the network. The factor BQ means Bad Quality. It is the proportion of the discontent rate which depends on the experience of the user. We will discuss this factor later by using the subscription model.

2.4 Subscription Model–Subscribe/Cancel Analysis

In this analysis, we address the problem of pricing in such a network/telecom service. There are many factors that impact the *MSR*: the factor BQ, service price, and competition/cooperation among service providers. The higher of BQ value, the fewer subscribers satisfy this service, the more is the number of subscribers not enjoying this service. Thus, the *MSR* will decrease in the next period (i.e., the next month in our scenario). In [6], a game theory analysis and Nash equilibrium are considered to find the solution to this issue. Briefly, if the service price offered by a service provider is too high, subscribers may churn to another service provider. However, if the service price is low, the total revenue of the service provider may not be maximized. For a WiDeMS service, we assume that the basic service fee (*B*) is 5\$ and the application service fee (*A*) is 0.001\$ per second. These fees are defined as variable so that we may adjust them to appropriate values.

In our study, we focused on the impact of service quality. Two factors were considered for the *MSR*: the rate of increase of subscribers entering the network and the WiDeMS service quality. In this paper, we focus on the packet delay, packet loss, and data rates. These factors are more sensitive when the link becomes congested. A previous paper [7] defined a computational model for digitized voice transmission using the E-model, R-factor, and Mean of Score (MOS). We use the method of calculating the R-factor as good-or-better (GoB) and poor-or-worse (PoW) to determine how the number of users who have had a poor experiences with WiDeMs. This number is determined as follows:

1. The WiMAX operator observes the mean inter packet delay.
2. The mean inter packet delay is mapped to the R-value by the equations (3-1), (3-27), and (3-28) given in [7].
3. Based on the R-value, the good-or-better (GoB) value and poor-or-worse (PoW) value is calculated by equations (B-1), (B-2), and (B-3) given in [7].

In this manner, the number of users who have had a poor experience because of packet delay and packet loss can be determined. We can also determine the increment or decrement of *MSR* for the next month. The correlations between the inter packet delay and the PoW and GoB values are shown in Fig. 5.

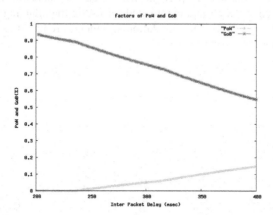

Fig. 5. Correlations between the inter packet delay and the PoW and GoB

The subscription model considers not only the packet delay but also the service blocking rate. WiMAX scheduling, including the resource allocation and admission control policies, has to be able to meet the QoS assurances for the WiMAX network. We measure the blocking rate for the overall connections. If the blocking rate is high, the BQ factor will increase. We argue that when the call blocking probability increases, subscribers will suffer from more call blockings and the level of user satisfaction will decrease. In WiDeMS, we define the leave threshold as 1%. If the block probability is higher than 2%, some subscribers will stop using WiDeMS.

The other factor impacting user satisfaction is the throughput (i.e., the data rate). In a WiMAX system, the data rate differs based on the modulation scheme and encoding rate. With higher modulation and encoding rates, the overall data rate will increase to 50–70 Mbps. The subscription model measures the average throughput of every FTP application. If subscribers suffer from lower data rates, they will leave in the next month if the data rate is insufficient. The leave probability table is shown below:

Table 3. Correlations between data rate and user experiment

Date rate	User experiment	Leave probability
>2 Mbps	Excellent	0%
500 Kbps~2 Mbps	Good	0%
100 Kbps~500 Kbps	Fair	10%
10 Kbps~100 Kbps	Poor	50%
<10 Kbps	Bad	90%

3 Simulation Environment of WiDeMS

In the network simulation stage, we simulated the WiDeMS framework. Our simulation tool was based on Qualnet simulation Software version 4.5 [9], which includes advanced wireless tools that support the IEEE 802.16e standard. The Qualnet platform is a PC-based software that is intended for use in the planning and designing of wired and wireless networks. The simulation arguments are based on the input criteria provided in its configuration file.

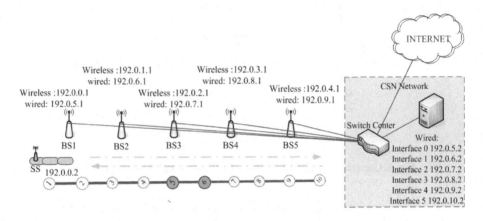

Fig. 6. Simulation scenario

As shown in Fig. 6, BS directly connects to the wired network and manages the admission control for each MS/SS. Each MS/SS is located on an MRT train. It connects to the WiMAX-BS by the 802.16 PHY/MAC protocol. In this configuration, because the MS/SSs are situated on MRT-trains, the passengers can link with the MS/SS through a WLAN or wired interface. In the WiDeMS, we assume that the MRT trains support a WLAN network. The distance between MRT stations is approximately 1.5 km. An MRT train moves to the next MRT station and stops for a while. It then moves to the next MRT station. The drive duration of an MRT train is 140 s while the stop duration is approximately 25 s. The maximum speed of the MRT train is up to 70 km/h.

Our framework assumes that the end-user subscriber uses the WiMAX service every day. The number of passengers and MSR will also be set in the configuration file. These input parameters and the configuration file are changed for different times and different cases. For example, the number of subscribers and the duration for each application will be reset for the next period.

In the operation flow, the overall process is divided into 20n parts for one day. n describes the MRT round trip periods in one day. First, the input arguments are stored in the simulator's configuration file (for example: the MAC/PHY setting, node placements/movements, and application descriptions). All of these input arguments will be inputted into the simulator. An MRT train starts from station 1, moves to station 10, and returns to station 1. For each round trip, an automatic function for the application model will add and remove VoIP/FTP applications in different time slots T $(0 < T < 20n - 1)$. At the end of the simulation, the simulation results are saved in our database, which can be used for evaluation in the evaluation stage. Table 4 shows the parameters and input arguments for the Qualnet simulator.

Table 4. Simulation parameters

Simulator parameter	Value
Channel bandwidth	20 MHz
BS frame interval	20 ms
BS TDD DL duration	10 ms
BS TTG/RTG/SSTG	10 μs/10 μs/4 μs
BS DCD broadcast interval	5 m
Propagation path loss model	TWO-RAY
PHY802.16 TX power	20.0
Handover RSS Trigger	−76.0 dbm
Wired link Bandwidth	100 mb/s
Simulation time	1400 s/per round of MRT

Supporting the QoS in a WiMAX-system is a difficult task because the characteristics of wireless links are highly variable and unpredictable. To cope with these issues, the QoS is usually managed at the medium access control (MAC) layer. In the Qualnet simulator, the QoS is supported by the bandwidth request, scheduling, traffic classifier, and UL/DL functions. However, for a researcher and designer, it is more difficult to build a system that can detect and maintain the different classifications of applications. Two different processes in our simulation were run in Qualnet. The first situation was not assigned a QoS guarantee, while the other was assigned a QoS guarantee. In the QoS-guarantee scenario, the UGS, rtPS, nrtPS, ertPS, and best effort classes were included in the WiMAX system. However, with no QoS-guarantee, all of the applications used the best effort class instead of the QoS supported classes. We called this situation the no QoS-guarantee scenario. On the other hand, the QoS-guarantee scenario considered all of the traffic types for all applications.

4   Simulation Results

In our simulation, WiDeMS was initialized and deployed in January 2010 with operations and management provided for 3 years (from January 2010 to December 2012). The numerical results and analyses are shown below:

a. Average delay on MRT

We collected VoIP packet transmission data for different MSRs, as shown in Fig. 7. In the case of an MSR of less than 0.06 with the no QoS guarantee, the average delay was less than 0.05 s. When the MSR approaches 0.1, the average delay for each user approaches 0.1–0.2 s, and even reaches 0.3 s at the peak rate. The high average delay not only resulted in a huge number of instances where receiving data of VoIP

packet transmit out of time threshold but also decreased the MSR because the users
experienced low quality of service. In the QoS-guarantee scenario, every VoIP session was assigned to the UGS class. All of the VoIP sessions received their packets
within the assured time delay. The average delay was close to 20 ms. This is suitable
for every UGS class application.

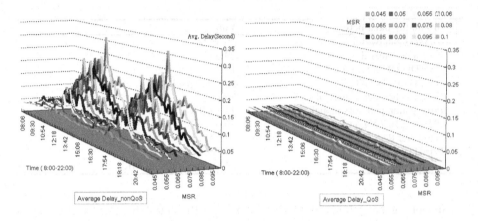

Fig. 7. VoIP packets average delay on MRT

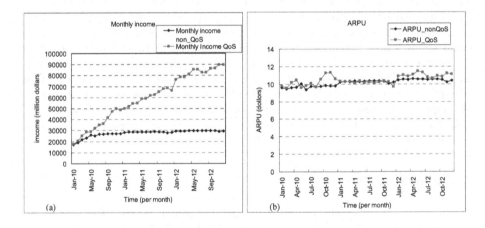

Fig. 8. Monthly income (a) and ARPU (b) for 3 years of operation

b. Monthly income

During the initial operating period, the profit is not very good. This is because the
MSR is initialized at 5% and increases by 1% every month. After one year of service,
the profit reaches $40000–45000. On the other hand, the quality of the WiDeMS is
not very good when the MSR is high (at about 10%–12%). So a portion of the subscribers will leave during the following month. This depends on the GoB values of the
subscription model. Finally, the monthly profit reaches a stable condition.

c. Average Monthly Revenue (ARPU) for Customers

Based on Equation (1) and (2), we focused on the overall subscribers' ARPU using formula (4).

$$ARPU = P(i) / N(i) \tag{4}$$

The results of the ARPU for the QoS guarantee and no QoS guarantee scenarios are close to 10. These stable ARPU results occur because every subscriber's applications are generated by the application model. The average service duration is close to the mean of the VoIP/FTP duration (i.e., VoIP call holding time and FTP transmit time). The ARPU reveals that most of the users use the WiMAX service if they agree to pay the WiMAX operator $10 every month.

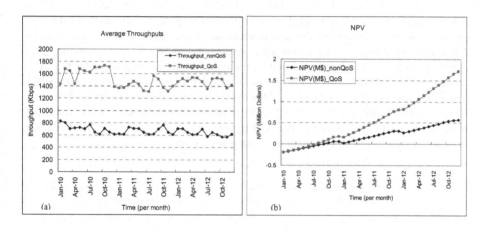

Fig. 9. Average throughput evaluation (a) and net present value (b) for 3 years of operation.

d. Average Throughputs

Fig. 9(b) depicts the FTPs' average throughputs for the QoS guarantee and no QoS guarantee scenarios. The average throughputs were based on all of the FTP sessions. It shows that the average throughputs were close to 1.2 Mbps–1.8 Mbps for the QoS guarantee, which was higher than the no QoS guarantee situation. This is because, for the no QoS guarantee scenario, all of the applications use the UL/ DL bandwidth for the BE class. The efficiency of the contention based strategy is lower than the pre-allotment in the MAC-scheduling since it wastes time dealing with the bandwidth request process. The best effort class needs every application to use the time slots to acquire the UL/DL bandwidth. Moreover, collisions may occur when the same application take the same random back off number if MSs/SSs need to transmit in the same time. This is the reason for the lack of bandwidth used and the allocation for the no QoS guarantee scenario.

e. Economic analysis

The results of the economic analysis are shown in Fig. 9(b). The total investment and profit were calculated for the QoS guarantee and no QoS guarantee scenarios. In the early part of the operation, the CAPEX and OPEX are about 0.2 million dollars. After a year, the NPV will not only balance but also increase every month. After three years, the overall profit with the QoS guarantee is close to 1.7 million dollars. But in the no QoS guarantee scenario, the profit is 0.5 million dollars. Since the quality is not good in the no QoS guarantee scenario, subscribers will leave because of bad experiences. However, the OPEX is deposited at the beginning of the year. Therefore, the NPV is reduced in January of every year.

f. Sensitivity analysis

To satisfy the uncertainty inherent in our assumptions, we used a method for testing the degree of sensitivity. The sensitivity analyses in the research included the OPEX, CAPEX, and service price level. For each parameter, 50% deviations from the base assumptions were defined as the upper (maximum) and lower limits (minimum) for the sensitivity analysis. The parameters were analyzed one-by-one, changing their values between the minimum and maximum and plotting the respective NPV values, as shown in Fig. 10(a). This graph shows that the NPVs are significantly more sensitive to changes in the service price. However, changing the service price level is a service problem, because when a WiMAX operator changes the service fee, it directly impacts the ARPU. In addition, the ARPU will sometimes affect the MSR and the monthly income when it is too expensive for a subscriber.

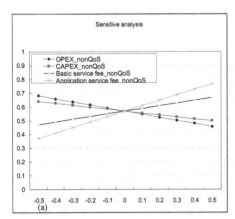 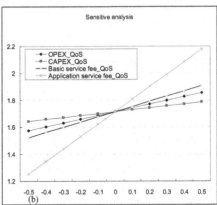

Fig. 10. Sensitivity analyses for OPEX, CAPEX, and service price level for no QoS guarantee (a) and QoS guarantee (b) scenarios.

Finally, this paper summarizes the results of the WiDeMS simulation. We calculated the figures for subscribers, avg. MSR, monthly income, ARPU, and NPV for every year. These statistics can be easily seen for every year.

Table 5. The result for the WiDeMS per year

Scenario	Period	Avg. subscribers	Avg. MSR	Monthly income	ARPU	NPV (M$)
WiDeMS_Non-	1st year	2522	0.07	24478.894	9.69	0.057746738
QoS	2nd year	2777	0.078	28600.691	10.29	0.307455031
	3rd year	2813	0.079	29541.852	10.50	0.56845726
WiDeMS_QoS	1st year	3274	0.091	35594.059	10.27	0.16172277
	2nd year	5067	0.157	60203.191	10.35	0.81676115
	3rd year	7207	0.20	83644.936	10.30	1.715783253

5 Conclusion and Future Work

It is possible that modifications for the input arguments can be seen from both the business and network perspectives. For example, if one or more factors, like the BS location, user model, application model, or MSR, were changed, different system performance and financial results might be seen. In our simulation results, the WiDeMS shows results whether the WiMAX operator supports a QoS guarantee or not.

Our deployment framework, models, and the performance of our simulation are suitable, not only for a WiMAX core network, but also for other technologies with high data rates and wide service coverage, e.g., LTE, 3.5 G, etc. The simulation results for the proposed WiDeMS show the forecast profit and the bottleneck of the network, which is the bandwidth overhead. Greater bandwidth leads to a higher data rate for subscribers. However, determining how to use a limited total bandwidth for all of the users is the critical issue for the deployment inside moving MRT trains.

The future work on the WiDeMS will focus on the following researches. The first thing is to put more than one SS/MS onto an MRT train. Each SS/MS can link to the same or a different BS(s). Every SS/MS allows a subscriber to use their cell phones or notepads by connecting to the BS. Furthermore, each SS/MS handles a different handoff threshold. This scenario may lead to load balancing/sharing effects because users simultaneously employ different BSs from the same MRT train. On the other hand, the deployment strategy should also be considered in the load balancing scenario for the WiDeMS. However, the use of a heuristic method in the research may take these factors into consideration for a future WiDeMS. In addition, by changing the basic assumptions and the station topology, other interesting simulation results will be illustrated in future work.

References

1. WiMAX forum, WiMAX deployment considerations for Fixed Wireless Access in the 2.5 GHz and 3.5 GHz licensed Bands, white paper (July 2005)
2. Lannoo, B., Verbrugge, S., Ooteghem, J.V., Quinart, B.: Business Scenarios for a WiMAX Deployment in Belgium. In: Mobile WiMAX Symposium (2007)
3. Grønsund, P., Engelstad, P.E., Johnsen, T., Skeie, T.: The Physical Performance and Path Loss in a Fixed WiMAX Deployment. In: International Wireless Communications and Mobile Computing Conference, pp. 439–444 (2007)
4. WiMAX forum, WiMAX System evaluation methodology version 2.1 (2008)

5. Smura, T.: Competitive potential of WiMAX in the broadband access market: a techno-economic analysis, Helsinki University of Technology,
   http://www.netlab.tkk.fi/u/tsmura/publications/
   Smura_ITS05.pdf
6. Niyato, D., Hossain, E.: A game theoretic analysis of service Competition and pricing in Heterogeneous wireless Access Networks. Wireless Communications, IEEE Transactions (December 2008)
7. ITU-T Recommendation G.107, The E-Model, a computational model for use in transmission planning (March 2005)
8. WiMAX forum: The Business Case for Fixed Wireless Access in Emerging Markets (2005)
9. Scalable network technologies: qualnet simulator 4.5,
   http://www.scalable-networks.com/products/developer/
   new_in_45.php

Triangular Tiling-Based Efficient Flooding Scheme in Wireless Ad Hoc Networks

In Hur[1], Trong Duc Le[2], Minho Jo[3], and Hyunseung Choo[1]

[1] School of Information and Communication Engineering
Sungkyunkwan University, South Korea
{oracl0307,choo}@skku.edu
[2] Digital Media and Communications Business
Samsung Electronics
letrongduc@gmail.com
[3] Graduate School of Information Management and Security
Korea University, South Korea
minhojo@korea.ac.kr

Abstract. Flooding is an indispensable operation for providing control or routing functionalities to wireless ad hoc networks. The traditional flooding scheme generates excessive packet retransmissions, resource contention, and collisions since every node forwards the packet at least once. Recently, several flooding schemes have been proposed to avoid these problems; however, these flooding schemes still have unnecessary forwarding nodes. In this paper, we present an efficient flooding scheme to minimize the number of forwarding nodes, based on a triangular tiling algorithm. Using location information of 1-hop neighbor nodes, our proposed scheme selects the nodes which are located closest to the vertices of an equilateral triangle which is inscribed in the transmission coverage as forwarding nodes. The most significant feature of our proposed flooding scheme is that it does not require any extra communication overhead other than the exchange of 1-hop HELLO messages. Simulation results show that our proposed scheme is so efficient that it has the capability to reduce the number of forwarding nodes such that it approaches the lower bound, hence, it alleviates contention and collisions in networks.

Keywords: Wireless Ad Hoc Networks, Broadcasting, Flooding, 1-hop Neighbor Information.

1 Introduction

Wireless ad hoc networks have been developed in recent years due to their self-creating, self-organizing capabilities, which do not require any permanent network infrastructure [1]. Common scenarios for wireless ad hoc networks include survivable, efficient, dynamic communication networks for emergency and rescue operations, disaster relief efforts, and similar tasks where a fixed infrastructure is not always easily available. In such networks, flooding is one of the most fundamental operations for propagating control messages to every node. In *pure flooding* (or blind flooding) first discussed in [2,3], every node in the network forwards a flooding message when it receives a flooding message for the first time. However, this simple mechanism generates excessive

C.S. Hong et al. (Eds.): APNOMS 2009, LNCS 5787, pp. 345–354, 2009.
© Springer-Verlag Berlin Heidelberg 2009

retransmission packets, resource contention, and collisions in the network because all nodes need to forward the flooding message at least once. This is referred to as the *broadcast storm problem* [4,5].

To reduce the excessive flooding traffic, several flooding schemes have been proposed recently [6,7,8,9,10]. However, these flooding schemes are inefficient in reducing redundant transmissions, require significant control overhead, or employ numerous unnecessary forwarding nodes. In particular, [7] and [8] require every node to collect and maintain 2-hop neighbor information, which will incur extra system overhead. Also, the accuracy of the information is poor, due to the high node mobility.

In this paper, we propose an efficient flooding scheme for minimizing the number of forwarding nodes, by employing a triangular tiling algorithm. In our approach, each node selects the nodes located at the vertices of an equilateral triangle as the forwarding nodes. If there is no node located at a certain vertex, the node closest to the vertex is nominated as the forwarding node. In this manner, the flooding procedure continues hop-by-hop until every node receives the message. Based on this idea, our proposed scheme can minimize the number of forwarding nodes more efficiently than previous work. The proposed scheme has the following contributions:

- Our proposed scheme fully exploits the geographic location information of 1-hop neighbors to efficiently minimize the number of forwarding nodes during the flooding process and to solve the broadcast storm problem in wireless ad hoc networks.
- Our proposed scheme does not require any extra communication overhead other than the exchange of 1-hop HELLO messages.
- For each node, the number of forwarding nodes is limited to the number of vertices of an equilateral triangle, thus, the broadcast collision and congestion are reduced.
- Since the time complexity of our proposed scheme for computing the forwarding nodes in each step is linear with respect to the number of nodes, it is the lowest of all the flooding schemes considered in this work.

In order to evaluate the performance of our proposed scheme, it is implemented along with other flooding schemes, using the *ns-2 simulator* [11]. Simulations are performed with various transmission ranges and numbers of nodes, to validate their effect on performance. The simulation results show that the proposed scheme significantly reduces the number of retransmissions while guaranteeing high message deliverability compared to other work.

The remainder of this paper is organized as follows: Section 2 reviews existing flooding schemes in wireless ad hoc networks. Section 3 describes our proposed flooding scheme based on triangular tiling. In Section 4, we discuss the performance of our flooding scheme via simulation and compare its performance with other several flooding schemes. Finally, we conclude our work in Section 5.

2   Related Work

A pure flooding (or blind flooding) is a typical example in the first category, which does not require any information of neighbors. It is the most popular form of broadcasting because of its simplicity. In pure flooding, every node in the network forwards a flooding message exactly once after receiving it for the first time. However, redundant transmissions in pure flooding may cause the broadcast storm problem, resulting in contention and collisions in the networks.

Edge Forwarding [6] that uses location information of 1-hop neighbor is an effective receiver-based flooding scheme that works by partitioning the transmission coverage. In Edge Forwarding, each node divides its transmission coverage into six equally sized sectors. When a node, say B, receives a new flooding message from its parent node, say A, B first determines which sector of A is currently in B. B's partition lines divide the sector into six sub-partitions. If there is at least one node in each sub-partition, B does not forward the message. By doing so, it reduces the forwarding nodes in flooding operation.

Another important scheme is flooding based on *connected dominating set* (CDS) [7,8]. A dominating set (DS) is a subset of nodes such that every node in the graph is either in the set or adjacent to a node in the set. A CDS is a connected DS and any routing in mobile ad hoc networks can be achieved efficiently via the CDS. To maintain CDS in the network, however, each node is required to collect and maintain 2-hop neighbor information. Maintaining 2-hop neighbor information for each node incurs extra overhead of the system. Also, the information can be hardly accurate when the mobility of the nodes in the network is high.

The authors in [9] propose an efficient flooding scheme called 1HI (1-hop Information) to minimize the number of the forwarding nodes in mobile ad hoc networks (MANETs). When a source node has a message to flood, based on the 1-hop neighbor information, it computes the boundary of its neighbor's coverage area and selects the neighbors that contribute to this boundary as forwarding nodes. After that the source attaches the list of the forwarding nodes to the message and broadcasts the message to its neighbors. Upon receiving a flooding message, if the message has already been received, the receiver node is discarded; otherwise, it checks if it is itself in the forwarding list. If yes, it computes the next hop forwarding nodes among its neighbors and relays the message in the same manner as the source does. Finally, the message reaches all the nodes in the network.

In [10], Le and Choo present an improved scheme for 1HI that used 2-Hop Backward Information (2HBI). When a node has a message that needs to be flooded to every node, it selects the forwarding nodes in the same manner as 1HI. Next, the node further optimizes the number of forwarding nodes based on the three optimization rules. After that, the node attaches this optimized list of the forwarding nodes into the message and broadcasts it. Consequently, 2HBI has fewer forwarding nodes than 1HI, thus, it can more effectively reduce contention and collisions than 1HI in networks.

3   The Triangular Tiling-Based Flooding Scheme

The goal of every efficient flooding scheme is not only to cover all the nodes in the network but also to minimize the number of forwarding nodes participating in the flooding procedure. Therefore the overlapped areas of the connected forwarding nodes must be minimized. To this end, we consider the triangular tiling to flood messages in wireless ad hoc networks. In geometry, the triangular tiling is a type of regular tiling of the Euclidean plane, as shown in Fig. 1(a). Initially, the entire plane is covered with equilateral

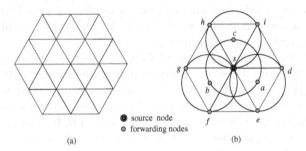

Fig. 1. (a) Triangular tiling (b) each node has three forwarding nodes located at the vertices of equilateral triangle

triangles and then transmission coverage areas are drawn that circumscribe them. Obviously, the plane can be fully covered by these coverage areas in this manner if each equilateral triangle has forwarding nodes at its vertices, as shown in Fig. 1(b). However, it does not ensure that forwarding nodes are exactly located at the vertices of equilateral triangles in wireless ad hoc networks. Thus, the forwarding nodes should be selected as the nodes which are located closest to the vertices.

We assume that all nodes in the network have the same transmission range R. Thus, the network assumed connected can be represented as a unit disk graph $G(V, E)$, where V is a set of nodes and E is a set of links. Each node v in V has a unique node ID, denoted by $id(v)$. Let $N(v)$ denote the set of neighbor nodes of v. That is, nodes in $N(v)$ are within the transmission range of v and can receive signals transmitted by v. Node v needs to know the information of its direct neighbors, including their IDs and their geographic locations. The 1-hop neighbor information can easily be obtained from the HELLO messages periodically broadcasted by each node. The location information of each node can be obtained via GPS or various distributed localization methods [12,13] when GPS service is not available.

The basic mechanism of our scheme involves guiding flooding such that it propagates along the vertices of equilateral triangles. Our proposed flooding scheme has three steps: 1) transmission coverage partitioning, 2) forwarding node selection, and 3) forwarding set optimization.

3.1   Transmission Coverage Partitioning

Each node divides its transmission coverage into three equal-size sectors, denoted as P_0, P_1, and P_2, respectively, as shown in Fig. 2(a). Let (x_u, y_u) denote the location of node u. Since u is a neighbor node of node s, node s determines the sector that contains node u as follows:

- If $x_s \geq x_u$ and $y_s < y_u$, then node u is in P_0 if $\frac{x_s - x_u}{y_u - y_s} \leq \sqrt{3}$; otherwise, node u is in P_1
- If $x_s > x_u$ and $y_s \geq y_u$, then node u is in P_1
- If $x_s < x_u$ and $y_s < y_u$, then node u is in P_0 if $\frac{x_u - x_s}{y_u - y_s} \leq \sqrt{3}$; otherwise, node u is in P_2
- If $x_s \leq x_u$ and $y_s \geq y_u$, then node u is in P_2

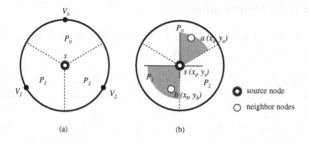

Fig. 2. Transmission coverage partitioning (a) partitioning and naming of node s' transmission range (b) node a and b are node s' P_0, P_1 neighbor nodes, respectively

For instance, in Fig. 2(b), node a is located in P_0 of node s since $x_s < x_a$, $y_s < y_a$ and $\frac{x_a - x_s}{y_a - y_s} \leq \sqrt{3}$. And node b is located in P_1 of node s since $x_s > x_b$, $y_s > y_b$. In this manner, every node in the network can determine the sector to which each of its neighbors belongs. After partitioning the transmission coverage into three sectors, each node computes the vertices of an equilateral triangle inscribed in its transmission coverage. These vertices are denoted as V_0, V_1, and V_2 as shown in Fig. 2(a). A vertex of equilateral triangle V_i is located at partition P_i, where $0 \leq i \leq 2$.

3.2 Forwarding Node Selection

Before flooding a message, the source node and every intermediate node must select its next-hop forwarding nodes from its neighbor nodes. For each partition, it selects only the node that is closest to the vertices of equilateral triangle as the forwarding node. In an empty partition, no forwarding nodes are selected.

Fig. 3 shows an example of the forwarding node selection step for each partition of node s. In this example, $N(s) = \{a, b, c, d, e, f\}$ and nodes a, b, and c are located in P_2 of node s. In this partition, s chooses node a which is closest to the vertex V_2 as the forwarding node. In the same manner, s selects the forwarding nodes d in P_0. Let $F(s)$ denote the set of forwarding nodes of s; accordingly, $F(s) = \{a, d\}$. Notice that node s does not select any forwarding node in P_1 since there is no neighbor of s located in this partition.

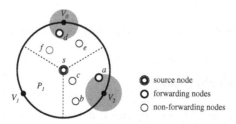

Fig. 3. Example of forwarding node selection

3.3   Forwarding Set Optimization

After selecting the forwarding set, every forwarding node optimizes its forwarding set based on two following optimization rules:

• **Rule 1:** Each node removes from its forwarding set the nodes that are covered by its parent node. By using rule 1, we can prevent the messages from propagating backward. Let us consider in Fig. 4 the case of forwarding node c, which has just received a flooding message from node s. The node c selects its forwarding nodes which are closest to the vertices of an equilateral triangle inscribed in its transmission coverage. Thus, the forwarding set of node c is $F(c) = \{d, e, s\}$. It is clearly shown that s and d had previously received a flooding message from node s. For this reason, node c no longer needs to consider the forwarding duty of nodes s and d. Consequently, by applying rule 1, nodes s and d can be removed from the $F(c)$. The resulting forwarding set of node c contains only node e.

• **Rule 2:** Each node that is closer to the parent node removes from its forwarding set the nodes that are in the overlapping coverage area of its covering area and covering areas of other forwarding nodes. This rule prevents different forwarding nodes from selecting the same node as their forwarding node. According to this rule, node e is removed from $F(c)$, because it is also in the coverage area of forwarding node a, which is farther from parent node s than node c. Then, the forwarding set of node c is now empty, which implies that there will be no next-hop forwarding node from node c.

After optimizing the forwarding set, the node whose forwarding set is empty abandons its retransmission role. As a result, node c abandons its role of forwarding node since its forwarding set is empty.

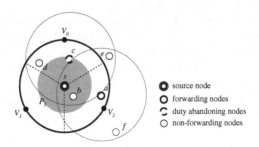

Fig. 4. Example of the forwarding set optimization

4   Performance Evaluation

By using the *ns-2 simulator* [11], we implement and compare our scheme with five representative flooding schemes in wireless ad hoc networks: *pure flooding, CDS-based flooding, Edge Forwarding, 1HI, 2HBI*. The simulation parameters are listed in Table 1. The popular two-ray ground reflection model is adopted as the ratio propagation model.

The MAC layer scheme follows the IEEE 802.11 MAC specification. We use the broadcast mode with no *RTS/CTS/ACK* mechanisms for all message transmissions. Each data packet with its attached information has a constant length of 256 bytes. The bandwidth of a wireless channel is set to *2 Mb/s* as the default. We consider the minimum CDS (MCDS) as a lower bound on the number of retransmissions since the number of forwarding nodes is no fewer than the number of MCDS in the network. Although computing MCDS is NP-hard, this is possible when we use the ratio-8 approximation algorithm [14]. Because every scheme being considered requires nodes to send HELLO messages to their 1-hop neighbors periodically, the cost of the HELLO messages are ignored in our performance evaluation.

We use three metrics to evaluate the efficiency of the flooding schemes: *ratio of forwarding nodes, number of collisions,* and *delivery ratio. The ratio of forwarding nodes* is given by the total number of transmissions that every forwarding node needs to cover the entire network over the total number of nodes in the network. *The number of collisions* is defined to be the total number of collided packets that each node encounters before it successfully receives the flooding message. And *the delivery ratio* is the ratio of the nodes that successfully received messages over the number of the nodes in the network for 100 seconds. We investigate the impact of two parameters on these metrics: *the number of nodes* and *the transmission range*, respectively.

4.1 Simulation Results with Increasing Number of Nodes

In this simulation, we vary the number of nodes from 200 to 1000 in a square area of $1000\ m \times 1000\ m$. The transmission range is fixed at $250\ m$. In the simulation, nodes are placed at random locations in the network and the results are computed as the average of 100 runs. Each simulation is run for 100 seconds, as shown in Fig. 5(c)

The performance of our proposed scheme is significantly better than the performance of *pure flooding, CDS-based flooding, Edge Forwarding, and 1HI*, and slightly better than the performance of *2HBI* in terms of the ratio of forwarding nodes as shown in Fig. 5(a). The curve of our proposed flooding scheme almost approaches the lower bound, since we select at most two forwarding nodes to forward flooding messages. And, the curve of our scheme decreases when the number of nodes increases, because

Table 1. Simulation parameters

Parameters	Value
Simulator	*ns*-2 (version 2.31)
MAC Layer	IEEE 802.11
Data Packet Size	256 bytes
Transmission Rate	2 Mb/s
Transmission Range	$100 \sim 300\ m$
Number of Nodes	$200 \sim 1000$
Number of Trials	100
Size of Square Area	$1000\ m \times 1000\ m$
Network Load	$10\ Pkt/s$

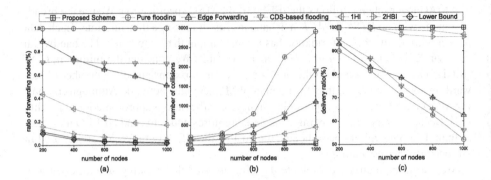

Fig. 5. Simulation results with increasing number of nodes on performance

there is a greater chance of finding forwarding nodes that are closer to the vertices of the equilateral triangular in a flooding operation. As shown in Fig. 5(a), when the number of nodes in our proposed scheme reaches 1000, only 2.2% of the total number participate in forwarding, whereas in Edge Forwarding, CDS-based flooding, 1HI, and 2HBI, 51.1%, 69.5%, 16.9%, and 4.4% participate in forwarding, respectively.

In Fig. 5(b), our scheme and 2HBI have a much lower rate of collisions compared with other schemes. This means that the broadcast problem can be alleviated in our proposed scheme. This is because it has fewer forwarding nodes and the senders always select the farthest forwarding nodes. In Fig. 5(c), the delivery ratio of our scheme is guaranteed as 100 % when the number of nodes varies from 200 to 1000, while the delivery ratio of other schemes are 52%–100%. Therefore, the delivery ratio of our scheme is significantly better than that of other schemes. Note that although collisions can occur, a node that misses flooding messages from a forwarding node still has a chance of receiving messages from another forwarding node.

4.2   Simulation Results with Increasing Transmission Range

In this simulation, we vary the transmission range from 100 m to 300 m in each simulation run and randomly place 500 nodes in a square region of 1000 m × 1000 m. The network load is set to *10 Pkt/s*. We study the performance with respect to the transmission range of each node. The simulation results are plotted in Fig. 6. Each simulation is run for 100 seconds, as shown in Fig. 6(c).

As the transmission range increases, the number of neighbors of each node increases. This has the same effect on the increase of network density as the increase of nodes in a fixed square region. Similar to the influence of varying the number of nodes, the performance of our proposed scheme is significantly better than that of all other schemes shown in Fig. 6(a). The curve of our scheme approaches the curve of the lower bound as the transmission range increases. When the transmission range reaches 300 m, only 3.8% of nodes participate in forwarding in our flooding scheme, whereas in Edge Forwarding, CDS-based flooding, 1HI, and 2HBI, the participation rate is 57.8%, 67.1%, 19.2%, 7.9%, respectively. This is because increasing the transmission range not only results

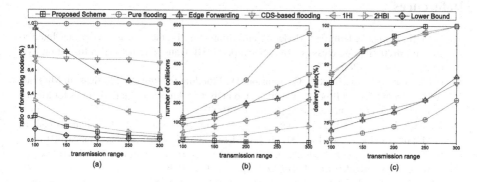

Fig. 6. Simulation results with increasing transmission range on performance

in a higher network density, but also faster flooding. This also means that the flooding operation can be done in fewer steps due to the large transmission range of nodes.

The curves in Fig. 6(b) show similar trends to Fig. 5(b). When the transmission range increases, there is a higher chance of collisions. On the other hand, the curve of our proposed scheme decreases since the forwarding nodes are selected to be as far away as possible. The delivery ratios of all compared flooding schemes increase when the transmission range increases, as shown in Fig. 6(c). The delivery ratio of our scheme is slightly worse than 1HI and 2HBI when the transmission range varies from 100 m to 150 m. Beyond a transmission range of 200 m, however, the performance of our scheme is better than 1HI, 2HBI in terms of the delivery ratio. This is because a node selects at most two forwarding nodes, so the connectivity between nodes is weaker than that of other flooding schemes in which the senders can select more forwarding nodes.

5   Conclusion

In this paper, we addressed the flooding problem, to minimize the number of forwarding nodes and to cover every node in the network. We presented an efficient flooding scheme that uses only 1-hop neighbor information and employs a triangular tiling algorithm to flood messages in the network. Our scheme has the lowest time complexity for selecting the forwarding nodes which is linear with respect to the number of nodes; it is on the order of $O(N)$, where N is the number of nodes. Simulation results show that our flooding scheme has fewer forwarding nodes, and collisions, while achieving a higher *delivery ratio* than the existing efficient flooding schemes.

Acknowledgment

This research was supported by MKE, Korea under ITRC IITA-2009-(C1090-0902-0046) and by MEST(Korea), under WCU Program supervised by the KOSEF (No. R31-2008-000-10062-0) and was supported by the Brain Korea program, the Ministry of Education, Science and Technology, the Korean government. Dr Choo is the corresponding author.

References

1. Woo, S., Hong, J., Kim, H.: Modeling and Simulation Framework for Assessing Interference in Multi-Hop Wireless Ad Hoc Networks. KSII Trans. Internet and Information Systems 3(1), 26–51 (2009)
2. Ho, C., Obraczka, K., Tsudik, G., Johnson, D.: Flooding for Reliable Multicast in Multihop Ad Hoc Networks. In: Proc. Int'l Workshop on Discrete Algorithms and Methods for Mobile Computing and Communications, pp. 64–71 (1999)
3. Jetcheva, J., Hu, Y., Maltz, D., Johnson, D.: A Simple Protocol for Multicast and Broadcast in Mobile Ad Hoc Networks (2001) Internet Draft: draft-ieft-manet-simple-mbcast-01.txt
4. Ni, S., Tseng, Y., Chen, Y., Sheu, J.: The broadcast storm problem in a mobile ad hoc network. In: Proc. ACM/IEEE MOBICOM, pp. 151–162 (1999)
5. Shin, M., Arbaugh, W.A.: Efficient Peer-to-Peer Lookup in Multi-hop Wireless Networks. KSII Trans. Internet and Information Systems 3(1), 5–25 (2009)
6. Cai, Y., Hua, K.A., Phillips, A.: Leveraging 1-Hop Neighborhood Knowledge for Efficient Flooding in Wireless Ad Hoc Networks. In: Proc. 24th IEEE Int'l Performance Computing and Communications, IPCCC (2005)
7. Dai, F., Wu, J.: An Extended Localized Algorithm for Connected Dominating Set Formation in Ad Hoc Wireless Networks. IEEE Trans. Parallel and Distributed Systems 15(10), 908–920 (2004)
8. Wu, J., Lou, W., Dai, F.: Extended Multipoint Relays to Determine Connected Dominating Sets in MANETs. IEEE Trans. Computers 55(3), 334–347 (2006)
9. Liu, H., Jia, X., Wan, P.J., Liu, X., Yao, F.F.: A Distributed and Efficient Flooding Scheme Using 1-Hop Information in Mobile Ad Hoc Networks. IEEE Trans. Parallel and Distributed System 18(5), 658–671 (2007)
10. Le, T.D., Choo, H.: Efficient Flooding Scheme Based on 2-Hop Backward Information in Ad Hoc Networks. In: Proc. IEEE Int'l Conf. on Communications (ICC), pp. 2443–2447 (2008)
11. The Network Simulator, www.isi.edu/nsnam/ns (cited on 10th May, 2007)
12. Meguerdichian, S., Slijepcevic, S., Karayan, V., Potkonjak, M.: Localized algorithms in wireless ad-hoc networks: location discovery and sensor exposure. In: Proc. 2nd ACM international symposium on Mobile ad hoc networking & computing, pp. 4–5 (2001)
13. Bulusu, N., Heidemann, J., Estrin, D., Tran, T.: Self-configuring localization systems: Design and Experimental Evaluation. ACM Trans. Embedded Computing Systems (TECS) 3(1), 24–60 (2004)
14. Wan, P.J., Alzoubi, K., Frieder, O.: Distributed Construction of Connected Dominating Set in Wireless Ad Hoc Networks. In: Proc. 21st IEEE INFOCOM, pp. 1597–1604 (2002)

Modeling Directional Communication in IEEE 802.15.3c WPAN Based on Hybrid Multiple Access of CSMA/CA and TDMA

Chang-Woo Pyo

National Institute of Information and Communications Technology, 3-4 Hikarino-Oka, Yokosuka, 239-0847, Japan

Abstract. This paper focuses on the performance of the hybrid multiple access of CSMA/CA and TDMA, namely *CSMA/CA-TDMA*, in the 802.15.3c WPAN. A contribution of this paper is to provide an analytical model to study the system throughput of CSMA/CA-TDMA. A key property of the 802.15.3c WPAN is the directional communication due to the directivity of the mmWave radio. Another contribution is to analyze the system throughput for directional communication based on CSMA/CA-TDMA in the 802.15.3c WPAN. Numerical results show that the large number of the directions supported by the PNC could be a high overhead in the network as well as CSMA/CA-TDMA has higher system performance rather than CSMA/CA in case of happening high collisions.

1 Introduction

There is a growing demand for high-speed wireless multimedia services for both personal as well as commercial purposes. Several research groups have conducted studies to realize high-speed wireless communications across the world. For realizing high-speed WLAN, IEEE 802.11n has been introduced to provide a maximum throughput of up to 400 Mbps [1]. IEEE 802.11ac [2] and IEEE 802.11ad [3] have been introduced for developing very-high-throughput medium access control (MAC), and for evaluating the IEEE 802.11n MAC with higher PHY rates.

For high-speed wireless multimedia services such as uncompressed high-definition TV (HDTV), instantaneous music, and image data transmissions, it is necessary that the data rate of wireless communications exceeds 1 Gbps. Thus, millimeter-waves (mm-Waves) have attracted considerable attention as powerful media for developing such high-speed wireless communications [4]. IEEE 802.15.3c has been introduced to standardize a mm-Wave wireless personal area network (we call 802.15.3c WPAN) operating in the unlicensed band of 57-66 GHz with a targeted data rate of over 1 Gbps [5].

This paper focuses on the performance of wireless multiple access in the 802.15.3c WPAN. The most famous wireless multiple access is contention-based CSMA/CA used in the 802.11 WLANs. On the other hand, the 802.15.3c WPAN

C.S. Hong et al. (Eds.): APNOMS 2009, LNCS 5787, pp. 355–364, 2009.

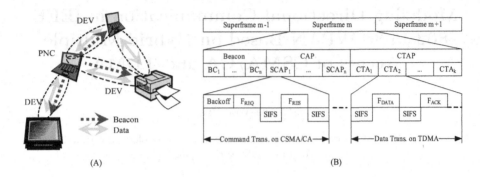

Fig. 1. (A) A pico-network and (B) a superframe construction for directional communication in the 802.15.3c WPAN

has applied a *hybrid multiple access* of contention-based CSMA/CA and contention-free TDMA, namely "*CSMA/CA-TDMA*", that supports the different channel usages [5][6]. In CSMA/CA-TDMA, CSMA/CA is mainly used for control signal transmission, while TDMA is used for high speed data transmission.

A contribution of this paper is to provide an analytical model to study the system throughput of CSMA/CA-TDMA. Lots of analytical models for CSMA/CA have been published [7]-[11]. However, these are not directly adopted for the throughput analysis of the 802.15.3c WPAN because the data transmission of the 802.15.3c WPAN is mainly performed by TDMA, then the throughput of TDMA is quite different from that of CSMA/CA. Another contribution is to analyze the system throughput for directional communication in the 802.15.3c WPAN. A key property of the 802.15.3c WPAN is the directional communication due to the directivity of mm-Wave. Thus, the 802.15.3c WPAN provides the "*directional CSMA/CA-TDMA*" mechanism, which enables to communicate among devices directionally as well as to get high communication gain. As our basic understanding, this paper is the first attempt to analyze the system throughput for the directional CSMA/CA-TDMA in the 802.15.3c WPAN.

The rest of the paper is organized as follows. Section 2 describes the overview of the 802.15.3c WPAN. Section 3 provides an analytical model to study the system throughput for the directional CSMA/CA-TDMA. The results of throughput analysis are shown in Section 4, and we conclude this paper in Section 5.

2   Overview of 802.15.3c WPAN

The 802.15.3c WPAN (referred to as a *piconet*) allows a number of independent devices to communicate with each other. A piconet consists of several components, as shown in Fig. 1 (A). The basic component of a piconet is the device (DEV). In particular, one device is required to play the role of a piconet coordinator (PNC). The PNC is responsible to permit devices to be members of the piconet, store their information for maintaining the piconet, announce the existence of the piconet, and synchronize communications among devices in the piconet. Here, we summarize the MAC features specified in the 802.15.3c WPAN draft standard [5].

2.1 802.15.3c WPAN MAC

Hybrid Multiple Access. The PNC provides multiple channel accesses per unit of superframe (SF), which is the basic timing of the piconet, as shown in Fig 1 (B). The SF comprises three major portions − beacon, contention access period (CAP), and channel time allocation period (CTAP). The beacon is broadcasted for SF timing allocation only by the PNC, while each device in the piconet accesses a channel either during the CAP or CTAP. The CAP and CTAP work on the different multiple access types of contention-based CSMA/CA and contention-free TDMA, respectively. The portions of SF are described in detail as follows:

− Beacon: The beacon broadcast by the PNC is used to set the timing allocations of SF and to transmit management information for the piconet.
− Contention Access Period (CAP): The contention access period (CAP) is used to communicate commands and/or any asynchronous data present in the SF. The multiple access method during the CAP is CSMA/CA.
− Channel Time Allocation Period (CTAP): The channel time allocation period (CTAP) is composed of channel time allocations (CTAs). The CTAs are used for commands, isochronous streams, and asynchronous data connections. The multiple access method during the CTAP is TDMA.

Directional CSMA/CA-TDMA. To perform directional communication, device detection shall be done before communication between the directional antenna capable devices. The followings are two categories of device detection for directional communication.

− PNC − DEV: The beacons are broadcast by the PNC to all available directions and the CAP is divided into several sub-CAPs (S-CAPs) to receive the feedbacks from the devices as shown in Fig. 1 (B). The number of S-CAP depends solely on the receive direction of the PNC since each S-CAP identifies the receiving direction of the PNC. In normal, the number of S-CAPs is the same as the number of beacons.
− DEV − DEV: The CTA is used for the device detection between two devices in order to reduce the signaling overhead in the CAP.

3 Analytical Model

A main contribution of this paper is the analytical evaluation for the system throughput upon the directional CSMA/CA-TDMA in the 802.15.3c WPAN.

3.1   S-CAP on CSMA/CA

Data transmission in CSMA/CA-TDMA is mainly performed at the CTAs. Then, the system throughput depends on the number of allocated CTAs. Before deriving the number of allocated CTAs, particularly, it is necessary to consider command transmission during the CAP because the CTA request command shall be successfully transmitted between devices and the PNC to allocate CTAs for communication. As mentioned before, the CAP in the SF is divided into several S-CAPs for directional communication. Thus, the CTA request transmission shall be completed in the S-CAP.

Each S-CAP is also worked on the multiple access of CSMA/CA. During the S-CAP, in addition, all devices sense the status of the shared wireless channel to transmit frames. The ideal wireless channel for transmission during the S-CAP is based on the assumption that all frames can be transmitted and received error free if there is no collision. Initially, any device can transmit a CTA request command to the PNC during the S-CAP. When the request command is successfully transmitted, the PNC allocates a CTA for the communication between the source and destination devices. However, when the transmission of the request fails due to a collision, the device waits for a random backoff time before retransmitting the request. Note that the collision avoidance by RTS/CTS adopted in the IEEE 802.11 WLANs cannot be applied in the 802.15.3c WPAN; hence, we do not consider RTS/CTS for the system throughput analysis.

Let us assume j devices are contending the channel at a given time and the average frame transmission probability of the devices is γ. The probabilities of the channel to be *idle* (i.e., the probability that none of devices transmits a frame, denoted by p_i), *success* (i.e., the probability that just one device transmits a frame, denoted by p_s) and *collision* (i.e., the probability that multiple devices transmit frames at the same time, denoted by p_c) can be derived from the Bernoulli process.

$$\begin{cases} p_i = (1 - \gamma)^j, \\ p_s = j \cdot \gamma \cdot (1 - \gamma)^{j-1}, \\ p_c = 1 - j \cdot \gamma \cdot (1 - \gamma)^{j-1} - (1 - \gamma)^j. \end{cases} \tag{1}$$

Upon following the random backoff procedure in CSMA/CA, the frame transmission probability of a device, γ, can be derived from the average number of transmissions made by a device during the average backoff time for the device to successfully transmit a frame. Then, γ can be presented by

$$\gamma = \frac{E[A]}{E[D]}, \tag{2}$$

where $E[D]$ is the average backoff time for a device to successfully transmit a frame when there are N channel contending devices, and $E[A]$ is the average number of transmission attempts made by the device during $E[D]$.

To derive the average backoff time, $E[D]$, we use the exponential backoff procedure defined in the 802.15.3c WPAN. From the limit of space, we do not

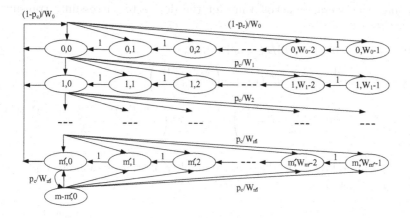

Fig. 2. State transitions of the backoff procedure for CSMA/CA

describe the details of the exponential backoff procedure, but the backoff procedure is the same as that defined in 802.11 WLANs. Figure 2 shows the state transitions of the exponential backoff procedure. The notation (i, W_i) indicates the backoff window size of $W_i = 2^i W$ at the ith backoff stage. The average backoff time for a device to successfully transmit a frame, $E[D]$, can be described as the average backoff time taken by the device to successfully transmit the frame during each backoff stage. At the initial backoff stage of $i = 0$, a device shall wait for a random backoff time of $W_0 = W$ for transmission. When the frame is successfully transmitted after the backoff time, the expected backoff time taken by the device, denoted by $E[D|i = 0]$, is given as

$$E[D|i = 0] = (1 - p_c)\frac{W}{2}.$$

When the frame is successfully retransmitted after s collisions, i.e., the frame is successfully transmitted at the backoff stage of $i = s$, the expected backoff time for the device, denoted by $E[D|i = s]$, is given as

$$E[D|i = s] = p_c^s(1 - p_c)\left(\frac{\sum_{l=0}^{s} 2^l W}{2}\right),$$

where $1 \le s \le m'$. In case of collisions after the maximum backoff stage of m', the backoff stage of the device returns to the maximum backoff stage of $i = m'$. The expected backoff time of the device during which the frame is successfully transmitted within the retransmission limit of m is given as

$$E[D_m] = E[D|i = m' + 1] + ... + E[D|i = m]$$
$$= p_c^{m'+1}\left(\frac{\sum_{l=0}^{m'} 2^l W + 2^{m'} W}{2}\right) + ... + p_c^m\left(\frac{\sum_{l=0}^{m'} 2^l W + (m - m')2^{m'} W}{2}\right).$$

Therefore, the average backoff time for the device to successfully transmit a frame, $E[D]$, is

$$E[D] = E[D|i = 0] + \ldots + E[D|i = m'] + E[D_m] \tag{3}$$

$$= (1 - p_c)\frac{W}{2} + \cdots + p_c^{m'}(1 - p_c)\left(\frac{\sum_{l=0}^{m'} 2^l W}{2}\right)$$

$$+ p_c^{m'+1}\left(\frac{\sum_{l=0}^{m'} 2^l W + 2^{m'} W}{2}\right) + \cdots + p_c^m\left(\frac{\sum_{l=0}^{m'} 2^l W + (m - m')2^{m'} W}{2}\right).$$

During the period of $E[D]$, the average number of transmission attempts made by a device, $E[A]$, can also be modeled as follows:

$$E[A] = (1 - p_c) + p_c(1 - p_c)2 + \cdots + p_c^m(m + 1) = \frac{1 - p_c^{m+1}}{1 - p_c}. \tag{4}$$

3.2   Number of Successful CTA Requests in S-CAP (N_{suc})

Let us assume that after a certain period of time in the piconet consisting of M devices, i CTAs are allocated by $2i$ devices, denoted by CTA_i, where none of the devices allow multiple communication links. In CTA_i, $2i$ devices of source and destination are active, while $M - 2i$ devices are idle in the piconet. Since the maximum number of CTAs cannot exceed the maximum number of possible communication links ($M/2$), the value of i ranges from 0 to $\frac{M}{2}$.

At the beginning of a S-CAP in CTA_i, on average $N = (M - i) \cdot P_{req}/N_{SCAP}$ devices send the CTA request commands to the PNC, where P_{req} is the probability of transmitting a frame by each device and N_{SCAP} is the total number of S-CAPs per SF. Note that the CTA requests transmitted from iP_{req}/N_{SCAP} communicating (active) devices are the requests to release the allocated CTAs, while the CTA requests transmitted from $(M - 2i)P_{req}/N_{SCAP}$ non-communicating (idle) devices are the requests to allocate new CTAs. After a certain time, one CTA request is served and the remaining $N - 1$ CTA requests are continued that the process of frame service may continue until all CTA requests are served.

In order to derive the number of successful CTA requests during the S-CAP, we investigate the channel status, which can either be idle or busy. The channel is idle only when there is no frame transmission, and it is sensed busy because of collisions or successful frame transmissions. In a given length of the S-CAP, T_{SCAP}, the average number of busy slots, n_{bs}, can be expressed as

$$n_{bs} = \frac{T_{SCAP} - n_{is} \cdot T_{is}}{T_{bs}}, \tag{5}$$

where n_{is} is the average number of idle slots, and T_{is} and T_{bs} are the duration of an idle slot and a busy slot, respectively. Since the channel is idle only when all contending devices are in the backoff stage, n_{is} can be approximated as the average backoff time a device experiences as derived in (3). Upon the 802.15.3c draft standard, T_{is} is given as $SIFS(2.5us)+pCCADetectTime(5us)$. Given the channel is busy in T_{SCAP} and implying at least one device is transmitting in the given slot, two events may occur. One event is collision resulting from multiple simultaneous frame transmissions and another is successful frame transmission when only one device transmits in that slot. Thus, T_{bs} can be represented by

$$T_{bs} = \overline{p_s}(T_{req} + SIFS + T_{ACK} + BIFS) + \overline{p_c}(T_{req} + BIFS), \qquad (6)$$

where BIFS is a backoff IFS defined as SIFS (2.5us) + pCCADetectTime (5us). T_{req} is the length of the CTA request frame and T_{ACK} is the length of the ACK frame. From the channel state transitions, the probability of successful frame transmission in a busy slot, $\overline{p_s}$, and the probability of frame collision in a busy slot, $\overline{p_c}$, are derived as

$$\overline{p_s} = \frac{p_s}{p_s + p_c} = \frac{n \cdot \gamma \cdot (1-\gamma)^{n-1}}{1 - (1-\gamma)^n},$$

$$\overline{p_c} = \frac{p_c}{p_s + p_c} = \frac{1 - (1-\gamma)^n - n \cdot \gamma \cdot (1-\gamma)^{n-1}}{1 - (1-\gamma)^n}, \qquad (7)$$

where n is the average number of contending devices during T_{SCAP} with the uniform distribution of $[N, 0]$.

During a transmission, since the two events of collision and success are independent of each other, the average number of busy slots that contribute to successful transmissions can be considered as a binomial random variable. The total successful command transmissions in the overall CAP containing more than one S-CAP, denoted by n_{suc}, can be derived as

$$n_{suc} = N_{SCAP} \times \sum_{l=0}^{n_{bs}} l \cdot \binom{n_{bs}}{l} (\overline{p_s})^l (\overline{p_c})^{n_{bs}-l}. \qquad (8)$$

3.3   Number of Allocated CTAs (N_{CTA})

As already mentioned, the CTA requests transmitted from the active devices are the requests to release allocated CTAs, while the CTA requests transmitted from the idle devices are the requests to allocate new CTAs. Therefore, during the CAP, when n_{suc}^{id} is larger than n_{suc}^{av}, the PNC needs to allocate $n_{suc}^{id} - n_{suc}^{av}$ additional CTAs. On the contrary, the PNC may release $n_{suc}^{av} - n_{suc}^{id}$ allocated CTAs. Then, n_{suc}^{id} and n_{suc}^{av} that are the number of idle devices and the number of active devices with the success slots, respectively, are given as

$$n_{suc}^{id} = \frac{M - 2i}{M - i} \cdot n_{suc}, \quad n_{suc}^{av} = \frac{i}{M - i} \cdot n_{suc}. \qquad (9)$$

On the basis of the number of requests at the SF containing i CTAs, the expected number of allocated CTAs for the following SF, $\overline{N_{CTA_i}}$, becomes

$$N_{CTA_i} = \begin{cases} i + (n_{suc}^{id} - n_{suc}^{av}), & n_{suc}^{id} > n_{suc}^{av}, \\ i - (n_{suc}^{av} - n_{suc}^{id}), & n_{suc}^{id} \leq n_{suc}^{av}. \end{cases} \qquad (10)$$

3.4   System Throughput (S)

The throughput of the piconet is successfully performed by data transmission in the allocated CTAs. In a given time of one SF, then the throughput is given as

$$S_i = \frac{N_{CTA_i} \cdot k \cdot B}{L_{sf}}, \tag{11}$$

where k is the number of transmitting frames per CTA, B is the payload of the frame and L_{sf} is the SF length consisting of the beacon, the CAP and N_{CTA} numbers of CTAs. The system throughput for a long term period is given by the average throughput on the allocated $CTAs$ being from 0 to $M/2$.

$$S = \frac{\sum_{i=0}^{M/2} S_i}{M/2} = \frac{\sum_{i=0}^{M/2} (N_{CTA_i} \cdot k \cdot B)/L_{sf}}{M/2}. \tag{12}$$

Table 1. Parameters for throughput analysis

Parameters	Values
Number of devices in a piconet (M)	1-20
Number of frames per CTA (k)	30, 50
Frame payload (B)	10KBytes
Data rate (R)	1.62Gbps (BPSK), 3.24Gbps (QPSK)
Length of beacon (T_{beacon})	50us
SIFS	2.25us
MIFS	0.5us
Length of command (T_{req})	22.2us
Length of ACK (T_{ACK})	17.8us
Length of frame (T_{frame})	Preamble (1.185us) + Header (0.44us) + Payload length (B/R)

4   Numerical Results

In order to study the system throughput of the 802.15.3c WPAN based on the hybrid multiple access of CSMA/CA-TDMA and the effect of channel inter-ference on the system throughput, the parameters are shown in Table I; these parameters are based on the 802.15.3c draft specification. The required CTA length depends on the service requirement of applications. That is, the CTA length is determined on the basis of the number of transmitting frames and the length of the frame. To simplify the analysis, we consider that the aver-age number of transmitting frames, μ, and the average length of the frames, T_{frame}, are constant for each device. T_{frame} can be obtained from the data frame structure consisting of the preamble, header and payload. Moreover, the frame transmission time per CTA, T_{CTA}, is $\mu(T_{frame} + SIFS + T_{ACK} + SIFS)$

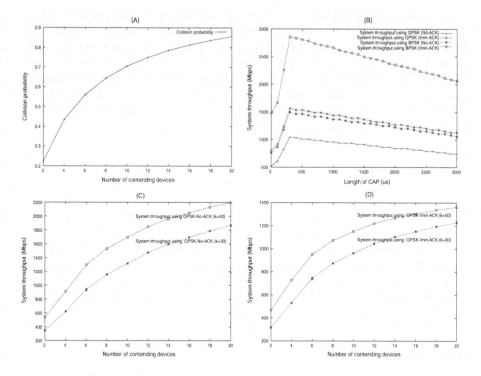

Fig. 3. (A) Collision probability in CAP, (B) System throughput with the length of CAP, (C) System throughput with the number of contending devices and No-ACK, (D) System throughput with the number of contending devices and Imm-ACK

if immediate ACK (Imm-ACK) is used, and $\mu T_{frame} + (\mu - 1)MIFS + SIFS$ if No-ACK is used. The length of SF is $L_{sf} = T_{beacon} + T_{CAP} + T_{CTA} \times N_{CTA}$.

Fig. 3 shows the numerical results of the performance analysis in the 802.15.3c WPAN. Fig. 3 (A) shows the effect of contending devices on frame collision in the fixed length of the CAP. We can see the probability of frame collision increases as an increase in the contending devices. This result is intuitively understandable because it causes the multiple access of CSMA/CA in the CAP. From the result, a large length of the CAP is expected to reduce frame collision, but we can see an interesting result in Fig.3 (B). The result shows that the effect of the length of the CAP on the system throughput. We can see that the system throughput increases with an increase in the number of contending devices. Meanwhile, we can also see that there is a sudden drop in the system throughput when the number of contending devises crosses a certain limit. This result indicates that the system throughput based on TDMA in the 802.15.3c WPAN is significantly affected by the length of the CAP based on CSMA/CA. It also means that the large number of the directions supported by the PNC could be a high overhead in the network.

Fig. 3 (C) and (D) show the effects of contending devices on the system throughput with respect to the ACK methods of No-ACK and Imm-ACK. The both figures show the system throughput increases as an increase in the contending devices regardless of the types of ACKs. The reason is because the overall times used for communication in the SF increases as the contending devices increase. These results show a significantly difference between CSMA/CA-TDMA and CSMA/CA. CSMA/CA-TDMA has higher system performance rather than CSMA/CA in case of happening high collisions.

5   Conclusions

This paper studied the performance of the hybrid multiple access of CSMA/CA-TDMA in the 802.15.3c WPAN. This paper provided a novel analytical model to study the system throughput of CSMA/CA-TDMA as well as to analyze the system throughput for directional communication based on CSMA/CA-TDMA in the 802.15.3c WPAN. From the obtained results, we found that the large number of the directions supported by the PNC could be a high overhead in the network as well as CSMA/CA-TDMA had higher system performance rather than CSMA/CA in case of happening high collisions.

References

1. IEEE 802.11 Working Group Homepage,
   http://www.ieee802.org/11
2. Status of Project IEEE 802.11 Task Group AC,
   http://www.ieee802.org/11/Reports/tgacupdate.htm
3. Status of Project IEEE 802.11 Task Group AD,
   http://grouper.ieee.org/groups/802/11/Reports/vhtupdate.htm
4. Ogawa, H.: Millimeter-Wave Wireless Personal Area Network (WPAN) and Its Standardization Activity within IEEE802.15. In: Proc. ICMMWT (2004)
5. IEEE 802.15.3c, http://www.ieee802.org/15/pub/TG3c.html
6. Pyo, C.W., et al.: MAC Enhancement for Gbps Throughput Achievement in Millimeter-Wave Wireless PAN Systems. In: IWCMC (2008)
7. Bianchi, G.: Performance Analysis of the IEEE 802.11 Distributed Coordination Function. IEEE JSAC 18, 535–547 (2000)
8. Wang, Y., Garcia-Luna-Aceves, J.J.: Collision Avoidance in Multi-Hop Ad Hoc Networks. In: Proc. IEEE International Symp. on MASCOTS (2002)
9. Tickoo, O., Sikdar, B.: Queueing Analysis and Delay Mitigation in IEEE 802.11 Random Access MAC based Wireless Networks. In: Proc. IEEE Infocom (2004)
10. Cai, L.X., Shen, X., Mark, J.W.: Voice Capacity Analysis of WLAN With Unbalanced Traffic. IEEE Tran. on Vehicular Technology 55(3) (May 2006)
11. Malone, D., Duffy, K., Leith, D.: Modeling the 802.11 Distributed Coordination Function in Nonsaturated Heterogeneous Conditions. IEEE/ACM Trans. on Networking 15(1) (February 2007)

Management of Upstreaming Profiles for WiMAX*

Sheng-Tzong Cheng[1] and Red-Tom Lin[2]

[1] Department of Computer Science and Information Engineering,
National Cheng Kung University, Taiwan
[2] Network and Multimedia Institute,
Institute for Information Industry, Taiwan

Abstract. In recent years many Internet video platforms, such as YouTube and
MySpaceTV, are developed for users to share their videos with others. However, they are unable to do that wirelessly in a real-time fashion due to the insufficient network bandwidth or transmission quality. Under such context, the
IEEE 802.16 standards have been proposed for high-speed communications. It
enables video streaming in real time for mobile users. In this paper, we propose
an architecture for mobile users to perform the upstreaming of their videos. The
users can upstream their real-time videos through their mobile devices and
wireless network technology. Due to the various network conditions and video
quality, we propose the multiple profiles upstream function. We provide adaptive forward error correction codes, automatic repeat request, and unequal error
protection for selection base on the hardware capability and the network states
to ensure the quality of service. Finally, we use NS-2 as simulation tool to verify our algorithm.

Keywords: WiMAX, QoS, FEC, ARQ, ULP, NS-2.

1 Introduction

H.264 has great advantage of coding efficiency compared with the successful prior
coding standards. It can save 64.46%, 48.80%, and 38.62% bit-rate compared with
that of MPEG-2, H.263, and MPEG-4 respectively under the same reconstructed
picture quality. But the high coding efficiency is acquired by heavy computation. It is
estimated that the complexity of H.264 encoder is about 5-10 times as that of MPEG-
4, and the complexity of H.264 decoder is about 2-4 times as that of MPEG-4. The
high coding complexity limits the application of H.264 in the domain of real time
video communication.

With the rapid growth of wireless technologies, the task of providing broadband
last mile connectivity is still a challenge. Such wireless solutions avoid the prohibitive
cost of wiring homes and businesses and allow a relatively faster deployment process.
Among the broadband access technologies that are being sought, WiMax (worldwide
interoperability of microwave access) is perhaps the strongest contender that is being
supported and developed by a consortium of companies.

---

* This research was supported by the Applied Information Services Development & Integration
project of Institute for Information Industry and sponsored by MOEA, R.O.C.

C.S. Hong et al. (Eds.): APNOMS 2009, LNCS 5787, pp. 365–374, 2009.

Nowadays, users do not only use broadband Internet just for connectivity and web surfing. Services like video on demand are becoming popular in the last mile. The widespread use and bandwidth demands of these multimedia applications are far exceeding the capacity of current 3G and wireless LAN technologies. Moreover, most access technologies do not have the option to differentiate specific application demands or user needs. WiMax is envisioned as a solution to the outdoor broadband wireless access that is capable of delivering high-speed streaming data. WiMax offers some flexible features that can potentially be exploited for delivering real-time services. In this paper, we explore the possibility of supporting upstream over WiMax and suggest means through which the quality of upstream can be improved. Specifically, the contributions of this paper are listed as follows:

- The quality of upstream depends on the loss of the packets . We try to recover as many dropped packets as possible and get the balance between loss rate and redundant packets in real-time.
- The weights of video packets from different frame are not equal. We will allocate the redundant packets to different frame. The number of redundant packets is based on the loss rate of video packets during a GOP and the ratio of redundant packets we assign to the frame bottom on the frame's weight.
- Forward error correction (FEC) is more appropriate than automatic retransmission request (ARQ) in streaming video applications , but ARQ could fix the packet loss more efficient within the limitation of delay time. We would activate the ARQ mechanism if the condition is completion.
- The hardware level of the device used by User would not afford all error correcting method. We propose the multiple profiles to fit to the different hardware and network conditions.

The rest of this paper is organized as follows: In Section 2, we provide an overview on the techniques that have been proposed to support streaming services over wireless channels. In Section 3, we simply introduce the features of personal upstreaming. In Section 4, we show the techniques of error correction we used and activate parts of them based on the conditions. In Section 5, we present the simulation model and results. Conclusions are drawn in Section 6.

2   Related Work

Supporting real-time applications over any wireless network poses many challenges, including limited bandwidth, coping with bandwidth fluctuations, and lost or corrupted data. Due to the growing popularity of streaming services over wireless networks, the problems have been well researched and many solutions have been proposed which combine audio and video processing techniques with mechanisms that are usually dealt with in the data link and physical layers. These approaches can broadly be classified into two categories: automatic repeat request (ARQ) and forward error correction (FEC). ARQ schemes provide high reliability when the channel is good or moderate. However, for error-prone channels, the throughput drops due to the increased frequency of retransmissions. Since error control techniques base on ARQ introduce delay that may be unacceptable, FEC is more appropriate in streaming video application, which arranges the data and redundancy bits so as to recover the original data even when not all of the bits are received.

Intelligent FEC packet adjustment has been an important issue in previous FEC studies [5]. For example, more redundant FEC packets have greater opportunity to recover erroneous packets, but those packets may overload wireless networks when traffic load is high; fewer FEC packets, on the other hand, may fail to recover erroneous packets when wireless error rate is high. Improving network performance by dynamically tuning FEC strength to the current rate of wireless channel loss is the objective of this paper.

Several unequal loss protection (ULP) schemes have recently been proposed for video over wireless and wired networks, but these schemes only consider a single aspect, i.e. either different layers of scalable coding or different frames in a GOP. On the other hand, Yang [1] takes into account the effect of the transmission error propagation, such as an ELEP model based FEC assignment algorithm. It is developed in order to minimize the decoder distortion. The work [1] only adds additional redundancy FEC codes and does not consider the restriction of the transmission bandwidth.

FEC and ARQ schemes are also used and applied to the MAC layer of WiMax [2]. These techniques do not contradict the MAC layer specifications that have been defined for WiMax. Under such a context, The novelty of our approach lies in the exploitation of the features of both VoIP and WiMax for improving the quality of VoIP calls over WiMax channels. Moon [3] implements a network adaptive selection of transport error control (NASTE) that selects or combines transport-layer error control mode(s) according to the target application and the channel status. NASTE effectively combines both end-to-end monitoring (E2EM) and cross-layer monitoring (CLM).

3 Personal Upstreaming

Satellite news gathering (SNG) is one of the first attempts for mobile users / units to perform worldwide newscasting. Mobile units are usually vans equipped with advanced, two-way audio and video transmitters and receivers, using dish antennas that can be aimed at geostationary satellites. Recently, with the advent of high quality, low latency video encoding for IP (H.264/MPEG-4 AVC or wavelet), lower-cost alternatives to expensive proprietary microwave solutions have entered the market. These are based on OFDM technology such as 802.16. We envision that the person SNG could be possible in which a mobile user could act as a mobile reporter in SNG by performing personal upstreaming.

The system architecture for Personal upstreaming has two parts as shown in Fig. 1. One part is the client side, including the mobile user providing the real-time video and the audience watching the contents provided by the mobile user. The other part is Streaming Service Center, including the IP Multimedia Subsystem (IMS) which allows users register and send message to users and the streaming servers which receive the contents from users and forward them to the audience.

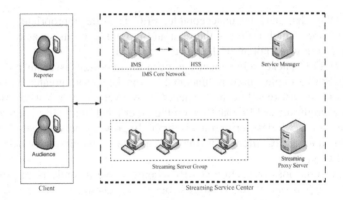

Fig. 1. System Architecture for Personal Upstreaming

The procedure for streaming the real-time video from User to Audience is called Push-to-Watch system. The steps are described as follows:

1.   A user sends a UPSTREAMING message to the IMS.
2.   The IMS forwards the UPSTREAMING message to a particular audience.
3.   A user upstreams the real-time video to the streaming server.
4.   The audience who wants to watch the video replies the message to IMS.
5.   The streaming server multicasts the video to the interested audience.

4   Hybrid Error Correction

If the quality of the source corrupted, the quality of the video sent to the Audience would be awful. In this paper, we propose several kind of error correction to ensure the quality of the vide source.

4.1   Adaptive FEC

In FEC approach, the source node transmits parity packets along with the original data packets. The receiver can accurately recover any lost data packets less than the parity packets. The amount of parity packets is determined at the time of FEC encoding.

The simultaneous, parallel transmission of multiple symbols also makes it possible to use error correction to reconstruct the contents of faulty carriers. These characteristics result in a stable connection with very low bit error rations (BER). The BPSK, QPSK, 16QAM, and 64QAM modulation modes are used, see in Fig. 2, and the modulation is adapted to the specific transmission requirements. The packet loss would be affected by the modulation, so monitoring the feature of MAC layer and using the result to determine the FEC strength might be meaningless. We decided to analyze the real loss information through the feedback channel of WiMax, then use the loss rate to generate the FEC strength and refresh it during each GOP.

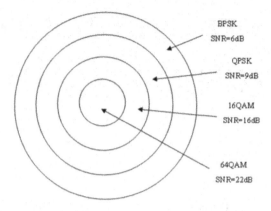

Fig. 2. The modulation models for WiMax

4.2 GOP-Base ULP

The three major picture types found in typical video compression designs are Intra coded frames, Predicted pictures, and Bi-directional predictive pictures. They are also commonly referred to as I frames, P frames, and B frames.

Intra coded frames are pictures coded without reference to any pictures except themselves. Often, I frames are used for random access and are used as references for the decoding of other pictures. Predicted frames require the prior decoding of some other pictures in order to be decoded. In older standard designs, P-frames use only one previously-decoded picture as a reference during decoding, and require that picture to also precede the P picture in display order. Bi-directional predicted frames require the prior decoding of some other pictures in order to be decoded. B frames are never used as references for the prediction of other pictures. As a result, a lower quality encoding can be used for such B frames because the loss of detail will not harm the prediction quality for subsequent pictures.

From the temporal dependency existing in a GOP, we can see that the earlier an error occurs in a GOP, the more frames will be corrupted. For example, an error in I-frame will corrupt all the following frames in the GOP whereas an error in the last frame of the GOP does not affect any other frames.

In MPEG-4, the relations between each frame of a GOP sequence IBBPBBPBBPBBPBB are shown in Fig. 3. If I frame is broken, the whole frames in the GOP would be effected. If P1 frame is broken, B1, B2, P1, B3, B4, P2,B5, B6, P3, B7, B8, P4, B9, B10, P5,B11 and B12 frame would be effected.

In H.264, P-frames can use multiple previously-decoded pictures as reference during decoding, and can have any arbitrary display-order relationship relative to the pictures used for its prediction. We can easily show the differences between Fig. 3 and Fig. 4. If P3 frame is broken, B1, B2, P1, B3, B4, P2,B5, B6, P3, B7, B8, P4, B9, B10, P5,B11 and B12 frame would be affected.

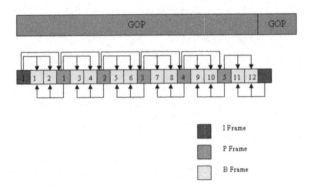

Fig. 3. The reference of the frames in MPEG-4

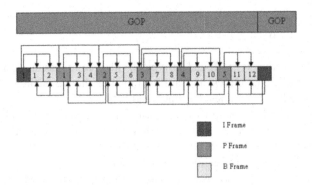

Fig. 4. The reference of the frames in H.264

4.3 ARQ Based on Transmission Delay

In wireless environment, the traditional FEC cannot perform well as in the wired network due to burst error characteristics. The traditional FEC uses consecutive data packets to make parity packets, but if errors occur in bulk, FEC would not get enough packets to recover the lost data packets. We need ARQ to assist the FEC to recover the burst loss in wireless networks.

In traditional video codec, the real time application would not allow the ARQ because the delay time of the retransmission is too long. In H.264, the full decompression in the GOP would need the I frame of the next GOP so we acquire more time to be the buffer. In others words, we can activate ARQ mechanism in real time upstream with H.264 codec.

We propose a simple strategy for ARQ with H.264 codec. We would monitor the delay time of network to decide either we should retransmit the packet or just ignore it. If the ARQ packet is received after the decoding time, it just wastes the bandwidth. When the ARQ is activated, the front of the GOP will get more protection than the rear of the GOP. The rear of the GOP has less chance to retransmit because it is closer to the decoding time. We would reduce the FEC strength of the front of GOP to achieve the fairness.

4.4 Multiple Profiles

The devices, used by the mobile users in personal upstreaming, are light and portable. We may not realize all methods we mentioned before because the capability of the device is restricted. We propose the conception of multiple profiles in Table 1. Each profile has its own error correcting mechanisms to fit in the limitations of the device and the network conditions.

- High Profile: The device might be high level laptop. It equips with the high frequency processor and the chip supporting H.264 encoding and the delay of the network is short.
- Medium Profile: The grade of hardware is the same with High Profile, but the delay of the network is long.
- Low Profile: The device might be a PDA with the camera so we used the simplest error correction mechanism.

Table 1. Multiple Profiles

Profile	Function
High	FEC、ARQ、H.264
Medium	FEC、H.264
Low	FEC、MPEG-4

5 Simulation

We use NS-2 with NIST WiMax module [4] and Evalvid [5] as the tools. NS-2 can simulate the traffic in WiMax and Evalvid can generate the traffic of the video stream. Table 2 shows the network environment. Many literatures have used Gilbert model to characterize wireless channel loss process. The Gilbert-Elliot error model is a two-state Markov model with geometrically distributed residence time. The Gilbert-Elliot error model predicts the state of the next packet by just looking at the previous received packet. Fig. 5 shows the Gilbert-Elliot error model state transition diagram. When the channel state is "Good", the packet loss rate is P_G. When the channel state is "Bad", the packet loss rate is P_B. In addition, P_{GB} stands for that the channel state changes to "Bad" from "Good" and P_{BG} stands for that the channel state changes to "Good" from "Bad". In Gilbert-Elliot error model, the average packet loss rate is $P_{average} = P_G \pi_G + P_B \pi_B$. We keep P_G as 0.04 and vary P_B to obtain the targeted average loss rate.

Fig. 6 shows the simulation procedures in this paper. First we get the YUV video sequence and encode it to the codec we want. Then we can get the packet information through the parser and adjust it for feeding it to NS-2 traffic simulation. After the traffic simulation, we can get the damaged packets and recover them. We can

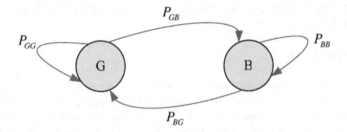

Fig. 5. State Transition Diagram for Gilbert-Elliot Error Model

Table 2. NS-2 Environment Settings

Parameters	Value
Channel Bandwidth	10M(bps)
Downlink / Uplink ration	3 : 7
Uplink Symbols	42
Modulation	OFDM_64QAM_3_4
Frame Duration	0.004
Packet size of Video Packet	500 (bytes)
Delay time	100ms

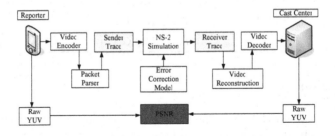

Fig. 6. State Transition Diagram for Gilbert-Elliot Error Model

combine the recovered packets and the video before transmitting to a damaged video. We decode the damage video to a new YUV video sequence. We can get the PSNR value by comparing the original YUV video sequence and the damaged YUV video sequence.

Because many signals have a very wide dynamic range, PSNR is usually expressed in terms of the logarithmic decibel scale. The PSNR is most commonly used as a measure of quality of reconstruction in image compression.

Two simulations with different error corrections were performed in our adaptive FEC, our adaptive FEC with ARQ and SFEC. SFEC means static FEC so the FEC strength is constant in SFEC and its protection is averaged. The first simulation is for H.264 and the second is for MPEG-4. The first simulation results are shown in Fig. 7. SFEC can provide better quality but make more redundant packets. Our mechanism can adjust the number of the redundant packets based on the network conditions and the quality is better than SFEC when the amounts of their redundant packets are the same. The simulation results for MPEG-4 are shown in Fig. 8. The MPEG-4 decoder we used may not have error correction so the video without protection is almost broken. The mechanism we proposed is still better than SFEC but the effect is less. The fact that B frames in the video encoded by MPEG-4 is more than the video encoded by H.264 might cause the results.

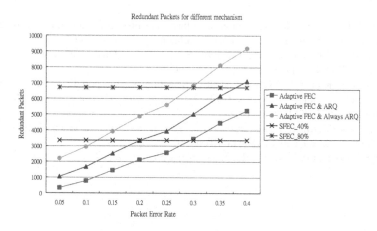

Fig. 7. H.264 Redundant Packets

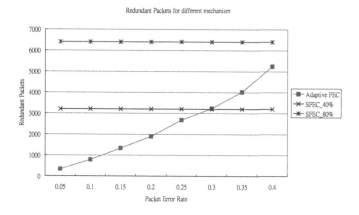

Fig. 8. MPEG-4 Redundant Packets

6 Conclusions

In this paper, we developed the application of upstream through IEEE 802.16 and proposed multiple profiles to adapt the upstream to different network conditions. Finally, we confirmed the efficiency through the simulation result. The number of

FEC packets, the allocation ratio of FEC packets and ARQ mechanism are initially set. We can keep arguing these parameters for optimizing them.

Our future work is to tackle the part between the Cast Center and the Audience. We can not only stream through the traditional CDNContext Delivery Network, but also integrate the Cast Center and the Audience into P2P network to utilize the bandwidth.

References

1. Yang, X.K.: Unequal loss protection for robust transmission of motion compensated video over the internet. Signal Process.: Image Commun. 18, 157–167 (2003)
2. Sengupta, S., Chatterjee, M., Ganguly, S.: Improving Quality of VoIP Streams over WiMax. IEEE. T. Computers 57, 145–156 (2008)
3. Moon, S.T., Kim, J.W.: Network-adaptive Selection of Transport Error Control (NASTE) for Video Streaming over WLAN. IEEE. T. Consum. Elec. 53, 1440–1448 (2007)
4. NIST: IEEE 802.16 Module for NS-2
   http://www.antd.nist.gov/seamlessandsecure/download.html
5. EvalVid: A Video Quality Evaluation Tool-set,
   http://www.tkn.tu-berlin.de/research/evalvid/

Architecture of Context-Aware Workflow Authorization Management Systems for Workflow-Based Systems

Seon-Ho Park, Young-Ju Han, Jung-Ho Eom, and Tai-Myoung Chung

Internet Management Technology Laboratory,
School of Information and Communication Engineering,
Sungkyunkwan University,
300 Cheoncheon-dong, Jangan-gu,
Suwon-si, Gyeonggi-do 440-746, Republic of Korea
{shpark,yjhan,jheom}@imtl.skku.ac.kr,tmchung@ece.skku.ac.kr

Abstract. Workflow technology is gaining major influence in many application domains. Especially, with the progress of the ubiquitous computing technology, a context-aware workflow has received considerable attention as a new workflow approach for a ubiquitous computing environment. This paper discusses security considerations for context-aware workflow systems and presents an architecture of a context-aware workflow authorization management (CAWAM) system. The CAWAM system offers management and enforcement of an authorization policy for context-aware workflow systems, and provides security-related context management functions. The CAWAM system utilizes the GCRBAC model for an access control service. In addition, the system supports a web-based policy management user interface.

1 Introduction

Workflow has broadly used in many application domains, such as manufacturing, banking, healthcare, office automation, etc. Recently, there are many attempts to adopt workflow models to ubiquitous computing [1–5]. Context-aware workflow has received considerable attention as a new context-aware model for ubiquitous computing environment. Context-aware workflow does not only use context-aware applications, but also enables workflow technology to be applied in new domains that are process oriented. For example, the health-care assistance service (in [1]) provides context-aware execution functions for the medical guideline and activities related with the execution which is managed by the workflow engine.

From a security perspective, context-aware workflow system(CAWS) is a new challenge for exploring security policies that are intended to control access to sensitive resources. A considerable number of studies have been conducted both on context-based authorization [6–11] and on workflow authorization [12–14]. However, an access control for context-aware workflow systems has not received much attention.

C.S. Hong et al. (Eds.): APNOMS 2009, LNCS 5787, pp. 375–384, 2009.

This paper explores access control considerations for CAWSs and presents the architecture of a CAWAM(Context-Aware Workflow Authorization Management) system to provide a context-based authorization management service for the CAWSs. We used the generalized context-role based access control (GCRBAC) model to apply the security-sensitive context information and the workflow-related data to the authorization policy. CAWAM architecture includes the web-based policy UI for ease of policy management.

The remainder of this paper is composed as follows. Section 2 illustrates the CAWSs and access control requirements for the CAWSs. Section 3 describes an overview of CAWAM system. Section 4 presents GCRBAC model. Section 5 illustrates detailed architecture of CAWAM system. In section 6 we explore related works. Finally, section 7 concludes this paper.

2   Access Control on Context-Aware Workflow Systems

Several studies have been made on the context-aware workflow systems. Ardissono et al. presented a framework integrating workflow management and context-aware action execution [1]. Their framework, so-called "CAWE (context-aware workflow execution)" enables to support context-dependent courses of actions and a runtime selection of an appropriate context-dependent part of a workflow. Westkaemper et al. [2] and Lucke et al. [3] studied on a "Smart Factory". The Smart Factory approach enables a multi-scale manufacturing by using the ubiquitous computing technologies and tools. Wieland et al. presented concepts for modeling context-aware workflows and showed how to realize them extending BPEL and using the Nexus Platform [4], and also presented the Smart Workflows [5] which is context-aware processes.

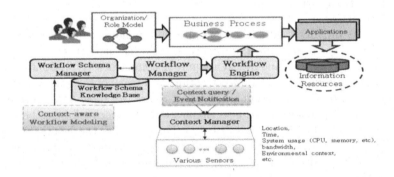

Fig. 1. Generalized architecture of Context-aware Workflow System

Fig. 1 illustrates the generalized architecture of CAWSs. We analyzed the characteristics on such CAWSs from an access control point of view, and found some requirements of an access control for the CAWS as follows.

Req.1. A user should have access rights that are required to do his/her job functions. (Least privilege principle)

Req.2. Users cannot be granted permissions to be mutually exclusive. (Separation of Duty principle)

Req.3. A user can be granted his/her access rights only if the current user context(e.g. location, position, social situation) is satisfied to particular context conditions.

Req.4. Accesses to an object are not allowed if the current context on the object is not satisfied to particular context conditions that are specified in according to process rules

Req.5. To perform some job functions belong to a workflow, the corresponding access rights should be available only during an execution of the job functions.

3   CAWAM System Overview

In this section, we present the CAWAM(Context-Aware Workflow Authorization Management) system which is an authorization management system for a CAWS. The CAWAM is designed to reflect the requirements described in section 2.2. The CAWAM provides policy management functions for configuration and enforcement of authorization policy based on a user's current context(e.g. a floor number, a room number, blood pressure, or social situation data such as "who you are with", etc.), an environment context(e.g. time of the day, date, temperature, air quality, etc.), an object-related context (e.g. an execution state of a process, a memory usage, a cpu usage, an I/O state of an object, etc.), and a task-related context (e.g. a cardinality of a task, dependencies within a workflow, control flows of a workflow, status of task instance in the task state transition diagram, etc.).

We explored several access control models to find an access control mechanism suitable for the CAWAM system, and decided to extend CRBAC(context-role based access control) model [11], because any model suitable for a context-aware workflow system does not yet exist. The CRBAC model has all advantages of a NIST RBAC model [15] because it is an extension of the RBAC model, and also enables a management of context-based access control policies. However, the CRBAC model does not suitable for a workflow system. So we improved the CRBAC model to fit an access control for workflow-based applications. An extended CRBAC model is described in section 4.

The CAWAM offers a context management function and a workflow-related data management function, as well as an access control function. For ease of policy management, CAWAM system also provides web-based interface for authorization policy configurations. The CAWAM system architecture consists of an access control(AC) part, a context management(CM) part, a workflow management (WM) part, and an authorization management (AM) part. Fig. 2 illustrates an overview of the CAWAM system. A detailed architecture of the CAWAM system is described in section 5.

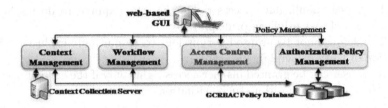

Fig. 2. CAWAM system overview

4 Generalized CRBAC Model

This section presents the generalized CRBAC (GCRBAC) model which is an extended CRBAC model to fit for context-aware workflow systems. The GCRBAC model is a more flexible and highly expressive access control model because it provides four types of context-roles which are used as generalized context modeling tools. The context roles consist of the user context role, the object context role, task context role and the environment context role which are used to utilize the user's current context, the object-related context, the task-related context, and the environmental context, respectively. Fig. 3 shows the GCRBAC model(b) and the CRBAC model(a), and details important notations of the GCRBAC model.

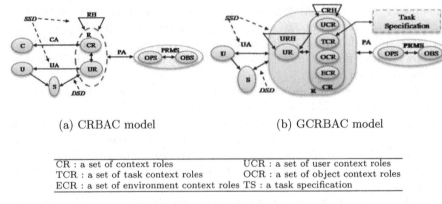

(a) CRBAC model   (b) GCRBAC model

CR : a set of context roles	UCR : a set of user context roles
TCR : a set of task context roles	OCR : a set of object context roles
ECR : a set of environment context roles	TS : a task specification

(c) Elements which are added in the GCRBAC

Fig. 3. CRBAC model(a), GCRBAC model(b), and important notations of added component for the GCRBAC model(c)

As the fig. 3 indicates, the GCRBAC model is natural extension of the CRBAC model, therefore many properties of the CRBAC model are used for the GCRBAC model.

Definition 1 [U, UR, P, OBJ, OP, UA, PA, RH]

- U, UR, R, P, OBJ, OP (users, user_roles, roles, permission, objects, operations)
- UA is the mapping that assigns a user_role to a user.(Formally: $UA \subseteq U \times UR$)
- PA is the mapping that assigns permissions to a role. (Formally: $PA \subseteq R \times P$)

Definition 2 [Context Roles (CR) and Context Information (CI)]. The context role is a GCRBAC component which is used to capture security-related information about the various types of context information for use in GCRBAC policies. A CR is a high level information which is logical and human-understandable, whereas a context information (CI) which is gathered from various types of embedded sensors in real physical world is low-level information.

Definition 3 [Roles (R)]. Role is the mapping that assigns a user_role to a context_role. (Formally: $R \subseteq UR \times CR$.)

Property 1 [Types and definition of CR]
All of the CRs are classified into four types: UCR, TCR, OCR, and ECR. **CR** is defined as a 4-tuple of $\langle UCR, TCR, OCR, ECR \rangle$.
$CR \subseteq UCR \times TCR \times OCR \times ECR$

Property 2 [Representation of CI]. CIs are represented using first order logic. Examples of the CI are $Time(10:00, 15:00)$, $Location(Park, entering, roomA)$, and so on. It is possible to construct more complex **CI** expression by performing boolean operations like conjunction($\vee$), disjunction($\wedge$) and negation ($\sim$) over basic **CIs**.

GCRBAC system has to offer management functions for support the mappings between CRs and CIs. GCRBAC model provides the prototype of such mapping functions.

Definition 4 [Mapping function mCR and mCI]. mCR is a total function that converts several CIs into a context role. mCI is a total function that converts a context role into a set of context information. The following formalizes the function.

- $mCR(ci : CI) \rightarrow 2^{CR}$
- $mCI(cr : CR) \rightarrow 2^{CI}$

Definition 5 [Task and Task Specification]. Task(**T**) is a set of related permissions and the basic unit of work process that a user has to do. Task Specification (**TS**) is a catalog that contains all tasks and their attributes.

Property 3 [Attributes of tasks]. If the type of task T_k is a type wT, T_k has following attributes :

- Task Instance Threshold (TIT): the maximum number of task instances which are able to be executed for a specific task at the same time

- *Task Type (TT): types of task*
- *Status of Task Instance (STI): status of task in a state transition diagram of a task process instance* (refer to (b) of Fig.4)
- *Task Activation Conditions (TAC): precondition for activation of a task T_k; TAC of T_k can be represented as $\{pT, TTs, BOs\}$ (pT is a set of prior tasks, TT represents pTs which are associated by a BO, BO is boolean operations which specify control pattern of pTs ($BO = \{\odot, \otimes, \oplus\}$))* (refer to (a) of Fig.4)

Property 4 [Types of tasks]. *All of the tasks are classified into two types.*

- *type wT: Tasks which belong to one or more workflow*
- *type nT: Tasks which is not related with workflow*

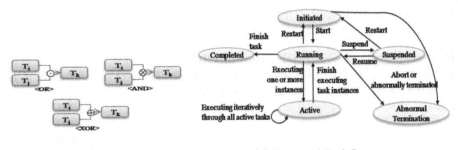

(a) Boolean operations   (b) Status of Task Instance

Fig. 4. Boolean operations which specify control patterns of prior tasks in the task activation condition(a); State transition diagram of a task process instance used in CAWAM system(b)

5   CAWSM Architecture

This section describes detailed architecture, functions and operations of core components of CAWAM system Architecture. CAWAM system consists of the access control component, context management component, the policy management component, workflow management component, and the web-based interface. Fig. 5 illustrates the architecture of CAWAM.

5.1   Access Control Part

The access control component consists of the security service manager(SSM) module, the access control engine (ACE) module, and the authorized session manager (ASM) module. The SSM provides a function of interface for authorization policy enforcement and manages information about overall user security such as a user identification management, a user authentication management, and an access control management. Although this paper does not describe the

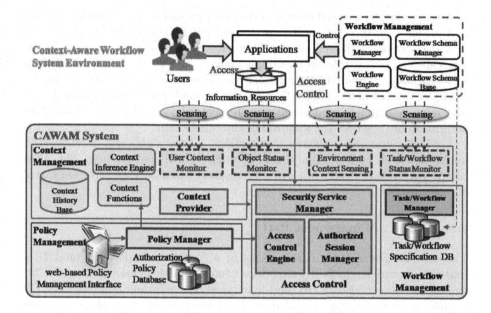

Fig. 5. CAWAM Architecture

user authentication and identification mechanisms to concentrate on authorization management, the CAWAM architecture includes user authentication management mechanism which is able to support the multi-authentication and the partial credential management. From a security management point of view, the identification and the authentication have to precede the access control. In this paper, we assume all users are already authenticated.

The ACE enforces the access control policies based on GCRBAC models. Input and output of the ACE are well defined where the former is the user access requests and the latter is either allowing or denying those requests. The request contains the activated user-role, the object that needs to be accessed, operation that needs to be performed, and task that can be performed by the user-role. When the ACE is received an access control request from the SSM, the ACE get appropriate policy rules which is related with the request from the policy manager module in the policy management part.

An authorization of GCRBAC is configured using access control transaction. A transaction specifies a particular action to be performed in the system. The transaction for configuration of the authorization rule is defined as following definition.

Definition 6 [Authorization Transaction of CAWAM]. *The authorization transaction of the CAWAM system is a 3-tuple in the form of $\langle$ ur, cr, p $\rangle$ where $ur \in UR$, $cr \in CR$, $p \in P$, (ur, cr) $\in$ **R**, and (r, p) $\in$ **PA**. Transaction $\langle$ ur, cr, p $\rangle$ indicates that any user u, where (u, ur) $\in$ **UA**, is authorized for permission **p** if only **cr** is activated.*

A policy database would consist of transactions listing, paired with a permission bit for each transaction. The permission bit indicates whether the associated transaction is allowed or prohibited. Each $\langle$ *transaction, permission_bit* $\rangle$ is called a policy rule. For example, policy rule would be represented as $\langle\langle Sales\_manager, OpenSalesReportCtx, Open\_SalesReportFile \rangle, ALLOW\rangle$. In this example, *Sales_manager*, *OpenSalesReportCtx*, and *Open_SalesReportFile* are user role, context role, and permission, respectively, and *ALLOW* means that permission bit is set for the approval of the transaction.

The ASM manages activated user sessions. Access rights can be allowed to users only if all context conditions are satisfied with policy rules. The ASM enforces activation/deactivation of user sessions based on the activation status of relevant CRs.

5.2   Context Management Part

The context management part consists of a context provider(CP) module, a context inference engine(CIE) module, a context functions management (CFM) module, a context history DB, and four types of monitoring modules. The CP module gives the access control part the context information processed. The CFM offers functions for management of the mapping between context information and context role. In addition the CFM provides various mapping functions which are created based on *mCI* and *mCR* in Definition 4. The CIE module provides reasoning functions to make an inference from low-level context data. The four types of monitoring modules are user context monitors, object status monitors, environment context sensors, and workflow status monitors. These monitors collect specific types of context data and give them the CIE module. CAWAM system includes a context collection server which manages context monitoring modules and 10 sensor motes which collect environmental data such as a temperature and an illumination. In the sensor network of our

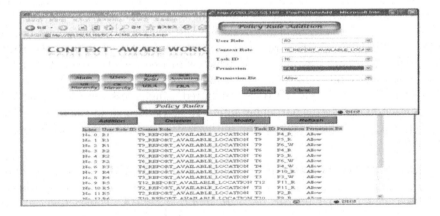

Fig. 6. A Screenshot of security policy configuration UI

system, the communication between the server and sensors is encrypted using skipjack algorithm for secure communication.

5.3 Policy Management Part

The policy management part includes a policy manager, a policy database, and a web-based user interface for policy management. The policy manager provides authorization policies to the access control part and the context functions management module in the context management part. The user interface is a web-based graphic user interface for the policy manager module. A policy administrator inputs simply the policy information to the user interface, then the policy manager module makes the policy rules using the information from the user interface. We implemented the user interface using the visual C# .Net in the Microsoft .NET framework. Fig. 6 is a screenshot of the policy management user interface.

6   Related Works

This section illustrates several access control mechanisms as the related works. Various access control models have been proposed for various kinds of information systems. Especially, role-based access control (RBAC) model[15] is important and popular access control model which is broadly used in many application domains. A large number of extensions to the RBAC model was proposed in order to the enforcement of more complex rules and the inclusion of context information in authorization decisions. Especially, with the development of the ubiquitous computing environment, temporal constraints and spatial constraints became the important aspects of access control. To address general time-based access control requirements, Bertino et al. propose a Temporal Role Based Access Control (TRBAC) model[6]. Covington et al. introduced the notion of environment role and proposed Generalized Role Based Access Control (GRBAC) model [7]. GRBAC model enhanced traditional RBAC by incorporating the notion of object roles, and environment roles, with the traditional notion of subject roles. The CRBAC model[11] introduced the notion of the context-role which facilitates configuration and management of context-related policies.

Various access control models have been studied for workflow security. The basic authorization model for workflows is the WAM (Workflow Authorization Model)[12]. The WAM is capable of configuring authorizations in a way that subjects gain access to required objects only during the execution of the task, thus synchronizing the authorization flow with the workflow. The WAM associates an authorization template with each task, which allows appropriate authorizations to be granted only when the task starts and to revoke them when the task finishes. Wu et al.[13] presents the authorization and access control mechanism for application data in workflow systems. They discuss key access control requirements for application data in workflow applications using healthcare examples. Thomas and Sandhu have introduced Task-Based Authorization Controls (TBAC), a family of models, in which permissions are actively (de)activated according to the current task/process state[14].

7 Conclusion

With the progress of the ubiquitous computing technology, a context-aware workflow has received considerable attention as a new workflow approach for the
ubiquitous computing environment. In this paper, we discussed security considerations for context-aware workflow systems and presented an architecture of
a context-aware workflow authorization management (CAWAM) system. The
CAWAM provides the management and enforcement functions for authorization
policy suitable for context-aware workflow systems, and provides the context
management functions and workflow-data management functions. In this paper,
we also presented the extended context-role based access control model for use
in the CAWAM system.

References

1. Ardissono, L., et al.: Adaptive medical workflow management for a context-
   dependent home healthcare asistance service. In: First International Workshop
   on Context for Web Services 2005, pp. 59–68 (2006)
2. Westkaemper, E., et al.: Smart factory - bridging the gap between digital planning
   and reality. In: 38th CIRP International Seminar on Manufacturing Systems (2005)
3. Lucke, D., et al.: Smart factory - a step towards the next genereation of manufacturing. In: 41st CIRP Conference on Manufacturing Systems, pp. 115–118 (2008)
4. Wieland, M., et al.: Towards context-aware workflows. In: Proc. of UMICS 2007
   (2007)
5. Wieland, M., et al.: Context Integration for Smart Workflows. In: Proc. of 6th
   PERCOM, pp. 239–242 (2008)
6. Bertino, E., et al.: TRBAC: A temporal role-based access control model. ACM
   TISSEC 4(3), 191–233 (2001)
7. Covington, M.J., et al.: A Context-Aware Security Architecture for Emerging Applications. In: Proc. of 18th ACSAC, pp. 249–258 (2002)
8. Bertino, E., et al.: GEO-RBAC: A Spatially Aware RBAC. In: Proc. of 10th ACM
   SACMAT, pp. 29–37 (2005)
9. Ray, I., Toahchoodee, M.: A spatio-temporal role-based access control model. In:
   Barker, S., Ahn, G.-J. (eds.) Data and Applications Security 2007. LNCS, vol. 4602,
   pp. 211–226. Springer, Heidelberg (2007)
10. Chen, L., Crampton, J.: On Spatio-Temporal Constraints and Inheritance in Role-
    Based Access Control. In: Proc. of ASIACCS 2008, pp. 156–167 (2008)
11. Park, S.-H., et al.: Context-Role Based Access control for Context-Aware Appli-
    cation. In: Gerndt, M., Kranzlmüller, D. (eds.) HPCC 2006. LNCS, vol. 4208,
    pp. 572–580. Springer, Heidelberg (2006)
12. Atluri, V., Huang, W.-K.: An authorization model for workflows. In: Martella, G.,
    Kurth, H., Montolivo, E., Bertino, E. (eds.) ESORICS 1996. LNCS, vol. 1146,
    pp. 44–64. Springer, Heidelberg (1996)
13. Wu, S., et al.: Authorization and Access Control of Application Data in Workflow
    Systems. Journal of Intelligent Information Systems 18, 71–94 (2002)
14. Thomas, R.K., Sandhu, R.S.: Task-based authorization controls (TBAC): A family
    of models for active and enterprise-oriented authorization management. In: The
    IFIP WG11.3 Workshop on Database Security, pp. 166–181 (1997)
15. Ferraiolo, D.F., et al.: Proposed NIST Standard for Role-Based Access Control.
    ACM TISSEC 4(3), 224–274 (2001)

Baseline Traffic Modeling for Anomalous Traffic Detection on Network Transit Points

Yoohee Cho[1], Koohong Kang[2], Ikkyun Kim[3], and Kitae Jeong[1]

[1] Network Lab. KT
463-1 Jeonmin-dong, Yuseong-gu Daejeon, South Korea
{yoohee,kjeong}@kt.com
[2] Dept. of Information and Communications Engineering, Seowon University
231, Mochung-Dong, Chongju, 361-742, South Korea
khkang@seowon.ac.kr
[3] Information Security Research Division, ETRI,
Gajeong-dong, Yuseong-gu, Daejeon, 305-700, South Korea
ikkim21@etri.re.kr

Abstract. Remarkable concerns have been made in recent years towards detecting the network traffic anomalies in order to protect our networks from the persistent threats of DDos and unknown attacks. As a pre-process for many state-of-the-art attack detection technologies, baseline traffic modeling is a prerequisite step to discriminate anomalous flow from normal traffic. In this paper, we analyze the traffic from various network transit points on ISP backbone network and present a baseline traffic model using simple linear regression for the imported NetFlow data; bits per second and flows per second. Our preliminary explorations indicate that the proposed modeling is very effective to recognize anomalous traffic on the real networks.

Keywords: Intrusion Detection, Anomaly, DDoS attack.

1 Introduction

1.1 Background and Related Works

Protecting network systems against recurrent and new attacks is a pressing problem. Seamless and secure traffic streams through the Internet are becoming increasingly important today because corporations, public organizations, and even individuals are more heavily depending on the Internet for their daily activities. One of the important components of the Internet security is intrusion detection system (IDS), which gathers and analyzes information from diverse sources of computers or networks, and then identify possible security breaches. Moreover, we usually refer 'intrusion detection' to a comprehensive term for detection actions that capture attempts to compromise the confidentiality, integrity, or availability of systems or networks.

Traditionally, intrusion detection systems have been classified into two categories; a signature-based system and an anomaly-based system [1]. Signature

C.S. Hong et al. (Eds.): APNOMS 2009, LNCS 5787, pp. 385–394, 2009.
© Springer-Verlag Berlin Heidelberg 2009

detection involves searching network traffic for some series of bytes or packet sequences known to be malicious. A key advantage of this detection method is that signatures are easy to develop and understand if we know what network behavior we're trying to identify. However, signature engines have some disadvantages because they only detect known attacks, and signatures must be created for every attacks, and novel attacks cannot be detected. The anomaly detection technique relies on the concept of a baseline for network behavior. This baseline is a description of acceptable network behavior, which is automatically self-learned or specified by the network administrators, or both. Events in an anomaly detection engine are caused by any behaviors that fall outside the predefined or acceptable model of network behaviors. An integral part of baselining network behavior is the engine's ability to dissect protocols at all layers. A disadvantage of anomaly-detection engines is the difficulty of defining rules. Each protocol being analyzed must be defined, implemented and tested for accuracy. Moreover, detailed knowledge of normal network behavior must be constructed and transferred into the engine memory for detection to occur correctly. On the other hand, once a protocol has been built and a behavior defined, the engine can scale more quickly and easily than the signature-based model because a new signature does not have to be created for every attack and potential variant.

Thottan et al. [2] briefly review the commonly used methods for anomaly detection. However, we simply classify them as two categories; one is the time series modeling approach, and the other is the statistical modeling approach. The first approach attempts to build diverse traffic profiles in time series, such as symptom-specific feature vectors or just traffic volume [2,3,4]. The second approach learns a statistical model of normal network traffic, and flags deviations from this model [5,6]. Models are usually based on the distribution of source and destination addresses and ports per transaction. In [5], packet header anomaly detector (PHAD) learns the normal range of values for 33 fields of the Ethernet, IP, TCP, UDP, and ICMP protocols. PHAD uses the rate of anomalies during training to estimate the probability of an anomaly while in detection mode. If a packet field is observed n times with r distinct values, there must have been r anomalies during the training period. If this rate continues, the probability that the next observation will be anomalous is approximated by r/n. Although there are a number of variations of these two approaches, we note some general and important shortcomings of them. In time series modeling approach, we have to big tune for many parameters to characterize the statistical behavior of abnormal traffic patterns. For example, we have to choose the values of six parameters in [4]. In statistical modeling approach, we have to train attack-free training data which will not always be available in a practical system [5]. Recently, much more sophisticated work has been done in this area. Sperotto [7] investigates how malicious traffic can be characterized on the basis of aggregated metrics, in particular by using flow, packet and byte frequency variations over time. In [8], they examined the characteristics of recent Internet traffic from the perspective of flows. and found that the frequent occurrence of flash flows highly affects the performance of the existing flow-based traffic monitoring systems.

1.2   Motivation and Contribution

Since year 2000, Internet service provider (ISP), which traditionally had provided only high speed transfer service to customers, have developed the various premium services to enhance the quality of service (QoS) on their network. Moreover some ISPs have additionally invested on security services for their customers such as network firewalls. In spite of the havily investments on security services during several years, ISPs cannot help putting budget into new security systems because of the persistent threats of DDoS attack and new worms. As a preprocess for many state-of-the-art attack detection technologies, baseline traffic modeling is a prerequisite step to discriminate anomalous flow from normal traffic. In this paper, we analyze the traffic from various network transit points on ISP backbone network and present a baseline traffic model using simple linear regression for the imported NetFlow data; bits per second and flows per second. Our preliminary explorations indicate that the proposed modeling is very effective to recognize anomalous traffic on the real networks. Even if the NetFlow data were used widely for the network management, planning, and billing, our research results give a chance to use it for network security. In particular, we investigated and analyzed the in-bound and the out-bound traffic of various access networks. Since we could not find any reports about the traffic of ISP networks, we note that the gathered data will be important for the network management and security technologies.

The rest of this paper is organized as follows. In Section 2, we summarize the network environment and basic traffic distribution to observe in this paper. we analyze the regression model of campus access network in Section 3, and then we compare the model with the one of small-and-medium scale business (SMB) access network traffic in Section 4. and finally, we conclude in Section 5.

2   Traffic Pattern on Network Transit Points

As mentioned earlier, baseline traffic modeling of in-out traffic pattern is very crucial to the acurate working of anomaly detection engines which are based on network traffic pattern. In this section, we investigate the traffic characteristics on network transit points connecting to different types of stub networks.

2.1   Network Environments

Figure 1 shows access networks as an experimental network used throughtout in this paper, where the network consists of two types of access network; the one connects to nine stub networks for SMB company of class C (/24) networks and the other connects to five stub networks for university networks of class C networks. In Fig. 1, each access router is enabled to export the Netflow and SNMP MIB data to the Netflow collector. For the Netflow configuration, we set the sampling rate at 1:500, and the value of inactive and active timeout at 15 seconds and 30 minutes respectively. We measured the full Netflow information every five minutes during one week.

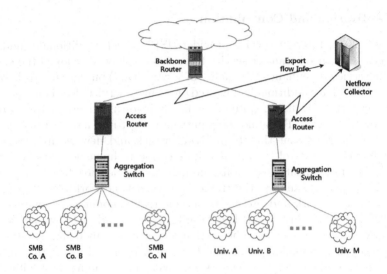

Fig. 1. Network view of Netflow information

2.2 General Traffic View

Figure 2 and Figure 3 show the aggregated BPS (Bits per Second) and FPS (Flows per Second) of the ingress and egress traffic at the access routers during a week, where a unit of X axis indicates five minutes duration and it begins on Sunday midnight (00:00). From the figures, we note that there are time-based and day-based repeated patterns of the volume. However, in case of SMB traffic, the traffic pattern on weekend is slightly different from the one of week-days. Therefore, the traffic patterns on different access routers might be obviously diverse with the connecting stub networks. We set up models revealing a

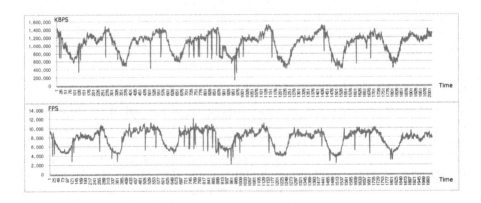

Fig. 2. Traffic relationship between BPS and FPS gathered from campus access networks during 7 days

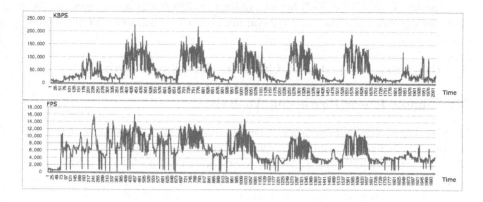

Fig. 3. Traffic relationship between BPS and FPS gathered from SMB(small and medium business) company access networks during 7 days

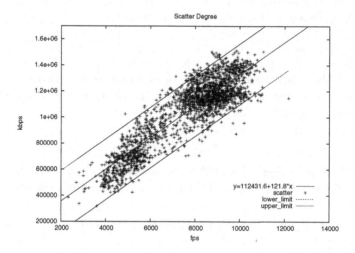

Fig. 4. Scatter diagram between BPS and FPS gathered from campus access networks

relationship between FPS and BPS gathered from both of the routers, and then we compare them. The scatter diagram, shown in Fig. 4, suggests that there is a strong statistical relationship between BPS and FPS, and the tentative assumption of the straight-line model $y = \beta_0 + beta_1 x + \epsilon$ appears to be reasonable.

From Fig. 2 and Fig. 4, we note that there is an outstanding features with the relationship between BPS and FPS in terms of the network traffic anomaly; (i) the BPS is much larger than the FPS, and (ii) the FPS is much larger than the BPS. The typical type of DDoS (Distributed Denial of Service) attack is a flooding attack, which generates larger traffic volume to the victims so that CPU and memory and the other network resources in the victims are exhausted. Such an overload of network traffic is related to the case (i). On the contrary, in the case of (ii), the TCP SYN attacks make the FPS much larger than the BPS what we expected because they try continuously to reconnect to TCP sessions.

3   Regression Model of Campus Access Networks

3.1   Simple Linear Regression Analysis

Regression analysis is a statistical technique for investigating and modeling the relationship between variables. As we metioned before, there is a linear relationship between FPS and BPS. Therefore, we can consider the simple linear regression model; that is, a model with a single regressor x and the relationship between the response y and x a straight line [9]. The simple linear regression model is

$$y = \beta_0 + \beta_1 x, \tag{1}$$

where the dependent variable y is BPS and the independent variable x is FPS. Regression coefficient β_0 is a constant or intercept, and β_1 is a slope of regression line. The goal of linear regression is to adjust the values of slope and intercept to find the line that best predicts y from x. More precisely, the goal of regression is to minimize the sum of the squares of the vertical distances of the points from the line. But, all observed data for y couldn't be explained with regression equation 1. That is, there is an error in predicting y for x, therefore, the prediction regression equation for i^{th} observed x_i is as follows.

$$\hat{y}_i = \beta_0 + \beta_1 x_i \tag{2}$$

Main point of use of regression model is to predict newly observed value y for a specific x. Given new x_0, $100(1 - \alpha)\%$ prediction interval is as follows,

$$
\begin{aligned}
\hat{y}_0 - t_{(\alpha/2, n-2)} \sqrt{MS_E \left(1 + \frac{1}{n} + \frac{(x_0 - \bar{x})^2}{S_{xx}}\right)} \leq y_0 \\
\leq \hat{y}_0 + t_{(\alpha/2, n-2)} \sqrt{MS_E \left(1 + \frac{1}{n} + \frac{(x_0 - \bar{x})^2}{S_{xx}}\right)}
\end{aligned}
\tag{3}
$$

where, $t_{(\alpha/2, n-2)}$ is t-distribution of $n-2$ degree of freedom, and MS_E is residual mean square, $S_{xx} = \sum_{i=1}^{n} (x_i - \bar{x})^2$, and $\bar{x} = \frac{1}{n} \sum_{i=1}^{n} x_i$. In this paper, we try to solve the regression coefficient of equation 2 and the residual mean square of the equation 3 using the SPSS (Statistical Package for the Social Science) program [10]. Figure 4 presents the upper and lower bound lines of the 95% confidence interval based on equation 3. We consider the points beyond the predict region as anomalous traffic, and the percentage of the confidence interval need to be tuned as the network type. Table 1 shows the regression coefficients based on campus network type traffic for a week, and the simple linear regression line is as follows,

$$\hat{y}_i = 112431.626 + 121.811 x_i \tag{4}$$

Table 1. Regression coefficients from universities traffic

Coefficients[a]

Model	Unstandardized Coefficient		Standard Coefficient			95% Confidence Interval for B	
	B	Std. Error	Beta	t	Sig.	Lower Bound	Upper Bound
1 (Constant)	112431.626	11209.969		10.030	.000	90447.246	134416.007
FPS	121.811	1.418	.887	85.878	.000	119.029	124.593

a. Dependent Variable : BPS

Table 2. Residual Analysis of universities traffic

ANOVA[b]

Model		Sum of Squares	df	Mean Square	F	Sig.
1	Regression	1.035E+14	1	1.03543E+14	7374.962	.000[a]
	Residual	2.821E+13	2009	14039771330.0		
	Total	1.317E+14	2010			

a. Predictors : (Constant), FPS
b. Dependent Variable : BPS

3.2 Verification of Regression Model

We are often interested in testing hypotheses and constructing confidence intervals about the model parameters. In order to verify the linear relationship between x (FPS) and y (BPS), we consider the hypothesis related to the significance of regression as follows,

$$H_0 : \beta_1 = 0$$
$$H_a : \beta_1 \neq 0$$

Failing to reject $H_0 : \beta_1 = 0$ implies that there is no linear relationship between x and y. Alternatively, if $H_0 : \beta_1 = 0$ is rejected, this implies that x is of value in explaining the variability in y. We can test the hypotheses using F-test, and note that the followings for the significance level α;

$$\text{If } F' \leq F(1 - \alpha; n - 2), \text{ conclude } H_0$$
$$\text{If } F' > F(1 - \alpha; n - 2), \text{ conclude } H_a,$$

where, we conclude that $F(.95; 1, 2008) < F(.95; 1, 1000) = 3.85$ from the F-distribution table. Because $F' = 7374.962 > F(.95; 1, 2008)$ from Table 2, null hypothesis H_0 is rejected, therefore, we verify that there is a linear relationship between FPS and BPS.

In a real problem, we don't get to observe the errors $e_1 \ldots e_n$; but we do observe the residuals $\hat{e}_1 \ldots \hat{e}_n$;. If the residuals appear to be roughly independent and normally distributed with mean 0 and constant standard deviation, we assume the same is roughly true for the errors. Hence we make histograms and normal probability plots to check normality of the residuals, make residual plots (residuals vs. fitted values, i.e., $\hat{e}_i$ vs. $\hat{Y}_i$) to see if the standard deviation of the residuals remains relatively constant as the fitted values change. Residual plots also help us verify whether a linear relationship between the mean of Y and X

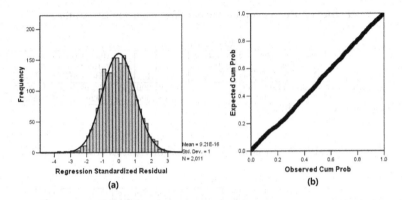

Fig. 5. (a) Histogram of regression standardized residual of the linear model for the university access network traffic. (b) Normal probability plot of standardized residual.

Table 3. Model Summary of campus access network traffic

Model Summary[b]

Model	R	R Square	Adjusted R Square	Std. Error of the Estimate	Durbin-Watson.
1	.877[a]	.786	.786	118489.54102	.329

a. Predictors : (Constant), FPS
b. Dependent Variable : BPS

seems appropriate. Residual plots can be used with other information to check the assumption that all Y values are independent of one another.

Figure 5 (a) shows the histogram of regression standardized residual of the linear model for the university access network traffic. As the figure, the pattern show here indicates no problems with the assumption that the residuals are normally distributed at each level of Y and constant in variance across levels of Y. The normal probability plot is a graphical technique for assessing whether or not a data set is approximately normally distributed. As Figure 5 (b), X axis means the observed cumulative probability and Y axis means the expected cumulative probability. We confirm that the residual is normally distributed, because the plot is evenly spread along the diagonal line. As the result of the verification, we confirmed that there is a linear relationship between FPS and BPS traffic parameters the university access network. Therefore, we can use the equation 3 as a baseline traffic with significance level to detect anomalous traffic.

4   A Model of SMB Access Networks

As shown Figure 3, the SMB traffic feature is a little different from the one of the campus access network. We try to apply the same approach to the SMB access network. Figure 6 shows the scatter diagram, regression line with the 95% significance level of two parameters (FPS and BPS) gathered from SMB access network for a week. The relation between FPS and BPS does not look linear,

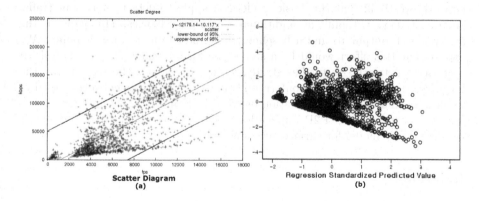

Fig. 6. (a) Scatter diagram between FPS and BPS gathered from SMB access networks. (b) Scatter plot of regression residual.

Table 4. Model Summary of SMB access network traffic

Model Summary[b]

Model	R	R Square	Adjusted R Square	Std. Error of the Estimate	Durbin-Watson.
1	.711[a]	.505	.505	31881.48212.599	.316

a. Predictors : (Constant), FPS
b. Dependent Variable : BPS

meaning that it would be hard to draw a line that passes very close to each of the point. Nonetheless, we could decide to model the relationship between FPS and BPS by a line because it is the easiest way to model a relationship. Strictly speaking, although the value (.329) of Durbin-Watson in the Campus access network model is not much different from the one (.316) in the SMB access network, the value of R-square is just .505, where an r-squared of 1.0 would mean that the model fits the data perfectly with the line going right through every data point, and also as Figure 6 (b), We notice that the residuals plot shows the residuals are not normally distributed - the residual scatter plot is not evenly sparse. Therefore, we noticed that the model of SMB access network does not conform to the linear regression model.

5   Conclusions and Future Work

In this paper, we analyzed the traffic of the two kinds of access networks and presented a baseline traffic model using the simple linear regression analysis between FPS and BPS parameters, which gathered from the transit points on ISP backbone network. We verified that there is a linear relationship between FPS and BPS traffic parameters of the university access network, whilest the model of SMB access network couldn't conform to the linear regression model. Although the baseline model in this paper is not perfect for every type of access network, we think that this basic model can apply to detect anomalous traffic on transit point without any additional security appliances. This paper is the first research output to find a baseline traffic model on the transit points. We expect that this effort would be a step-stone for anomalous traffic analysis in large scale network. As further strudy, we need to study the reason why the linear relationship could not conform to traffic of SMB access network, and also, we need to find out how the relation between FPS and BPS is linear for various tcp/udp ports such as P2P applications. After that, we need to find better baseline traffic model for various access networks.

References

1. Foster, J.C.: IDS: Signature versus anomaly detection (2005),
   http://searchsecurity.techtarget.com
2. Thottan, M., Ji, C.: Anomaly detection in ip networks. IEEE Transactions on Signal Processing 51, 2191–2204 (2003)
3. Barford, P., Plonka, D.: Characteristics of network traffic flow anomalies. In: Proceedings of the 1st ACM SIGCOMM Workshop on Internet Measurement, pp. 69–73. ACM, New York (2001)
4. Brutlag, J.D.: Aberrant behavior detection in time series for network monitoring. In: LISA 2000: Proceedings of the 14th USENIX conference on System adminis-tration, pp. 139–146. USENIX Association (2000)
5. Mahoney, M.V.: Network traffic anomaly detection based on packet bytes. In: SAC 2003: Proceedings of the 2003 ACM symposium on Applied computing, pp. 346–350. ACM, New York (2003)
6. Anderson, D., Frivold, T., Valdes, A.: Next-generation intrusion-detection expert system (NIDES). Technical Report SRI-CSL-95-07, Computer Science Laboratory, SRI International (1995)
7. Sperotto, A., Sadre, R., Pras, A.: Anomaly characterization in flow-based traffic time series. In: Akar, N., Pioro, M., Skianis, C. (eds.) IPOM 2008. LNCS, vol. 5275, pp. 15–27. Springer, Heidelberg (2008)
8. Kim, M.S., Won, Y.J., Hong, J.W.: Characteristic analysis of internet traffic from the perspective of flows. Computer Communications 29, 1639–1652 (2006); Moni-toring and Measurements of IP Networks
9. Montgomery, D., Peck, E.A.: Introduction to Linear Regression Analysis, 2nd edn. John Wiley and Sons, Inc., Chichester (1992)
10. SPSS: (Spss manual), http://www.spss.com/

IP Prefix Hijacking Detection Using Idle Scan*

Seong-Cheol Hong[1], Hong-Taek Ju[2], and James W. Hong[1]

[1] Dept. of Computer Science and Engineering, POSTECH, Korea
{pluto80,jwkhong}@postech.ac.kr
[2] Dept. of Computer Engineering, Keimyung University, Korea
juht@kmu.ac.kr

Abstract. The Internet is comprised of a lot of interconnected networks communicating reachability information using BGP. Due to the design based on trust between networks, IP prefix hijacking can occurs, which is caused by wrong routing information. This results in a serious security threat in the Internet routing system. In this paper, we present an effective and practical approach for detecting IP prefix hijacking without major change to the current routing infrastructure. To detect IP prefix hijacking event, we are monitoring routing update messages that show wrong announcement of IP prefix origin. When a suspicious BGP update that causes MOAS conflict is received, the detection system starts idle scan for IP ID probing so that distinguish IP prefix hijacking event from legitimate routing update.

Keywords: BGP, IP Prefix Hijacking, Routing, Security.

1 Introduction

The Internet is a decentralized network comprised of many interconnected networks. Each network communicates reachability information using BGP (Border Gateway Protocol). The BGP is the de-facto inter-domain protocol that maintains a table of IP networks or prefixes. It designates network reachability among Autonomous System (AS) and there are more than 30,000 ASes in the Internet routing system [8]. The routers maintain and update their own routing table according to the routing information exchanged via BGP.

However, the Internet routing infrastructure is vulnerable to attacks due to lack of BGP security guarantees. The Internet was designed to provide communication on the basis of trust between networks. BGP also does not guarantee any security properties such as the authenticity of origin information and path attributes. IP prefix hijacking is the one of BGP security attacks, which a BGP router, for malicious purpose or by misconfiguration, announces an IP prefix that the router does not own. It results in reachability problem and communication failure in the Internet. IP prefix hijacking incidents are often reported on the NANOG mailing list [1].

* This work was partly supported by the IT R&D program of MKE/IITA [2008-F-016-02, CASFI] and WCU program through the KSEF of MEST, Korea [R31-2008-000-10100-0].

C.S. Hong et al. (Eds.): APNOMS 2009, LNCS 5787, pp. 395–404, 2009.
© Springer-Verlag Berlin Heidelberg 2009

To mitigate the impact of wrong routing information, some BGP extensions have been proposed, such as Secure BGP (S-BGP) [2], Secure Origin BGP (soBGP) [3]. Maybe these solutions can solve well-known BGP security problems, but it is difficult to deploy in practical network because the digital signature techniques that is used by S-BGP and soBGP cause high overhead and these improved protocols require changes to the existing protocol.

Many previous studies have proposed a method to detect IP prefix hijacking events [4, 5, 14, 15, 16, 17]. These are easily deployable solutions using passive monitoring or active probing. But some of these approaches use the routing registry information, such as IRR (Internet Routing Registry) databases, which can be outdated. The IP prefix hijacking events must be distinguished from legitimate routing updates because both cases cause MOAS (Multiple Origin AS) change.

To detect IP prefix hijacking event, we are monitoring routing update messages that show wrong announcement of IP prefix origin. When IP prefix hijacking occurs, there would be two networks having same IP address space in the Internet. Because the basic route selection process is to select routes with the shortest path, only the ASes close to the attacker AS are likely polluted. Our work focuses on fingerprinting two ASes having same IP prefix to distinguish IP prefix hijacking event from legitimate routing update. Our goal is to propose an easily deployable method that satisfies all of the following requirements: No modifications to routing protocol and current routers and performing the detection process without AS cooperation.

The organization of this paper is as follows. Section 2 describes the related works dealing with IP prefix hijacking. Section 3 defines the problem at hand and describes our solution approach. The experiment results are discussed in Section 4. Finally, Section 5 concludes the paper with possible future work.

2   Related Work

IP prefix hijacking is caused from an attack on the inter-domain routing protocol. The RPSEC (Routing Protocol Security Requirements) working group proposed a lot of Internet-Drafts about a scheme to improve routing protocol security, for examples, general security threats and requirements to routing protocols [6, 7]. Path attributes and Network Layer Reachability Information (NLRI) authentication is one of the requirements. This provides a means to verify and assure peering relationships and prefix advertisements against unauthorized announcements.

One of the BGP security architecture is S-BGP [2] which employs three security mechanisms – Public Key Infrastructure (PKI), optional BGP transitive path attribute and IPsec. S-BGP requires working with the Internet registries and ISPs to set up the PKT. However, PKI causes high overhead and requires a wide deployment in the Internet registries, router vendors and ISPs. The other proposal is soBGP [3] which is a deployable mechanism for validating the authorization of the BGP data. Its design goal is to be able to attain security profit without the participation of every AS and configure the level between security and overhead. While S-BGP and soBGP may be able to solve the security problems of routing protocol, it is not easy for both solutions to deploy in the current Internet infrastructure.

Zhao et al. [9] first explained MOAS conflict meaning that multiple origin ASes announce the same IP prefix. Originally, a unique AS number is allocated to each AS for use in BGP routing. However, many cases such as use of static routing and private AS number cause MOAS in the Internet [14]. When looking only the BGP update message, we cannot find any difference between legitimate MOAS and IP prefix hijacking.

Lad et al. [5] propose the method which monitors occurrence of new origin ASes in real time and notify the prefix owners that a suspicious update occurs. However, this method needs to rely on mutual cooperation between ASes. Karlin et al. [10] propose a system that automatically delays the use and propagations of suspicious routes. Introducing delay gives the human operators and systems time to investigate the suspicious route. The routers identify suspicious routes by consulting a table of trusted routing information learned from the recent history of BGP update messages. This method has some false positive cases which can legitimately occur: Provider change and occurrence of previously unseen provider.

In this paper, we propose a real-time detection method of IP prefix hijacking events. Our contribution in this paper is to propose an easily deployable detection method without AS cooperation.

3 Proposed IP Prefix Hijacking Detection Method

In this section, we propose a method to detect an illegal BGP update message. IP prefix hijacking events have some common characteristics such as MOAS and invalid route in a BGP message. Using these characteristics, we can identify problematic update messages and detect hijacking activities. In this study, we focus on the following objectives.

- Without changing BGP routing infrastructure
- Do not rely on mutual cooperation

The first objective means that the proposed detection approach must be easily deployable. The second requirement infers that the detection method should be effective without any AS cooperation.

3.1 IP Prefix Hijacking

IP prefix hijacking is a well-known security threat that corrupts the Internet routing tables. Each AS uses BGP to advertise its own prefixes to communicate with other ASes, but BGP does not provide any mechanisms to authenticate routing announcements. Therefore, a malicious router can announce wrong routing information to target prefix on the Internet without any authentication process. Sometimes, malicious users use IP prefix hijacking to get IP addresses on purpose to do spamming or DDoS attack.

IP prefix hijacking can occur on purpose or by accident in several ways. Many previous studies have classified IP prefix hijacking in detail [12, 16, 17]. We briefly explain the three types of IP prefix hijacking. Regular prefix hijacking occurs when the attacker AS announces a prefix that it does not actually own. As its wrong announcement is propagated, the Internet becomes to be polluted. Because the routers prefer the shortest AS path to forward traffic, not all of ASes in the Internet are polluted. Subprefix hijacking happens when the attacker AS announces a more specific prefix than what may be announced by the true origin AS. Most ASes are impacted by this announcement because the priority of more specific IP prefix is higher in route selection process. Lastly, IP prefix interception is that the attacker AS forwards the hijacked traffic to the origin AS. In this case, the victim cannot recognize the occurrence of prefix interception.

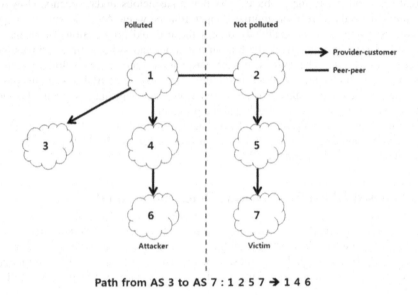

Path from AS 3 to AS 7 : 1 2 5 7 → 1 4 6

Fig. 1. Example of IP prefix hijacking: polluted and unpolluted ASes

Our approach focuses on the regular prefix hijacking. Fig. 1 shows an example of IP prefix hijacking with AS relationships. We suppose that the attacker AS is 6 and the victim AS is 7. When the attacker announces the IP prefix that the victim actually originated, this malicious routing information is propagated in the Internet. Typically ISPs can filter the announcements from their downstream ASes containing invalid IP address space, but previous hijacking incidents shows that it may not be applied by misconfiguration. With the given shortest AS path preference in routing, networks (AS 1 and AS 4 in Fig 1.) close to the attacker AS are polluted by the malicious announcement. AS 3 also receives the announcement and must decide whether the update is applied to routing table.

Because the routing tables of ASes near AS 3 are polluted, it cannot directly reach AS 7, but the unpolluted ASes can still arrive at the victim AS. A detection system using the information from multiple BGP monitoring points can recognize a MOAS conflict caused by IP prefix hijacking. However, it requires that monitoring points are located in both polluted and unpolluted ASes. That is, appropriate probing locations must be selected so that probing packets should reach two conflicted origin ASes through the different AS paths.

As mentioned above, we focus on the objective that we should avoid multiple vantage points. This requires additional techniques to properly detect IP prefix hijacking. Single vantage point cannot find any difference between legitimate MOAS and IP prefix hijacking.

We design the IP prefix hijacking detection algorithm using idle scan technique. Our algorithm identifies a suspicious BGP update message and verifies whether it is the IP prefix hijacking event.

3.2 Approach

In this section, we describe our solution approach. Fig. 2 shows an overview of our detection system in deployment. The detection system connects to the BGP router in observer AS to monitor BGP update messages and its routing table.

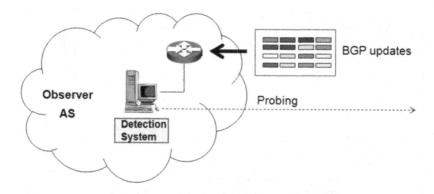

Fig. 2. Overview of the detection system in deployment

A BGP update message consists of withdrawn routes, the reachability information in the NLRI field and the AS_PATH attribute. The NLRI field indicates the IP address space about the destination AS and the AS_PATH attribute has the AS level path to reach the announced address space. With the comparison between update message and routing table, we can observe a suspicious event, especially MOAS conflict. When a suspicious BGP update that causes MOAS conflict is received, the system starts idle scan so that distinguish IP prefix hijacking event from legitimate routing update.

A probing technique called reflect-scan for fingerprinting the victim network is proposed [12]. This method is derived from the TCP idle scan technique described in [13]. The reflect-scan focuses on detecting subprefix hijacking cases, but it is applied to regular prefix hijacking in our approach.

The idle scan technique is used for completely blind port scanning that attackers can scan a target with sending a packet to a dummy host instead of the target. We utilize this technique on the purpose to reach the victim AS because we cannot directly arrive at the victim AS when IP prefix hijacking occurs as in Section 3.1. The key idea is to use the sequential IP ID increment property in IP packet and allow the unpolluted AS to forward the traffic to the victim AS.

The proposed technique is explained in Fig.3. First, the detection system selects a host (*Ha*) in the suspicious IP prefix, which satisfies the property to assign IP ID packets incrementally. Also, *Ha* should be idle because other traffic except for IP ID probing packets can interfere with the scan logic. And, the detection system should select a host (*Hr*) in the previous hop AS which is a just previous AS in the destination of AS_PATH to the target IP prefix. For example, if AS_PATH is 'a b c d', the previous hop AS is 'c' to the target AS 'd'. This host should be alive and in service with open TCP port. The web server that always opens the HTTP port is a good candidate for *Hr*.

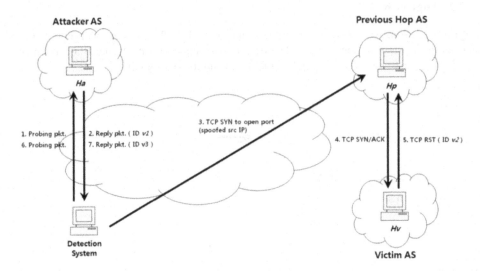

Fig. 3. Idle scan for IP prefix hijacking detection

After selecting the hosts, the detection system starts IP ID probing. The system sends a probing packet to *Ha* and records IP ID value in the reply packet. If a spoofed TCP SYN packet in which the source IP and *Ha*'s IP are same is sent to *Hr*, then *Hr* would response with a TCP SYN/ACK packet to *Ha*'s IP. When IP prefix hijacking occurs, *Ha* and *Hv* should be different. *Hv* that receives an unsolicited SYN/ACK packet will respond with a TCP RST. Therefore, one more probing to *Ha* can verify whether the received BGP update is the IP prefix hijacking event, because the IP ID difference between step 2 and step 7 is only one (that is, $v3 = v1 + 1$). In case of legitimate updates, *Ha* and *Hv* is the same host, and the IP ID difference is likely two or more ($v3 = v2 + 1 = v1 + 2$).

The probing packets used in step 1 and 6 do not need to be only TCP SYN/ACK packets like TCP idle scan technique. The proposed method requires the target hosts having predictable IP ID numbers for outgoing IP packets. To satisfy this requirement, we can select the protocol of probing packet that is expected to reply with incremental IP ID generation. More details are given in the next section.

The target hosts should be likely idle to reduce the false detection rate. To increase the detection accuracy, we can try to send multiple probing packets at step 1, 3, and 6. If the target is not as busy as well-known web server, we can sufficiently infer the occurrence of IP prefix hijacking as sending many probing packets.

4   Experiment Results

In this section, we present our experiment results to validate the proposed method and also discuss some of the obstacles.

We divide the validation process into three steps – correctness, feasibility and effectiveness. The correctness validation is that we test whether the proposed method detects the IP prefix hijacking events correctly. We should examine if the method can be used on real network, and the effectiveness of the proposed method should be measured from the performance point of view. In this paper, we carry out the correctness test and other validations are remained for future work.

4.1   Experiment Environment and Analysis

We performed an experiment to validate the correctness of our proposed method. Fig. 4 shows the experimental test-bed. We constructed the test-bed network which consists of routers and hosts using Linux machines. IP prefix hijacking condition is made by manipulating the routing table directly. The attacker and victim hosts can be operated on various operating.

This test-bed simulates the IP prefix hijacking case as described in Section 3.1. We suppose that Net6 is attacker AS and attempts to steal the IP prefix owned by Net7. After IP prefix hijacking event by Net6, Net1 and Net4 are polluted, but Net2 and Net5 are not polluted because the route selection process selects the shortest path. Fig. 4 shows that the routing tables are finally updated by the malicious action of Net6.

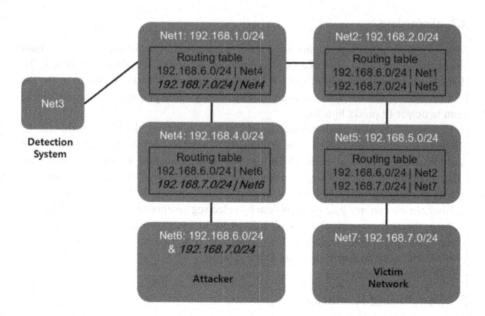

Fig. 4. Test-bed for correctness validation

The detection process is performed in Net3. We installed various operating systems (OS) on attacker AS Net6, such as Windows, Linux and etc., and checked for the IP ID probing process. As a result, the IP ID difference between the reply packets to two continuous probing packets was only one. Therefore, we conclude that idle scan technique is effective for detecting IP prefix hijacking events.

Table 1. IP ID generation pattern for different operating systems

Operating Systems	Reply Packets	
	TCP RST	ICMP
Windows	Incremental	Incremental
Linux	Zero	Incremental
Solaris	Incremental	Incremental
Router	Incremental	Random

Table 1 summarizes IP ID generation pattern of reply packets on various operating systems. We can use TCP probing for idle scan on the cases of most OS types except for Linux. Linux replies with zeroed IP ID packets in response to TCP SYN/ACK packets. In this case, we derive other protocols as reply packets, such as ICMP packets caused by UDP port scan to closed port. Linux also replies with sequential IP ID generation to ICMP packets. We can select appropriate probing packets to guarantee that the target systems assign IP ID packets incrementally on a global basis.

4.2 Discussions

When IP prefix hijacking occurs and an AS is impacted by that event, the AS cannot reach the victim network because the neighbor networks are already polluted. To distinguish a hijacking event from a legitimate update, we should be able to probe the attacker and victim AS. Idle scan can be a solution for that purpose and we utilize idle scan to detect IP prefix hijacking.

However, idle scan needs an appropriate target host having the property that the IP ID sequence generation happens incrementally. Therefore, we must perform port scan and OS identification on each AS in the routing table to find a candidate host. Another concern is that spoofing packets cannot be forwarded the target machine by egress filtering in ISPs. The higher service providers such as Tier-1 ISP are less likely to filter, so we can mitigate this problem by selecting an appropriate previous hop AS.

Also, the real network consists of a diversity of systems and devices, so we may suffer from unexpected responses in performing IP ID probing. We will solve this problem in future by feasibility tests in the Internet.

5   Concluding Remarks

In this paper, we have presented our algorithm of IP prefix hijacking detection. For detecting hijacking events, our algorithm relies on common characteristics of IP prefix hijacking such as MOAS and invalid route in BGP message. Our goal is to accurately identify hijacking events and distinguish them from the valid BGP updates. The proposed system does not require any protocol changes and multiple vantage points.

For future work, we will perform feasibility and effectiveness validation for the proposed method. The key problem is how to apply and validate the method in the real network. We will improve the algorithm to complement the reachability difficulties to the victim network.

We also plan to examine more IP prefix hijacking strategies and improve our algorithm that can detect various hijacking types. Furthermore, it is necessary to investigate the clear differences between IP prefix hijacking and valid updates.

References

1. North America Network Operator's Group, NANOG: NANOG Mailing Lists, http://www.nanog.org/mailinglist/
2. Lynn, C., Mikkelson, J., Seo, K.: Secure BGP (S-BGP) (June 2003); IETF Draft: draft-clynn-s-bgp-protocol-01.txt
3. Weis, B.: Secure Origin BGP (soBGP) Certificates (July 2004); IETF Draft: draft-weis-sobgp-certificates-02.txt
4. Kruegel, C., Mutz, D., Robertson, W., Valeur, F.: Topology-based Detection of Anomalous BGP Messages. In: Vigna, G., Krügel, C., Jonsson, E. (eds.) RAID 2003. LNCS, vol. 2820, pp. 17–35. Springer, Heidelberg (2003)
5. Lad, M., Massey, D., Pei, D.: PHAS: A Prefix Hijacking Alert System. In: Proceedings of the 15th USENIX Security Symposium, Vancouver, B.C., Canada, August 2006, pp. 153–166 (2006)
6. Barbir, A., Murphy, S., Yang, Y.: Generic Threats to Routing Protocols (October 2004); IETF Draft: draft-ietf-rpsec-routing-threats-07
7. Christian, B., Tauber, T.: BGP Security Requirements (March 2006); IETF Draft: draft-ietf-rpsec-bgpsecrec-04
8. BGP Routing Table Analysis Reports, BGP Reports, http://bgp.potaroo.net/
9. Zhao, X., Pei, D., Wang, L., Massey, D., Mankin, A., Wu, S.F., Zhang, L.: An Analysis of BGP Multiple Origin AS (MOAS) Conflicts. In: Proceedings of the 1st ACM SIGCOMM workshop on Internet Measurement, San Francisco, USA, November 2001, pp. 31–35 (2001)
10. Karlin, J., Forrest, S., Rexford, J.: Pretty Good BGP: Improving BGP by Cautiously Adopting Routes. In: Proceedings of the 14th IEEE International Conference on Network Protocols, Santa Barbara, California, USA, November 2006, pp. 290–299 (2006)
11. Ballani, H., Francis, P., Zhang, X.: A Study of Prefix Hijacking and Interception in the Internet. ACM SIGCOMM Computer Communication Review 37(4) (October 2007)
12. Hu, X., Mao, Z.M.: Accurate Real-time Identification of IP Prefix Hijacking. In: Proceedings of the IEEE Security and Privacy, Oakland, California, USA, May 2007, pp. 3–17 (2007)
13. Insecure.Org., TCP Idle Scan (-sI), http://nmap.org/book/idlescan.html
14. Tahara, M., Tateishi, N., Oimatsu, T., Majima, S.: A Method to Detect Prefix Hijacking by Using Ping Tests. In: Ma, Y., Choi, D., Ata, S. (eds.) APNOMS 2008. LNCS, vol. 5297, pp. 390–398. Springer, Heidelberg (2008)

15. Mao, Z.M., Rexford, J., Wang, J., Katz, R.H.: Towards an Accurate AS-level Traceroute Tool. In: Proceedings of the 2003 Conference on Applications, Technologies, Architectures, and Protocols for Computer Communications, Karlsruhe, Germany, August 2003, pp. 365–378 (2003)
16. Zheng, C., Ji, L., Pei, D., Wang, J., Francis, P.: A Light-Weight Distributed Scheme for Detecting IP Prefix Hijacks in Real-time. ACM SIGCOMM Computer Communication Review 37(4) (October 2007)
17. Zhang, Z., Zhang, Y., Hu, Y.C., Mao, Z.M., Bush, R.: iSPY: Detecting IP Prefix Hijacking on My Own. In: Proceedings of the ACM SIGCOMM 2008 conference on Data Communication, Seattle, USA, pp. 327–338 (2008)

A PKI Based Mesh Router Authentication Scheme to Protect from Malicious Node in Wireless Mesh Network*

Kwang Hyun Lee and Choong Seon Hong**

Department of Computer Engineering Kyung Hee University
Seocheon, Giheung, Gyeonggi, 446-701 Korea
khlee@networking.khu.ac.kr, cshong@khu.ac.kr

Abstract. Wireless mesh network can increase service coverage by low-cost multiple-paths between a source and a destination. But because of the wireless nature, wireless mesh network is vulnerable to diversified threats. Especially attacks by malicious nodes can decrease the performance or ravage the network. Many protocols, based on Public Key Infrastructure (PKI), have been proposed to ensure security however; these protocols cannot efficiently prevent the cases of compromised nodes. In this paper, we propose a mesh router authentication scheme to solve this problem of existing protocols. Although our scheme increases the end-to-end delay, the proposed mesh router authentication scheme can guarantee secure wireless mesh network communication from malicious nodes.

Keywords: Wireless mesh network, Security, Wormhole attack, PKI.

1 Introduction

Wireless mesh network [1] can be deployed more efficiently in a wide network area. In case of Wi-Fi to extend network area, the cost of deployment will be highly increased [2]. But a network can be extended in an efficient and a low-cost way using wireless mesh network in comparison to the Wi-Fi network.

Wireless mesh network structure is similar to MANET (Mobile Ad-hoc Network) [3]. Wireless mesh network and MANET have to apply policy of high level security because of the ad hoc nature of these networks. But MANET cannot apply heavy security policy due to the performance limitation. On the contrary, wireless mesh network has the processing ability to apply heavy security policies. In this paper, we propose mesh router authentication scheme for adapting optimized algorithm for security in wireless mesh network. By using the proposed mesh router authentication

* This research was supported by Basic Science Research Program through the National Research Foundation of Korea(NRF) funded by the Ministry of Education, Science and Technology (R01-2008-000-20801-0).
** Corresponding author.

C.S. Hong et al. (Eds.): APNOMS 2009, LNCS 5787, pp. 405–413, 2009.

scheme, delay time is increasing. We are able to detect malicious nodes in the network and delete them from the network.

This paper is organized as follows. In Section 2, we give a brief overview of wireless mesh network and security threats due to malicious nodes. We describe proposed mesh router authentication scheme in Section 3. In Section 4, we explain the scenario for the attack prevention scheme. The performance evaluation is presented in Section 5. Finally, the conclusion and future works are provided in Section 6.

2 Background

Wireless mesh network consist of one or more mesh routers and mesh clients. The mesh clients are mobile ad hoc nodes with scarce resources while the mesh routers have no mobility, and they communicate each other using high bandwidth radio links. The mesh routers are used to forward the mesh client packets to the mesh gateway. The mesh gateway connects the mesh clients to the Internet. As the mesh routers have functions of self-configuration, self-healing and self-organization, mesh routers can compose or modify routes by themselves.

The wireless mesh network can be classified as three types. First type is Infrastructure/Backbone wireless mesh network. It provides the basic routing backbone consisting of only mesh routers. Second type is Client wireless mesh network. Client wireless mesh network seems like simple ad-hoc networks. It consists of just client that also work as routers or forwarding nodes. The last type is Hybrid wireless mesh network. This type has the combined characteristics of both Infrastructure/Backbone and Client wireless mesh network.

2.1 Problem Statement

Wireless mesh network is vulnerable to diversified threats. A malicious node can launch an attack during the mesh network initialization. Especially, Wormhole attack, which can easily breakdown the network and affects its performance significantly. The following provides a brief description of the Wormhole attacks [4].

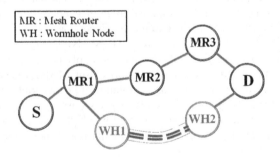

Fig. 1. An example of Wormhole attack

Wormhole attack is different from other network attacks. The majority of the network attacks consist of single node. However, in an attacker uses two nodes and makes an optimized tunnel between them. This can be achieved by using high bandwidth links, wired line, or encapsulation. This optimized link appears tempting for other nodes of the network. As a consequence, network nodes start to use the optimized, but compromised link for data forwarding. At first sight, Wormhole tunnels appear to be good, as they provide fast links for data forwarding. However, Wormhole nodes can do sniffing, modifying and dropping of data packets. Therefore, nodes have to establish the network without Wormhole nodes.

3   Proposed Mesh Router Authentication Scheme

If there is no authentication scheme, malicious nodes can execute an attack such as Wormhole. Therefore, if mesh routers are authenticated for each communication, a malicious node cannot harm the network. In this section, we explain our proposal for mesh router authentication based on PKI (Public Key Infrastructure) to protect the network from malicious nodes in wireless mesh network.

3.1   Assumptions

We maintain some assumptions, as follows:
- There is a CA (Certification Authority), which has the list of public key of all mesh routers.
- All mesh routers have hash function and public key of the CA.
- Both sender and receiver can listen to the transmission of each other.
- Routing protocol used is DSR (Dynamic Source Routing) [5].

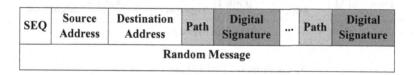

Fig. 2. Proposed RREQ message format

3.2   Proposed PKI Based Mesh Router Authentication Scheme

During the network initialization, each mesh router broadcasts the RREQ message. The format of RREQ packet is shown in Fig. 2. Our scheme includes a digital signature to authenticate mesh router. If mesh router receives the RREQ message, the mesh router creates the digital signature with the help of private key and random message. The created digital signature is then added in the RREQ message. The RREQ message is then rebroadcasted for route discovery. Using the digital signature, we can authenticate the mesh router.

Our scheme uses two ways for the authentication, first scheme works during initialization of the network. On the other hand, second schemes works after any route update. In first method, intermediate nodes authenticate the mesh router. In second method, intermediate mesh routers just check the broadcast time of next hop node by WPA (Wormhole Prevention Algorithm) [6].

3.2.1 Mesh Router Authenticated by Intermediate Mesh Router

An intermediate mesh router which called MR1, broadcasts the RREQ message to configure path and wait for receiving RREQ message from the next hop mesh router (by overhearing). The MR1 then uses a WPT (Wormhole Prevention Timer) to detect the Wormhole node. The WPT can be calculated by following equation:

$$WPT = \frac{2 \times Transmission\ Range(TR)}{The\ propagation\ speed\ of\ a\ packet} \tag{1}$$

If the RREQ message is arrived within WPT, the intermediate mesh router (MR1) encrypts the path and next hop node's digital signature by CA's public key, and sends the RAREQ (Router Authentication Request) message to the CA. Fig. 3 shows the RAREQ message format. The RAREQ message is used for confirmation of mesh router's digital signature by the CA. The CA compares the RAREQ with the digital signature of the mesh router and then, the CA sends the RAREP message that has the result of the comparison to the requester mesh router. Using this authentication way we can find the malicious mesh router in a faster way even though this scheme has large overhead.

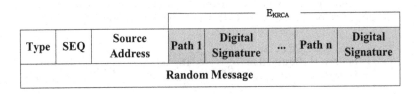

Fig. 3. Proposed RAREQ message format

3.2.2 Mesh Router Authenticated by Source Mesh Router

The intermediate mesh router overhears the next hop mesh router's RREQ message. At this time, if the RREQ message is not arrived within WPT, the intermediate mesh router just drop the RREQ message as it was not sent to confirm the router authentication. If the source mesh router's RREQ message is arrived at the destination mesh router, it encrypts paths and digital signatures of source and unicast the RREP message into reverse direction. After the source mesh router receives the RREP message, it sends the encrypted paths and digital signatures to the CA using the request message (RAREQ) message.

As we mentioned above, the CA compares encrypted path nodes with their digital signatures present at the CA. And then the CA replies the RAREP (Router Authentication Reply) message that contains comparing results.

Fig. 4 shows the RAREP message format. As shown in Fig.4, the path and the comparing results are encrypted using the source node's public key. In this case, the network traffic is lower in amount as compared to the case of the intermediate authentication. However if malicious mesh router broadcasts the RREQ message, the source mesh router waits for receiving the destination router's RREP message.

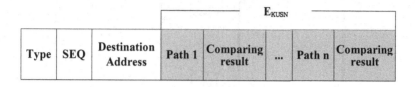

Fig. 4. Proposed RAREP message format

3.3 The Mesh Router Reliability Table

The mesh router maintains a mesh router reliability table by using authentication results with the neighbor mesh routers. Because mesh networks use the multi-path routing protocol [7], the mesh router reliability tables are maintained by comparing results information of neighbor mesh routers. If the networks have a malicious mesh router, the malicious mesh routers are detected by its neighbor nodes. Then neighbor nodes report to other node. The mesh routing reliability table stores a value against each neighbor router.

This value indicates the number of times the RREQ was heard by a neighbor node without hearing the forwarded RREQ message by the same router node. This value is maintained to identify the malicious neighbor router nodes. For instance, if any router received the RREQ message and did not forward the RREQ message, we can doubt this node as a malicious node. Each time this happens, the value is increased. If any routing path has this value more than the threshold value, the mesh routers is removed from the routing path in the routing table.

4 Protecting Scenario of Proposed Scheme

Proposed scheme supports more robust security than the traditional scheme. The traditional scheme has threats such as the case of malfunction mesh router or attack from an authenticated mesh router. But these attacks can be prevented by rechecking the mesh router in case of already authenticated mesh routers. In this section, we explain a scenario of our proposed scheme.

4.1 In Case of Wormhole Node Just Can Forward the RREQ Message

As shown in fig. 5, we can prevent the malicious mesh router that has no broadcasting function. If router A needs to configure the path, the router A broadcasts the RREQ message with the RM (Random Message). At this time, if the wormhole nodes have no broadcasting function, the RREQ message cannot attach router A within WPT.

RM : Random Message

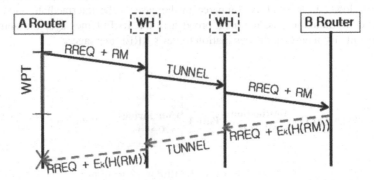

Fig. 5. In case of wormhole node just can forward the RREQ message

4.2   In Case of Wormhole Node Can Broadcast RREQ Message

If the wormhole node can broadcast the RREQ message, the traditional scheme (ex. WPA) cannot prevent the wormhole node. However our scheme can prevent the wormhole node as follows.

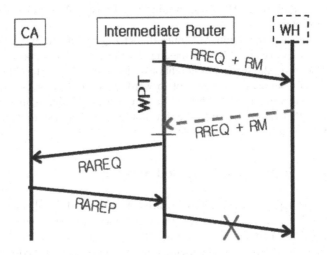

Fig. 6. Mesh router authenticated by intermediate mesh router

In case of the mesh router is authenticated by the intermediate mesh router, the intermediate mesh router sends the RAREQ message to the CA. After receiving the RAREP message from the CA, the Intermediate mesh router decides whether the path is removed or not. Fig 6 shows a simple example of preventing wormhole node.

Fig 7 shows the case of the mesh router authenticated by the source mesh router. If the destination mesh router receives the RREQ message, the destination mesh router encrypts paths and the digital signature by using CA`s public key. And the destination mesh router unicast the RREP message and encrypted message through reverse path to the source mesh router. The destination mesh router encrypts path and the digital signature because if the destination mesh router cannot encrypt the message, the malicious nodes can modify the digital signature. Therefore, in the proposed scheme; during the transmission process, encrypted message cannot be modified by the malicious nodes. The CA decrypts this message and after confirming the mesh router and signature, the destination mesh router sends the RAREP message to the source node.

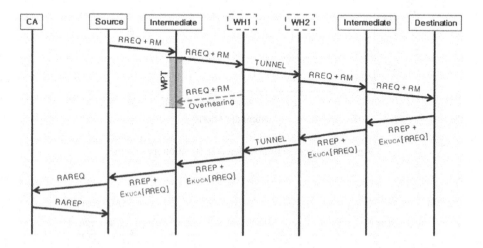

Fig. 7. Mesh router authenticated by source mesh router

5 Simulation Result

Parameter	Value
Simulation code	C++
Network form	Grid topology
The number of nodes	10~60
Total number of simulation	100

To evaluate the performance of our scheme we perform the simulation in C++. We assume that network topology is grid topology and mesh routers have uniform distance in between them. We select the source mesh router and the destination mesh router randomly. We perform our simulation 100 times. We increase nodes to check changing value depending on the number of nodes.

As shown in Fig. 8, x-axis denotes the number of nodes and y-axis denotes time (in milliseconds) of finding destination mesh router. Two lines show the traditional DSR and our proposed scheme, respectively. As the numbers of nodes are increased, the time to detect the destination mesh router is also increased. As shown in the fig. 8, our scheme works slower than the traditional scheme. But it provides better security, as it uses public key encryption between the source router node and the destination router

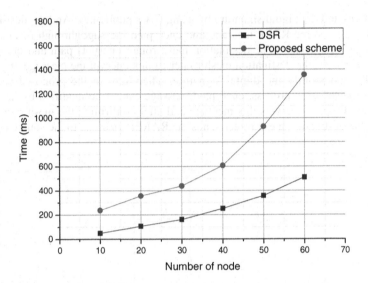

Fig. 8. The average time to detect destination mesh router

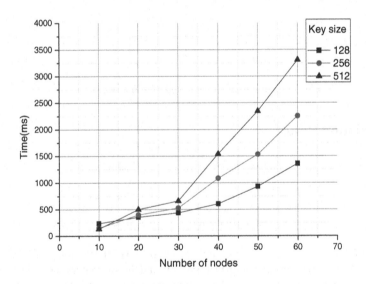

Fig. 9. The average time to detect destination mesh router with the increase of number of nodes for different key sizes

node. The DSR protocol does not use any kind of such encryption or authentication. So, the worm whole attacks are easily possible. Therefore, as our scheme has more delay but it is a trade-off between the security and delay.

Fig. 9 shows the time delay in detection of the destination router nodes with the increase of number of nodes for different key sizes. We perform the simulation using 128-bit, 256-bit and 512-bit public key encryption. We can see that the delay is increased with the use of higher bit key but the difference is not significant. As shown in Fig.9, when the number of nodes is smaller then the delay between using different keys is not significant. However, as the numbers of nodes increase, the delay is increased significantly. With this graph we can find the optimal value of number of nodes for a certain size of the public key. As the mesh nodes have the processing capability of performing the public key cryptography. And the number of router nodes in a wireless mesh network is not high. Hence, using higher bit keys seems to provide a good and secure solution for the wormhole attack problem.

6 Conclusion

We need to apply more robust scheme for secure networking because wireless mesh network have high performance,. In this paper, we propose a suitable enhanced scheme for wireless mesh network using a PKI based mesh router authentication scheme. Our scheme uses mesh routers' digital signatures to authenticate each other when detecting a path and a Route-Reply forwarding timeout to detect the malicious nodes, which causes the wormhole attacks. Therefore, our proposed scheme can prevent not only wormhole attack but also other attack from malicious nodes. Our proposed scheme has weak point that it has large overhead. We are currently working to make our scheme a lightweight scheme for wireless mesh network.

References

1. Akyildiz, I.F.: A Survey on Wireless mesh network. IEEE Radio Communications (September 2005)
2. Ben Salem, N., Hubaux, J.-P.: Securing Wireless mesh network. Wireless Communications, IEEE 13(2), 50–55 (2006)
3. Nandiraju, D., Santhanam, L., Nandiraju, N., Agrawal, D.P.: Achieving Load Balancing in Wireless mesh network Through Multiple gateways. In: Mobile Adhoc and Sensor Systems (MASS), October 2006, pp. 807–812 (2006)
4. Glass, S., Portmann, M., Muthukkumarasamy, V.: Securing Wireless mesh network. Internet Computing, IEEE 12(4), 30–36 (2008)
5. Johnson, D.B., Maltz, D.A.: Dynamic Source Routing in Ad Hoc Wireless Networks. Mobile Computing (1996)
6. Choi, S., Kim, D.-y., Lee, D.-h., Jung, J.-i.: WAP:Wormhole Attack Prevention Algorithm in Mobile Ad Hoc Networks. In: 2008 IEEE International Conference on Sensor Networks, Ubiquitous, and Trustworthy Computing, June 11-13, pp. 343–348 (2008)
7. Nandiraju, N.S., Nandiraju, D.S., Agrawal, D.P.: Multipath Routing in Wireless mesh network. In: Mobile Adhoc and Sensor Systems (MASS), October 2006, pp. 742–746 (2006)

A Personalization Service in the IMS Networks

Youn-Gyou Kook and Jae-Oh Lee

School of Information Technology, Korea University of Technology and Education
Cheonan, Chungnam, Korea 330-708
{ykkook,jolee}@kut.ac.kr

Abstract. In this paper, we present the personalization service in the IP Multimedia Subsystem (IMS) networks. It is the customized service that the service provider serves the optimal content on a suitable user's situation within the various services. To serve that, the service provider has to develop the various types based on the analysis of user's profiles and usage patterns that are recorded on the server. We have implemented the IMS infrastructure and simulated the personalization service on that.

1 Introduction

The IP Multimedia Subsystem (IMS) will be the next generation networks suggested by 3GPP (3rd Generation Partnership Project) [1]. The IMS is the convergence communication environments that are made up of wired and wireless networks based on IP to provide the multimedia services which include a voice, audio, video, large scale data and others to the customer [8, 11]. The service providers need new paradigm services to serve the customers.

It is necessary that the service provider provide the personalization service which is customized on analyzing user profiles and usage patterns. That could be acquired from the Session Initiation Protocol (SIP) messages and Home Subscriber Server (HSS) [8, 9]. And the service provider needs an open Service Delivery Platform (SDP) to provide the digital convergence services and the new paradigm services like a customized service and a 3rd party service which is developed by the other service developers [4, 7, 12].

Therefore, this paper presents the implementation of IMS infrastructure and the personalization service. This service is composed of the analyzed elements that are service type, service timestamp, used time, used devices and user's private information. The elements can be detailed and extended by the service provider's policy for providing the customized service to the user.

The remainder of this paper is organized as follows. The next section presents personalization services in the IMS. Section 3 describes the implementation of our IMS infrastructure and the simulation of the personalization service on our scenario.

2 The Personalization Service in the IMS

In these days, the requirement of the user is transforming the passive service environments in which uses the listed services by the service provider into the interactive

C.S. Hong et al. (Eds.): APNOMS 2009, LNCS 5787, pp. 414–417, 2009.

and conventional ones in which the user has the opportunity to request the customized services. It is necessary that these service environments can be supported to provide the digital convergence services, large scale services and individual specific services. The IMS networks is suited to be in that environments because that is made up of wired and wireless networks based on All-IP and is able to provide the various personalization services.

The service provider to serve the customized services based on the IMS infrastructure has to possess many services of all kinds and to develop the various service types. He must consider optimization, customization and profit problems when he develops and provide the services. And he has to extend his roles to being 'the service enablers' who can support the 3rd party services, it means that he can develop a new paradigm service, support the 3rd party services and orchestrate the partial services of an existing service, a new service and the 3rd party service.

However, the user faces complexities of selecting the appropriate services. It is difficult to select one from his interesting contents. So the service provider has to consider the customized services when he generates and orchestrates a new service. It could be solved the problem by the personalization service based on analysis of user's profiles and usage patterns. 'The Service Provider' manages the services that are composed of the existing service enablers and the new ones to guarantee the stability in the IMS environments. 'The Service Adapter' analyzes the user's profiles and usage patterns and acquires the elements of the personalization services. 'The Service Generator' generates the new services and orchestrates the partial services within the 3rd party services based on Web services.

We can analyze the customer's information that are transported by the SIP protocol and recorded the user's log on HSS in the IMS. The SIP protocol includes the used device information, status, timestamp, expiration and others. Figure 1 shows a SIP message on the IMS.

```
INVITE sip:jolee@kut.ac.kr SIP/2.0
Via: SIP/2.0/UDP 220.68.70.190:5000 comp=sigcomp; branch=z9hg4bk9hah
Max-Forwards:70
Route:<sip:220.68.70.190:5000 comp=sigcomp>
        <sip:220.68.70.190:4000;lr>
From:Infotel <sip:infotel@kut.ac.kr>
To:JaeOhLee<sip:jolee@kut.ac.kr>
Call-ID:23fi571ju
CSeq:8348 INVITE
Content-Type:application/sdp
Content-Length:187
```

Fig. 1. A SIP message on the IMS

And the HSS is in charge of that: the management of user's mobility, the support of call and session, the creation of user's security information, user's authority, message integrity, message encryption, the management and store of user's information. So we analyze the user's profiles and the usage patterns including his interesting services and so on. And we acquire the elements of the personalization services. The elements can be analyzed in detail by the service provider's policy, but this paper presents the basic elements. The elements of the personalization services follow as that:

1) **Service Type:** It is a kind of service that is served by the service provider. These are voice communications, MMS, SMS, e-mail, game, web browsing, searching, notifications, advertisement service, and so on.

2) **Service Timestamp:** It is the timestamp that the user accesses to use the services. To be the various time slots that user accesses the service, it orders the set of time-stamps on which based the length of time used or the most frequently accessed and indicated the priority timestamp. The service timestamps are composed of the time slots ranged from TS_1 to TS_{24} slot.

3) **Service Span:** It is the time that the user used each service. We could know user's royalty and favor of a service and define a service size of the personalization services. The service span is the average span of using each service in a day.

4) **Service Device:** It is a kind of device that the user accesses to use the service. We could know the user's device information from the SIP messages.

5) **Service User:** It is the user's profiles such as his age, sex, job and so on. It is used to recommend what, when and how service is fit for him.

We generate and orchestrate the personalization services based on the analyzed elements of the user's profiles and usage patterns. And we can define to serve the what, when and how of service.

3   Implementation of the Personalization Service

In this section, we describe the scenario and implementation of the personalization services in the IMS infrastructure. As described in Part 2, the personalization services have the structure of the meta-data that are composed of a service type, service size and service time.

The IMS infrastructure is implemented in JAIN SIP 1.2 API and Apache for Web-service. We designed to route a specified IP address and Port, and simulated the personalization services in a user's local area networks. Figure 4 shows the implemented IMS infrastructure that is P-CSCF node, I-CSCF node and S-CSCF node. P-CSCF (proxy-call/session control function) node executes the role of proxy and user agent, transports the SIP messages to I-CSCF node. I-CSCF (interrogating-CSCF) node acquires the address of S-CSCF from HSS and routes the received SIP messages from the other networks as like a P-CSCF through S-CSCF node. S-CSCF (serving-CSCF) node manages the customer's information at HSS, interacts with SDP.

We have implemented the application service with Java and WSDL2Java. And the mobile service has been implementing Java with the stub generator. To generate the customized service, we considered the elements of a personalization service as above presented. Figure 1 shows the application and mobile service on the mobile emulators.

4   Conclusion

The IMS is the convergence communication environments that are made up of wired and wireless networks based on IP to provide the multimedia services which include a voice, audio, video, large scale data and others for the customer. It is necessary that the service delivery platform is open to include the 3rd party service and the customer specific service. So, the service provider makes efforts to possess the many and various services and develop a new paradigm service that converge the digital contents with another service environments. However it is difficult that the user use on his fit the services because of the more and various services.

In this paper, we described the personalization service in the implemented IMS infrastructure to provide the digital convergence service. Moreover we presented the architecture of an open service delivery platform and analyzed the elements of the personalization services based on the customer's profiles and usage patterns. On generating the personalization services, we considered the extracted elements is composed of service type, service timestamp and service size. The elements of the personalization services can be analyzed in detail by the policy of the service provider. The future work includes the monitoring and security of service in the IMS.

References

[1] http://www.3gpp2.org/, TFC3131 (3GPP2-IETF)
[2] http://java.sun.com/webservices/docs/1.3/tutorial/doc/index.html
[3] http://www.analysysmason.com/
[4] http://www.parlay.org/en/products/
[5] http://www.serviceoriented.org/web_service_orchestration.html
[6] Pichot, A., Audouin, O., Alcatel Research: Grid services over IP Multimedia Subsystem. In: BroadNet 2006, IEEE, October 2006, pp. 1–7 (2006)
[7] Devoteam Group: Service Delivery Platform: The Key to Service Convergence, Devoteam white paper (October 2007)
[8] Camarillo, G., Garcia-Martin, M.A.: The 3G IP Multimedia Subsystem (IMS), 2nd edn. Wiley, Chichester (2006)
[9] Hurtado, J.A., Martinez, F., Caicedo, O., Ramirez, O.: Providing SIP services support on Mobile networks. In: 4th ICEEE (September 2007)
[10] Kim, S., Min, J., Lee, J., Lee, H.: W3C. "Mobile Web Initiative" Workshop (November 2004)
[11] Cho, J.-H., Lee, J.-O.: The IMS/SDP structure and implementation of presence service. In: Ma, Y., Choi, D., Ata, S. (eds.) APNOMS 2008. LNCS, vol. 5297, pp. 560–564. Springer, Heidelberg (2008)
[12] Ren., L., Pei., Y.Z., Zhang., Y.B., Ying, C.: Charging Validation for Third Party Value-Added Applications in Service Delivery Platform. In: Network Operations and Management Symposium, NOMS 2006, 10th IEEE/IFIP (October 2006)
[13] Maes, S.H.: Service Delivery Platforms as IT Realization of OMA Service Environment: Service Oriented Architectures for Telecommunications. In: WCNC 2007, pp. 2885–2890. IEEE Communications Society, Los Alamitos (2007)
[14] Boll, S.: Modular Content Personalization Service Architecture for E-Commerce Applications. In: WECWIS 2002, June 2002, pp. 213–220. IEEE Computer Society, Los Alamitos (2002)
[15] Chou, W., Li, L., Liu, F.: Web Services for Communication over IP. IEEE Communications Magazine, 136–143 (March 2008)
[16] Cheng, Y., Leon-Garcia, A., Foster, I.: Toward an Autonomic Service Management Framework: A Holistic Vision of SOA, AON, and Autonomic Computing. IEEE Communications Magazine, 138–146 (May 2008)

Dimensioning of IPTV VoD Service in Heterogeneous Broadband Access Networks*

Suman Pandey[1], Young J. Won[1], Hong-Taek Ju[2], and James W. Hong[1]

[1] Dept. of Computer Science and Engineering, POSTECH, Korea
{suman,yjwon,jwkhong}@postech.ac.kr
2 Dept. of Computer Engineering, Keimyung Univ., Korea
juht@kmu.ac.kr

Abstract. The IPTV subscription rate has increased steadily since the introduction of IPTV Video on Demand (VoD) services. We have developed a model to determine the optimum deployment strategy for IPTV delivery network from the IPTV service providers' perspective. The analysis technique in this paper helps us determine the best deployment scenarios to support a certain number of customers within the tolerant boundary of Quality of Experience (QoE) measures. We define important QoE measures. The QoE will help service providers make deployment decisions including the number of servers, distance of servers from a community, and desirable access network bandwidth capacity.

Keywords: IPTV, VoD, QoE, IPTV network dimensioning.

1 Introduction

ITUT defines IPTV as multimedia streaming over IP networks with reasonable quality assurance [10]. Some IPTV service providers own full networking infrastructure such as AT&T and Verizon. Others need not necessarily own the full infrastructure such as AOL, Apple and Google; they might lease some part of the infrastructure or rely on the current Internet. Software product such as Microsoft Mediaroom [11] is used as the IPTV service provisioning platform by these telcos to provide IPTV services. IPTV video services can be broadly classified as live TV and video on demand (VoD). In future, a significant fraction of the IPTV traffic, up to 90%, may be due to VoD services [6]; and VoD services are resource intensive. However, the planning and deployment of VoD services in access networks (AN) has not been thoroughly studied. Mathematical models for IPTV network deployment [1, 4] have been proposed; however, they do not consider the heterogeneous aspect of network and actual deployment strategies. Furthermore, there are studies concentrating only on some specific technology deployment such as Ethernet based WDM networks with ring topology [2], Ethernet over SONET, and Ethernet over Fiber technology [3]. The network cost models for VoD services [4] have focused on video distribution strategies for reducing

---
* This work was partly supported by the IT R&D program of MKE/IITA [2008-F-016-02, CASFI] and WCU program through the KSEF of MEST, Korea [R31-2008-000-10100-0].

C.S. Hong et al. (Eds.): APNOMS 2009, LNCS 5787, pp. 418–422, 2009.

network cost based on hit ratio and cache size. The benefits of a P2P delivery mechanism for IPTV VoD servers have been explored [5], however they proposed bandwidth modeling only for P2P. In this study, we developed mathematical models to dimension IPTV network from a service provider perspective.

IPTVs can have a complex network architecture as illustrated in Fig. 1 [3, 6]. A typical IPTV network is composed of super headend (SHE) and video hub offices (VHOs) at the national core. VHOs are connected to multiple video source servers (VSOs) which constitute the metropolitan area network (MAN) core, and VSOs are connected to central offices (COs) which are closer to the user communities. VSOs' edge routers route and aggregate local loop traffic from passive optical networks (PONs) ANs or from digital subscriber line access multiplexers (DSLAMs). There are video on demand (VoD) servers at each of these hierarchical levels. COs and VSOs are connected to the ANs. ANs can have multiple outside plants (OSP) connected to the CO. Each OSP may have a co-located DSLAM. These DSLAMs can be connected with the remote DSLAMs which will be directly connected with the end user communities. The IPTV infrastructure is logically divided into three main parts: client domain, network provider domain, and service provider domain [1]. IPTV service providers need to lease services from all these domains, for example, obtaining sustainable bandwidth within the client and network provider domains, and leasing optimum number of content servers. IPTV service providers provide the audio-video services to different kinds of users having different access network technology including fiber to home (FTTH), fiber to building (FTTB), xDSL or cable network. In such a heterogeneous environment, IPTV service providers need to consider heterogeneous network conditions while making optimum deployment strategy.

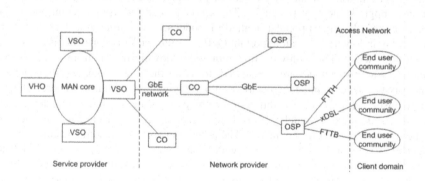

Fig. 1. IPTV network architecture

This paper focuses on determining the optimum network deployment strategy for VoD services in a heterogeneous networking environment. We define essential QoE measures, and build an analysis model for each QoE. We can state our problem as follows: "Using user QoE & user requirements help service providers determine the number of servers, access network bandwidth requirements, and distance from server to users to fulfill IPTV VoD service in a heterogeneous access network".

2 Proposed Method

In this section, we explain our definition of community, the selected QoE metrics and overall design. A community is a group of IPTV subscribers sharing the similar characteristics, such as physical closeness, distance from the IPTV server, and the type of network access technology in use. A clustering-based algorithm can be implemented to delimit a community. Depending on the QoE measure, the community chooses the best servers for VoD data delivery. The community provides user viewing behavior, number of users in the community, mean link capacity, mean request rate during peak viewing hours, mean session duration, and mean distance from the community to the server. The network and server properties and their related metrics are driven from the network service and storage providers.

The network performance highly affects users' QoE for audio video services. Thus, while designing or dimensioning a network for video traffic, QoE is an important measure. ITU-T [10] divides QoE into two categories: subjective QoE and objective QoE. The subjective QoE includes emotions, linguistic background, attitude, and motivation. The objective QoE includes information loss and delay. Objective includes service factors, transport factors and application factors. In this paper when we refer to QoE we mean objective QoE. We utilize three objective QoE for this work, first is server's waiting time, second one way minimum delay and third is access network bandwidth consumption.

Server waiting delay is an important objective QoE measure. Communities generate service requests and servers fulfill these requests. As the load on a server increases the performance decreases; this causes delay in serving the requests. This QoE metric can help in determining whether IPTV service providers need to increase the number of server or upgrade the server capacity. The server is modeled using queuing theory [1]. The Erlang C [9] is most popular method to model server waiting time. One-way minimum delay in low traffic is an important QoE measure for VoD. One way is considered because we need to know only the time required to download VoD from server to community. Low traffic hours are considered, to avoid the effect of network congestion. This way we can determine the behavior of underlying network infrastructure. This is a deterministic delay, which is addition of the processing time of intermediate devices and end host, and propagation delay [8] of heterogeneous network. Heterogeneous network properties can be input to this model. The access network bandwidth consumption also affects user viewing quality; high bandwidth consumption causes an increase in latency and poor application performance at the end user. As we already discussed the AN can be heterogeneous. The download and upload bit rate of the network may vary depending on the different access technologies used [7]. While modeling the bandwidth consumption these parameters should be considered.

The inputs from the community, network providers and service providers are passed to the analysis models, which calculate three QoE measures, and help in analyzing the best deployment strategies (Fig 2.). All the QoE simulation models are independent of each other. The desired QoE values are provided in Service Level Agreement (SLA) from the IPTV service provider to the subscriber. For example SLA can define server waiting probability < 0.5 s, bandwidth occupancy < 0.5 and one way delay < 0.005 s. Service providers should meet minimum QoE conditions, and upgrade their networks whenever the QoE is not satisfied.

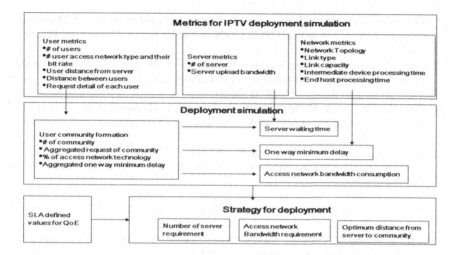

Fig. 2. Overall design showing interaction of input metrics with the models and optimization strategies for deployment

3   Conclusion and Future Work

The proposed modeling technique allows us to determine the optimum deployment conditions for a given number of potential IPTV VoD users while satisfying the pre-defined QoE measures. By analyzing user requirements and network configurations, we calculated substantial objective QoE such as server waiting time, one way minimum delay, and access network bandwidth consumption. The expected QoE conditions are compared, and used to guide deployment decisions considering the following parameters: the number of VoD servers, the physical distance of server from community, and AN bandwidth capacity. For future work, we will develop a more sophisticated model that considers content popularity and multiplexing aspects of network dimensioning. We will also explore various delivery mechanisms such as P2P, and multicasting for delivering VoD as well as live TV. We will develop models that consider various delivery mechanisms.

References

[1] Agrawal, D., Beigi, M.S., Bisdikian, C., Lee, K.W.: Planning and Managing the IPTV Service Deployment. In: 10th IFIP/IEEE International Symposium on Integrated Network Management, Munich, Germany, May 21-25, pp. 353–362 (2007)
[2] Wauters, T., Colle, D., Pickavet, M., Dhoedt, B., Demeester, P.: Optical Network Design for video on demand services. In: Optical Network Design and Modeling Conference, Milan, Italy, February 2005, pp. 250–259 (2005)
[3] El-Sayed, M., Hu, Y., Kulkarni, S., Wilson, N.: Access Transport Network for IPTV Video Distribution. In: Optical Fiber Communication Conference, March 5-10 (2006)

[4] Thouin, F., Coates, M.: Video-on-demand server selection and placement. In: Mason, L.G., Drwiega, T., Yan, J. (eds.) ITC 2007. LNCS, vol. 4516, pp. 18–29. Springer, Heidelberg (2007)

[5] Chen, Y.F., Huang, Y., Jana, R., Jiang, H., Rabinovich, M., Rahe, J., Wei, B., Xiao, Z.: Towards capacity and profit optimization of video-on-demand services in a peer-assisted IPTV. Multimedia Systems 15, 19–32 (2008)

[6] Simsarian, J.E., Duelk, M.: IPTV Bandwidth Demands in Metropolitan Area Networks. In: 15th IEEE Workshop on Local and Metropolitan Area Networks, NJ, USA, June 10-13, pp. 31–36 (2007)

[7] Won, Y.J., Choi, M.J., Hong, J.W.K., Hwang, C.K., Yoo, J.H.: Measurement of Download and Play and Streaming IPTV Traffic. IEEE Communications Magazine 46, 154–161 (2008)

[8] Bovy, C.J., Mertodimedjo, H.T., Hooghiemstra, G., Uijtervaal, H., Mieghem, P.V.: Analysis of end-to-end delay measurements in Internet. In: PAM, Colorado, USA, March 25-27 (2002)

[9] Adan, I., Resing, J.: Queueing Theory, Department of Mathematics and Computing Science. Eindhoven University of Technology (2002)

[10] ITUT- IPTV Focus group proceedings (2008), http://www.itu.int/ITU-/IPTV/

[11] Microsoft media room, http://www.microsoft.com/mediaroom/

Remote Configuration Mechanism for IP Based Ubiquitous Sensor Networks

Kang-Myo Kim, Hamid Mukhtar, and Ki-Hyung Kim

Ajou University, San 5, Woncheon-Dong, Yeongtong-Gu,
Suwon, South Korea
{rlaay,hamid,kkim86}@ajou.ac.kr

Abstract. Wireless Sensor Networks (WSNs) are becoming increasingly important because of their reduced cost and a range of real world applications. IP
connectivity to WSNs has enabled ubiquity of devices. 6LoWPAN networks are
the realization of these IP based Ubiquitous Sensor Networks (IP-USNs). Management of such low power and constrained networks is a crucial problem. Remote configuration is an important management aspect where the devices are
configured on runtime. In this paper we propose a remote configuration mechanism for IP-USNs.

Keywords: 6LoWPAN, sensor network management, remote configuration.

1 Introduction

IEEE standard 802.15.4 [1] has emerged as a strong technology for wireless sensor
networks to morph Personal Area Networks (PANs) into Low power Personal Area
Networks (LoWPANs). LoWPANs are characterized by low data rates, low power
consumption, low cost, autonomous operations and flexible topologies.

In order to fully realize a pervasive or ubiquitous environment, LoWPANs must be
connected to the Internet Protocol (IP)-based networks. This integration would make
the resource sharing possible within both networks, maximizing the utilization of
available resources. The motivation for IP connectivity, in fact, is manifold. Firstly,
the pervasive nature of IP networks allows the use of existing infrastructure. Secondly, IP based technologies, along with their diagnostics, management and commis-
sioning tools, already exist, and are proven to be working. Thirdly, IP based devices
can more easily be connected to other IP networks, without the need for translation
gateways etc. Internet Engineering Task Force (IETF) [2] is standardizing the transmission of IPv6 over LoWPANs through a working group known as 6LoWPAN [3].
6LoWPANs are envisioned to play a major role in the future pervasive paradigm.

The 6LoWPANs are a suitable choice for an infrastructure to provide ubiquitous
and robust connectivity to users. To keep such networks always operational, robust
network management architecture is needed.

In this paper, we propose a remote configuration mechanism for IP based ubiquitous
sensor networks. The architecture enables the efficient configuration of sensor devices
from any client over the internet. Moreover it defines the information exchange

C.S. Hong et al. (Eds.): APNOMS 2009, LNCS 5787, pp. 423–426, 2009.

mechanisms between the entities. The configuration is done through an acknowledged connectionless service in order to save energy and reduce the memory footprint. The architecture is implemented and tested and it helps in efficient and easy remote configuration.

2 Related Works

A management framework for Heterogeneous Wireless Sensor Networks is proposed in [4]. The framework defines configuration as a component by using a configuration module on the network manager and the sensor devices. However the implementation and evaluation details of the protocol are not available. WSNView [5] is a configuration and management tool for wireless sensor networks. The tool displays the configuration information and configuration objects can be modified with the help of WSNView. However the evaluation details of this protocol are also not available.

3 Configuration Management of IP-USNs

On one hand, 6LoWPANs are IPv6 networks; while on the other hand, these are sensor networks that comprise a large number of nodes with extremely limited resources. Therefore, the management solutions for traditional IP networks cannot be applied directly because of the resource constraints. For example, a 6LoWPAN node may run out of energy causing a fault in the network. This node failure is a 'design feature' of sensor networks as compared to other networks where it is less expected.

As another feature, the applications, when designed for traditional networks may have restrictions in terms of performance and response time as compared to the hardware limitations, when designed for sensor networks. The traditional networks run a diversity of applications as compared to LoWPANs where the network is generally executing a single application in a cooperative fashion.

Furthermore, as an inherent WSN characteristic, 6LoWPAN could possibly be a data centric network which is different than traditional IP network behavior. On the contrary, because of IP support, there is a possibility that LoWPANs support a variety of services making it further complicated for network management operations.

3.1 Characteristics of 6LoWPANs

6LoWPANs are characterized by their inherently small packet size and low bandwidth. The IEEE802.15.4 based physical layer allows the packet size of only 127 bytes and a maximum data rate of 250 kbps.

The devices within 6LoWPANs are considered to be low power with low memory and processing capability. They are deployed in large numbers and tend to be unreliable in operation and availability. The devices may go to sleep mode in order to conserve energy for long periods.

4 Remote Configuration Architecture

In this section we describe the remote configuration architecture. The architecture consists of three components i.e. Sensor server, Authentication server and the Remote Client.

4.1 Sensor Server

Sensor server is application working on IP-USN sensor node. Remote client connects to sensor server and configuration data is exchanged. Sensor server consists of four modules i.e. communication, parsing, core and data gathering module.

4.2 Authentication Server

Authentication for access control in case of remote configuration is a necessary consideration. The remote client which needs to access the sensors needs a password to connect to each sensor node. A key Ku is generated from the password and it is sent with a nonce NA. The nonce provides protection against reply attacks. An adversary can easily replay old messages if the counter were not present in the Message authentication code (MAC). The authentication server generates the MAC through NA and Ku and compares it. Upon reception of an authentication request, the authentication server checks the client's access privileges. If the client is a legitimate with access privileges to the particular sensor nodes, the server gets the allowed list of IP addresses of the client from the gateway and sends the list to the remote client. The communication between authentication server and the sensor server is done with the help of a key Km. The authentication and data freshness in this case is also done with the help of MAC and nonce. At this stage the remote client sends a key establishment request to the authentication server with the list of nodes that it wants to connect. The server generates a session key Krc and shares that with the required nodes and the remote client. The key can be used for authentication of the remote client as well as the encryption of data if required.

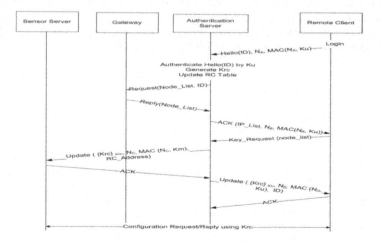

Fig. 1. Authentication Sequence

4.3 Remote Client

The manager can connect to sensor server through remote client. It is an application which can execute on any computer in the internet. It gets the list of available sensors from the authentication server and configures the sensor nodes remotely over the internet. Moreover it performs logging of all the authentication and configuration events. The configuration requests are sent over UDP as unicast packets therefore remote client maintains a reliability timer. Every configuration request is coupled with an acknowledgement. If the acknowledgement is not received then the request is sent again. Upon reception of the configuration request, the sensor server checks the session key and if the client is authenticated then the configuration is successful.

5 Conclusion and Future Works

This paper presents a remote configuration mechanism for IP based Ubiquitous Sensor Networks. We describe the components and the communication interfaces for the remote configuration architecture. Moreover we implemented the proposed architecture. Our future work focuses on the GUI design for such configuration, security and efficiency of the authentication scheme.

References

1. 802.15.4-2003, IEEE Standard, Wireless medium access control and physical layer specifications for low-rate wireless personal area networks (May 2003) Internet Engineering Task Force, http://www.ietf.org/
2. IPv6 over Low Power WPAN Working Group,
   http://www.ietf.org/html-.charters/6LoWPAN-charter.html
3. Ruiz, L.B., Norgueira, J.M.S., Loureiro, A.A.F.: MANNA: Management Architecture for Wireless Sensor Networks. IEEE Communications Magazine 41(2) (February 2003)
4. Chen, J.-L., Lu, H.-F., Lee, M.-Y.: WSNView System for Wireless Sensor Network Management. In: IMSA 2007 (2007)
5. Policy-based Management of Ad-hoc Enterprise Networks. In: HP Openview University Association 9th Annual Workshop, HP-OVUA (June 2002)

Asymmetric DHT Based on Performance of Peers

Kazunori Ueda[1] and Kazuhisa Kawada[2]

[1] Electronic and Photonic Systems Engineering, Kochi University of Technology,
185, Miyanokuchi, Tosayamada, Kami, Kochi, Japan
ueda.kazunori@kochi-tech.ac.jp
[2] NTT Software Corporation

Abstract. Recently, a distributed hash table (DHT) has been used to construct peer-to-peer (P2P) overlay networks. DHT has high scalability for the number of peers, and peers can efficiently retrieve contents from the overlay network. Since DHT algorithms are used to construct symmetric network models, the communication cost of maintaining the routing table is low for peers that have a large computing or network resource and vice versa. We propose an asymmetric DHT network model that can be used to allocate appropriate jobs to a peer according to its performance.

1 Introduction

This logical network is termed "overlay network." Overlay networks such as Gnutella that are constructed by popular P2P applications are not unstructured. On the other hand, a network distributed system termed distributed hash table (DHT) is structured. DHT enables fast search of contents or peers. However, the role of each peer is the same role, irrespective of its performance; in other words, the network is symmetric. Therefore, in some cases, frequently accessed contents are allocated to peers that do not have enough resources such as data storage.

We proposed the basic concept of a method used to construct an asymmetric network on the basis of the performance of peers [1]. In this study, we describe a method used to implement an asymmetric DHT based on Kademlia. To verify the effectiveness of the asymmetric DHT, we carry out simulations and draw a comparison between the number of control packets generated by asymmetric Kademlia and by original Kademlia.

2 Kademlia

Kademlia [2] is a well-known DHT algorithm. An overlay network of Kademlia is constructed using a binary tree. The routing table in Kademlia is defined as k-bucket. The number k of k-bucket represents the capacity of k-bucket. Each peer has multiple k-buckets of each "prefix length" that represents how many first bits are the same between IDs of peers. This "prefix length" can be determined by the "NOT XOR" operation performed between a peer and other peers. For example, the peers from "01100" to "01111" are included into the prefix length "2" k-bucket of the peer "01010," because the result determined by the "NOT XOR" operation performed between the peer

C.S. Hong et al. (Eds.): APNOMS 2009, LNCS 5787, pp. 427–430, 2009.

"01010" and the peers from "01100" to "01111" is "110xx", and this means that the first "two" bits of the peers' IDs are the same.

In general, it is difficult to implement a DHT system because of its complexity. However, since the Kademlia algorithm is very simple, it can be easily implemented. Therefore, Kademlia is used as a method for constructing an overlay network for several P2P file-sharing applications.

3   Asymmetric DHT

3.1   Basic Concept of Asymmetric DHT

Many existing DHT algorithms are used to construct a symmetric overlay network. In other words, in an overlay network, the role of each peer is the same. In addition, in some cases, peers do not have enough resources to manage frequently accessed contents. To solve this problem, we propose the basic concept of constructing an overlay network using an asymmetric DHT algorithm. In the case of an overlay network that is constructed using the asymmetric DHT algorithm, many routing packets are transferred to peers that have sufficient resources. In routing, if possible, peers attempt to transfer routing packets to those peers that have sufficient resources. As a result, the total number of messages used for constructing the overlay network is reduced. A method used for constructing an overlay network that is based on our basic concept of asymmetric DHT takes into account the resources of peers, such as CPU resource, storage resource, and network resource. A high-performance peer is a peer that has sufficient resources, and a low-performance peer is a peer that does not have sufficient resources. We cannot make sweeping statements about the suitable resource of a peer because the resource depends on the characteristics of each P2P application.

In an overlay network constructed using the asymmetric DHT algorithm, the size of the routing table of each peer is not the same. The size of the routing table depends on the distribution of high-performance peers in the overlay network. We define "tablesize" as the number of high-performance peers in the routing table of a low-performance peer. A low-performance peer can add a peer into its routing table unless the number of high-performance peers in the routing table is less than the tablesize. As a result, the routing tables of most low-performance peers will be small if high-performance peers are present in the overlay network. The size of the routing table of a high-performance peer is the same as that of the original DHT. Since the routing table of a low-performance peer is small and incomplete, low-performance peers attempt to transfer requests to high-performance peers by looking up a high-performance peer in its routing table. In an overlay network, the role of high-performance peers is the same as that of a hub peer, and high-performance peers receive more number of requests than low-performance peers. Since a high-performance peer has the same size of routing table as compare to original Kademlia, routing from a source peer to a destination peer is achieved.

3.2   Asymmetric DHT Applied for Kademlia

The basic concept of asymmetric DHT can be applied to most DHT algorithms. To apply the concept to existing DHT algorithms, it is necessary to sort peers into a group

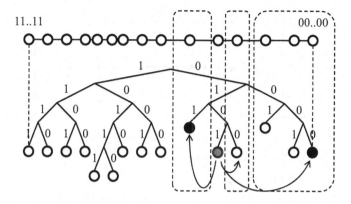

Fig. 1. Routing table in asymmetric Kademlia (tablesize = 2)

that consists of high-performance peers or a group that consists of low-performance peers. The size of the routing table of a low-performance peer also changes. We applied this basic concept of asymmetric DHT to Kademlia, because Kademlia is a simple DHT algorithm and used for the base system of many P2P file-sharing applications. In the following explanations, we describe a system known as "asymmetric Kademlia" that is based on the basic concept of asymmetric DHT for Kademlia.

In the asymmetric Kademlia, the number of peers included in the k-bucket of a low-performance peer depends on the distribution of high-performance peers. Figure 1 shows the k-bucket of a low-performance peer 1010. In Kademlia, while a peer updates the peer list in the k-bucket, it leaves those peers from the k-bucket with priority that have been present in the overlay network for a long time. On the other hand, in asymmetric Kademlia, a peer leaves high-performance peers from the k-bucket with priority. In the case of routing in asymmetric Kademlia, basically, the role of high-performance peers is the same as that of a hub peer, i.e., high-performance peers transfer a query message or a request message. A peer attempts to transfer a message to a high-performance peer that is included in the k-bucket of the peer. High-performance peers transfer the message to other high-performance peers if possible.

4   Performance Evaluation

To verify the efficiency of asymmetric DHT, we compared asymmetric Kademlia with the original Kademlia by carrying out simulations. We used Overlay Weaver as a simulation tool [3]. Simulations were carried out with following scenario: place 10000 peers in a network, connect each peer to the initial peer, random peers put files according to DHT algorithm. (The number of files is 10000.), random peers request and get a random file. The effectiveness of asymmetric Kademlia depends on the distribution of high-performance peers. We carried out simulations to estimate the total number of messages in an overlay network using the asymmetric Kademlia. Valuable parameters are "tablesize" and "p". Tablesize represents the capacity of the k-bucket. The tablesize

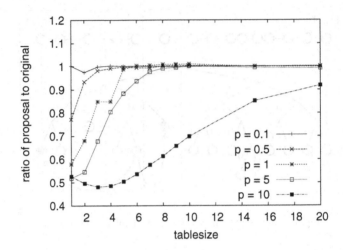

Fig. 2. Ratio of asymmetric Kademlia to original Kademlia (number of messages)

varies from 1 to 20. We change the percentage "p" of high performance peers from 0.1% to 10%. Figure 2 shows the ratio of the number of messages in the original Kademlia to those in the asymmetric Kademlia. In this figure, basically, the number of messages can be reduced by applying our proposed concept for Kademlia. From Figures 2, we conclude that the tablesize of the k-bucket should be approximately set within the range 4-8 so that our basic concept of asymmetric DHT can be applicable without considering the distribution of high-performance peers.

5   Conclusion

We proposed the basic concept of an asymmetric DHT algorithm. This algorithm enables use to construct an overlay network, in which high-performance peers have more prominent roles than low-performance peers, and save control packets for maintenance of routing table. Results of simulation showed that the total number of control packets in overlay networks constructed using the asymmetric Kademlia is less than those in networks constructed using the original Kademlia.

References

1. Kawada, K., Ueda, K.: Asymmetric-DHT based on node capabilities. In: Technical report of IEICE. ICM, March 2009, vol. 108, pp. 69–74 (2009)
2. Maymounkov, P., Mazieres, D.: Kademlia: A Peer-to-peer Information System Based on the XOR Metric. In: Druschel, P., Kaashoek, M.F., Rowstron, A. (eds.) IPTPS 2002. LNCS, vol. 2429, pp. 53–65. Springer, Heidelberg (2002)
3. Shudo, K., Tanaka, Y., Sekiguchi, S.: Overlay Weaver: An overlay construction toolkit. Computer Communications 31, 402–412 (2008)

Load Balance Based on Path Energy and Self-maintenance Routing Protocol in Wireless Sensor Networks

Yao-Pin Tsai[1,2], Ru-Sheng Liu[1], and Jun-Ting Luo[1]

[1] Department of Computer Science and Engineering, Yuan Ze University
[2] Department of Computer and Communication Engineering, Technology and Science Institute of Northern Taiwan
ybchai@tsint.edu.tw, csrobinl@saturn.yzu.edu.tw,
jtl815@gmail.com

Abstract. A wireless sensor network consists of one or more hubs and lots of sensors. Sensors have capability of sensing, wireless communication and data processing. Therefore, how to extend the network lifetime is an important issue. In this paper, we propose a routing protocol which uses the water flow idea and allocates the loading based on the lowest energy in the paths. Our simulation result shows that the approach can extend the lifetime efficiently.

Keywords: load balance; path energy; self-maintenance; routing protocol; wireless sensor networks.

1 Introduction and Related Works

Malfunction and limited-energy are the primary problems of wireless sensor networks. Therefore, how to make the network messages reach their target hubs successfully for network users to take corresponding action is an important issue. In EAR (Efficient and Reliable Routing Protocol) [1], a routing protocol is proposed for a network to transmit messages in more efficient energy-saving manner and to send packets with more reliable communication.

2 LBPS

In this paper, we propose a routing protocol called load balance based on path energy and self-maintenance routing protocol in wireless sensor networks (LBPS in its abbreviated). The LBPS methods exchange of RREQ and RREP packets is replaced with periodical Hello messages and the overall path. The energy is taken into consideration of load control in order to balance the energy consumption of the nodes and to replace the unfair method the energy of next hop is the only consideration. The method can be divided into two main concepts, water flow routing and efficient path energy.

C.S. Hong et al. (Eds.): APNOMS 2009, LNCS 5787, pp. 431–434, 2009.

2.1 Water Flow Routing

Distance is denoted as the number of nodes in a path from a node to a hub. In order to make packets be transmitted follows the principle that Water flows downwards, each node will send a Hello message for a period of time. The message concludes the value of Distance to the hub for constructing routes.

In initial stage, the value of *Distance* of the hub is set as 0 and those of the other sensor node as infinity (∞). When a node receives a Hello message, it will update its own neighbor list table and estimate whether or not it has to update the value of its *Distance*. The node will select a *Distance* with minimum value plus 1 from its neighbor list table and compare it with its current *Distance*. There are two conditions as follows in which *Distance* will be changed.

(1)   Condition 1: If the value of *Distance* plus 1 is smaller than its current one, then update Distance as the smaller value.
(2)   Condition 2: If the value of *Distance* plus 1 is larger than the current one, it means that the route of next hop may lose. The initial stage gets started and *Distance* is set as ∞ to find a new path.

In the duration of two periodic Hello messages, immediate Hello message is used for speed up the response to topology changes. There are two conditions in which Immediate Hello message is sent.

(1)   The value of *Distance* is changed.
(2)   A node receives a Hello message with *Distance* of infinity.

It means that the path of the neighbor node may be lost. If so, flow the principle of condition 2 as above to update the value of *Distance*.

In this paper, neighbor tables will be updated when there is a change in the neighborhood and then the minimum *Distance* of these neighbor nodes is calculated again to generate new *Distance*. If the new value is different from the original one, the value of Distance is updated and an Immediate Hello Message is sent right away.

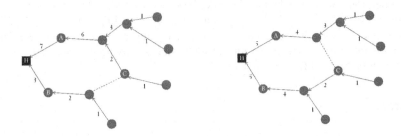

Fig. 1. Load balance of sensor nodes

In the left part of Figure 1, to consider the overall energy consumption of the nodes, numbers near arrows denote their quantity of load. From the load balance point of view, the load of Node A is 7 while the load of node B is only 3. For a long time, the energy consumption of Node A will be much larger than Node B. If the routing path of Node C can be changed under reachable communication range, then the loading of

Node A and B, as the right part of Figure 1, will be the same and the energy consumption can be balanced to improve huge energy consumption issue on some node.

2.2 Load Balance Strategy

An example of load balance based on path energy is shown in Figure 2, in which there are only two paths can be selected and the upstream load comes from K nodes in total. The purpose of this paper is to distribute loading according to path energy from a node in a path to the hub in order to balance the energy consumption of nodes in the network as far as possible. Here, we assume that the path energy is J_m and J_n, respectively. The load of Node m comes from i nodes and that of Node n comes from $K-i$ nodes, where $i : K - i = J_m : J_n$. A node will select a node with minimum *Distance* as its next hop. The loading is distributed based on path energy of all paths available. In Figure 2, the upstream load comes from K nodes. M and N denote percentage of load that is distributed to Node m and n, respectively. Therefore, Hello message includes $\{m : MK, n = NK\}$. After the ratio calculation is done, whenever a node has to transmit a packet, it will generate a uniform random number and determine the next hop according to the random number. In Figure 3, a random number, ranging from 1 to K, is generated. If the random number is locates between M and K, then the packet will be sent to Node m. For a long time, the load distribution will approach ideal balance.

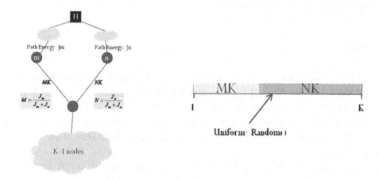

Fig. 2. Percentage of load   **Fig. 3.** Random path selection

3 Simulation and Analysis

The transmission power consumption is calculated by the radio model the same as that in LEACH [2]. At first, we compared the overall energy consumption and path length of EAR and LBPS. Figure 4 shows that average overall energy consumption of LBPS is smaller than EAR.

In Figure 5, the simulation terminates when 50% nodes can't communicate with the hub. As we can see , the packet delivery ratio of LBPS is better than EAR. LBPS utilizes periodic and real-time Hello message to fast response to changes of network topology and update routing tables of nodes at the right moment to keep latest information as far as possible.

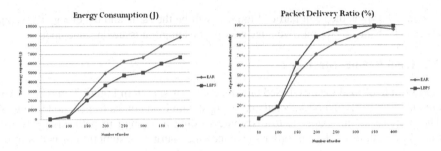

Fig. 4. Comparison of overall energy consumption   **Fig. 5.** Packet delivery ratio

4   Conclusion

In LBPS, if a node has more than one path with the same path length, the path is selected according to path energy in the manner of probability. For a long time, traffic
load of each path can be almost distributed much more balanced based on the percentage of path energy. In the future, the path can be selected according to energy
percentage of each path. This method is not in the manner of probability but the path
is determined by upstream nodes directly. By this method, the packet route and load
balance can be control more precisely.

References

1. Loh, P.K.K., Long, S.H., Pan, Y.: An Efficient and Reliable Routing Protocol for Wireless
   Sensor Networks. In: Sixth IEEE International Symposium on a World of Wireless Mobile
   and Multimedia Networks (WoWMoM 2005) (2005)
2. Heinzelman, W., Chandrakasan, A., Balakrishnan, H.: Energy-Efficient Communication
   Protocol for Wireless Microsensor Networks. In: Proc. 33rd Hawaii Int'l. Conf. Sys. Sci.
   (January 2000)
3. Muruganathan, S.D., Ma, D.C.F., Bhasin, R.I., Fapojuwo, A.O.: A centralized energy-
   efficient routing protocol for wireless sensor networks. IEEE Communications Magazine,
   8–13 (March 2005)
4. Brownfield, M.I., Mehrjoo, K., Fayez, A.S., Davis IV, N.J.: Wireless sensor network en-
   ergy-adaptive mac protocol. In: IEEE 3rd CCNC, January 2006, pp. 778–782 (2006)
5. Zhang, W., Jia, X., Huang, C., Yang, Y.: Energy-aware location-aided multicast routing in
   sensor networks. In: IEEE Conference On Wireless Communication, Networking and Mobile Computing, September 2005, pp. 901–904 (2005)

Performance Evaluation of VoIP QoE Monitoring Using RTCP XR

Masataka Masuda[1], Kodai Yamamoto[2], Souhei Majima[1], Takanori Hayashi[1], and Konosuke Kawashima[3]

[1] NTT Service Integration Laboratories
[2] NTT Network Service System Laboratories
NTT Corporation,
3–9–11 Midori-cho Musashino-shi, Tokyo 180-8585, Japan
[3] Tokyo University of Agriculture and Technology
2–24–16 Naka-cho Koganei-shi, Tokyo 184-8588, Japan

Abstract. In next generation networks (NGNs), not only bearer quality management but also quality of experience (QoE) management is needed. One means of QoE monitoring is IETF RFC3611 RTCP XR which is a protocol for reporting the quality information measured at VoIP terminals. In our previous study, we propose a QoE monitoring method of VoIP using the XR. However, XR is not a fully mature technology, and some report parameters are vaguely defined to implement on a VoIP terminal. We focus on the XR parameters of network factors, loss rate, round trip delay, and jitter. We clarified the applicability to QoE monitoring of these XR parameters based on experimental results.

Keywords: RTCP XR, Quality, QoE, VoIP, Monitoring.

1 Introduction

The primary features of next generation networks (NGNs) are QoS control for networks, and public switched telephone network replacement support for VoIP networks. In conventional networks, the bearer quality is managed. In NGNs, it is also essential for the end user to manage Quality of experience (QoE) [1]. The differences in VoIP terminals' implementations strongly affect the QoE. Therefore, it is essential to monitor the QoE in networks on a call-by-call basis [2], and a quality measurement function should be implemented at VoIP terminals.

QoE monitoring technologies that are embedded in VoIP terminals are being studied with keen interest [3] One of these technologies is IETF RFC3611 RTCP XR, which is a protocol for reporting the quality of information measured at VoIP terminals. We proposed a QoE monitoring method of VoIP by using XR, and clarified the parameters which are needed for monitoring QoE [4]. Some XR parameters are vaguely defined to implement on a VoIP terminal. The definitions of report parameters have not been extensively evaluated in terms of their applicability to QoE monitoring. We evaluate the applicability to QoE monitoring using XR parameters of network factors; loss rate, round trip delay, and jitter.

C.S. Hong et al. (Eds.): APNOMS 2009, LNCS 5787, pp. 435–439, 2009.

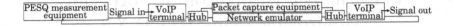

Fig. 1. Configuration of VoIP system

2   QoE Monitoring Technology of VoIP

QoE monitoring technologies that are embedded in VoIP terminals are standardized in RFC3611 XR. XR is an extended protocol of RTCP for reporting the quality information measured at VoIP terminals. RFC3611 defines the packet format and many report parameters, but the applicability of the report is not defined. In our previous study, We proposed QoE monitoring method using XR [4]. The scenario represents the case where the operator of a carrier network responds to a user complaint. The operator can monitor the current QoE from end user terminals, identify factors causing degradation of QoE.

To enable the use of this QoE monitoring method, we focus on network and terminal factors in various report parameters of XR. However, the report parameters are currently implemented on VoIP terminals using various definitions, for instance, RFC3611 does not clearly define the report parameters.

3   Definition of Network Parameters

RFC3611 defines the report format of network factors; loss rate, round trip delay (RTD), and jitter. The loss rate is the fraction of RTP data packets from the source lost. This value is calculated by dividing the total number of packets lost by the total number of packets expected, multiplying the quotient by 256, and taking the integer part. The RTD is the round trip time between RTP interfaces, expressed in milliseconds. This value is measured using RTCP, and the time of receipt of the most recent RTCP packet, minus the last sender report (LSR) time reported in its SR, minus the delay since last SR (DLSR) reported in its SR. The jitter is the transit time between two packets "above" the sequence number interval. All jitter values are measured as the difference between a packet's RTP timestamp and the reporter's clock at the time of arrival, measured in the same units. RFC3611 does not clearly show this above sequence number and the jitter calculation equation, and the jitter packet format must be a positive number, but a negative number may be calculated in the above definition.

4   Performance Evaluation

4.1   Experimental Conditions

We conducted a performance assessment experiment to demonstrate the effectiveness of QoE monitoring using XR. We evaluate two commercial VoIP terminals equipped with XR. The configuration for the VoIP system under test is outlined in Fig. 1. The SR and XR are sent together as a compound packet, and the transmission interval is 5 seconds, which is recommended for a fixed minimum interval. The reporting range of an XR packet is implemented as follows:

- **Terminal A:** The RTP packets received at the interval between receptions of the XR packets (interval),
- **Terminal B:** The RTP packets received at the interval between the beginning of the RTP session and the transmitting of the XR packet (accumulate).

The network conditions are emulated by an network emulator, and the number of conditions (combinations of packet-loss rate (0–10%), delay (10–150 ms) and jitter (0–20 ms)) was 54. We repeatedly fed speech samples with a total time of 120 seconds into the system under all testing conditions. We evaluated the performance of QoE monitoring using XR (the XR values) and compared them with the actual values.

4.2 Performance Requirement

To evaluate the performance of QoE monitoring, we used the root-mean-square error (RMSE), which was calculated using the difference between the XR values and the actual values. We set the performance requirement as follows: "The RMSE in estimating quality should be less than or equal to the mean of the 95% confidence intervals in subjective speech quality assessment." This means that a QoE monitoring algorithm accurately estimates subjective quality comparable with the statistical ambiguity of subjective assessment results. In our previous study, we found that the mean for the 95% confidence interval in subjective quality assessment under similar testing conditions was about 0.29 [4]. When this was mapped onto the R-scale, it became about 11.5 at minimum. The R value is a quality evaluation metric, and calculated using ITU-T Recommendation G.107 "the E-model" [5]. The loss rate and the RTD can be mapped onto R-scale.

The jitter cannot be mapped onto R-scale. Therefore we set the performance requirement, "The degradation of listening quality is discoverd from the jitter." In this experiment, we discoverd the degradation of listening quality by using perceptual evaluation of speech quality (PESQ) [6]. In the test condition at maximum jitter, which does not affect degradation of the PESQ value, the jitter value is 1.144 ms (average), and 1.720 ms (maximum). In the test condition at minimum jitter, which affects degradation of the PESQ value, the jitter value is 1.837 ms (average), and 2.466 ms (maximum). The value of difference between the previous two conditions is the boundary. Therefore, we set the performance requirement as follows: "The RMSE value should be less than or equal to 0.689 ms (the mean jitter) or 0.746 ms (the maximum jitter)."

4.3 Experimental Results

To evaluate the performance of loss-rate measurement using integer mapping, we compared the XR value with the actual value. The actual value is the loss rate, which is measured using RTP packet capture. Figures 2 and 3 plot the

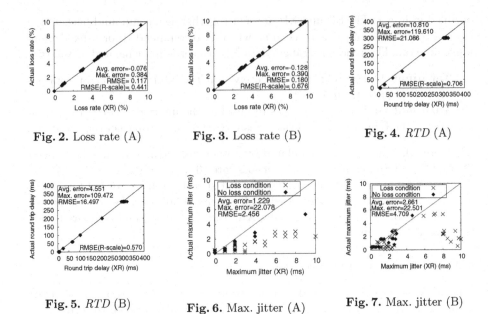

Fig. 2. Loss rate (A)   **Fig. 3.** Loss rate (B)   **Fig. 4.** *RTD* (A)

Fig. 5. *RTD* (B)   **Fig. 6.** Max. jitter (A)   **Fig. 7.** Max. jitter (B)

performance of loss-rate measurement using XR, and show the RMSE. A plot signifies the average loss rate under testing conditions. The measurement error is calculated, the XR value minus the actual value. Figures 2 and 3 show that both terminals satisfy the requirement, i.e., the RMSE was 0.441 and 0.676, respectively. The XR values are less than the actual values, because the decimals are dolloped when the measured loss rate maps onto the integer.

The XR value of *RTD* is measured using RTCP, but the transmission delay of RTP packets must be measured to monitor the QoE. Therefore, we evaluate the performance of the estimation of the RTP transmission delay using the XR value. We define the actual *RTD* which is calculated by the average of one-way delay of RTP packets plus the average of one-way delay of RTP packets which are transmitted in reverse direction of the same time. The RTP packets are the same target packets for measurement as those of XR. Figures 4 and 5 plots the performance of *RTD* measurement by XR. Figures 4 and 5 show that both terminals satisfy the requirement, i.e., the RMSE was 0.706 and 0.570, respectively. In the test conditions which is large delay variation, the maximum estimation error was large depending on the measurement method using RTCP.

We define jitter as the relative delay between two packets always above sequence number one to use under various conditions: packet loss and packet reordering. If there is no packet above sequence number one, the packet is excluded from the jitter calculation. The jitter packet format in XR must take a positive number, therefore the actual jitter is an absolute value and calculated by

$$J_n = |(t_{b(n)} - t_{a(n)}) - (t_{b(n-1)} - t_{a(n-1)})|, \tag{1}$$

J_n where is the jitter of nth packet, $t_{a(n)}$ is the transmitting-time of nth packet, $t_{b(n)}$ is the receiving-time of nth packet.

The mean jitter measurement by the both terminals satisfies the requirement, i.e., the RMSEs were 0.395 and 0.086. The measurement error was less than 1 ms. Figs. 6 and 7 plot the performance of the maximum jitter measurement with XR. The plot symbols are classified based on the testing conditions, which does or does not consist of the lost packets. The maximum jitter measurement by the both terminals does not meet the requirement, i.e., the RMSEs were 2.456 and 4.709. Therefore, the degradation of listening quality is not detected using XR value of the maximum jitter. Figures 6 and 7 show that the measurement error is large in the test conditions, including the packet loss.

5   Conclusion

We evaluated the performance of QoE monitoring by using XR. We will study QoE monitoring performance by using the terminal factors in the future.

References

1. ITU-T Rec. P.10/G.100 Appendix 1, Definition of QoE (January 2007)
2. ETSI TIPHON TS101 329-5 Annex E, Method for determining an equipment impairment factor using passive monitoring (November 2000)
3. IETF RFC3611, RTP control protocol extended reports (RTCP XR) (November 2003)
4. Masuda, M., et al.: End-to-end quality management method for VoIP speech using RTCP XR. IEICE Trans. on Communications E90-B(11)(November 2007)
5. ITU-T Rec. G.107, The E-model (March 2005)
6. ITU-T Rec. P.862, Perceptual evaluation of speech quality (PESQ) (February 2001)

Research on Service Abstract and Integration Method in Heterogeneous Network Platform

Fei Ma[1,2], Wen-an Zhou[1], Jun-de Song[1], and Guang-xian Xu[2]

[1] Beijing University of Posts and Telecommunications, Beijing.100876, China
[2] Liaoning Technical University, Huludao.125105, China
mafee2007@gmail.com, zhouwa@bupt.edu.cn, jdsong@bupt.edu.cn

Abstract. With the development of information and communication technologies, network convergence is inevitable trend. Because there are many heterogeneous networks in our daily lives, the heterogeneous network platform should provide the universal interfaces, regardless of the diversity of heterogeneous networks. A service abstract and integration method is proposed with corresponding interfaces in the paper. According to the application requirements, the platform can gain series of common service capability components from bottom heterogeneous networks by the abstract method. These service capabilities can be encapsulated to provide open interfaces with web service. The common services are also integrated to support new services or applications by the integration method. An illustration shows that the method is feasible on the platform to some extent.

Keywords: service abstract method; service integration; common service; heterogeneous network platform.

1 Introduction

In order to support the applications of modern service industry including e-commerce, digital medical, tele-education etc, the heterogeneous network platform should provide the universal interfaces, regardless of the diversity of heterogeneous networks. In reference [1], three approaches including SIP Servlet, Java and BPEL were used to build a composite service with analysis on developer experience. In [2], a new value-added service creation pattern for convergence network, capability injection was presented. In [3], the author summarized the applications of web services in Parlay family and focused on interaction mechanism of Parlay API based on web services on the basis of analysis of these technologies. In [4], a non-functional feature based service capability feature (SCF) discovery mechanism was proposed to meet the requirements of the client applications and to discover the SCF.

We can see from the researches above, Parlay X is based on telecom networks not including Internet networks and cannot satisfy the application requirements of modern service industry. In this paper, the heterogeneous network platform extends the applications of parlay X web service over Internet domain, which is composed of telecom and Internet networks.

C.S. Hong et al. (Eds.): APNOMS 2009, LNCS 5787, pp. 440–443, 2009.

2 System Platform Architecture

In this section, adaptation layer is proposed to provide open interfaces regardless of diversity of heterogeneous networks. As the intermediate layer between heterogeneous networks and common service integration layer, the network adaptation layer is composed of basic service sub-layer and adaptation service sub-layer, which is described in Fig.1. In basic service sub-layer, common service capabilities are abstracted and encapsulated with standard interfaces, which are easy to be controlled. Thus, it can guarantee the openness of network capabilities for the development of modern service industry. In adaptation service sub-layer, in order to adapt service capacities to network resource distribution, it is divided into three planes including management plane with functions of authentication and user capability management, control plane with functions of intelligent handoff and capability adaptation, and data plane with functions of load balancing and protocol conversion.

Service integration layer connects network adaptation layer with corresponding interfaces, which involves service integration unit, E-payment sub-platform and interactive service sub-platform. It is used to support the applications by integrating the common service from bottom network.

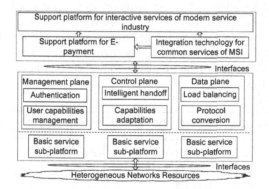

Fig. 1. The architecture of platform

3 Service Abstract and Integration Method

In this paper, the service abstract method adopts a combination way, from the top down and from the bottom up. At first, these upper-layer applications are researched and analyzed to get the common requirements for network capability. And then the common requirements are mapped into lower layers. Based on these requirements, network adaptation layer of this platform abstracts the network common service capabilities from bottom-layer heterogeneous networks such as call control, user state/terminal information, connectivity management, account management and charging capability etc. The way from the top down is used to specify the requirements to get the common service capabilities. Thereafter, these common service capabilities are encapsulated into a set of sub-platforms in light of the common requirements including SMS sub-platform, soft switching sub-platform and soft switching sub-platform etc.

And the service integration layer may take further integration for common service capabilities and sub-platforms. The way form the bottom up is to integrate common services for upper applications. The "black box" method is adopted in definition of these sub-platforms, which provides interfaces for the users regardless of internal execution.

After common service capabilities are researched and analyzed from heterogeneous networks, standard interfaces are defined in form of software components with formal description language. According to the requirements, there are some new interfaces on basis of parlay X web service for instance voice call interface with capability adaptation functions. After basic service sub-platforms and interfaces are defined, they are encapsulated with Web Service and issued over adaptation layer. When the application needs the sub-platform, the service is bound and called after lookup.

With the development of technologies, new value-added services emerge in great numbers. Service abstract method can extract the basic common service capabilities from existing services to generate the service units. When a user needs a new service or value-added service, it can be assembled with service integration method according to user requirement description. The process of service integration is described in Fig. 2. At first, the service providers issue the services and register them in service register server. When a user has the service request, the description document of service requirements should be submitted to the common service integrated system. And the integrated system queries the services in service registration server and receives the return information in step 4 and 5. The integration system will integrate the common service and define the service interfaces. Furthermore, it binds the service and registers the integrated service into the registration server as a new service and finally returns the results to the user.

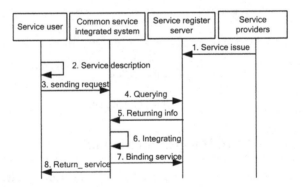

Fig. 2. Common services integration process

4 Service Illustration

With the method and interfaces above, an illustration is given under the circumstance of network adaptation layer and multi-terminals, which is described in Fig.3. Firstly, the application server periodically gains the state information of users and terminals in step 1-4. Assume that the terminal 2 of user A initiates a call to user B by the network adaptation server. The server automatically selects the optimal terminal 2 for user B, and the connection is established in step 7. When the terminal 2 of user B disconnects for some malfunction reason, the server will adapt another terminal to keep the service continuity in step 8-10. Similarly, the terminal 2 of user A may be switched to the terminal 1 in step 11-13. Thus, the terminal 1 of user A is connected with the terminal 1 of user B for call process. The adaptation server receives the hang-up message from user A, and then it will notify user B and terminate the connection in the end.

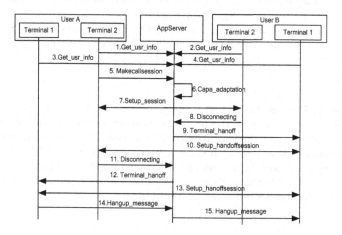

Fig. 3. Service interfaces illustration

5 Conclusions

In order to support the applications of modern service industry, this paper has presented the architecture of heterogeneous network platform based on adaptation layer. A service abstract method is proposed to get a set of SCF components and sub-platforms, which can be integrated to support new services by service integration method. A definition method of service interfaces is given in form of web service. At last, service application is given as an illustration, which indicates the feasibility of the service abstract method and corresponding interfaces.

References

1. Yuan, Y., Wen, J.J., Li, W., Zhang, B.B.: A Comparison of Three Programming Models for Telecom Service Composition. In: The Third Advanced International Conference on AICT 2007, May 13-19, pp. 1–10 (2007)
2. LI, J.-l., Sen, S., Yang, F.-c.: A Converged Network Based Value-Added Service Creation pattern. Journal of Beijing University of Posts and Telecommunications 32(suppl.), 35–39 (2009)
3. Yang, X., Su, S., Chen, J.: Research on Parlay API based in web services. Modern Science & Technology of Telecommunications 3, 26–28 (2005)
4. Peng, J., Yang, F.-c.: Non-functional feature based Parlay SCF discovery mechanism. Journal on Communications 27(12) (December 2006)

Security and Reliability Design of Olympic Games Network

Zimu Li[1], Xing Li[1], Jianping Wu[1], and Haibin Wang[2]

[1] Network Research Center, Tsinghua University, 100084 Beijing, China
{lzm,xing,jianping}@cernet.edu.cn
[2] Technology Department, The Beijing Organizing Committee for the
Games of the XXIX Olympiad, 100083 Beijing, China
wanghaibin@beijing2008.cn

Abstract. High security and reliability is the essential design principle of the Olympic Games Network. This paper depicts the organization and its functionality of security and reliability design in 29th Olympic Games Network, such as architecture, 2-level DMZ, A/B systems, IP encoding and routing/switching. In addition, it describes the network optimization to meet the special requirements of Beijing MSTP MAN. In this paper, some experience of realizing a high security and reliability network is provided for reference.

1 Introduction

Games Network is the basic system of Olympic Games results distribution, information publication, media transmission and games management. As a critical system, high reliability and security of Games Network is the essential requirement for a successful Olympic Games. This paper gives an overview of the reliability and security design of 29th Beijing Olympic Games Network.

2 Network Architecture

29th Olympic Games Network is composed of more than 50 venues around Beijing, Tianjin, Qingdao, Shanghai, Shenyang, Qin Huangdao and HongKong. Many critical applications, such as Timing & Scoring, Results Diffusion System (IDS), Commentator Information System (CIS), Games Management System (GMS) and etc are running on it. The architecture is shown in Fig. 1.

In Fig.1, it shows that Games Network is typical 2-core architecture that two core switches are deployed in PDC (Primary Data Center) and SDC (Secondary Data Center, a total backup of PDC) respectively and connected by three high-speed Ethernet channels. Except few none-competition venues, most venues connect PDC and SDC through different geographic leased lines. In most venues, two aggregation switches are deployed and backed up each other. All venue access switches connect with aggregation switches by two uplinks. In venues, each server has two NICs connecting with two different access switches by "teaming". Therefore, from servers to core

C.S. Hong et al. (Eds.): APNOMS 2009, LNCS 5787, pp. 444–447, 2009.
© Springer-Verlag Berlin Heidelberg 2009

switches, each level of the network has two switches and two uplinks respectively. Any single device failure would not affect the Games.

Fig. 1 shows that the under layer transmission system is composed of ADM (Add/Drop Multiplexer) device and MSTP (Multiple Service Transportation Platform) which is the most popular MAN technology in recent years. MSTP is compatible with SDH and provides multiple types of Ethernet interface that can be connected with core switches in PDC and SDC directly without any additional device. By this way, it looks like all venues LANs are connected together with a virtual switch and thus complexity of management is reduced.

3 Security Architecture and 2-Level DMZ

Security architecture is in accordance with network architecture except firewalls are highlighted. Fig.2 gives an overview of Games Network Security Architecture.

In Fig.2, all venues without external leased line connect each other by MSTP ring. As MPC (Media and Press Center) and IBC (International Broadcast Center) provide service mainly to journalists that need communication with their home news agencies by leased lines and VPN, MPC and IBC are protected by firewalls so that only the permitted VPN traffic are allowed to get out to the pre-determined destinations.

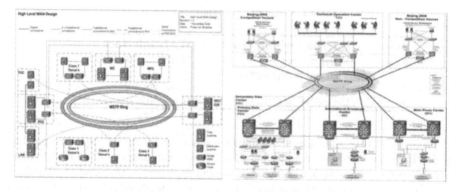

Fig. 1. Games Network architecture   **Fig. 2.** Network Security Architecture

To ease management and improve physical security level, all global critical servers of Olympic Games are deployed in PDC/SDC. Some of the servers only provide service to internal clients and some others provide service to Internet clients. Hence, a 2-level DMZ (Demilitarized Zone) is constructed to protect these servers in PDC/SDC. In DMZ, from inside to outside direction, the 1st level firewall ensure servers in Internal DMZ can only be accessed by internal clients whereas Internet clients can access only servers in External DMZ by the 2nd level firewall. In addition, the 2-level firewall guarantees that data can only be transmitted from internal servers to external servers. The 2-level firewall architecture and single direction traffic filtering prevent network attacks from both outside and inside at the utmost level and gives the servers maximum protection.

4 A/B Redundant System

Redundant system based on hot-standby technology is an effective measurement to improve reliability and availability. From Network Architecture, we can see that most venues have two aggregation switches, A switch and B switch, which backed up each other. The two aggregation switches are connected by PortChannel and uplinking to core switches in PDC and SDC by separate fiber routes. Within venues, all access switches connect with A switch and B switch respectively.

To minimize modification of client's configuration, HSRP is configured on A switch and B switch so that client can access network by the virtual gateway of HSRP even if any single switch or link is down. Thus, by A/B redundant system, high reliability and availability achieved.

5 IP Address Encoding

IP address encoding is an important component of network design. As Olympic Games Network is an isolated network, it is possible to encoding the network from the top down according to IPv4 address format like "A.B.C.D". Table 1 gives the encoding scheme within venues and the WAN interface address encoding scheme.

Table 1. Venue and wan interface IP encoding

Byte	Interface within venue	WAN Interface
A	Venue number	100(PDC), 200(SDC)
B	Vlan number/application type	254
C	Device type	Venue number
D	Device No. of C byte	1(PDC/SDC), 2(venue)

From Table 1, IP addresses are defined by venues and applications (corresponding with VLAN number) restrictively. It is such an easy thing from an IP address to identify the corresponding switch, venue, VLAN and application. This kind of IP address encoding makes the network structure very clear and simplified network management and routing policy greatly.

6 Routing and Switching

The routing & switching of Games Network is composed of layer-3 OSPF routing and layer-2 VLAN switching. Core switches in PDC/SDC and aggregation switches in venues run OSPF routing protocol as Area 0. Aggregation switches and access switches in venues run VLAN protocol and each venue belongs to an OSPF Total Stub Area. All servers in venue forward packets to the virtual gateway which has been configured on both aggregation switches. On the whole, Games Network has a simple routing and switching system so that reliability and security can be achieved without much trouble. Nevertheless, 29th Olympic Games Network indeed has some special requirements because of the MSTP MAN.

6.1 Physical Link Failure Detection

For that ADM is the under layer MSTP transportation device of Games Network and all venue switches connect with ADM by Ethernet interfaces, it is hard to detect immediately that there is a link failure. To avoid this routing problem, Fast Hello is used to detect the transportation link failure as soon as possible. Every other 1 second, one Hello packet will be sent out to his peer OSPF neighbor to see if there is a link failure occurred. By this way, Games Network solves the MSTP link failure-detection problem almost perfectly.

6.2 Fast Convergence and Loop Free

For Games Network, fast convergence and real time responding are essential requirements for results publication and news transmission. At the same time, routing paths should be considered carefully because there are so many redundant devices and links exist and loops may exist. So many technologies such as Rapid-PVST, Root Guard, Port Fast, BPDU Guard and Loop Guard as well as Fast Hello are used to solve this problem.

7 Conclusion

29th Olympic Games Network is a large-scale network which is composed of more than 1100 network devices and 50+ venues in 7 cities around China. More than 400 engineers work together to guarantee the network operation. Besides the technologies this paper describes in above chapters, many other technologies such as storage backup, remote monitoring, authentication and authorization, logging alarm and so on are also deployed in Games Network. All of these technologies and experienced engineers together provide the assurance of the "truly exceptional Games" in history.

Acknowledgments. Thanks Atos Origin for the Olympic Games IT operation and the technical documents.

References

1. Marc, G.: OG08STECSAR00121-SFA-Master Systems Architecture, Atos Origin, 200728210 (2007)
2. Pieter, J.: OG08STECSAR00412-SFA-Network Architecture, Atos Origin, 2007211212 (2007)
3. Gao, C.: OG08SSECSAR00220-APP-Games Security Architecture, Atos Origin, 20071228 (2007)
4. Pieter, J.: OG08STECNET00117-SFR-Games IP Addressing Scheme, Atos Origin, 20082527 (2008)
5. Pieter, J.: OG08STECREQ00315-SFA-Beijing 2008 Venue Classification and WAN Bandwidth Requirements, Atos Origin, 20080507 (2008)

An Implementation of the SDP Using Common Service Enablers

Jae-Hyoung Cho and Jae-Oh Lee

Information Telecommunication Lab,
Dept. of Electrical and Electronics Engineering,
Korea University of Technology and Education, Korea
{tlsdl2,jolee}@kut.ac.kr

Abstract. The structures and systems of network and communication are independently constructed with existing based-IT service and 3G services without considering common functions between them. Consequently, the study of standardized Service Delivery Platform (SDP) structure and its related interfaces are needed to provide the methodology of existing or new services in an efficient way. Therefore, in this paper, we suggest the structure of common service enablers (i.e., PoC, Presence) for the SDP and show the way of interaction between them. Finally, we will describe an implementation of the components of the SDP using them.

1 Introduction

The IMS is an architectural framework for delivering Internet Protocol (IP) multimedia to mobile users. As it is such a framework that provides access to the content of Internet and mobile services anywhere and anytime with guaranteed Quality of Service (QoS) and manageability. Currently, in network business and communication, structures and systems are independently constructed with existing IT services and 3G services without considering common functions between them. In order to solve these problems, it is necessary that the study of the SDP provides the methodology of existing or new services to users and easily develops to service providers. This anticipant platform enables the IP Multimedia Subsystem (IMS) structure to provide the control of service consistently and integrate wired and wireless network to meet those demands. The IMS calls for the creation of the SDP that can develop new service based on existing service of IMS with valued additional services [1].

The SDP simplifies service interactions through service providers and prevents the overlap of resources through common enablers. Because of these capacities, the SDP can provide new services and reduce processes of service development and complications. Also the SDP extends the IMS service layer with containing functions such as third-party service registration, device management, network service, network gateway and so on. In this paper, we define that whole of enablers is the SDP and will suggest the structures of PoC and Presence Service for the SDP/IMS. In section 2, we will descript the structure of PoC and Presence service, and then in section 3, we will show prototyped SDP that is composed of two common service enablers (PoC and Presence enablers).

C.S. Hong et al. (Eds.): APNOMS 2009, LNCS 5787, pp. 448–452, 2009.

2 Related Works

PoC is a walkie-talkie type of service. Users press (and hold) a button when they want to say something, but they do not start speaking until their terminal tells them to do so. There are several incompatible PoC specifications at present. Most of them are not based on the IMS, but consist of proprietary solutions implemented by a single vendor. As a result these PoC solutions generally cannot interoperate with equipment from other vendors. This situation prompted Open Mobile Alliance (OMA) to create the PoC working group to start working on the OMA PoC service. OMA's PoC service is based on the IMS [2]. The reason for PoC service deployment in the IMS is that PoC server itself does not need to handle such as SIP signaling routing, discovery and address resolution services, Session Initiation Protocol (SIP) compression, Authentication Authorization Accounting (AAA), QoS control, and so on [3].

Presence is one of basic services that are likely to become omnipresent. On the one hand, the presence service is able to provide an extensive customized amount of information to a set of users. On the other hand, third-party services are able to read and understand presence information, so that the service provided to the user is modified (customized) according to the user's needs and preference expressed in the presence information [4]. Presence is the service that allows a user to be informed about the reachability, availability, and willingness of communication of another user. The presence service is able to indicate whether other users are online or not and if they are online, whether they are idle or busy (e.g., attending a meeting or engaged in a phone call). Additionally, the presence service allows users to give details of their communication means and capabilities (e.g., whether they have audio, video, instant messaging, etc. capabilities and in which terminal those capabilities are present).

3 An Implementation of Prototyped SDP Model

The figure 1 describes the deployment of enablers. Enabler is located between Application Service and IMS-Core. In this structure, Presence Enabler includes to function of PoC server (aggregation proxy function, XDMS function). Presence Enabler (PE) includes aggregation proxy function in PoC. XDMS function is included to PE that replaces XDMS with PIDF [5].

Figure 2 describes that application services are used common service enablers two cases. Case 1 creates services using orchestration between common service enablers. Intra-bus which is composed of the SDP between service enablers is SOA/WS and inter-bus is used to SIP. Orchestration executes about service orchestration and interaction. Service Orchestration is used to create fast service in IP domain and service interaction controls about events of signaling path in the network. This way is implemented by SDP SOA. This way needs modeling tool that is able to compose a useful service. Between service enabler orchestration is used Business Process Engineering Language (BPEL) and enterprise [6]. BPEL supports two different types of business processes. Firstly, executable processes allow us to specify the exact details of business processes. They can be executed by an orchestration engine. In most cases BPEL is used for executable processes. Secondly, abstract business protocols allow us to specify the public message exchange among common service enablers. They do not

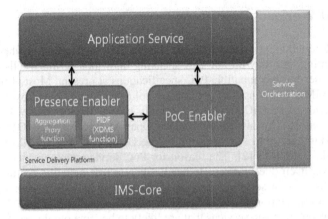

Fig. 1. Overview of Structure between the SDP and Application Service

include the internal details of process flows and are not executable. BPEL is used to SOA/WS technology in between inner service enabler or between other service enabler interfaces. The protocol is used to bus structure implement to abstract access on destination. Case 2 creates application service that calls to want enablers using SIP. If service provider or user want to use common service enablers (PoC or Presence enabler), they is able to use through pushing button. This case does not consider orchestration, but reuses common service enablers.

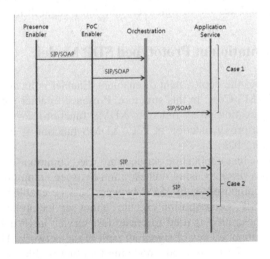

Fig. 2. Two Cases of Using Common Service Enablers

The figure 3 represents the structures of presence and PoC enablers in detail. When SIP message accesses from the IMS, the roles of *Presence Server (PS)* accomplish to register presence information and upload registering information to *ResourceList*

Pidfparser defines related to information (e.g., presence, location and status of device, user information) of user and entities using Presence Information Data Format (PIDF). *ResourceList* is implemented possible to store resource information of user and entities. Later on, it will be extended to interact with the Home Subscriber Server (HSS) for AAA and QoS. If the user that sends SUBSCRIBE message or PUBLISH message for registration in *ResourceList* is existence in *ResourceList*, PM re-uploads partly modified presence information or creates PIDF document about presentity. Presence Manager (PM) makes suitable SIP messages to process SIP messages from PS [7].

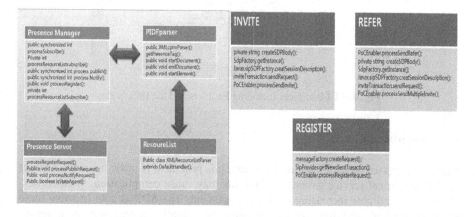

Fig. 3. The Structures of Presence and PoC Enabler in Detail

PoC process three kinds of SIP messages (REGISTER, REFER, INVITE message). INVITE creates "*sdpbody*" and initializes to dialog for furthering run the media session between users. After creating "*sdpbody*", using "*inviteTranscation.sendRequest( );*", PoCE sends Invite message to the user. So the user presses call button to send an INVITE request to other users and waits for acceptance of the callee(s).

REFER is used to add another users. In our implementation, is used as multiple-REFER. Since multiple REFER are typically used within the context of an application, the user generating it has generally an application-specific means to discover the result of the transactions initiated by the enabler. Consequently, multiple-REFER is normally combined with an extension that eliminates the implicit subscription that is usually linked to REFER.

The figure 4 describes implemented IMS nodes (P-CSCF, S-CSCF, I-CSCF), presence enabler and PoC enabler. The S-CSCF contains user's addresses as well as presence enabler and PoC enabler. Presence information was implemented by XML-based as PIDF, RIDF and is able to extend for including more presence information. This implementation is common service enablers (e.g., presence, PoC) through unify common resource and was used to messenger service.

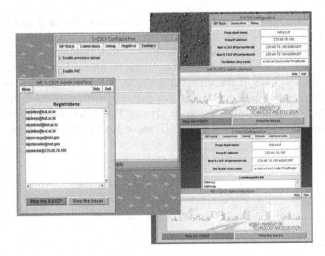

Fig. 4. GUI Interfaces Including Enablers and the IMS Nodes

4 Conclusions

Currently, service development method, which is provided, is made up of the form of separated vertical service structures. And this may cause complication, intensiveness of resource and cost increasing of the management. The SDP is able to solve these problems through unifying common resource. By simplifying the interaction among enablers, the SDP reasonably provides new innovative services for the market and makes service management problem easier by reducing the complication of service development. In this article, we suggest structure of enabler and prototyped SDP. In the future, our enablers need to develop including various user information (to extend PIDF, RIDF) and we will develop another enablers that are able to include another common resource (for multimedia service).

References

1. Cho, J.-H., Lee, J.-O.: Deploying application services using service delivery platform (SDP). In: Ata, S., Hong, C.S. (eds.) APNOMS 2007. LNCS, vol. 4773, pp. 575–578. Springer, Heidelberg (2007)
2. OMA, PoC-V2_0, (August 2008)
3. RFC 3859, A Presence Event Package for Session Initiation Protocol (SIP) (August. 2004)
4. 3GPP TS 23.141 v7.2.0, Presence Service; Architecture and functional description(Release 7) (September 2006)
5. OMA, Presence_SIMPLE-V1_1 (January 2008)
6. Adams, R., Boisseau, M., Bowater, R.: Service Orchestration Implementation Options. In: IBM 2004 (2004)
7. Cho, J.-H., Lee, J.-O.: The IMS/SDP structure and implementation of presence service. In: Ma, Y., Choi, D., Ata, S. (eds.) APNOMS 2008. LNCS, vol. 5297, pp. 560–564. Springer, Heidelberg (2008)

Decreasing Access Control List Processed in Hardware

Takumichi Ishikawa and Noriaki Yoshiura

Department of Information and Computer Science, Saitama University
255, Shimo-ookubo, Sakura-ku, Saitama City, Saitama Prefecture, Japan
Tel.:+81-48-858-3498, Fax:+81-48-858-3498
ishikawa@fmx.ics.saitama-u.ac.jp, yoshiura@fmx.ics.saitama-u.ac.jp

Abstract. Access control list (ACL) is one of the most important things in computer network security. While cheap router or PC processes ACL by software, network equipment such as Layer 2 or 3 switch processes ACL by hardware because there is a speed limit in software process ability. The hardware process of ACL can handle high speed network packet, however, this capability limits ACL configuration such as the limit of the number of rules in ACL. This paper proposes the software that decreases the number of rules in ACL to satisfy the limit of hardware. This paper also evaluates this software by experiment in which this software is applied to practical ACL.

1 Introduction

Nowadays, computer security becomes more important because of the spread of computer network. Firewalls are one of the equipment which protects computer from computer network attacks such as portsweep, brute force attack and so on. The firewalls check whether the IP header of a packet matches predefined rules. The set of predefined rules is called "Access Control List" (ACL), which is a sequence of such rules. There are two ways of processing ACL; one is a software processing and the other a hardware processing. The software processing, theoretically, has no number limit of the rules of ACL. However, the amount of time to process ACL for each packet is the total time taken to compare the packet to each rule until it matches some rule. If many packets match low priority rules, ACL processing takes much time. The priorities of ACL rules can be arranged in order to decrease the taken amount of time, but optimization of ACL rules for minimum amount of time for processing is NP-complete[1].

The hardware processing of ACL is realized by hardware circuit such as Content Addressable Memory (CAM). In the hardware processing, all rules of ACL are compared with a packet at the same time and the amount of time taken in processing a packet is constant[2,3]. However, there is a limit of the number of rules in ACL because the hardware circuit size decides the number of rules which it can handle. Another disadvantages of hardware processing are that hardware processing of ACL costs more than software one and that processing ACL consumes electric power according to the size of hardware circuit for ACL. It follows that the small size of hardware circuit is preferred. However, high speed processing of ACL is needed for smooth high speed network communication.

C.S. Hong et al. (Eds.): APNOMS 2009, LNCS 5787, pp. 453–457, 2009.

By the background described above, this paper proposes the algorithm of decreasing the number of ACL rules within the limit of hardware circuit in keeping the meaning of the ACL rules. This paper also implements this algorithm and evaluates this implementation by using practical ACL rules.

2   The Algorithm and the Implementation of Decreasing ACLs

This paper handles the ACL which is practically used in Saitama University Japan. Since the university uses OmniSwitch in Alcatel Lucent as core Layer 3 switch, this paper uses the syntax of OmniSwitch as that of ACL[4]. ACL is handled by hardware circuit, which limits the configurable size of each element [5]. Therefore, the users of such switches have to keep the size of ACL entries within the limits. Handling many entries of ACL requires decreasing the entries of ACL within the limits. This paper proposes the algorithm which decreases ACL entries until the entries is not more than the limits of hardware.

Decreasing the entries of ACL must keep the meaning of the entries of ACL. The decreasing procedure in this paper starts with finding two rule entries with the same priority and action, and checks the condition entries used in such two rule entries. The condition entries in ACL can consist of many elements according to the OmniSwitch manual, however, it is difficult to target all elements in decreasing the entries. Therefore, this paper focuses network group in OmniSwitch ACL because there seems to be many overlaps in network groups. In the process that network administrators write the ACL rules, the descriptions of network group are liable to overlap, that is, several descriptions of network group means the same network address or group. The reason of overlap is considered as follows. Network administrators usually write the entries of ACL for each subnetwork in the organization which the administrators belong to. It is because each subnetwork decides its own network policy and the network administrators receives the network policies from each subnetwork. It is difficult that the administrators try to rearrange these network policies because human error may change the meaning of ACL. Thus, this paper proposes the algorithm which decreases the ACL entries by rearranging network groups of source and destination addresses in the conditions of ACL. We explain several possible cases of decreasing the entries of ACL as follows:

- **The case that source addresses are the same and the destination addresses are not in the two conditions**
  Table 1 shows this case example, in which a packet from network group N1 to network group N2 or N3 is to be passed. There is a possibility that N2 and N3 are merged into one network group and third, fifth and eighth lines are removed.
- **The case that the destination addresses are the same and the source addresses are not in the two conditions**
  This case is similar to the previous case.
- **One network group includes the other network group**
  Table 2 shows this case example, in which the source address of one condition includes the source address of the other condition or the destination address of one condition includes the destination address of the other condition. Since network group N1 includes network group N2, second, fifth and eighth lines can be removed.

Table 1. In the two conditions, the source addresses are the same

1 policy network group N1 IPaddressA/mask IPaddressB/mask
2 policy network group N2 IPaddressC/mask IPaddressD/mask
3 policy network group N3 IPaddressE/mask IPaddressF/mask
4 policy condition C1 source network group N1 destination network group N2
5 policy condition C2 source network group N1 destination network group N3
6 policy action A1 disposition accept
7 policy rule R1 precedence 10 condition C1 action A1
8 policy rule R2 precedence 10 condition C2 action A1

Table 2. One network group includes another network group

1 policy network group N1 133.38.0.0 mask 255.255.0.0
2 policy network group N2 133.38.1.0 mask 255.255.255.0 133.38.2.0 mask 255.255.255.0
3 policy network group N3 133.37.0.0 mask 255.255.0.0
4 policy condition C1 source network group N1 destination network group N3
5 policy condition C2 source network group N2 destination network group N3
6 policy action A1 disposition accept
7 policy rule R1 precedence 10 condition C1 action A1
8 policy rule R2 precedence 10 condition C2 action A1

Each of the above three cases does not imply that a redundant network group can be removed immediately because the redundant network group may be used in another condition. Removing it may make some conditions change. Therefore, it is necessary to check whether the candidate of removed network groups appears in another conditions. Therefore, merging two network groups requires checking the effect on the other ACL entries. However, it takes much time to check this effect when trying to merge network groups appearing in many places in all condition entries of ACL and the amount of taken time is impractical in some cases.

The algorithm of this paper tries to merge the network groups which appears in up to four places in all condition entries in order to decrease ACL entries within practical amount of time. There are several cases that a network group appears in more than four places and the algorithm of this paper cannot handle such cases.

2.1   Implementation

We implement the algorithm by script language Perl. This software obtains the file including ACL as input and outputs the file including ACL in which entries are decreased. As input, this software also obtains the sizes of hardware such as the limits of rule, condition and network group entries. This software checks whether the number of each kind of entries is over the limit. If all kinds of entries are within the limits, this software terminates and outputs input ACL. If there is a kind of entry which is more than the maximum, this software stores each entry of "rule", "condition" and "network group" into array and checks whether there are network group entries which can be merged. If such entries exist, this software checks whether each of such entries appears in up to four places in all conditions and whether the ACL entries decrease when merging such network group entries to one network group entry. This paper evaluates the software by experiment. In the computer used in this experiment, OS is Vine Linux 4.1 CR, CPU is Intel(R) Pentium(R) 4 CPU 1.8 GHz and the size of memory is 512 Mbyte. This experiment uses the ACL which is really used in Saitama University Japan and supposes that the size of hardware for ACL is one eighth of the real OmniSwitch ACL, that is, the maximum numbers of rules, conditions and actions are 256, and that of network groups is 128. This experiment also handles the case of one over sixty-four size of hardware to show the capability of this software.

Table 3 shows the results in the cases of one eighth size and one over sixty-four size of hardware. In the case of one eighth, this software outputs the revised ACL entries within the sizes of hardware, on the other hand, in the case of one over sixty-four, this software cannot output the entries within the sizes of hardware. The time taken by this process is 11.45 seconds in the case of one eighth size, that is 19.16 seconds in the case of one over sixty-four size. The used memory size in both cases is about 3.5 Mbyte.

Table 3. The result of the experiment

	Sample ACL Size	1/8 hardware scale	1/64 hardware scale
Number of policy rules	176	**140**	**103**
Number of policy conditions	176	**140**	**103**
Number of network groups	153	**128**	**108**

3   Conclusion

This paper proposed the algorithm which decrease the entries of ACL within the sizes of hardware. It also implemented this algorithm and experiments it by using the ACL entries which are used in Saitama University Japan. This experiment showed that this implementation can decrease the ACL entries within the sizes of hardware. The algorithm in this paper is not complicated or outstanding, however, the experiment result shows that the software proposed in this paper is effective for decreasing ACL entries. There are several reasons why this software is effective; one of them seems to be that network administrators hardly write very complicated ACL because it is difficult to write such an ACL. Another of them seems to be that it is not a big problem that this software cannot handle the network groups which appear in more than four places in all conditions; the network groups which appear in many places in all conditions stand out and network administrators may arrange ACL entries by by focusing the network groups. By this experiment, we find several future works and one of them is to handle the order of the ACL entries.

References

1. Grout, V., McGinn, J., Davies, J.: Real-time optimisation of access control lists for efficient Internet packet filtering. J Heuristics (2007)
2. Alessandri, D.: Access Control List Processing in Hardware, Diploma Thesis, Zurich, Switzerland: ETH, Electrical Engineering Department (1997)
3. Ata, S., Hwang, H., Yamamoto, K., Inoue, K., Murata, M.: Managment of Routing Table in TCAM for Reducing Cost and Power Consumption, IEICE technical report, information networks (2007)
4. Alcatel-Lucent, http://www.alcatel-lucent.com/wps/potal/
5. Alcatel: OmniSwitch 7700/7800 OmniSwitch 8800 Network Configuration Guide (2005)

A Novel DNS Accelerator Design and Implementation

Zhili Zhang[1,2,*], Ling Zhang[2], Dang-en Xie[2], Hongchao Xu[2], and Haina Hu[2]

[1] Department of Computer Science, Xuchang University, Xuchang ,461000, China
[2] Computer Network Center, Xuchang University, Xuchang, 461000, China
Tel.:+86-374-2968339
zzl@xcu.edu.cn

Abstract. As a key technology to reduce DNS resolution time, DNS perform-ance optimization becomes a heated issue in the field of research. This paper proposes a novel DNS accelerator which based on the invalid TTL renewal pol-icy to attain the quicker query speed. Compared with traditional server of DNS, the accelerator of DNS has much more excellent function at cache hit-rate, net-work traffic, CPU load and query-response time by the performance test to the concrete website. The new DNS accelerator has a magnificent accelerating rate.

Keywords: DNS; TTL; Renewal policy; Accelerator.

1 Introduction

The traditional DNS has some fatal weaknesses when applied to the Internet, such as slow enquiries, long latency of user-perceived, slow update of root servers' records, large network traffic, configuration error-prone and so on [1]. According to the survey of 535,000 domains and 164,000 traditional DNS servers recently, we find that 79% of the domain names depend on two or less root servers [2], which leads to large query flow, long latency and slow response of users.

Taking into consideration the fact that the traditional DNS has long resolution time, many times of round-trip and large query flow, some authorities proposed the following solutions, such as pre-resolving [3], DNS pre-fetching based on the popu-larity, Pre-fetching using piggyback [4] and Pre-cache DNS records.

In this paper, a new DNS lookup accelerator is designed by improving the above-mentioned methods. Here we simply describe the procedures of the accelerator. When the client sends a request, it is resolved at the local DNS server at once. If the domain name exists in the local server's cache, a response will be replied to the client, or an iterative query happened. Meanwhile, the query is captured by a special module, and the query is sent to the database server. Then the query happened in the database server and finally the results are sent to the local DNS server. By this method, it re-duces the query-failed times and the UDP packets, so it saves the query time. The experiment shows the DNS accelerator can make the query speed faster by accelerat-ing the conversion of domain names to IP addresses.

---

* Corresponding author.

C.S. Hong et al. (Eds.): APNOMS 2009, LNCS 5787, pp. 458–461, 2009.

2 Design Principle and Critical Techniques of the DNS Accelerator

2.1 Design Principle of the DNS Accelerator

We add a database server to save domain names and IP addresses which the local DNS server has just resolved. It preserves the IP address of the authority domain name server corresponding to the local DNS server. When clients send out requests, the local DNS server resolves them immediately. If the domain name exists in the caches of the local DNS server, it issues a reply; otherwise, it sends out an iterative lookup. At this time, the lookup is captured and sent to the database server by the corresponding module. We look up the domain name in the database server as a result and return it to the local DNS server.

The DNS lookup accelerator is composed of four interrelated components: DNSCapture, DNSDBLookup, DNSDBSave and DNSDBRefresh. DNSCapture is installed on the DNS server. It listens to the UDP port 53 and hands over the DNS request and response packets to DNSDBLookup and DNSDBSave separately; DNSDBLookup is responsible for searching the database server and reverting enquiries results to the DNS servers; DNSDBSave stores query answers of the external DNS in the database. The three above-mentioned components are encapsulated into a procedure in the form of thread. We adopt a thread pool to design DNSDBLookup in order to deal with large numbers of concurrent enquiries.

DNSDBRefresh is an independent process. It regularly updates DNS records in the database according to the renewable strategy. Therefore, it can renew expired records in the database and maintain the effectiveness of the pop records.

2.2 Critical Techniques of System Implementation

(1) DNSCapture listens to the UDP port 53 by using LIBPCAP in the DNS server. DNSCapture concerns only about two types of data packets: if the source address of DNS lookup packets is the local machine and the destination port is 53, the DNS server cache does not answer the question and hands it over to external DNS servers.

DNSCapture captures it this time and transmits it to DNSDBLookup; otherwise , the database does not have the answer and the DNSCapture hands it over to DNSDBSave. DNSDBSave saves it in the database server.

(2) Once DNSDBLookup receives query requests which forwards from DNSCapture, it starts the searching for the answer in the database. If the answer exists, DNSDBLookup picks up the answer and amends the head mark of reply packets, then calls LIBNET library and returns the results of the enquiries to the DNS server; In addition, DNSDBLookup modifies the cumulative number of visits in caches. If the answer does not exist, DNSDBLookup does not deal with it. External DNS servers still reply to these packets of DNS lookup.

(3) External DNS servers send out the answer to the local DNS server; DNSCapture captures it and hands it over to the DNSDBSave. The DNSDBSave receives it and detects its format. And the DNSDBSave detects the query name of its mark whether it is consistent with the query name of the packet in HASH. If it is matched,

the DNSDBSave saves it in the database. Otherwise the DNSDBSave does not save it (to avoid DNS spoofing).

(4) The DNSDBRefresh regularly selects these domain records from the database and sends it to the external domain name server directly. This time we get the data form the database directly, needn't use iterative domain analysis. We look up the authority DNS server of this domain only once, this domain can be resolved. It will eliminate many times of iteration, shorten the refresh time and reduce unnecessary network traffic. They often have a high prevalence. After they are renewed, we can achieve quicker query speed.

3   Performance Test and Analysis of the Results

3.1   Test Analysis of the Hit-Rate

The data in Figure 1 were accumulated by DNSDBCache. The gray line indicates the number of DNS servers clients look up; the black line shows the number of external enquiries when there is no answer in DNS caches; the white line indicates the number when there is no answer in the database (secondary) cache. Figure 2 give the hit-rate of double-caches.

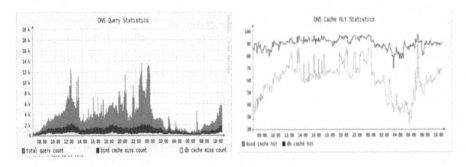

Fig. 1. Statistics of DNS lookup traffic   **Fig. 2.** The hit-rate of DNS cache and database cache

We can find that the hit-rate of DNS is improved greatly after adding the database (secondary) cache from figure 2. The average hit-rate is up to more than 95 percentage, due to the following two reasons. (1) Because bind sets memory as its cache, it only preserves limited DNS records. We set the database as cache, which can preserve more DNS records and reduce the swap as a result of lack of space at the same time. (2) We adopt the renewable strategy of cache. The expired records can be updated timely and saved in the cache to improve the hit-rate. If the memory of DNS server is smaller, the contrast of hit-rate will be more obvious.

3.2   The Number of Database Records and Test Analysis of the TTL

The black area in figure 3 shows the change trend of numbers of database records and the white area represents the number of records that need updating. We can see that the number of records changes smoothly, indicating that the database cache can meet users' requirements for the DNS lookup. There are a few records that need updating in every cycle, at most 5000 records. We can finish updating these records in a short time. Figure 4 shows that the distribution and change trend of the TTL in the DNS cache. We can see that from this chart the elimination strategy of cache can better control the number of records in the database.

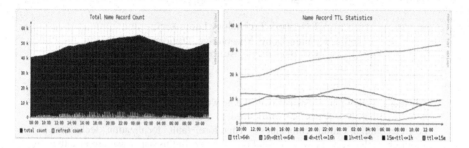

Fig. 3. The change trend of the database records **Fig. 4.** The distribution and the change trend of the TTL

4 Conclusions

This paper puts forward the lost efficacy TTL-based renewal policy to attain a quicker query speed. Compared with the traditional DNS server, the DNS accelerator is excellent in its performance at cache hit-rate, network traffic, CPU load and query-response time by the performance test to the concrete website. The new DNS accelerator has a good accelerating rate and optimizes the performance of DNS. Several major works should be further studied: (1) Apply the DNS accelerator to more complex and larger-scale commercial websites on the IPv6 DNS; (2)Further test should be done so as to make it a more advanced, powerful, reasonable and practical DNS accelerator on the Internet.

Acknowledgments. This work is support by NSF of Henan Province (No.2009A520024) and the higher Education Information Project of Henan Province (No.2008xxh021).

References

1. Mockapetris, P.: RFC1034 Domain Names– Concepts and Facilities (November 1987)
2. Mockapetris, P.: RFC1035Domain Names–Implementation and Specification (November 1987)
3. Li, D., Wu, J., Cui, Y.: Research on the Structures and Resolutions of Internet Namespaces. Journal of Software 16(8), 1445–1455 (2005)
4. Cohen, E., Kaplan, H.: Prefetching the means for document transfer: A new approach for reducing web latency. Computer Networks 39(4), 437–455 (2002)

Service Fault Localization Using Functional Events Separation and Modeling of Service Resources

Jinsik Kim, Young-Moon Yang, Sookji Park, Sungwoo Lee, and Byungdeok Chung

KT Network Technology Laboratory,
463-1, Jeonmin-Dong, Yuseong-gu, Daejeon, Korea
{jinsikkim,xavier,suji,sungwoo,bdchung}@kt.com

Abstract. Today many services over the networks, e.g., IPTV(Internet Protocol TV), web-based services are introduced to extend the business area of tel-cos(telecommunication companies). Therefore fault localization technique should be enhanced to cover the application layer. In this paper, we propose the rule-based alarm correlation method with functional event separation. And we also suggest modeling strategy for service resources and functional dependencies to enable those techniques. The proposed service fault localization method enables end to end service fault localization including application layer with less complexity and we prove it by implementing the pilot system for representative fault cases of real IPTV services.

Keywords: Service Fault Localization, Alarm Correlation, Resource Modeling.

1 Introduction

Telcos begin to introduce new businesses, e.g., IPTV or web-based portal site for expanding their business area. So some derivative changes should be considered from those expansions and one of them is the enhancement on network fault localization technique [1]. For service fault diagnosis, there are some researches. Hanneman [2] proposed hybrid event correlation architecture which used rule based and case based method and applied it to web hosting services. Cheng [3] suggested state transition model for service fault localization.But service fault localization techniques still should be enhanced for handling large sized complex IT services. We have already proposed network fault localization system [4]. In this paper we expand the method to cover application layer using functional event mapping and service resource modeling. And we implement fault localization pilot system for IPTV services using those methods.

2 The Characteristics of Service Fault Localization

We define the service fault as the whole malfunctions and faults which make an effect on QoS (Quality of Service). So it includes the faults of whole service-related resource like network elements, server hardware, database daemons, etc. The application layer has more characteristics in addition to the network layer's [5].

C.S. Hong et al. (Eds.): APNOMS 2009, LNCS 5787, pp. 462–465, 2009.
© Springer-Verlag Berlin Heidelberg 2009

First one is the inconsistency of alarm events among service resources. For instance, locations of alarm events of the network elements are almost standardized like rack, shelf, slot, unit and port. That's because the network layer is composed of similar equipments. But for the application layer, one application is usually different from the others.

Second one is that the relations and function flows among the applications are complex. Network layer usually accomplishes the function of transferring the data from one place to another. But IT services are composed of lots of flows among applications. Therefore we need to handle complex dependencies on analyzing service fault events.

Third one is the analysis complexity of the events itself. It means that there are multiple meaning on service alarms compared to the network alarms and so it should be handled more in detail according to it's functional situation.

3 Proposed Service Fault Localization and Pilot Implementation

3.1 System Overview

System diagram for service fault localization is shown on the left diagram of Fig.1. Through event normalization step, collected events are converted into analysis-ready status events. Function mapping is the key part of normalization step. We construct the function and flow database of services, and all the alarm events are mapped into proper functions by mapping table which the domain experts have defined. Finally on alarm correlation step, it extracts root cause event by applying alarm correlation rules from domain experts based on the dependencies and resource data of service resource database.

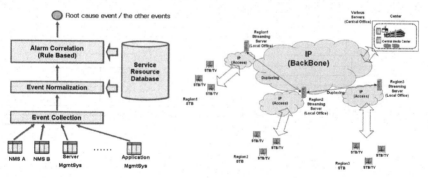

Fig. 1. Service fault localization system diagram & fault localization target IPTV networks and applications diagram. MgmtSys stands for management system.

3.2 Resource Modeling and Functional Event Mapping

IT services are provided through the various interactions among many service resources, so service resource modeling is essential for the alarm correlation of service to analyze the events based on service dependency flows.

Main concern of service resource modeling for service alarm correlation is that it should have affordable detail for both the event analysis and the low maintenance cost.

To satisfy the above requirement, resource modeling is focus on functional resource dependencies. If there is a fault on a service resource, it means that there is a problem on certain service function or flow. For instance, if the user authentication function is accomplished in order of A/B/C servers, the flow order is same and the user authentication related alarms can be mapped into user authentication function. So if there is an user authentication mapped alarm, we can track it on the user authentication function flow. So we model the dependencies of the functional flow of resources and it prevents the users to input whole the APIs(Application Program Interfaces) of the applications on the resource database.

The above functional resource modeling method is combined with functional alarm event mapping on event normalization step, So the functions, service resource flows of the functions and event function mapping table are the analysis base of the proposed method.

And it is also necessary to define the inter-layer structure on the upper of network layer. Typical layers for servers and applications are CPU, HDD/MEM, NIC(Network Interface Card), process external and process internal. If CPU is broken, all the other layers are damaged. The layer structure is shown on resource internal component of Fig.2. The last concern is duplex of resources. Most of the large-size IT services should handle massive access requests from the customers, so servers and applications have the structure of duplex or the architecture of regional request distribution. It is described on resource multiplexing table of Fig.2.

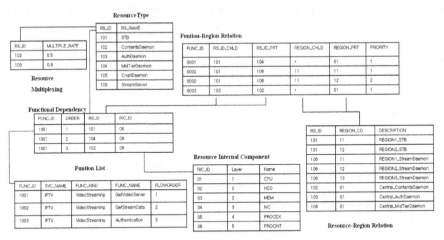

Fig. 2. Resource modeling example for IPTV application resources of right diagram of Fig.1. Video streaming function is composed of three sub functions of function list table. Each functions have flows through several resources on functional dependency table.

3.3 Pilot Implementation and the Result

IPTV service of KT is provided through QoS enabled IP backbone, access network and several kinds of applications. It is shown on the right diagram of Fig.1. In this pilot system, we analyze the 10 representative service flows of IPTV and the flows are divided into 16 separate function flows like IP allocation of STB, contents purchasing, video streaming, user authentication, etc. One of them is shown on Fig. 2. There are many fault localization methods[5][6] but we expand the rule-based alarm correlation technique and RCEE strategy[2]. 60 types of rules are extracted by domain experts. Those rules handle temporal relation, layer priority, functional dependency, time priority, event grouping strategy, alarm analysis mapping information utilizing, event status modification, etc. We build up 28 representative fault scenarios for various fault location from optical cable cut to application error. For each case, the related alarm events are grouped and root cause event is selected properly. So we can distinguish the fault location for representative fault cases with the proposed system.

4   Conclusions

We proposed service fault localization system for IT service based on service resource modeling and functional event mapping method for the application layer. Service fault localization complexity is minimized by the function separate analysis. It is possible by enhancing event normalization process of alarm correlation by adding functional event mapping process. For those processes functional dependency data of service resource database should be managed after modeling them.

We also implemented the pilot system for IPTV service based on the proposed method and showed that the proposed method can successfully used for fault localization on representative cases of IPTV service malfunctions.

References

1. Gardner, R.D., Harle, D.A.: Methods and systems for alarm correlation. In: GLOBECOM 1996, vol. 1, pp. 18–22 (1996)
2. Hanemann, A., Marcu, P.: Algorithm Design and Application of Service-Oriented Event Correlation. In: Business-driven IT Management 2008, pp. 61–70 (2008)
3. Cheng, L., Qiu, X.S., Zhang, N.: A Service-Oriented Alarm Correlation Method for IT Service Fault Localization. Communications and Networking in China, ChinaCom apos, 1–5 (2006)
4. Kim, J., Yang, Y.M., Park, S., Lee, S., Chung, B.: Fault Localization for Heterogeneous Networks Using Alarm Correlation on Consolidated Inventory Database. In: Ma, Y., Choi, D., Ata, S. (eds.) APNOMS 2008. LNCS, vol. 5297, pp. 82–91. Springer, Heidelberg (2008)
5. Steinder, M., Sethi, A.S.: A survey of fault localization techniques in computer networks. Science of Computer Programming 53, 165–194 (2004)
6. Hanemann, A.: A Hybrid Rule-Based/Case-Based Reasoning Approach for Service Fault Diagnosis. In: Proceedings of the 20th International Conference on Advanced Information Networking and Application, vol. 2, pp. 734–740 (2006)

Towards Fault Isolation in WDM Mesh Networks

Chi-Shih Chao and Szu-Pei Lu

Department of Communications Engineering
Feng Chia University, Taiwan 40724, ROC
cschao@fcu.edu.tw

Abstract. In this paper, what we did on modeling of fault propagation behavior in WDM mesh networks and creation of leader major alarm domain concept are introduced and utilized as a basis for the development of our collaborative WDM network fault isolation approach. This approach fully takes the individual advantages of our two developed WDM network fault isolation methods to handle different fault situations (single- and multi-fault situations).

Keywords: alarm clustering, passive probing, diagnostic power, dependency matrix, SRLG.

1 Introduction

To let our work be well done, what we have done previously in the field, along with the concepts behind, are introduced and fully utilized as the basis of the development of an effective WDN network fault isolation approach. This approach takes advantages of two self-made WDM network fault isolation methods – the passive probing-based fault isolation method and the alarm clustering-based fault isolation method – to deal with different fault situations. By working collaboratively, we found our combinational approach can be feasible to the fault isolation job. The rest of this paper is organized as follows. In Section 2, what we have performed on fault isolation in WDM networks is described. They include how to carve the correct and proper fault propagation behavior in WDM mesh networks, which profitable information can be employed to narrow down the size of the suspicious faulty area, etc. Based on these achievements as well as concepts, Section 3 introduces our two developed WDM network fault isolation methods with their corresponding advantages and disadvantages: one is good at single fault isolation and the other one is at multi-fault situation. At last, Section 4 provides further discussion as a brief conclusion.

2 Summary of Our Previous Works

While modeling the fault behavior for fault isolation, one thing is worth being noticed: The management system can receive different categories and amounts of alarms by varying the fault occurrence location and time in the monitored DWDM network. For precise specification of the fault pattern in path-protected WDM networks, the alarm dependence relationship [1] is introduced to classify caused alarms, based on

C.S. Hong et al. (Eds.): APNOMS 2009, LNCS 5787, pp. 466–469, 2009.

their "importance" for the corresponding fault, into two groups: *major alarms* mean they are always-present whenever and wherever a fault occurs, and *minor alarms* may or may not appear when a fault occurs. By this classification, we can filter minor alarms out of the alarm collection database to effectively analyze/attain the fault behavior, and hence improve the management system's feasibility and scalability.

In addition, in our work, the leader major alarm domain concept is employed [1]: Since the first alarm in the alarm propagation pattern can reveal the most possible location and beginning time of a fault, we can use the first major alarm to supplant the whole alarm string for fault diagnoses. Then, by employing the alarm domain theory, we can deduce a significantly smaller suspicious element set. To further enhance our fault diagnoses, some other information should be considered in the process of fault isolation to keep downsizing the suspicious faulty element set. The first type of such information which can facilitate this work is the dynamic light session information in the diagnosed WDM mesh network while faults occur. This type of information indicates that those elements, which are within the territory of the leader major alarm domain and meantime are traveled through by one of the workable light sessions during the period of faults, should be treated as the profitable information source (or evidence) for fault diagnosis. Since these elements can function well in the diagnosed network when a fault occurs, we should remove them from the suspicious element set.

Moreover, fault diagnosis in WDM mesh networks should also consider the physical network construction (or physical layout). In our work, SRLGs (Shared Risk Link Groups) for the physical construction of WDM mesh networks are taken into account. For some reasons (e.g., cost, administration, geography, environment, or else), it can be found that two or more sessions share the same conduit/PVC pipeline/cable at some link sections [2] in most of cases, where these sessions are viewed as being within the same SRLG. By employing the concept of SRLG, some suspicious elements can be exempted from inspection for fault diagnosis due to their sharing of "healthy SRLGs."

3 WDM Network Fault Isolation Methods

3.1 Passive Probing-Based Fault Isolation

In Fig. 1(a), there are five OXCs and five light sessions (or links) are established concurrently. Based on the concept of SRLG, network elements N1, N2, N3, N4 can be thought of as being at the same risk group because the failure of any one of them will cause the light session from OXC_0 to OXC_1 malfunction. Accordingly, we can derive five SRLGs from five existing light sessions. Integrating the SRLG concept with Brodie's active probing method, a passive probing dependency matrix which is used for the single fault isolation is built as shown in Fig. 1(b) where the five SRLGs (or the five light sessions) are employed as "passive" probes. For instance, the first row in the matrix with the value of (1111000000) indicates that the light session (or named SRLG1) traveling through network components N1, N2, N3, and N4 is used as "or passively becomes" one of the probes for fault isolation.

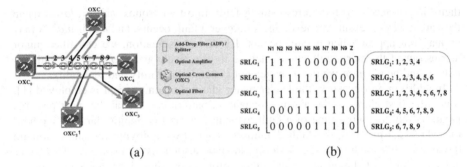

(a)   (b)

Fig. 1. (a) A WDM Example Network and (b) The Passive Probing Dependency Matrix for Single Fault Isolation

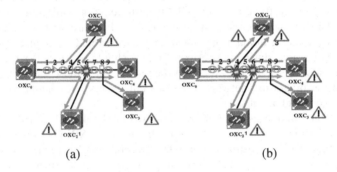

(a)   (b)

Fig. 2. Failure on (a) N6, and (b) N4 and N6

In case of the failure of N6 (Fig. 2(a)), OXC_1, OXC_2, OXC_3, and OXC_4 will emit a "No Signal" alarm separately; in fact, four trimmed "No Signal" alarm strings starting from these four OXCs would show up individually. By our leader major alarm domain concept, only the leader alarms of these four alarm strings would be taken into account in our fault isolation methods. To isolate the fault, we first get the complementary values of $SRLG_1$ which remains normal while the failure occurs. Then we execute the logic AND operation on all of "passive" SRLGs in the matrix, i.e., $\overline{SRLG_1} \cdot SRLG_2 \cdot SRLG_3 \cdot SRLG_4 \cdot SRLG_5$ and find that the resulting value of this operation is 0000010000 (the 6th bit is 1), which indicates the failure of N6.

3.2 Alarm Clustering-Based Fault Isolation

Despite the success of the adaption of Brodie's method to our passive probing-based fault isolation in WDM networks, there exist two fatal drawbacks:

1. The passive probing-based dependency matrix in Fig. 1(b) is not feasible for dealing with the case of multiple faults. For example in Fig. 2(2), suppose N4 and N6 fail simultaneously. It can be found that there would be five "No Signal" alarm strings starting from OXC_1, OXC_2, OXC_3, and OXC_4. If the dependency matrix in Fig. 1(b) is used, an unacceptable diagnostic result comes out: (0000000000), by logically ANDing all the rows of the dependency matrix. The result shows that

there is no fault being found; yet, interestingly, neither can it be assured that the network is with no problems (∵ Z bit is 0, not 1).

2. For this, Brodie *et al.* proposed their improvements on handling the situation of multi-faults: they expand the dependency matrix for single fault diagnosis to multi-fault diagnosis. For isolating two simultaneous faults, columns in the single-fault dependency matrix will be mutually logic ORed and the results will be appended to the original single-fault matrix. Nevertheless, this method is proved that it can easily make the dependency matrix for multi-fault diagnosis too large to be accommodated while computing.

To conquer the above two deadly drawbacks, the alarm clustering concept is used [3]. With the alarm clustering concept, the corresponding alarm domains for the five "No Signal" alarms in Fig. 2(b) can be drawn as: {N1~N4}, {N1~N6}, {N1~N8}, {N4~N9}, and {N6~N9}. By using the best explanation of alarm clustering, we figure out {N4, N6} as the best explanation for the alarm cluster formed by the five alarms. However, if the passive probing-based method is used, the logic ANDed result on the single-fault dependency matrix in Fig. 2(b) is (0000000000). Since it is not an appropriate diagnostic result, then the two-fault dependency matrix will be used to execute the logic AND operation on each matrix rows to check if a proper two-fault diagnosis result can be obtained. If it can't, next, the three-fault dependency matrix will be used to operate in the same fashion. We will continue in this way until a combination of faults comes up to be able to properly explain the alarm collection.

4 Further Discussion

It seems that the alarm clustering-based method has advantages over the passive probing-based method on mult-fault isolation. However, the alarm clustering-based fault isolation also has it own shortcomings in practice:

1. Although lots of efforts have done for a near-optimal algorithm with the polynomial computing complexity, alarm clustering has been proved that it is of NP-completeness [3] to get the best explanation.
2. In some cases, it is also possible that the alarm clustering-based fault isolation can make the same misdiagnosis as the passive probing-based method does. Let us consider the case of the failure on N5 and N6 on Fig. 1. The best explanation for the caused alarms would be {N6} — which is the same as the passive probing-based method with a single-fault dependency matrix does. In this case, both of these two methods make the same incorrect diagnosis.

References

1. Chao, C.S., Yang, D.L., Liu, A.C.: A Time-Aware Fault Diagnosis System in LAN. In: Proceedings of the 2001 IFIP/IEEE International Symposium on Integrated Network Management (IM 2001), Seattle, USA, May 14-18, pp. 499–512 (2001)
2. Perros, H.G.: Connection-Oriented Networks: SONET/SDH, ATM, MPLS, and Optical Networks, 1st edn. Wiley, Chichester (2005)
3. Chao, C.S., Liu, A.C.: An Alarm Management Framework for Automated Network Fault Identification. Computer Communications 27(13), 1341–1353 (2004)

Factors Influencing Adoption for Activating Mobile VoIP

MoonKoo Kim[1], JongHyun Park[1], and JongHyun Paik[2]

[1] ETRI, Gajeongno 138, Yuseonggu, Daejeon, Korea
{mkkims,tephanos}@etri.re.kr
[2] TTA, 267-2 Seohyeon-dong, Bundang-gu, Seongnam, Gyonggi-do, Korea
jhpaik@tta.or.kr

Abstract. The introduction of the voice service to WiBro and the commercialization of 4G services will lead not only to the supersession of the existing voice market by m-VoIP but also to a greatly expanded market through the introduction of various value added services. For the purposes of this study, a market survey of Koreans was carried out as part of a methodical study of the factors influencing m-VOIP acceptance. Based on the findings of the study, the implications for the promotion m-VoIP in Korea are presented as the result.

Keywords: Mobile VoIP, WiBro, 4G, Factors Influencing, Wireless LAN.

1   Introduction

The wired Internet telephone has had a great impact on the existing wired telephone, and is understood to have the potential to accelerate the replacement of existing services. Despite the increasing interest in and technical progress of m-VoIP, few studies have attempted to systemize worldwide case studies for in-depth analysis of the factors which influence service acceptance. For the purposes of this study, a market survey of Koreans was carried out as part of a methodical study of the factors influencing m-VOIP acceptance. Based on the findings of the study, the implications for the promotion m-VoIP in Korea are presented as the result. The implications of the study results will be presented as measures to promote the m-VoIP market in Korea and to strengthen the competitiveness of the relevant participants.

2   Analysis of the Factors Determining Acceptance of m-VoIP Service in Korea

2.1   Study Overview and Analysis of m-VoIP Market Demand

The Communication Economy Research Team of the Electronics and Telecommunications Research Institute conducted a market survey of 500 people residing in the Seoul Metropolitan area to study the factors which have an impact on demand for m-VoIP and its acceptance. The survey analyzed the respondents' m-VoIP usage

C.S. Hong et al. (Eds.): APNOMS 2009, LNCS 5787, pp. 470–472, 2009.
© Springer-Verlag Berlin Heidelberg 2009

intention, the fee level, and the factors which influence acceptance in addition to the statistical characteristics of the population.

As shown in Fig. 2., 33.2% of all respondents declared that they intend to use m-VoIP, indicating that the market potential of the service is high. The m-VoIP market demand was particularly high among males, respondents in their teens~30's, Seoul residents, high school students, white-collar workers, blue-collar workers, and heavy mobile phone users. In general, m-VoIP demand was high among consumers with heavy mobile phone usage, with the differentiating factor being occupation rather than age group.

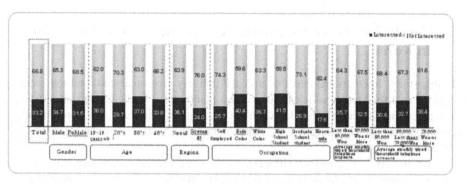

Fig. 1. m-VoIP Usage Intentions of Residents of the Seoul Metropolitan Area

2.2 Analysis of the Factors Influencing m-VoIP Acceptance

Fig. 2. shows the results of the study of those factors which have an impact on m-VoIP acceptance. The most significant factors with an impact on m-VoIP acceptance are the usage fee and call quality.

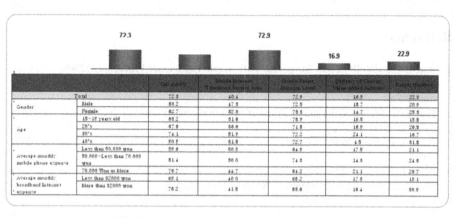

		Call quality	Mobile Internet Telephone Quality Info	Mobile Phone Message Level	Delivery of Various Value-added Services	People Matching
	Total	72.5	40.4	72.5	18.8	22.8
Gender	Male	80.2	47.3	72.5	19.7	20.9
	Female	62.7	42.8	70.5	14.7	25.8
Age	15~18 years old	66.2	91.8	78.9	18.5	15.8
	20's	67.8	36.8	71.8	18.8	26.8
	30's	74.1	51.8	72.3	24.1	16.7
	40's	80.8	31.8	72.7	4.5	51.8
Average monthly mobile phone expense	Less than 50,000 won	58.8	50.8	64.8	17.8	21.1
	50,000~Less than 70,000 won	81.1	90.0	74.8	14.8	21.8
	70,000 Won or More	78.7	44.7	84.3	21.1	25.7
Average monthly broadband Internet expense	Less than 32000 won	67.1	40.0	86.2	17.6	19.1
	More than 32000 won	78.2	41.8	85.8	18.4	50.8

Fig. 2. Factors Influencing m-VoIP Acceptance

Coverage, handsets, and value added services also ranked high. Respondents in their teens and heavy mobile telecommunication service users were the most sensitive to the usage fee, while call quality had a significant influence on acceptance among respondents in their thirties and forties, males, and intermediate mobile phone users. Males and people in their thirties were sensitive to coverage, while females and respondents in their twenties and forties were sensitive to handsets. Males, respondents in their thirties and heavy mobile telecommunication users were sensitive to value-added services. In other words, the key factors influencing m-VoIP acceptance are the usage fee and call quality. Furthermore, different genders, age groups and mobile telecommunication usage level were sensitive to different factors.

3   Conclusion: Direction for the Revitalization of m-VoIP Market in Korea

In this study, international cases of m-VoIP delivery were analyzed, and the service demand and the factors which influence acceptance were studied based on a market survey of the general public. Based on the findings of this study, the recommended direction for the revitalization of Korea's m-VoIP market may be described as follows:

First, it is likely that an m-VoIP market will be created and greatly expanded due to the emergence of 4G technology or technology that is similar to it, as mobile telecommunication technology including Korea's WiBro advances.

Second, in other countries, m-VoIP has mostly been offered under business partnerships between MVNO/wired operators and late-coming mobile telecommunication carriers rather than by mobile telecommunication carriers offering it as their own business model.

Lastly, in certain countries, mobile WiMax providers have offered the m-VoIP service from the commercialization phase to secure a competitive edge by using the technology.

References

1. Bae, D.C.: MVoIP Leader Skype. In: VoIP World Conference Proceeding (2008)
2. Kim, H.D.: Current Status of and Predictions for Mobile VoIP. In: ETRI Seminar presentation (2008)
3. Han, J.H.: Mobile VoIP and its Prospect in Convergence Paradigm. In: ETRI Seminar presentation (2008)
4. Park, J.H., Kim, M.K.: Characteristics, Prospects and Case Study of m-VoIP. In: ETRI (2008)
5. Park, J.H., Kim, M.G., Park, H.J.: Prediction of Wired/Wireless VoIP Demand and the Characteristics of Usage Intentions, Weekly Technology Trend No. 1275, IITA (2006)
6. http://www.fring.com
7. http://www.bolter.com

A Study on SDH/OTN Integrated Network Management Method

Kei Karato[1], Shun Yamamoto[1], Tomoyoshi Kataoka[1],
Yukihide Yana[2], and Yuzo Fujita[2]

[1] NTT Network Service Systems Laboratories,
Nippon Telegraph and Telephone Corporation
9-11, Midori-cho 3, Musashino-shi, Tokyo, 180-8585, Japan
{karato.kei,yamamoto.shun,kataoka.tomoyoshi}@lab.ntt.co.jp
[2] Network Service Solution Business Group, NTT Software Corporation
19-3, Naka-machi 1, Musashino-shi, Tokyo, 180-0006, Japan
{yana-y,fujita-y}@po.ntts.co.jp

Abstract. An efficient management method for the Optical Transport Network (OTN) and Synchronous Digital Hierarchy (SDH) integrated network is described. We propose layer boundary section by focusing on the connection terminal point function on the OTN layer and, a management method for a fully connected OTN/SDH network.

Keywords: SDH, OTN, Integrated management, overlay-network.

1 Introduction

1.1 Examination Target

We propose a layer boundary section for SDH and OTN layers. In addition, by using the idea of our proposed layer boundary section, we discuss a management method for an example topology network of SDH and OTN layers called Ring-in-Ring.

1.2 Integrated Network Management Model

In ITU-T G.805 Figure 3 [1], a function model with the SDH layer defined as Client layer (Client) and Server layer (Server) relationship. This function model can be applied to integrated network models consisting of SDH layer and OTN layer, so this examination is based on this Client/Server function model. For OTN architecture model see G.803 Figure I.1 [2] and SDH architecture models see, G.872 and other our previous research results [3].

2 Layer Boundary Section

2.1 Composition Patterns

To extract all layer boundary sections, we examine the pattern of network composition for Synchronous Digital Hierarchy (SDH) and Optical Transport Network (OTN)

C.S. Hong et al. (Eds.): APNOMS 2009, LNCS 5787, pp. 473–476, 2009.

layers. Based on function model that is Client/Server relationship, SDH and Client represent relation of Client/SDH (pattern 2), and OTN and Client represents relation of Client/OTN (pattern 1). Moreover, the relation of Client/Server is SDH and OTN. Therefore, ether SDH and OTN represent Server, and relation of Client/SDH/OTN is possible (pattern 3, 4). Pattern 4 is that one direction is Client/SDH/OTN, and other direction is Client/OTN pattern. Four composition patterns have been extracted from the combination of Client/Server function model (Figure 1).

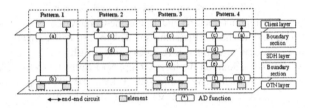

Fig. 1. AD functional position in boundary section

2.2 Adaptation Function and Layer Boundary Section

For the integrated network, it is necessary that the connection termination point is accurately defined, because it depends on the position of the connection termination point, and the function of layer boundary section is different. In G.805, the Adaptation (AD) function is defined between Client and Server. At the AD function, the Client layer data converts to the Server layer data. The termination point is defined as the AD function. Three kinds of boundary sections between layers (Client-OTN, Client-SDH, and SDH-OTN) are extracted. By placing the AD function at the ends of the boundary sections, six kinds of layer boundary sections are created (Figure 2).

Client/OTN Boundary section	
(a) The AD function is mounted on the Client side	(b) The AD function is mounted on the OTN side
Client — AD — OTN	Client — AD — OTN
OTN Physical Section(OPS)	Client/OTN physical Section(COS)
Client/SDH Boundary section	
(c) The AD function is mounted on the Client side	(d) The AD function is mounted on the SDH
Client — AD — SDH	Client — AD — SDH
SDH Physical Section(SPS)	Client/SDH physical Section(CSS)
Client/OTN Boundary section	
(e) The AD function is mounted on the SDH	(f) The AD function is mounted on the OTN
SDH — AD — OTN	SDH — AD — OTN
OTN/SDH physical Section(OSS)	SDH/OTN physical Section(SOS)

Fig. 2. Logical names of layer boundary sections

3   Management Method for SDH/OTN Integrated Network

3.1   Overlay Network Topology

Our management method for the SDH/OTN integrated networks is described by using the results of our proposed boundary section in the preceding chapter.

In an integrated network, SDH and OTN layer topologies are simple in some case and very complex in others, for example SDH and OTN overlay network. This network has never before been considered because each layer was separately managed in the current network management method.

Ring-in-Ring topology is one of the most complex topologies of overlay networks for composition Fig 1, pattern 3. In Ring-in-Ring topology, the OTN Section is the section between network elements (NEs) composed OTN ring topology and NEs that compose the SDH ring topology. Therefore OTN topology is a physical ring, but SDH topology is a logical ring because the SDH section consists of two boundary sections, two OTN NEs and, one OTN section.

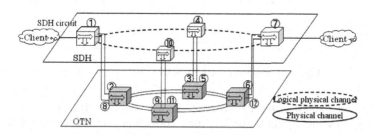

Fig. 3. SDH/OTH integrated network model called Ring in Ring

3.2 Examination of Management Method

Our management method for the Ring-in-Ring topology network (Figure. 3) was examined with respect to the SDH circuit management.

The route of the SDH circuit becomes Client-①-②-③-④-⑤-⑥-⑦-Client and Client-①-⑧-⑨-⑩-⑪-⑫-⑦-Client. There are alternate routes between the SDH and OTN layers.

In this Ring-in-Ring topology network, the current management method that SDH topology representing one sub-network and OTN topology representing another sub-network is impossible, because one SDH circuit consists of multi OTN circuits. Therefore not only the current method of managing sub-network but also the method for managing the circuit route was examined.

The four sections consists of NEs ①-⑩, ⑩-⑦, ①-④, and ④-⑦ of the SDH topology are called Multiple Sections (MSs). The four sections consists of NEs ⑧-⑨, ⑪-⑫, ②-③, and ⑤-⑥ of the OTN topology are called Optical Transport Sections (OTSs). One MS consists of one OTS, for example section ①-⑩ consists of section ⑧-⑨. Then the OTN ring-topology is divided into one OTS point-to-point topology, and SDH ring-topology becomes one virtual ring-topology. Next, the boundary sections ①-⑧, ⑨-⑩, ⑩-⑪, ⑫-⑦, ①-②, ③-④, ④-⑤, and ⑥-⑦ of the OTN and SDH layers are allocated a logical section with the AD function named SDH/OTN physical Section (SOS). Therefore, a nest relationship exists between all the logical sections and can be managed on the SDH virtual ring-topology. This SDH virtual ring is managed as one sub-network. The result of allocating a logical name to this composition is shown in Figure. 4.

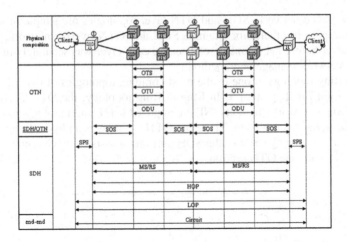

Fig. 4. Management method: SDH virtual ring topology

Besides this pattern, using the proposed logical section solves all 35 pattern network topologies, for example the NEs that composing the OTN ring topology are connected to the NEs that composing a part of the SDH ring topology.

References

1. ITU-T Recommendation G.805, Generic functional architecture of transport networks (March 2003)
2. ITU-T Recommendation G.803, Architecture of transport networks based on the synchronous digital hierarchy (SDH) (March 2003)
3. Yamamoto, S., Karato, K., Karato, T., Yana, Y., Fujita, Y.: A study of a network management system that consist of multi-layer. In: Proceedings of the 2008 IEICE Society Conference (1999)

Group P2P Network Organization in Mobile Ad-Hoc Network*

Rim Haw, Choong Seon Hong[**], and Dae Sun Kim

Department of Computer Engineering Kyung Hee University
Seocheon, Giheung, Yongin, Gyeonggi, 446-701 Korea
{rhaw,cshong,dskim}@khu.ac.kr

Abstract. This paper proposes an effective file searching scheme named Group P2P to reduce the number of message transmission. In the Group P2P, peers distinguish Parent Peer from Child Peer. Also, we propose another scheme to manage Child Peers for reducing overhead in the group as joining and leaving.

Keywords: Group P2P, Group, Parent-child relationship.

1 Introduction

Peer-to-Peer (P2P) [1] is one of the most killer applications in the recent network techniques. Specially, P2P research is frequently progressed in the Mobile Ad-Hoc Network. But, recent P2P method has been studied under wired environment, which is not appropriate for the mobile environment. Typical mobile P2P techniques for Ad-hoc Network are ORION (Optimized Routing Independent Overlay Network) [2] and DHT (Distributed Hashing Table) [3]. However these techniques may occur a query message increasing problem and increasing overhead, when a user wants to find a file or contents in P2P.

In this paper, we focus on reducing overhead and the number of query messages in file searching sequence. So, we organize group P2P network using peers which distinguish Parent Peer (PPR) from Child Peer (CPR) in Mobile Ad-hoc Network. PPR manages CPRs' information, and each P2P group communicates with each other. For file searching, instead of using multi-broadcast transmission for P2P, we use PPR to get information of file. It is helpful for us to avoid occurring overhead. Our scheme also helps to manage peer join and leave in the mobile environment.

2 Proposed Scheme

Fig. 1. (a) shows our proposed a parent-child peer group. The group range is decided by PPR's RSSI. And the group boundary is decided to compare CPR's group ID in each group edge. A PPR manages a CPR and registers CPR's file information to the

---

* This work was supported by the IT R&D program of MKE/IITA. [2008-F-016-02, CASFI].
** Corresponding author.

C.S. Hong et al. (Eds.): APNOMS 2009, LNCS 5787, pp. 477–480, 2009.

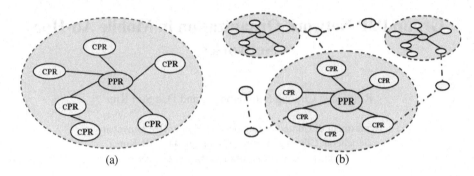

Fig. 1. (a) Proposed parent-child peer group (b) P2P network using proposed group

table in the peer group. Also each group communicates using the PPR and is organized the new P2P network.

Fig. 1 (b) shows P2P networks using group. We assume the PPR of each group itself knows local information and the neighbor PPR's local information for the peer group communication. If the new CPR joins in the P2P group, it makes a parent-child relationship with the recent PPR. At that time, the CPR checks a candidate of the PPR which has the smallest number of children and the strongest RSSI (Received Signal Strength Indication) [4] to choose PPR. Then the CPR joins a chosen PPR's group newly.

$$P_r = \left(\frac{K}{4\pi R} \right)^2 \cdot G_t \cdot G_r \cdot P_t \tag{1}$$

$$R = \frac{K}{4\pi} \sqrt{\frac{(G_t \cdot G_r \cdot P)}{P_r}} \tag{2}$$

Table 1. Parameters for RSSI Formula

Parameter	Description
P_r	Receiver side electric power [w]
P_t	sender side electric power [w]
K	Used wavelength (c/f) [m]
R	Distance between sender and receiver [m]
G_t	Sender side antenna electric power gain [dB]
G_r	Receiver side antenna electric power gain [dB]

To calculate distance between the PPR and the CPR, we can use RSSI. All peers can calculate RSSI by formula as shown in formula (1), (2) and Table. 1 [5]. If RSSI is strong, distance between PPR and CPR is close to each other. And if the number of children are small, group's depth is low. If the CPR leaves a group, the CPR sends a secede message to the PPR and the CPR. If the PPR and the CPR receive the leaving message, they erase the CPR from their file management table. And the CPR, which wants to leave group, sends a leaving message to other CPR, if the CPR leaving group joins other PPR. The new CPR wants to join the recent CPR. The new CPR sends its resource information and its ID to the recent CPR and gets PPR's ID. The recent CPR receives new CPR's information and updates its file management table. Then the recent CPR sends update information to its PPR. In this procedure, the PPR can get resource information from all CPR which joins group. Therefore, if the CPR wants to get some files information, the CPR sends a query message to its PPR and can get information. However file information, which is wanted by CPR, may not be in its PPR's file management table. Then the PPR sends a query message to the neighbor PPRs.

2.1 Group P2P File Searching Scenario

Fig.2. shows P2P file searching scenario. The CPR "A" in the left group peer "A" sends a query message to its PPR to find a file "★". First, the PPR "A" checks whether a file "★" is in its file management table or not. Then if a file does not exist in PPR "A" file management table, the PPR "A" broadcasts a query message to neighbor PPRs. This procedure is repeated until finding file "★".

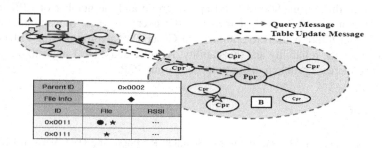

Fig. 2. Group P2P File Searching Scenario

3 Performance Evaluation

To evaluate the performance of our group P2P we perform the simulation using C++ codes. We increas the number of peers and compared with the number of message transmissions of both recent ORIOIN and group peer.

Fig. 3. (a) shows to compare the number of query message transmission and Fig. 3. (b) shows the overhead for updating the file management table as the number of joining and leaving peers changes between recent ORION and proposed group P2P. As shown in Fig. 3, we confirm decreased Query messages and reduced overhead in the group P2P.

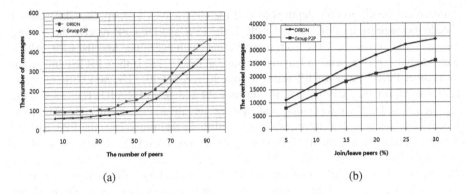

Fig. 3. (a) The number of query message transmission for the file searching (b) The overhead for updating the file management table

4   Conclusion and Future Works

In this paper, we introduce group P2P, which makes a relationship like parent-child, for reducing the overhead in mobile P2P environment. We can consider the number of query message transmissions are more decreased than recent ORION's transmission method in P2P network. However, our proposal increases cost of build and maintains the table according to method of build up the group and the number of CPR. If the number of groups is increased, it increases the query message transmission in the group P2P. Our future work is to optimize our Group P2P algorithm using PPR- CPR, and to improve the reliability and the stability. Also we adapt DHT algorithm to solve the message increase problem in our proposed Group P2P.

References

1. Aberer, K., Punceva, M., Hauswirth, M., Schmidt, R.: Improving Data Acess in P2P Systems. IEEE Internet Computing (February 2002)
2. Klemm, A., Lindemann, C., Waldhorst, O.P.: A Special-Purpose Peer-to-Peer File Sharing System for Mobile Ad-hoc Networks. In: IEEE Semiannual Vehicular Technology Conference 2003 FALL (October 2003)
3. Ding, G., Bhargava, B.: Peer-to-peer File-sharing over Mobile Ad hoc Networks. In: International Workshop on Mobile Peer to Peer Computing 2004 (IWMP2P 2004) (March 2004)
4. Wu, R.-H., Lee, Y.-H., Tseng, H.-W., Jan, Y.-G., Chuang, M.-H.: Study of characteristics of RSSI signal. In: ICIT 2008 (April 2008)
5. IEEE Std. 802.15.4-2006, IEEE Standard for Information technology — Telecommunications and information exchange between systems — Local and metropolitan area networks — Specific requirements - Part 15.4: Wireless Medium Access Control (MAC) and Physical Layer (PHY) Specifications for Low-Rate Wireless Personal Area Networks (WPANs) (2006)

Dynamic Reconstruction of Multiple Overlay Network for Next Generation Network Services with Distributed Components

Naosuke Yokoe[1], Wataru Miyazaki[1], Kazuhiko Kinoshita[1], Hideki Tode[2],
Koso Murakami[1], Shinji Kikuchi[3], Satoshi Tsuchiya[3], Atsuji Sekiguchi[3],
and Tsuneo Katsuyama[3]

[1] Department of Information Networking, Osaka University, Japan
{yokoe.naosuke,miyazaki.wataru,kazuhiko,murakami}@ist.osaka-u.ac.jp
[2] Department of Computer Science and Intelligent Systems,
Osaka Prefecture University, Japan
tode@cs.osakafu-u.ac.jp
[3] Cloud Computing Lab., Fujitsu Laboratories Limited, Japan
{skikuchi,tty,sekia,KatsuyamaTsuneo}@jp.fujitsu.com

Abstract. Because of rapid improvement in information and network technologies, many kinds of network services are created. In this paper, the concept of a new service platform is presented. Services are organized by components, and replications of components are distributed to many computers. Thanks to these replications, users can continue services by using another component as a substitute in case of failure. It is necessary to ask another component to find a substitute when a component fails. However, depending on which component to ask, the original version of the proposed method has a possibility to find only low-performance components. It is caused by constructing overlay network only once and keeping it as it is. We propose a new dynamic reconstruction method for overlay networks. Simulation experiments show how proposed method improves performance of our framework.

1 Introduction

With the recent improvement in information and network technologies, people can utilize many kinds of services with the Internet. Many companies and individuals are looking for the flexible and easy way to organize services. There are many kind of solutions to organize complex function by integrating remote components, such as CORBA, Java RMI, and Web Service [1]. Web service realize the best flexibility among them. However, it gains less reliability because of its infrastructure, the Internet. There are some approaches to overcome the weakness [2]. We propose a new platform for orchestrating services with remote components. In our proposal, many kinds of components organize services and each component has its replications. Replications are distributed to many computers on the network, and because of them, application servers can find new component and continue services even if one component gets failed. We have already described two methods for finding substitute in [3].

C.S. Hong et al. (Eds.): APNOMS 2009, LNCS 5787, pp. 481–485, 2009.

2   Model of Network Service Platform with Distributed Components

At the beginning, let us explain special terms for our platform. The word "component" means one application that can provide a service itself, and could be a part of more complex services. Every component needs to be installed on a computer, called as "node". We assume the "application server" that connects users with components. The application server uses one or more components to create its own service, and each user gets services via the application server.

We introduce "search component" to our platform for fault-tolerance. Search component is in charge of finding substitute in case of failure, and every component has its own search component. When an application server α decides to use component x_i, the address of search component x_s is also informed to α. x_s should be selected from the neighbors of x_i. In this case, "neighbors" means the components which have overlay link with x_i.

We can estimate performance of components with the following two factors. $P_{xi} = \frac{u_i \times l_x}{p_i}$, $L_{\alpha i} = RTT_{\alpha i}$. P_{xi} is a time to process a service with component x_i on node i, and $L_{\alpha i}$ is a network latency between x and application server α. u_i is the amount of accesses to node i per unit of time. Note that u_i is proportional to the total number of application servers using components on node i. l_x is capability required to process component x_i, and p_i means total capability of node i. $L_{\alpha i}$ is calculated directly from round-trip time between component x_i and application server α. Finally we can estimate service time of component x_i as follows; $S_{xi} = \omega \times P_{xi} + (1 - \omega) \times L_{\alpha i}$.

When x_i fails, server α asks x_s to find a substitute for a_i. Then x_s sends request message to all neighbors, and get replies that contain information about their performance. Furthermore, x_s calculates round-trip time from these sessions, and estimates S_{xi} for each neighbor.

Next, we describe component-based overlay networks. We assume that neighbors exchange information about their performance via these networks. Given the overhead of exchanging that kind of information in a certain period, we set a limit on the number of overlay links per node.

Now we describe how to construct those overlay networks. Assume that component x_i has been installed on node i. Since x_i does not know where other replication is initially, it needs to ask someone. We introduce a "Service Broker" for that role, which knows the locations of each replication. Component x_i asks the service broker for addresses of replications, and tries to connect with x_j firstly. If $D_i \geq ULD_i$ in this moment, where D_i is the total number of node i and ULD_i is its limitation, node i needs to disconnect one of its overlay links. If destination node j also has too many overlay links, do the same as node i. After removing a link or if D_i is less than ULD_i, x_i establishes the connection with x_j and tries next. These process will continue until component x_i connects overlay links the same number as average degree of node i. We denote average degree as $\bar{d}_i$, and it is calculated by D_i/n_i, where n_i is the number of components on node i. By connecting links up to $\bar{d}_i$, each components on one node can get even degrees. It ensures fairness of substitute-finding between components.

3 Dynamic Reconstruction of Overlay Network

3.1 Motivation

Let us show a sample scenario with Fig. 1. a_5 and a_{10} are search components of components around them, and numbers below the nodes denote Service Time of each components. When component a_5 needs to find a substitute, it can find components with relatively better performance. In contrast, a_{10} can find only low-performance components with high value of S_{xi}.

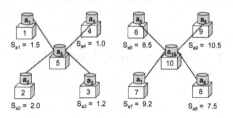

Fig. 1. An example of unfairness between two search components

This problem is caused because the overlay network had been constructed statically while the value of S_{xi} is always changing along with use. In this section, we propose a new scheme to reconstruct overlay networks dynamically.

3.2 Performance Function for Search Component

Search component x_j selects the substitute based on service time (S_{xi}). As mentioned before, service time is calculated with the process time and the network latency. Additionally, the process time on the node increases along with the number of application servers which have access to it. This means we can roughly define our performance function with the number of application servers.

The set $C = \{a_1, a_2, \cdots, a_m\}$ means neighbors of search component a_j. $N = \{n_1, n_2, \cdots, n_m\}$ denotes nodes where each component of C is working on. $S = \{s_1, s_2, \cdots, c_m\}$ is the numbers of application servers that have access to each node of N. We define our performance function as $f_j = \sum_{i=1}^{m} s_i$. High value of f_j indicates that a_k can select the substitute from the neighbors with short service time. Low value means an opposite.

3.3 Overlay Reconstruction Method

As a basic idea, we intend to make f as even as possible between search components. When application server α starts to use component a_1, also the address of search component a_{s1} is informed to α. We initiate our reconstruction focusing on this a_{s1}. First of all, we calculate performance function f_1 of a_{s1}. To smooth performance functions of all search components, we need to find the search component a_{sn} with f_n, which maximizes the value of $|f_n - f_1|$. Component a_{sn} should be selected from search components of components around a_1. Next, we sort N_1 and N_2 (node sets of a_{s1} and a_{sn}) this way: If f_1 is smaller than f_n, sort N_1 ascending with S_1 and sort N_2 descending with S_2. If f_1 is larger than f_n, sort N_1 and N_2 in an opposite manner. Then we seek N_1 and N_2 respectively, and look for two components that can exchange overlay network links.

Here "exchange" means disconnect links $a_{s1} \leftrightarrow a_{t1}$ and $a_{s2} \leftrightarrow a_{t2}$, then connect $a_{s1} \leftrightarrow a_{t2}$ and $a_{s2} \leftrightarrow a_{t1}$. If we can find the pair $\{a_{t1}, a_{t2}\}$, we simulate to exchange overlay links $\{a_{s1} \leftrightarrow a_{t1}, a_{s2} \leftrightarrow a_{t1}\}$. Then new performance functions $\{f'_1, f'_n\}$ are calculated. If the value of $|f'_n - f'_1|$ is smaller than $|f_n - f_1|$, exchange will be executed. By this process, search performance of a_{s1} and a_{sn} gets even.

4   Performance Evaluation

To evaluate the proposed method, we made a computer simulation with the following scenario. First, 100 nodes organize scale-free network. Each group of 20 nodes has the same capability: first group has 2.0, second 4.0, and so forth (6,0, 8.0, 10.0). We assume network latency of links between nodes as 1 unit time. ULD is set up to 100 times of capability. 30 replications per each 300 kinds of components were distributed to nodes randomly. All components use the same capability, 0.006 per access. There came 300 application servers, and they chose 5 components for service. We assumed one application server got γ accesses from users per unit time. It is assumed that all nodes failed respectively in the simulation, and substitute-finding method was invoked. Our model can be regarded as processor sharing model because many components are running on the same computer. We can calculate Response Time of component x_i and parameter ρ used for R_i as follows: $R_i = \frac{1/\mu_i}{1-\rho}$, $\rho = \sum_i \frac{\lambda_i}{\mu_i}$. Where λ_i is the number of application servers which use component i, multiplied by the number of accesses to the servers per unit time. μ_i is the capability of node, divided by the capability used for the service process of component i.

Let us show the evaluation result when access rate γ changed as 1 through 16. Fig. 2 shows the average of Response Time and Fig. 3 shows its variance.

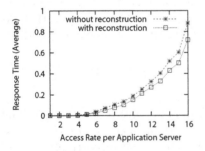

Fig. 2. Response Time (Average)

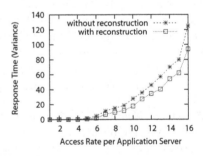

Fig. 3. Response Time (Variance)

Both Fig. 2 and Fig. 3 indicate that the proposed reconstruction method reduces the average and the variance of Response Time. It is because the proposed method makes performance of search component even, and each search component can find a good-performance substitute in case of failure.

References

1. Weerawarana, S., et al.: Web Services Platform Architecture: SOAP, WSDL, WS-Policy, WS-Addressing, WS-BPEL, WS-Reliable Messaging, and More. Prentice Hall, Englewood Cliffs (2005)
2. Hwang, S.-Y., et al.: Dynamic Web Service Selection for Reliable Web Service Composition. IEEE transactions on services computing 1(2) (April-June 2008)
3. Miyazaki, W., et al.: An Efficient Failure Recovery Scheme for Next Generation Network Services besed on Distributed Components. In: APNOMS (September 2008)

EECHE: An Energy-Efficient Cluster Head Election Algorithm in Sensor Networks*

Kyounghwa Lee, Joohyun Lee, Minsu Park, Jaeho Kim, and Yongtae Shin

Room 407 Information Science B/D, Soongsil University,
Sangdo5-dong Dongjak-gu Seoul, 156-743, South Korea
{khlee,jhlee,mspark}@cherry.ssu.ac.kr, jhkim@keti.re.kr,
shin@ssu.ac.kr

Abstract. One of the most important considerations in designing sensor nodes in a wireless sensor networks is to extend the network lifetime by minimizing an energy consumption with limited resources. In this paper, we propose an Energy-Efficient Cluster Head Election (EECHE) algorithm in sensor networks The proposed algorithm divides cluster area into two perpendicular diameters, and then elects cluster head by the density of member nodes and the distance from cluster head. Through the simulation experiments, we showed that our algorithm has improves the performance of cluster head election and provides more energy-efficient.

Keywords: wireless sensor network, clustering, energy-efficiency.

1 Introduction

A Sensor networks consist of large number of small, relatively inexpensive and low-power sensors that are connected to the wireless network[1]. Since each node only has limited energy resource and the battery recharge or replacement is impractical, sensor networks with a energy-aware design becomes important to achieve the desired life-time performance. The mean of energy-aware is that the energy spent on delivering packets from a source to a destination is minimized [2].

Recently, various clustering technique to reduce an energy consumption of sensor nodes have been developed. One of the most well-known clustering approaches is LEACH (Low-Energy Adaptive Clustering Hierarchy) [3] a clustering-based protocol that utilizes a randomized rotation of local cluster head to evenly distribute the energy load among the sensors in the network. But LEACH can not have chance to cluster head and consume their energy evenly with only probability. Unlike LEACH, where nodes self-configure themselves into clusters, LEACH-C (LEACH-Centralized) [4] utilizes the base station for the cluster formation. But, at the beginning of each round, LEACH-C causes on additional overhead to receive information from each node

---

* This research has been supported by a grant from the Korea Electronics Technology Institute 2008.

C.S. Hong et al. (Eds.): APNOMS 2009, LNCS 5787, pp. 486–489, 2009.
© Springer-Verlag Berlin Heidelberg 2009

about their location and energy level for the centralized cluster formation algorithm. HEED (Hybrid Energy-Efficient Distributed clustering) [4], that periodically elects cluster heads according to a hybrid of the node residual energy and a secondary parameter. But, HEED does not guarantee the number of elected cluster head.

In this paper, we propose an Energy-Efficient Cluster Head Election algorithm (EECHE) in the sensor networks. The proposed scheme is dividing cluster area into two perpendicular diameters to get four quadrants, then elects following cluster head by density and distance.

The structure of the paper is organized as follows: Section 2 describes an overview of the EECHE algorithm for sensor networks; Section 3 describes our application scenarios for EECHE; Sections 4 presents our evaluation and performance results; and Section 5 concludes this paper.

2 The EECHE Scheme

2.1 Network Model

During the phrase of cluster initialization, we assume the following properties about the sensor networks.

- The sensor nodes are quasi-stationary.
- All nodes have similar capabilities and can communicate using the same transmission power, but energy consumption is not uniform.
- Nodes are location-aware, i.e. equipped with GPS-capable antennae.

2.2 The EECHE Algorithm

The each steps of proposed EECHE algorithm are described as follows.

[step 1] **Local Grouping** divides cluster area into two perpendicular diameters to get four quadrants.

[step 2] **Compare the node density** that is the number of cluster members in each quadrant and select candidate quadrants.

[step 3] **Compare the node distance** that is from the nearest cluster head in candidate quadrants and select following cluster head.

Following cluster head determination, the nodes that are picked as cluster heads advertise their status to their immediate neighbors. Other nodes may also participate in the advertising process. Hearing these advertisement messages, each sensor node chooses the nearest cluster head and registers itself as a cluster member leading to the formation of clusters.

Figure 1 illustrates the Local Grouping that is divide cluster area into two perpendicular diameters at position of initial cluster header. M is a size of sensor field, R is a communication radius of initial cluster head, $C(x,y)$ is a position of initial cluster header, $C'(x',y')$ is a center of inscribed circle.

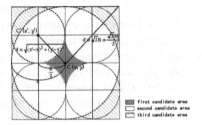

Fig. 1. Local Grouping by divide cluster area

2.3 Application Scenario

Figure 2 illustrates four cluster head election scenarios by EECHE algorithm, in which the energy consumption for communication between cluster head and member node is different by giving change in header position. Starting from a simple scenario, we describe our scheme how EECHE algorithm can elect following cluster head to support sensor application and inter-working.

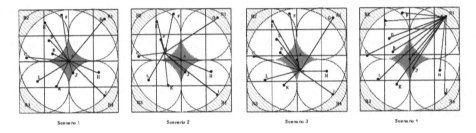

Fig. 2. Application scenarios for cluster head election

Sinario1 is example that create group within the radius of the first cluster header. It is the ideal form as control groups that set initial cluster header in center of radius. In this scenario, node *A* is a cluster head. Scenario 2 is example that changes cluster head according to EECHE algorithm. In *R2* area that have the most nodes of cluster area (thus density is high), the node *B* that is nearest from initial head elect next cluster head. In this scenario, node *B* is a cluster head. Scenario 3 is example that changes cluster head according to HEED algorithm that elect head by only consider communication distance and the remainder energy. In this scenario, node *J* is a cluster head. Scenario 4 is example that changes cluster header according to LEACH algorithm that elect cluster head by only probability without consideration about cluster form. In this scenario, node *G* is a cluster head.

3 Evaluation

We adopt the energy consumption model given in [5] for transmitting and receiving data. We simulated with the parameters of the system parameter that Radius of the region, R is 25m, Length of each packet l is 1,000bits, Electronics energy E_{elec} is 50nJ/bit, Transmitter energy ε_{fs} is 10pJ/bit/m^2, Amplier energy ε_{ms} is 0.0013pJ/bit/m^4.

$$E_{Tx}(l,d) = \begin{cases} l \times (E_{elec} + \varepsilon_{fs} x^2) & x < d_0 \\ l \times (E_{elec} + \varepsilon_{mp} x^4) & x \geq d_0 \end{cases} \tag{1}$$

$$E_{Rx}(l) = l \times E_{elec} \tag{2}$$

In order to calculate the energy consumption, we compare the different clustering protocol. Table 1 shows the energy consumption due to communication cost of once between whole node and cluster head by position of cluster. From the result, except scenario A that uses to control group, we can show that energy consumption of proposed algorithm is most efficient.

Table 1. Energy consumption for communication of once

CH	Protocol	A	B	C	D	E	F	G	H	I	J	K	L	TOT
A	initiate	0	1.7	9	4	9.2	9.7	15	5.3	13	0.7	4.2	7.3	78.8
B	EECHE	2	0	3.4	1	4.8	7.8	21	13	23	4	5	4.8	89.8
J	HEED	3.8	3.2	12	8.6	15	16	18	3.8	8.8	0	2.6	6.7	98.5
G	LEACH	16	20	40	19	19	10	0	14	29	18	34	42	261
K		4	5.3	8.4	11	20	25	35	11	13	2.6	0	1.9	136

4 Conclusion

In this paper, we propose an EECHE algorithm which improves the performance of clustering. The proposed algorithm divides cluster area into two perpendicular diameters, and then elects cluster head by the density of member nodes and the distance from cluster head.

In order to evaluating the performance of proposed scheme, we have compared the communication cost and energy consumption. Through the simulation experiments, we showed that the EECHE algorithm is energy-efficient more than about 10% than HEED, and more than about double or triple times than LEACH in some cases.

References

1. Bandyopadhyay, S., Coyle, E.J.: An energy efficient hierarchical clustering algorithm for wireless sensor networks. In: INFOCOM 2003, April 2003, vol. 3, pp. 1713–1723 (2003)
2. Xin., G., YongXin., W., Nas, L.F.: An Energy-Efficient Clustering Technique for Wireless Sensor Networks, pp. 248–252. IEEE Computer Society, Los Alamitos (2008)
3. Heinzelman, W.R., Chandrakasan, A., Balakrishnan, H.: Energy-efficient Communication Protocol for ireless Microsensor Networks. In: Proceedings of HICSS (2000)
4. Younis, O., Fahmy, S.: HEED: A Hybrid, Energy-Efficient, Distributed Clustering Approach for Ad Hoc Sensor Networks. IEEE Trans. Mobile Computing 3(4), 366–379 (2004)
5. Bandyopadhyay, S., Coyle, E.J.: Minimizing communication costs in hierarchically-clustered networks of wireless sensor. Computer Networks 44(1), 1–16 (2004)

Boundary-Aware Topology Discovery for Distributed Management of Large Enterprise Networks

Yen-Cheng Chen[1] and Shih-Yu Huang[2]

[1] Department of Information Management
National Chi Nan University, Puli, Nantou 545, Taiwan
ycchen@ncnu.edu.tw
[2] Department of Computer Science and Information Engineering, Ming Chuan University,
Taoyuan 333, Taiwan
syhuang@mcu.edu.tw

Abstract. Most of today's network topology discovery algorithms discover network topology in a best effort fashion. To facilitate distributed network topology discovery in case of multiple network management systems used for distributed management of large corporate networks, this paper introduces boundary awareness in network topology discovery and presents three practical heuristic boundary rules to help determine the boundary of the network to be managed. Implementations of these boundary rules are also presented.

Keywords: Topology discovery, boundary-awareness, distributed management, configuration management, SNMP.

1 Introduction

In recent years, several network topology discovery algorithms have been proposed [1]-[10]. SNMP-based algorithms are commonly used in commercial network management systems. SNMP-based algorithms find new topology information by retrieving the Management Information Base (MIB) provided in the routers. Most topology discovery algorithms find out subnets outward from the management station in an ad hoc manner. Best effort searches of subnets are performed until no new topology information is discovered. Conventional topology discovery algorithms have their own criteria to restrict the range the discovery. Known criteria are *exhaustive list of IP addresses, DNS zone transfer data* [5], *maximum hop count*, and *number of nodes*. However, these methods didn't originate from the perception of network boundaries. They can be regarded as criteria to stop the execution of a topology discovery algorithm.

In general, a network is administrated by an enterprise or organization. The network within an administrative domain is usually also the scope of network management. Therefore, it is important to control the discovery of network topology within an administrative domain. This paper will focus on the boundary-awareness of topology discovery for distributed management of large enterprise or ISP networks. We will consider the common characteristics of most enterprise networks and then present

C.S. Hong et al. (Eds.): APNOMS 2009, LNCS 5787, pp. 490–493, 2009.

three heuristic boundary rules useful for them. By these boundary rule tests, a topology discovery algorithm can know whether a boundary is reached.

2 Proposed Boundary Rules for Topology Discovery

Domain Boundary: Given one domain name, network topology discovery will be performed only within the network corresponding to the given domain or its certain sub-domain. The domain boundary rule is designed to limit the topology discovery within the network in an administrative domain. By the domain boundary rule, whenever a new subnet is found, we should further look for nodes with host names within the given domain. If a subnet contains nodes with host names within the given domain, the subnet will be included in the discovered topology. A network boundary may be found when a new discovered subnet contains no nodes within the given domain and contains nodes belonging to other domains.

LAN Boundary: Many enterprise intranets are constructed by one or more LANs connected by WAN links. The LAN boundary rule is to restrict the topology discovery within a LAN. By this rule, whenever a subnet is found during topology discovery, we must know whether it is a LAN. If the subnet is a LAN segment, it is within the scope of the network topology discovery. Otherwise, a boundary is reached.

Excluded Address Boundary: Given a list of IP addresses, the boundary of the network to be discovered is the set of subnets containing one of the IP addresses in the list. In contrast to the exhaustive list of all possible IP addresses to be discovered, a list of representative IP addresses is provided to explicitly specify all the possible boundaries of the network to be discovered. Each IP address in the list is used to denote an exterior subnet directly connected to the network to be discovered. By this rule, whenever a subnet is discovered and the subnet contains any one IP address of the list, the discovery has reached the boundary of the network.

3 Boundary-Aware Topology Discovery

A generic boundary-aware topology discovery algorithm, independent of the boundary rules adopted, consists of eight steps, described as follows.

1. Prepare information for a boundary rule and get the handler of the boundary rule.
2. Add the IP address of the default router of the management station into list *rList*.
3. Get the first element *r* of *rList* and remove it. If not any element is in *rList*, stop.
4. Find the subnets directly connected to router *r* and put them into the *dcSubnets* set.
5. For each subnet *s* in *dcSubnets*, determine whether *s* is within the scope of the network. If yes, put *s* in *iSubnets* and add (*s*, *r*) into *topology*.
6. If *s* is within the scope to be discovered, find the all active nodes in *s*.
7. Find all the next hop routers of *r*. The ipRouteNextHop variable in the ipRouteTable table of MIB II [11] can provide the required information.
8. For each next hop router *nhr* in a subnet in *iSubnets*, add *nhr* into *rList*. Go to 3.

In the following, we will discuss how to perform the boundary test.

Domain Boundary Test: The network administrator should first specify a domain name, denoted by dn. The domain boundary test consists of four steps:

1. For each subnet s in set $dcSubnets$, perform steps 2 and 3.
2. Find the all network nodes found in s.
3. For each node d, perform sub-steps 3.x.
   3.1: Get domain name of d, denoted by dn_d.
   3.2: If dn_d exists and is in domain dn, add s into the $iSubnets$ set and go to step 4.
   3.2: If dn_d exists and is in a domain different from dn, add s into the $oSubnets$ set.
4. Update $iSubnets$ by adding the subnets that are in $dcSubnets$ but not in $oSubnet$.

LAN Boundary Test: The ifType variable in ifTable, defined in MIB II [11], indicates the type of a network interface. It also implies whether the corresponding network is a LAN. The proposed test is described as follows.

1. Send SNMP get-next-request messages to router r to get the ipAddrTable table.
2. For each subnet s in set $dcSubnets$, perform the following steps.
3. By searching the ipAddrTable, determine the interface index ind, i.e. the ipAdEntIfIndex variable, of the network interface attached to subnet s.
4. Send a SNMP get-request message to router r to retrieve the ifType variable in the entry of the ifTable table with ifIndex = ind.
5. If the value of ifType is in the LAN type list, add s into $iSubnets$.

Excluded Address Boundary Test: A small list of IP addresses is prepared in advance. Each IP address in the list is an address outside of the network to be discovered. The detail of the test is described as follows.

1. For each subnet s in set $dcSubnets$, perform the following steps.
2. Get the subnet mask of s, denoted by sm.
3. For each IP address nip in list $boundAddrList$, perform steps 4 and 5.
4. Obtain nid by computing the bitwise AND of nip and sm.
5. If nid is identical to s, then mark s as an excluded subnet.
6. If s is not an excluded subnet, add s into set $iSubnets$.

4   Experiments

We conducted three experiments on the campus network of Ming Chuan University, with two campuses located in Taipei and Taoyuan respectively. These two campus LANs are connected by leased lines. Fig. 1 illustrates the campus network of Ming Chuan University. Our experiments were made to discover the Taoyuan campus network. By applying the domain boundary rule to discover networks in domain 'tj.mcu.edu.tw', we successfully found the networks in the Taoyuan campus, as shown in the lower part of Fig. 1. By applying LAN boundary rule, the discovery program found T1 links and stopped the discovery of networks beyond T1 links. Accordingly, the networks within the Taipei campus were not included in the discovered topology. By applying excluded address boundary rule, the algorithm didn't discover networks beyond the given IP addresses in the exclusive address list.

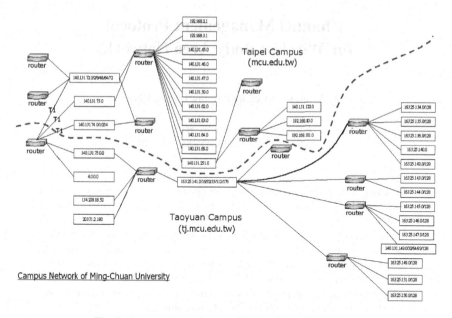

Fig. 1. Campus network map of Ming Chuan University

References

1. Breitbart, Y., Garofalakis, M., Martin, C., Rastogi, R., Seshadri, S., Silberschatz, A.: Topology Discovery in Heterogeneous IP Networks. In: Proceedings of IEEE Infocom 2000, vol. 1, pp. 265–274 (2000)
2. Govindan, R., Tangmunarunkit, H.: Heuristics for Internet Map Discovery. In: Proceedings of IEEE Infocom 2000, vol. 3, pp. 1371–1380 (2000)
3. Lin, H.C., Lai, S.C., Chen, P.W., Lai, H.L.: Automatic Topology Discovery of IP Networks. IEICE Transactions on Information and Systems E83-D(1), 71–79 (2000)
4. Lin, H.W., Wang, Y.F., Wang, C.H., Chen, C.L.: Web-based Distributed Topology Discovery of IP Networks. In: Proceedings of 15th International Conference on Information Networking, pp. 857–862 (2001)
5. Sianwalla, R., Sharma, R., Keshav, S.: Discovering Internet Topology,
   http://www.cs.cornell.edu/skeshav/papers/discovery.pdf
6. Lowekamp, B., et al.: Topology Discovery for Large Ethernet Networks. In: Proc. of ACM SIGCOMM 2001 (2001)
7. Barford, P., et al.: On The Marginal Utility of Network Topology Measurements. In: Proc. of ACM/SIGCOMM Internet Measurement Workshop (2001)
8. Spring, N., Mahajan, R., Wetherall, D.: Measuring ISP Topologies with Rocketfuel. In: Proc. of ACM/SIGCOMM 2002 (2002)
9. Lakhina, A., Byers, J., Crovella, M., Xie, P.: Sampling Biases in IP Topology Measurements. In: Proc. of IEEE Infocom (2003)
10. Emma, D., Pescape, A., Ventre, G.: Discovering topologies at router level. In: Magedanz, T., Madeira, E.R.M., Dini, P. (eds.) IPOM 2005. LNCS, vol. 3751, pp. 118–129. Springer, Heidelberg (2005)
11. McCloghrie, K., Rose, M.: Management Information Base for Network Management of TCP/IP –based internets, RFC 1213 (1991)

Channel Management Protocol
for Wireless Body Area Networks

Wangjong Lee and Seung Hyong Rhee

Kwangwoon University, Seoul, Korea
{woorihope,rhee}@kw.ac.kr

Abstract. Wireless body area networks are operated within 3m around a human body. The 400 MHz frequency band for the medical implanted communication (MICS) has already been used for other service and consists of 10 channels with 300 kHz bandwidth each. Although wireless BANs consider interference mitigation techniques (LBT and AFA), they are not able to guarantee a reliable transmission. In this paper, we suggest a new reservation mechanism for multi-channel management in wireless BANs. Our mechanism achieves performance enhancements and coexistence with conventional techniques. Simulation results also show that can be achieved efficient channel utilization.

Keywords: wireless BANs, multi-channel management, coexistence.

1 Introduction

Wireless body area networks (WBANs), which enable wireless communications within 3m around a human body, are under development by the IEEE 802.15.6 Task Group. Wireless BANs have considered various frequency bands: 2.4GHz ISM band, a wide frequency band for UWB and 400 MHz band for the MICS (medical implanted communication service). The MICS frequency band that consists of 10 channels with 300 kHz bandwidth has already been used for the Metaids (meteorological aids service). Therefore, this frequency band has need of a channel management mechanism. Wireless BANs manage multi-channels by using *listen before talk (LBT)* and *adaptive frequency agile (AFA)*. These contention based mechanisms prevent devices from causing interferences by sensing and hopping, but cannot guarantee a reliable transmission because any device cannot maintain the channel occupancy.

In this paper, we propose a multi-channel management mechanism that improves both throughput and reliability. This mechanism is based on the reservation mechanism. The coordinator manages data channels through the control channel. It also provides means for coexistence with other conventional mechanisms. We perform simulations to show that our mechanism achieves the performance enhancement.

2 Beacon Based Multi-channel Management Scheme

The current MICS frequency band consists of 10 channels with 300 kHz bandwidth. In our proposed mechanism, one is used as the control channel and other channels are

C.S. Hong et al. (Eds.): APNOMS 2009, LNCS 5787, pp. 494–497, 2009.

used for data transmissions. After scanning all channels, the coordinator sets one unused channel to the control channel. It allocates other unused channels to inbody devices by sending beacons based on a superframe through the control channel. Fig. 1 shows the channel plan and the superframe which is the basic timing structure of the control channel. The superframe is composed of 9 beacon slots which have durations of less than 5ms. During the beacon slot, the coordinator transmits a beacon frame that contains information for the channel reservation. By listening to the control channel, devices know which channels are used and reserved for communications between the coordinator and devices.

To start a communication between the coordinator and a device, the coordinator transmits a beacon via the control channel. On receiving its beacon, the device sends the ack frame in reply to the beacon and transacts an interface to the allocated data channel. When the coordinator receives the ack frame, it also transacts to the same data channel. After communicating on the data channel, the coordinator returns to the control channel. If the coordinator wants to communicate with other devices, it senses all data channels and sends beacons again. The device returns to the control channel or falls into a sleep state according to its duty cycle.

Because all devices know information of the channel allocation by listening to the control channel, it is important to transmit beacons on the control channel. But *listen before talk (LBT)* allows LBT devices to occupy channels when channels are idle for more than 5ms. If there are no beacons during some adjacent beacon slots, LTB devices can occupy the control channel. To prevent LBT devices from occupying the control channel, we propose the beacon repeat mechanism. The coordinator transmits the beacon and checks whether there are beacons to be sent or not. If there are no beacons for two consecutive slots, the coordinator copies the previous beacon and transmits the beacon during the second slot. This prevents the idle duration from exceeding more than 5ms.

Fig. 3 (a) shows that there are adjacent empty beacon slots due to coordinator 1 translating to its data channel. In the first superframe, two coordinators transmit beacons without a series of empty slots. Coordinator 2 receives the ack frame during the third beacon slot, and moves to the allocated data channel. This process leads two continuous slots to being empty in the second superframe. To prevent two adjacent empty slots, coordinator 2 sends its beacon during the forth slot instead of the fifth slot. The beacon slot shift hereby prevents LBT devices from occupying the control channel (Fig. 3(b)).

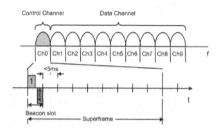

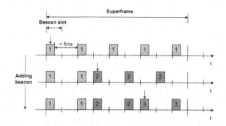

Fig. 1. Channel plan and superframe structure   **Fig. 2.** Beacon repeat by a coordinator

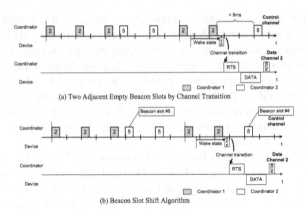

(a) Two Adjacent Empty Beacon Slots by Channel Transition

(b) Beacon Slot Shift Algorithm

Fig. 3. Beacon slot shift

The coexistent problem may occur on the data channel. If the device transmits no data for more than 5ms on the allocated data channel, the channel can be occupied by LBT devices. Because the device transmits data frames without sensing process, LBT devices and the device may cause collisions. Fig. 4 shows the RTS/ACK mechanism. After the transaction, the coordinator sends RTS within 5ms until the device sends data frame. Upon the device receiving the RTS, it transmits data frame and receive the ack frame in reply to the data frame. If the coordinator has a request for more data frames it must send the RTS within 5ms. When the device receives no RTS for more than 5ms, the device considers that data transmissions finished and releases the data channel.

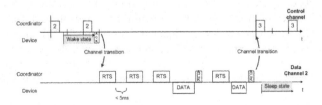

Fig. 4. RTS/ACK on data channel

3 Simulation Analysis

We have implemented our proposed mechanism in the ns-2 network simulator with the multi-channel module. In our simulations, listening times set to a smaller value than the value of RFID or sensor to organize the high-performance environment. Parameters for simulations are summarized in Table 1.

Table 1. Parameters for simulations

Attribute	Value
Number of channels	10 channels
Bandwidth of a channel	300 kHZ
Number of flows	2 flows
Listening Time	0.1 ms / 0.5
Traffic types	CBR

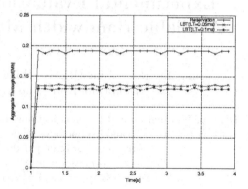

Fig. 5. Simulation results

To compare aggregate throughputs on MICS environments, we performed simulations on multi-channel environments. Fig. 5 shows simulation results for the conventional mechanism and our proposed mechanism. In the reservation simulation, devices transmit data frames on each independent channel without collisions after monitoring the control channel. These two channels are independent of each other. The LBT device has a low probability of collision because of channel sensing. The sensing process, however, causes more overhead. The performance of the LBT simulation is affected by the listening time. As the listening time is longer, the channel efficiency is decreased. To analyze effects of the listening time, we used 0.1ms or 0.5ms.

4 Conclusion

We have proposed a multi-channel management mechanism for IEEE 802.15.6 wireless BANs in order to achieve performance enhancements and coexistence. This mechanism is the channel reservation mechanism in which the coordinator assigns channels to devices. It also provides the backward compatibility by preventing reserved channels from being idle for listening time of the LBT technique. Simulation results show that our mechanism improves performance.

References

1. Zhen, B., Patel, M., Lee, S., Won, E., Astrin, A.: TG6 Technical Requirements Document (IEEE P802-15-08-0644-09-006). IEEE 802.15 WPAN document, IEEE (2008)
2. Tzamaloukas, A., Garcia-Luna-Aceves, J.J.: Channel-Hopping Multiple Access. In: IEEE International conference on Communications 2000, vol. 1, pp. 18–22. IEEE Press, New Orleans (2000)
3. Lee, W., Rhee, S., Kim, Y., Lee, H.: An Efficient Multi-Channel Management Protocol for Wireless Body Area Networks. In: The International Conference on Information 2009, Korea Institute of Information Scientists and Engineers, Chian Mai, Thailand (2009)

Experimental Evaluation of End-to-End Available Bandwidth Measurement Tools

Young-Tae Han[1], Eun-Mi Lee[1], Hong-Shik Park[1], Ji-Yun Ryu[2],
Chin-Chol Kim[2], and Yeong-Ro Lee[2]

[1] Dept. of Information and Communications, Korea Advanced Institute of Science
and Technology, Daejeon, S. Korea
{organon,boricha,parkhs}@kaist.ac.kr
[2] National Information Society Agency, Seoul, S. Korea
{rjy,cckim,lyr}@nia.or.kr

Abstract. Accurate bandwidth measurement is essential for network management, monitoring, and planning. Various active probing-based strategies and tools are have been developed for estimating available bandwidth. However, the test result of each tool could be different according to the strategies and tools even in the same network environment. The purpose of this paper is to give comprehensive information to users with interpretive comparison and performance test. The comparison and test with regard to measurement efficiency, and traffic load of probing packets. including IGI/PTR, pathload, pathChirp, spruce, and Iperf, are performed in dynamic network environments.

Keywords: Active measurement, available bandwidth, performance measurement, network monitoring.

1 Introduction

Accurate bandwidth measurement is essential for network management, monitoring, and planning. Various active probing-based strategies and tools are have been developed for estimating available bandwidth. However, the test result of each tool could be different according to the strategies and tools even in the same network environment not only because the dynamic network environment causes variations of test results but also because these tools are not standardized. Therefore, users should be preacquainted with differences of performance and characteristics of them. The purpose of this paper is to give comprehensive information to users with interpretive comparison and performance test. To provide comprehensive information about performance of existing tools for end-to-end available bandwidth measurement, the comparison and test of these tools, including IGI/PTR [1], pathload [2], pathChirp [3], spruce [4], and Iperf [5], is performed in terms of measurement efficiency, and traffic load.

The remainder of this paper is organized as follows. Section 2 summarizes bandwidth measurement tools and describes test environments and section 3 shows the results. Finally, section 4 concludes this paper.

C.S. Hong et al. (Eds.): APNOMS 2009, LNCS 5787, pp. 498–501, 2009.

2   Tools and Test Scenarios

Several open source tools for estimating available bandwidth are selected as following table 1.

Table 1. Available Measurement Tools

Tool	Author	Version	Method	Protocol
pathload	Jain	1.3.2	SLoPS	UDP
pathChirp	Ribeiro	2.4.1	SLoPS Chirp	UDP
IGI/PTR	Hu	2.1	SLoPS	UDP
Spruce	Strauss	0.2	PGM	UDP
Iperf	NLANR	2.0.4	Achievable TCP throughput	TCP/UDP

Due to the page limit , our test environment is not presented but our test environments is similar to [6]. To generate cross traffic, AX4000 is exploited. The tests are performed in the following scenarios:

(1) All measurement tools are tested without any cross traffic and all probing packets are traced and analyzed using tcpdump in terms of the number, traffic volume, and packet length distributions of probing packets. (2) All measurement tools are tested with cross traffic; 50, 100,150, 200, 250, 300, and 350Mbps. CBR cross traffic is injected along the path and generated packets are 1500-byte length of TCP. (3) All measurement tools are tested while increasing delay. The delay is increased by 50ms up to 400ms. (4) All measurement tools are tested while increasing packet loss. The packet loss rates of path are 1, 3, 5, and 10%.(5) Performance evaluation of the single and parallel mode of Iperf in lossy environment is performed. The network delay and packet loss are imposed with TC [7] at the linux-based network emulator.

3   Test Results

Table 2 shows workload characteristics of each tools. Time is the duration between the first probing packet and the last probing packet passed on the path. Packets are the total number of observed probing packets. V_p is the total volume of probing packets. To show efficiency and accuracy, we defined two metrics; the average probing rate, $R_p(V_p/T)$ and measurement error, $\varepsilon(|A_e - A_{ee}|/A_e)$ where A_{ee} is actual end-to-end available bandwidth (400Mbps) without cross traffic.

Fig. 1 shows the test results when network delay gradually increased. Only Iperf is affected by delay because TCP throughput is significantly affected by round trip time (RTT). According to [8], TCP throughput is being decreased when RTT is being increased.

Fig. 2 and 3 show test results of single and parallel modes of Iperf without any network congestion such as delay and loss.In the lossy network, we observed that

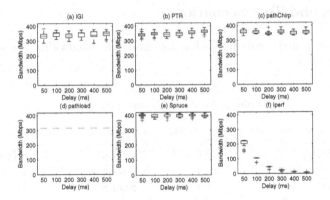

Fig. 1. Test results with delay increment

Table 2. Summary of probing packets

Tool	Time (second)	Packets	V_p (bytes)	Result (A_e) (Mbps)	R_p (Mbps)	ε
IGI/PTR	1	540	3,983,120	IGI-345	30.389	0.139
				PTR-351		0.122
pathload	1	946	564,512	315	4.307	0.213
pathChirp	120	39,618	40,489,332	348	2.574	0.130
Spruce	9	200	300,000	405	0.254	0.011
Iperf	10	510,518	60,963,944	407	46.512	0.020

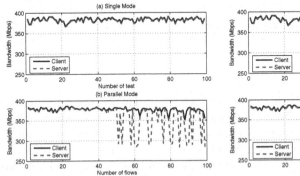

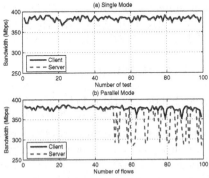

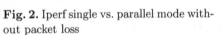

Fig. 2. Iperf single vs. parallel mode without packet loss

Fig. 3. Iperf single vs. parallel mode with packet loss

the performance of the parallel mode is unstable. According to our observation and [9], the loss rate of p is the dominant factor of the throughput of aggregate TCP flows. Thus, the parallel mode upon TCP is not appropriate for estimating bandwidth in the lossy network.

4 Conclusion

Several available bandwidth measurement tools have been evaluated . To verify the influence of delay and loss, tests were performed in dynamic network environments. Intrusiveness with respect to Iperf is the highest probing rate and this high probing rate may cause loss of cross traffic. According our observation, the TCP-based tool is more sensitive to delay, and packet loss than the UDP based tools. The delay is more critical than packet loss within TCP based tools. In UPD based tools, some tools such as pathChirp use sampling based estimation methods. In this case, the measurement performance is deteriorated by the packet loss. Thus, to develop more accurate and non-intrusive tools, more complex and dynamic environments should be considered.

Acknowledgments. This work was supported in part by the MKE, Korea, under the ITRC support program supervised by the IITA (IITA-2009-(C1090-0902-0036)).

References

1. Ningning, H., Steenkiste, P.: Evaluation and characterization of available bandwidth probing techniques. IEEE Journal on Selected Areas in Communications 21(6), 879–894 (2003)
2. Jain, M., Dovrolis, C.: End-to-end available bandwidth: measurement methodology, dynamics, and relation with tcp throughput. IEEE/ACM Trans. Netw. 11(4), 537–549 (2003)
3. Ribeiro, V., Riedi, R., Baraniuk, R., Navratil, J., Cottrell, L.: Pathchirp: Efficient available bandwidth estimation for network paths. In: Passive and Active Measurement Workshop, vol. 4 (2003)
4. Strauss, J., Katabi, D., Kaashoek, F.: A measurement study of available bandwidth estimation tools. In: Proceedings of the 3rd ACM SIGCOMM conference on Internet measurement, Miami Beach, FL, USA, pp. 39–44. ACM Press, New York (2003)
5. Iperf, http://sourceforge.net/projects/iperf/
6. Montesino-Pouzols, F.: Comparative analysis of active bandwidth estimation tools. In: Passive and Active Network Measurement, pp. 175–184 (2004)
7. TC, http://tldp.org/HOWTO/Traffic-Control-HOWTO/index.html
8. Padhye, J., Firoiu, V., Towsley, D., Kurose, J.: Modeling tcp throughput: a simple model and its empirical validation. In: Proceedings of the ACM SIGCOMM 1998 conference on Applications, technologies, architectures, and protocols for computer communication, Vancouver, British Columbia, Canada, pp. 303–314. ACM Press, New York (1998)
9. Hacker, T.J., Athey, B.D., Noble, B.: The end-to-end performance effects of parallel tcp sockets on a lossy wide-area network. In: Parallel and Distributed Processing Symposium., Proceedings International, IPDPS 2002, Abstracts and CD-ROM, pp. 434–443 (2002)

A Routing Management among CSCFs Using Management Technology

Ji-Hyun Hwang, Jae-Hyoung Cho, and Jae-Oh Lee

Information Telecommunication Lab,
Dept. of Electrical and Electronic Engineering
Korea University of Technology and Education, Korea
{korea4195,tlsdl2,jolee}@kut.ac.kr

Abstract. The IP Multimedia Subsystem (IMS) in ALL-IP based next generation communication environment can offer network/service providers various
multimedia services (e.g., voice, audio, video and data etc). The IMS is worthy
of notice in that it is possible to make wired/wireless convergence and quick
service development. The IMS needs an efficient Session Initiation Protocol
(SIP)-based routing among Call Session Control Functions (CSCF) in charge of
call and session processing. In this paper, we propose a routing management
among CSCFs using management technology.

1 Introduction

The IMS in the ALL-IP based next-generation communication environment is located
in the core technology. The reason why the IMS platform is attracted is the factor that
could rapidly develop and modify services and enable the customer to provide various
contents such as propagation of Voice over Internet Protocol (VoIP) among corporations, rapid progress of video transmitting/receiving, account management by a mobile phone, web-browsing and e-mail etc. The IMS can create services and has the
advantage that can control services by using internet-based technology regardless of
types of transmission network. Accordingly, the IMS improves price competitiveness
of services, makes easy interaction with various 3rd party applications based on efficient session management functions and permits the expansion of business areas
through interaction among services. In this paper, we propose a routing management
among CSCFs using management technology for efficient SIP Routing among CSCFs
in charge of call/session processing in the IMS network with FCAPS that is network
management function [1, 2].

2 The IMS Management System

There are a number of network management protocols (e.g., SNMP CORBA, JAVA
RMI, XML/HTTP, and so on). However, the standard group such as 3GPP and ITUT
was not concretely standardized interface between the CSCF and the management
system. In this paper, we will use the SNMP protocol for dynamic SIP routing in that

C.S. Hong et al. (Eds.): APNOMS 2009, LNCS 5787, pp. 502–506, 2009.
© Springer-Verlag Berlin Heidelberg 2009

setup and management of vendor's equipment are convenient [3]. The management system is able to obtain information related to performance of many P-CSCFs, S-CSCFs and I-CSCFs in the own home network. It periodically performs by polling through a Get-Request message and a Get-Response message. The CSCF agents report an extraordinary event which the CSCFs can generate to the management system by sending a Trap message. Figure 1 explains the IMS management system architecture.

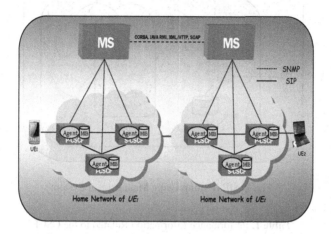

Fig. 1. The IMS Management System Architecture

The IMS session establishment is represented in figure 1. Given that the *UE1* and the *UE2* were registered, the *UE1* sends the SIP INVITE method included Session Description Protocol (SDP) to the P-CSCF in order to initiate the IMS Session with the *UE2*. In a static routing [2], it sends the INVITE method to the S-CSCF that was selected by the HSS. But, when the existing S-CSCF becomes overloaded in the viewpoints of the CSCF performance, the management system applies the dynamic routing algorithm to propose this paper. In other words, the management system is required to modify the SIP routing for the appropriate S-CSCF based on performance information criteria. Consequently, the P-CSCF can send the INVITE method to the S-CSCF that is selected by management system.

2.1 A Dynamic SIP Routing Algorithm Using Management System

Figure 2 denotes the forwarding procedure of a SIP message from the I-CSCF to the S-CSCF. When a lot of S-CSCFs (e.g., S-CSCF-A, S-CSCF-B, S-CSCF-C) exist in the IMS network, the management system notifies the I-CSCF of the appropriate S-CSCF's SIP URI (i.e., by sending a Set-Request message). And then, the I-CSCF can transmit the SIP message to the selected S-CSCF. The management system has CSCF's performance information, routing information and so on (i.e., the information is included in the management database). Finally, it is possible that the I-CSCF can decide the appropriate S-CSCF by performing the SIP routing algorithm of the management system criteria. One of the most important points for accomplishing dynamic SIP routing is to select management information objects. Major criteria on performance information in the dynamic SIP routing algorithm are expressed in Table 1.

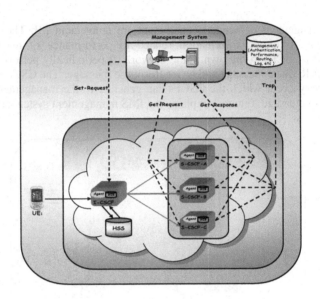

Fig. 2. SIP Routing Using Management System

Table 1. Performance Information Related to the CSCF

Priority	Performance List	Evaluation Criteria
1	System Down	On/Off
2	CPU Utilization	Tc
3	Memory Utilization	Tm
4	Throughput	High value is good.
5	Response Time	Low value is good.

The Threshold of CPU (Tc) Utilization and the Threshold of Memory (Tm) Utilization can be properly altered according to the decision-making procedure of management system. Throughput indicates processing ability of the CSCF and Response Time means transmission delay time of the CSCF. There is various performance information related to the CSCF. In current, we consider System down, CPU Utilization, Memory Utilization, Throughput and Response Time. The algorithm using Table 1 information is represented in figure 3.

The Management system periodically polls performance information from the CSCF agents by sending a Get-Request message and a Get-Response message. The Get-Request message Time (Gt) denotes time to send the Get-Request message. The Period of Time (Pt) means a cycle unit to send the Get-Request message. When Pt is shorter, each CSCF becomes overloaded in the viewpoints of network traffic. When Pt is longer, the management system is difficult to extract correct performance information. Therefore, it is remarkably important to set Pt in that the CSCF agent exceeds

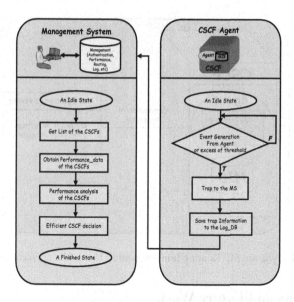

Fig. 3. A Dynamic SIP Routing Algorithm

Threshold (*Tc, Tm*) or generates an extraordinary event. Given that there are many CSCFs in the IMS network, the management system forwards a Get-Request message according to the *Pt*. And then, the management system can extract performance information from the CSCF agents, analyze performance information and modify routing information (i.e., routing information is included in the management database). Finally, the CSCFs can dynamically route to the appropriate CSCF. When the CSCF agents exceed Threshold (*Tc, Tm*) or generate an extraordinary event, the CSCF agents report a Trap message to the management system. And then, the management system save descriptions about the trap message (i.e., the descriptions are recorded on a Log information in the management database).

3   An Implementation Environment

Figure 4 shows the architecture of CSCF that performs the dynamic routing of a SIP message using the management system.

In figure 4, a management agent in CSCF periodically provides the management system with related performance information (i.e., by sending a Get-Response message). Therefore, the management system can offer the CSCF efficient routing information by using the dynamic SIP routing algorithm. That is, the management system uses Dynamic Routing Decision Module. Early routing information uses Java class file that is *ProxyRouterClass* which is defined in *JSIP* [6]. Therefore, the management system performs dynamic routing function of the SIP message by controlling Routing Action Module to the CSCF agent (i.e., through the SNMP).

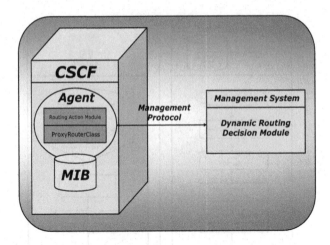

Fig. 4. Dynamic SIP Routing Implementation Using Management System

4 Conclusion and Future Work

To dynamically route the SIP message among CSCFs that take charge of part related with call and session processing in the IMS platform, we proposed a routing management among CSCFs using management technology instead of existing static routing. We designed a decision-making algorithm based on mainly F (Fault), P (Performance) in FCASPS that is network management function. A further direction for this research will be to extend the algorithm which can perform the dynamic routing capability by considering much more performance information of the CSCF.

References

1. Camarillo, G., Gracia-Martin, M.A.: The 3G IP Multimedia Subsystem (IMS), 2nd edn. Wiley, Chichester (2006)
2. 3GPP TS 23.228, IP Multimedia Subsystem (IMS), v.6.5.0 (June 2006)
3. Case, J.: A Simple Network Management Protocol (SNMP), IETF RFC 1157 (May 1990)
4. Rosenberg, J., et al.: SIP: Session Initiation Protocol, IETF RFC 3261 (June 2002)
5. 3GPP, IP Multimedia (IM) Subsystem Cx and Dx Interfaces; Signaling flows and message contents. TS 29.228, 3rd Generation Partnership Project (3GPP) (June 2005)
6. https://developer.opencloud.com/devportal/devportal/apis/jainsip/1.2/docs/index.html
7. Hwang, J.-H.: Routing Algorithm for Dynamic Management of CSCF Agent in the IMS Platform. In: Korea Network Operations and Management(KNOM)-Review (November 2008)

The Network Management Scheme for Effective Consolidation of RFID-Tag Standards

Yusuke Kawamura and Kazunori Shimamura

Kochi University of Technology, Japan
yusuke.kawamura@p-lab.jp

Abstract. The RFID tag sensor networks have been developed as the new infrastructure which might bring the next generation ubiquitous networking. However, the RFID network has difficulty of the effective mutual communication with another RFID network, because plural standards coexist for RFID tags. To solve this difficulty, the RNI Standard has been proposed as the network management method which enables the consolidation among the RFID tag standards. The RNI Standard originally proposed was unfortunately insufficient for the smooth communication. This study analyses the problems of the RNI Standard. And the Operational Platform based on the standard is newly proposed with the verification of it's expandability and practicality. The proposed scheme introduces the RFID tag separation processing environment. The proposed management scheme would be useful for the coming applications utilizing the plural RFID tag standards.

Keywords: RFID tag sensor network, RFID tag standard, RFID network integration standard, RNI Standard Operation Platform.

1   Introduction

The RFID (Radio Frequency Identification) tag is a smart device facilitating the wireless telecommunication function within an IC tip. The sensor network utilizing the RFID tags has been developed as an important basic infrastructure to realize the coming ubiquitous networked society in near future. However, two or more RFID tag standards which regulate the content of the storage region of the RFID tag. They are EPC, ucode and etc.. Because of the current plural standards and network environments, it is difficult for the RFID network of some standard to be connected to the other RFID network. Therefore, a serious obstacle now exists for the ubiquitous network society to utilize a RFID effectively.

1.1   RFID Network Integration Standard

The RFID Network Integrated Standard (RNI standard)[1] is proposed in order to solve the standard problem of the present RFID tags. The RNI standard makes it possible for the RFID network applications to use the plural RFID

C.S. Hong et al. (Eds.): APNOMS 2009, LNCS 5787, pp. 507–510, 2009.

tag standards in the environment unified. The schemes and features of the RNI standard could be explained as follows. The RNI standard aims to allow one RFID tag to be written down some RFID tag information of the different standards. The data within the RFID tag is described as the XML file. The sensor network based on the RNI standard accepts the XML data from the RFID tag interprets the sentences structure of the data and furthermore transmits it to the application server appropriate. The RNI standard can solve the problem of the standards in it's protocol. While the RNI standard is effective for the applications combining different standards RFID tags, it has some problems to be covered for the implementation. The management network which uses RNI Standard[2] lacks the structure to determine a usage of the plural RFID tag information. Therefore, the management network cannot use information in the RFID tag effectively. Moreover, there is a problem concerning to the speed of the response when the RFID tag is read. The management network takes too much time in operation compared to the services using the other electronic tags. Consequently, the basic scheme of this management network should be improved for the implementation for the industrial applications.

2   RNI Standard Operational Platform

The RNI Standard Operational Platform is proposed to solve the problematical point of management network which uses RNI Standard.

2.1   RNI Standard Management Platform Model

The RNI Standard Management Platform model proposed newly is shown in the figure [1]. RNI Standard Operational Platform consists of three facilities. They are RFID tag reading environment, RFID tag information decipher system and RFID application server. The RFID tag reading environment consists of parent RFID_R/W (Reader and Writer) and child RFID_R/W. Child RFID_R/W is located around parent RFID_R/W. Child RFID_R/W reads the RFID tag outside the intention of the user. And it transmits the RFID tag data to the Integration Standard Management System. The Integration Standard Management System generates an XML file from the RFID tag data received and performs sentence structure interpretation. The provided RFID tag information is saved in an RNI database. Next, Parent RFID_R/W transmits the RFID tag identification number of the read RFID tag due to the user's intention to the RFID application management system. The RFID Application Management System gets the RFID tag information which has made beforehand on the basis of the RFID tag authentication number from the RNI database. The RFID Application Management System gets the connection information of the RFID application server from Cache server or ANS (Application Name System). Where the Cache server is installed for shortening the processing time of searching the location information from ANS. And it transmits RFID tag information to an RFID application server by referring to the connection information. RNI Standard Operational

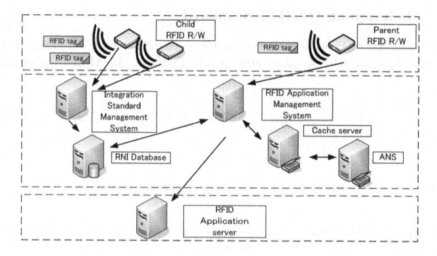

Fig. 1. RNI Standars Operation Platform model

Platform aims improvement of the response speed by performing preparation of RFID tag data by reading operation of child RFID_R/W beforehand. The user recognizes only the time from the reading of RFID tag which the user intends to the beginning of the application server. The user does not recognize the time for the generation and sentence structure interpretation of the XML file.

2.2   Operation Flow of the RNI Standard Management Platform

The flow from reading the RFID tag of child RFID_R/W to activation of the RFID application is shown below.

1. Child RFID_R/W reads a RFID tag and transmits the RFID tag data to the Integration Standard Management System.
2. The Integration Standard Management System generates an XML file from the RFID tag data received and performs sentence structure interpretation processing.
3. The RNI database stores RFID tag information provided by sentence structure interpretation processing.
4. Parent RFID_R/W reads a RFID tag and transmits the identification number in RFID tag data to the RFID Application Management System.
5. The RFID Application Management System collects RFID tag information from the RNI database by referring the received identification number.
6. The RFID Application management system collects the connection information of the application server which will be connected from the Cache server or ANS.
7. The RFID Application Management System transmits RFID tag information to the RFID application server by referring to the connection information.
8. The RFID application server activates the application.

3   Redesigning the RNI Standard

The data structure of the RNI Standard was redesigned as shown in figure [2] to use it on the RNI Standard management Platform. An ARM (Application

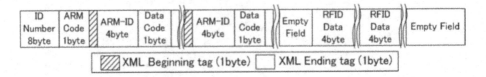

Fig. 2. The data structure of the RNI standard

Relationship Management) cord is added to distinguish the usage of the RFID tag. The usage defined by the ARM code is the following four kinds.

- A general usage
  One RFID tag can use only one RFID application.
- A joint ownership usage
  The information which Application of plural similar services utilizes in one RFID tag is described.
- A cooperation usage
  The information which RFID application of plural various services utilizes in one RFID tag is described.
- A compound usage
  Many information that one RFID application uses for one RFID tag is described.

4   Conclusion

In this research the problem regarding the speed when the management network uses original RNI Standard is solved with the parent and child structure of RFID_R/W in RNI Standard Operational Platform. In addition, as for the direction for uses of the RFID tag, the joint ownership usage, a cooperation usage and a compound usage are enabled by a new RFID network unification standard.

References

1. Nozaki, T., Shimamura, K.: Evaluation of utility by the application of using the RFID Network Integrated StandardSubsequences. In: Internationl Conference on Next Era Information Networking (NEINE 2008), pp. 19–21 (2008)
2. Kawamura, Y., Nozaki, T., Shimamura, K.: A caching method for RFID application to shorten data referring time in IP network. In: Internationl Conference on Next Era Information Networking (NEINE 2008), pp. 105–108 (2008)

ROA Based Web Service Provisioning Methodology for Telco and Its Implementation

WonSeok Lee, Cheol Min Lee, Jung Won Lee, and Jin-soo Sohn

Application Service Department, Central R&D Laboratory, Korea Telecom
{alnova2,cmlee,jwl,jssohn}@kt.com

Abstract. A web service is a software system designed to support interoperable machine-to-machine interaction over a network. Recently it is in the limelight being used for implementing Service Oriented Architecture (SOA). Traditionally, the web service for SOA is provisioned based on Simple Object Access Protocol (SOAP). In the advent of web 2.0, Representational State Transfer (REST) which focused on using the Web as it is by following its basic principles is come into being. As REST is becoming more popular, the design paradigm melted in REST makes Resource Oriented Architecture (ROA) an alternative or a complement to SOA. Telco has traditionally heavy transaction systems and network control systems are managed in a closed manner. So it is not easy to provide web services. In this paper, we propose a ROA-based web service provisioning methodology and a platform architecture for Telco. We also explain the implementation of the ROA-based web service platform in KT.

Keywords: ROA, SOA, Web 2.0, REST, SOAP.

1 Introduction

Web service is a subset of service and exposes a set of APIs to the client that can be accessible on the Web [1]. It is used to implement SOA, which includes three parts: Interaction Protocol, Service Description, and Service Discovery. Different protocols are designed to satisfy these parts. SOAP is defined for the message transportation. Web Services Description Language (WSDL) document describes a Web Service's interfaces, data type and provides users with a point of contact. Universal Description Discovery and Integration (UDDI) offers users a unified and systematic way to find service providers through a centralized registry of services [2] SOA is a kind of service-oriented distributed computing technology. SOA emphasizes re-usable service components and the interoperation between them. It uses web technology to provide a communication between heterogeneous service components. SOA using Web service applies the concept of services to Web. So it never uses key success factors of web but just use web as a transportation method. It is the biggest disadvantage that SOA is buried under a load of specifications in an effort to be platform agnostic and portable. Another disadvantage is that services are not discoverable. Furthermore, it is not suitable for simple interaction because SOAP needs a XML parser.

C.S. Hong et al. (Eds.): APNOMS 2009, LNCS 5787, pp. 511–514, 2009.

The term of REST were introduced in 2000 in the doctoral dissertation of Roy Fielding [3], one of the principal authors of the Hypertext Transfer Protocol (HTTP) specification. It is one of the architectural styles for distributed hypermedia systems that is suitable design style for ROA. ROA is considered to be a radical approach compared to SOA since ROA applies the concept of web to services [4]. Web is exploratory by nature and operated by well defined protocols and these are key points of the success and popularity of web.

Telco has the open API technology called Parlay [5] and it is extended to offer web services based on SOAP. But in the web 2.0 environments, more adaptable and agile web service providing method is needed. So we propose ROA-based web service provisioning methodology for Telco and explain the implementation of the ROA-based web service platform in KT.

2 ROA Based Web Service Provisioning Methodology for Telco

Telco has the open API technology which uses the unconstrained architecture using Gateway like PINT (PSTN/IN-Internet Internetworking) and uses the standard like Common Object Request Broker Architecture (CORBA) or SOAP. The latter is called Parlay or Parlay X. ROA is the current architecture of choice to overcome disadvantages of SOA. Furthermore Telco has network resources that can control network and it has customer resources based on the service agreement. It is more suitable to be defined as resources not as services. So ROA fits well to integrate Telco's resources.

We can devise resources like Fig. 1. In this figure a customer information can be access by URL like /Customer/{Customer ID}. The service agreement of this customer can be addressable like /Customer/{Customer ID}/Subscribe. We can get the XML or JSON message representing the subscribed information using GET method to /Customer/{Customer ID}/Subscribe. We can make a new resource by issuing POST method with a subscribe information to /Customer/{Customer ID}/Subscribe if we want to insert a new subscription transaction. POST method returns a subscription ID identifies that subscribing transaction. If we issues GET method to /Customer/{Customer ID}/Subscribe/{Subscription ID}, we can get the information of that subscription.

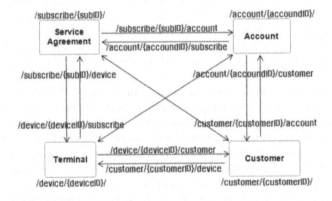

Fig. 1. Resources of Telco

If someone want to call this customer in a SNS service, we can get available contacts using the GET method on /Customer/{Customer ID}/AvailableContacts. This operation will returns devices that is available to call and hyper links like /Customer/{Customer ID}/device/{device ID}. We can make a call or send a short message by using the POST method with caller information to those hyper-links.

As Fig. 1 shows, available resources of Telco can have various URLs and one resource can be referenced by other resource. We can access account by /account/{account ID} or /subscribe/{Subscriber ID}/account/. So we can easily access a resource from other resource.

3 Implementation of the ROA Based Web Service Platform in KT

Previously, KT provides SOAP APIs for web and service profiles to service sites. These APIs includes authentication and billing. It also provides the information that is needed by a business. This is implemented 3-tied system, which has a database that stores profiles and a middleware that has business logics and control transaction and web servers that receives SOAP request.

As the client that uses web services is increasing and service requests are becoming diverse, it is difficult to manage SOAP specification and service discovery. So we implemented a ROA based web service platform. We used ruby on rails because by using it we can make web services with easy and agility. We also develop adapters named ActiveTPMonitor and ActiveLink that can incorporate legacy systems. Fig. 2 shows our web service platform.

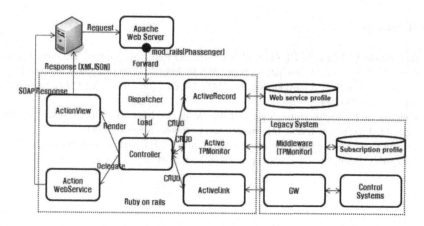

Fig. 2. Implementation of the ROA based web service platform

ActiveRecord is a database model that is provided Ruby on rails package. For accessing the subscription profile through middleware, we implemented ActiveTPMonitor class that is corresponding to ActiveRecord. We can access the subscription profile through ActiveTPMonitor as the same interface that is used by ActiveRecord. To interconnect control systems, we abstract network protocol like TCP or SOAP to

ActiveLink. This is also corresponding to ActiveRecord. By these classes, we can abstract the inner state and actions that is mapped to legacy systems.

Allowing that REST API is suitable interface for ROA, we have to maintain previous SOAP interface for compatibility. So we modify ActionWebService to make mapping information and translate SOAP API to REST API. We made REST/SOAP hybrid web service platform in that services are translated into resource and its method.

4 Conclusions

In this paper, we describe ROA based web service provisioning methodology for Telco and its implementation. We defined Telco's resources as URL and discussed how to use it. We proposed a system architecture that can integrate legacy systems and expose its functions by URL to 3rd party. Our methodology and implementation examples can be used as a reference model to build ROA based web service for Telco. There is a skeptical point of view about offering Telco's resources as web services to 3rd parties to make mash-up services. But the web is the general trend and we cannot go against that trend. The web is gradually sinking into our every life and services using the web and mash-up between them will become more active. Telco has a point of contact with real customers and can control the telecommunication network. These can be very valuable resource to both Telco and Companies that provides web 2.0 services. Mixing these two domains can make innovative services that converge online to offline and virtual to real.

References

1. Richardson, L., Ruby, S.: RESTful Web Services. O'Reilly Media, Inc., Sebastopol (2007)
2. Curbera, F., Duftler, M., Khalaf, R., Nagy, W., Mukhi, N., Weerawarana, S.: Unraveling the web services web and introduction to SOAP, WSDL, and UDDI. IEEE Internet Computing 6(2), 86–93 (2002)
3. Fielding, R.T.: Architectural Style and the Design of Network-based Software Architectures, PhD thesis, UC Irvine, Information and Computer Science (2000)
4. Overdick, H.: The Resource-Oriented Architecture, IEEE Congress on Services (2007)
5. The parlay group, http://www.parlary.org/en/index.php

Attack Model and Detection Scheme for Botnet on 6LoWPAN*

Eung Jun Cho, Jin Ho Kim, and Choong Seon Hong[**]

Dept. of Computer Engineering, Kyung Hee University, Korea
{ejcho,jhkim}@networking.khu.ac.kr, cshong@khu.ac.kr

Abstract. Recently, Botnet has been used to launch spam-mail, key-logging, and DDoS attacks. Botnet is a network of bots which are controlled by attacker. A lot of detection mechanisms have been proposed to detect Botnet on wired network. However, in IP based sensor network environment, there is no detection mechanism for Botnet attacks. In this paper, we analyze the threat of Botnet on 6LoWPAN and propose a mechanism to detect Botnet on 6LoWPAN.

Keywords: Botnet, 6LoWPAN, Attack Model.

1 Introduction

Attack types are varied with the development of the computer technology. In the past, attacks were launched done for just taking pride, and there was no other purpose. On 25th of January, 2003, in Korea, there was a serious attack against the Internet, it was a Slammer worm. The Slammer worm used the weak point of MS-SQL, and disabled entire network till next morning. Nowadays, attackers require money to stop attacking commercial site. According to this shifting of the attack paradigm, new kind of attack will be possible when new kind of technology is developed.

In this paper, we analyze attack case of Botnet [1][2], which is the most powerful attacking tool nowadays on 6LoWPAN (IPv6 over Low power WPAN)[3], and propose a detection mechanism of Botnet on 6LoWPAN.

2 Attack Model of Bonet on 6LoWPAN

In this section, we introduce an attack model of Botnet on 6LoWPAN. A sensor node provides low computation power, limited battery life and low bandwidth compared to that of a PC. If an attacker wants to launch DDoS(Distributed Denial of Service) attack with sensor nodes, the attacker has to infect much more number of sensor nodes than personal computer. However, although the attacker infects enough number of sensor nodes to launch DDoS attack, they cannot continue DDoS attack because of

---

[*] This research was supported by the MKE, Korea, under the ITRC support program supervised by the IITA (IITA-2009-(C1090-0902-0016)).

[**] Corresponding author.

C.S. Hong et al. (Eds.): APNOMS 2009, LNCS 5787, pp. 515–518, 2009.

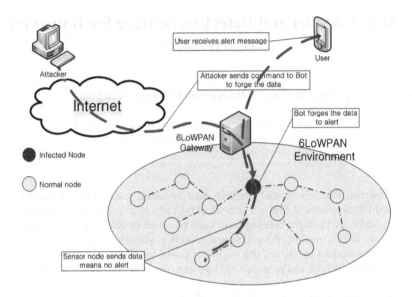

Fig. 1. Attack model of Botnet on 6LoWPAN

the limited battery. On the other hand, the attacker can utilize one characteristic to achieve his/her goal; acts as a route of the communication. With this characteristic, the attacker can forge the data packet which is passed through infected sensor node.

Figure 1 shows the example how the attacker forges the data packet which flows to user. At first, the attacker sends command to a bot (infected node) to forge the data. After this step, the bot can try to forge the data of a packet which is passed through infected node. A user requests the temperature data to the specific sensor node. Then the sensor node replies to user that temperature is "20". When the packet, which includes temperature data passes through infected node, the bot forges the data to "10". Finally, the user receives the forged temperature data. So the user can turn on the heater or turn off the air conditioner to make temperature normal. If this situation is in a hospital, it can be a serious accident.

3 Detecting Mechanism for Botnet on 6LoWPAN

In this paper, we assume the following rules. First, sensor nodes use only TCP to provide the service. Second, one 6LoWPAN provides only one service. Third, procedure of the communication between sensor node and user is typical. In 6LoWPAN environment, all packets, which flow from IP network to the 6LoWPAN, have to pass through a 6LoWPAN gateway. With this characteristic, we propose a detection mechanism of Botnet on 6LoWPAN. Figure 2 shows the Botnet detection module which should be installed on the 6LoWPAN Gateway. To analysis the traffic data, we store the data from the packet that passes through the 6LoWPAN Gateway.

Control field check module calculates the sum of TCP control field during a connection. Characteristic of traffics are almost same because sensor network provides limited service. However, in case of infected node, more traffic is required to maintain Botnet and update bot module. Due to this characteristic of Botnet, the sum of TCP control field value can be different during a connection. *Packet Length check* module calculates the average packet length during a connection. Application data of packets which are sent by user and attacker are quite different. As mentioned above, bots have to transmit more data to update module and maintain Botnet. So, packets, which are sent and received by infected node, are longer than normal one. *Activity check module* counts the number of connections established by sensor node. The attacker cannot know where the bot node is. Therefore the bot has to report its state and address information to the attacker. *Activity check module* counts these connections. *Bot Analysis module* uses above information to detect whether a node on 6LoWPAN is a bot or not. First, from *Value(C)*, *Bot Analysis module* analyzes all traffic of a node to find the malicious traffic. If the malicious traffic is found, it analyzes *Value(L)* and *Value(A)* also. All values are enough to determine that there is the bot on 6LoWPAN, *Bot Analysis module* makes the alert message to manager. The following is specific operation code of *Bot Analysis module*.

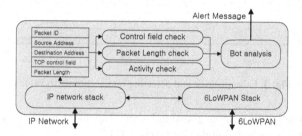

Fig. 2. Additional module on 6LoWPAN Gateway to detect Botnet

```
Value_c : value from Control field check module
value_l : value from Pakcet Length check module
value_a : value from Activity check module
value_ci : value from Control field check moudule of ith node
t : value of threshold
for(i=0; i < number of nodes; i++)
{
   if(value_ci < average value_c of other node - t_1 &&
      value_ci > average value_c of other node + t_1)
    if(value_ai> average value_a &&
       value_li < average value_l - t_2 &&
       value_li > average value_l + t_2)
     Makealert();
}
```

4 Evaluation

Table 1 shows parameter and value to simulate our mechanism. And Figure 3 shows detecting rate according to value *t*, and rates of false positive according to value *t*. We vary the number of nodes to compare. To decrease rates of false positive, value *t*

Table. 1. Simulation Parameters

Parameter	Value
Random range of packet length	10~11
Number of nodes	Variable
Packet generation rate	0.5 packet per every 1ms
Random range of ACK value	5~7
Test time	400000ms
The number of simulation	100

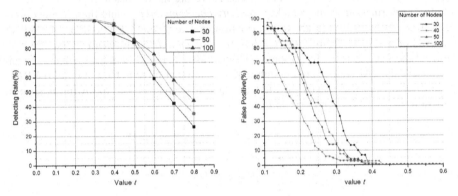

Fig. 3. False positive and detecting rate according to value t

should be pre-defined by doing pre-simulating. And more number of nodes makes lower false positive rates and higher detecting rates.

5   Conclusion and Future Works

In this paper, we explain the threat of Botnet attack on 6LoWPAN, and propose a detection mechanism for Botnet on 6LoWPAN. To apply our mechanism in the real sensor network, we need to consider about 6LoWPAN environment having thousands of sensor nodes and supporting multiple tasks. To reduce overheads of 6LoWPAN, we will consider sampling mechanism to analyze traffic as future works. Also, we will calculate optimal value of t.

References

1. Puri, R.: Bots & Botnet: An Overview,
   https://cours.ift.ulaval.ca/fileadmin/cours/20064_2772A/
   public/Botnet.pdf (August 2003)
2. Holz, T., Steiner, M., Dahl, F., Biersack, E., Freiling, F.: Measurements and Mitigation of Peer-to-Peer-based Botnets: A Case Study on Storm Worm, April 9 (2008)
3. Kushalnagar, N., Montenegro, G., Schumacher, C.: IPv6 over Low-Power Wireless Personal Area Networks (6LoWPANs): Overview, Assumptions, Problem Statement, and Goals, IETF RFC 4919 (August 2007)

An Efficient Strategy for Achieving Concurrency Control in Mobile Environments

Salman Abdul Moiz[1] and Lakshmi Rajamani[2]

[1] Research Scientist, Centre for Development of Advanced Computing, Bangalore, India
[2] Professor, CSE, University College of Engineering, Osmania University, Hyderabad, India
salmanmca@gmail.com, rlakshmi2006@yahoo.com

Abstract. In a mobile database environment, multiple mobile hosts may update the data items simultaneously irrespective of their physical locations. This may result in inconsistency of data items. Several Concurrency Control techniques have been proposed to eliminate the inconsistency of data items. Timeout based strategies are proposed in the literature to reduce the starvation problem in mobile environments with reduced rollbacks. However in each attempt the time of execution of the transaction is unnecessarily wasted though the timer value is known. In this paper we propose a strategy which executes a transaction only when the sufficient time is available for execution or the remaining time of execution is comparatively less. Experimental results show better throughput, and less waiting time for individual transactions.

Keywords: Mobile Host, Fixed Host, Transaction, Timer, Threshold, Rollback.

1 Introduction

In a mobile computing environment, data may be accessed anytime and anywhere with or without network connectivity. The characteristics of mobile environment make data accessibility a challenging issue.

A crucial consideration in mobile databases is the ability to simultaneously access the data items irrespective of the physical locations of mobile users. To handle the concurrency control issue, various concurrency control techniques have been proposed in literature which are usually based on three mechanisms viz, locking, timestamps and optimistic concurrency control. Though these techniques are well suited for the traditional database applications they may not work efficiently in mobile database environments.

A Mobile Host which is in a disconnected mode may lock the data item for indefinite amount of time leading to starvation. To solve this problem various timeout mechanisms are proposed [1,2]. However in these mechanisms the transaction is executed even when the time for execution of a transaction is higher. In this paper we propose a strategy which increases the throughput of the system at the same time the waiting time of a transaction may decrease. The idea behind this technique is that if the transaction takes a larger time & the timer value is sufficiently small, the transaction shouldn't be executed.

C.S. Hong et al. (Eds.): APNOMS 2009, LNCS 5787, pp. 519–522, 2009.
© Springer-Verlag Berlin Heidelberg 2009

The remaining part of this paper is organized as follows: Section 2 summarizes the survey of existing techniques, section 3 describes the architecture of mobile environment, section 4 specifies the proposed concurrency control strategy, section 5 specifies the performance metrics and section 6 concludes the paper.

2 Related Work

Multiple mobile hosts may access the same data items leading to inconsistency. Several valuable attempts to efficiently implement the concurrency control mechanisms have been proposed. Even though some of the proposals address all the operational requirements, only a subset of performance issues is addressed.

The two phase locking protocol in not suitable for mobile environments as it requires clients to continuously communicate with server to obtain locks and detect the conflicts [3]. A Timeout based Mobile Transaction Commitment Protocol provides non-blocking, however it faces the problem of time lag between local and global commit. The Mobile 2PC protocol preserves 2PC principle, however it assumes that all communicating partners are stationary with permanent available bandwidth[4].In timer based strategies, if the timer value is small compared to the time of execution of a particular application or service my a mobile host, it will still be executed and the timer period is wasted. The time for execution of a transaction can be predicted from earlier successful transactions.

The proposed concurrency control strategy helps in increasing the commit rate and throughput of the transactions executed concurrently.

3 Mobile Database Architecture

The Mobile Database architecture (Fig.1) consists of two major entities i.e. a Mobile Host (MH) and a Fixed Host (FH). Transactions initiated at mobile host may be executed at fixed host or mobile host or the execution of the transaction is distributed

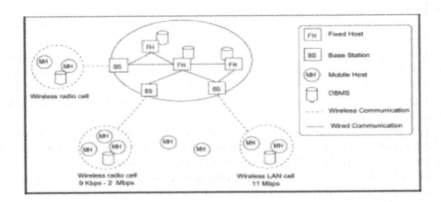

Fig. 1. Mobile Environment [4]

among mobile host and fixed host respectively. Mobile host may be disconnected to save bandwidth and battery. Hence disconnections are treated as normal situations and not as failures [2].

In this paper we assume that the mobile host will initiate the transaction, the respective data items are locked at fixed hosts and the required data items, after execution of transaction the results are integrated with fixed host.

4 Concurrency Control Strategy for Mobile Environments

In the timeout based mechanisms the commit, rollback or abort decision is made at when the timer expires. In the proposed strategy the decision regarding the state of the transaction is made at the beginning.

A mobile host locks the data items on fixed host (if available). An initial timer t is set and the transaction starts executing. If the time of execution specified by fixed host is less than or equal to the current timer value t, then the transaction is executed.

If the time for execution is slightly greater than the timer value, we proceed to execute the transaction based on the remaining time of execution. If the transaction is not completed within time interval t, the timer value is increased by a small factor δ, so that the transaction whose time of execution is slightly greater than t also gets executed. However if it is detected at a later stage that the transaction is taking more time it may be rolled back but it has to be executed within the maximum timer value(Threshold), otherwise it may be aborted.

5 Performance Metrics

Table 1 represents the expected time of completion of individual transaction maintained at fixed host. The results give a comparison of timeout based techniques and the proposed strategy.

Table1. Mobile Hosts requesting for similar data items

Mobile Host	Expected time of completion
M1	3
M2	4
M3	12
M4	6
M5	2
M6	5
M7	8

Table 2. Comparison of Proposed Strategy and Time stamped strategies

Dynamic Timer Adjustment Strategy					Proposed Concurrency Control Strategy						
Mobile Host	t	CT*	Status	t=t+δ	Mobile Host	RT	Decision	t	CT*	Status	t=t+δ
M1	3	3	Commit	3	M1	Nil	√	3	3	Commit	3
M2	3	6	Rollback	4	M2	25%	√	3	7	Commit	4
M3	4	10	Rollback	5	M3	66.7%	×	4	-	Rollback	5
M4	5	15	Rollback	6	M4	16.7%	√	5	13	Commit	6
M5	6	17	Commit	6	M5	Nil	√	6	15	Commit	6
M6	6	22	Commit	6	M6	Nil	√	6	20	Commit	6
M7	6	28	Rollback	7	M7	25%	√	6	28	Commit	7
M2	7	32	Commit	7	M3	**	√	7	40	Commit	8
M3	7	39	Rollback	8							
M4	8	45	Commit	8							
M7	8	53	Commit	8							
M2	8	61	Rollback	9							
M2	9(>T)	-	Aborted	-							

Right section summary:

*CT: Completion time   *RT: Remaining Time **Commits**: 7 **Rollbacks**:1   **Aborts**: 0 **Total Time**: 44msec

Comparative Results: The waiting time in transaction queue is reduced and The throughput has increased and the number of commits have also increased.

Left section summary:

*CT: Completion time **Commits**: 6, **Rollbacks**:5, **Aborts**: 1, **Total Time**: 61 msec

6   Conclusion

A transaction throughput would increase if the possible time of its execution is compared as against the current timer value. If the timer values is very small as compared to the time of execution of a transaction it may be executed later, this increases the throughput and increase the success rate of the transaction.

References

1. Choi, H.J., et al.: A Time stamp Based optimistic Concurrency Control for handling Mobile Transactions. In: Gavrilova, M.L., Gervasi, O., Kumar, V., Tan, C.J.K., Taniar, D., Laganá, A., Mun, Y., Choo, H. (eds.) ICCSA 2006. LNCS, vol. 3981, pp. 796–805. Springer, Heidelberg (2006)
2. Kumar, V., et al.: TCOT, A Timeout based Mobile Transaction Commitment Protocol. In: IIS9979453 (2004)
3. Nouali, N., et al.: A Two Phase Commit Protocol for Mobile Wirelss Environment. In: 2005 16th Australasian Database Conference, ADC 2005, vol. 39 (2005)
4. Serran-Alvarado, P., et al.: A Survey of Mobile Transactions Distributed & Parallel Databases, vol. 16, pp. 193–230. Kluwer publishers, Dordrecht (2004)

Design of Intersection Switches for the Vehicular Network*

Junghoon Lee[1], Gyung-Leen Park[1,**], In-Hye Shin[1], and Min-Jae Kang[2]

[1] Dept. of Computer Science and Statistics
[2] Dept. of Electronic Engineering
Jeju National University, 690-756, Jeju Do, Republic of Korea
{jhlee,glpark,ihshin76,minjk}@jejunu.ac.kr

Abstract. This paper proposes an efficient message switch scheme on the vehicular telematics network, especially for the intersection area where routing decision may be complex due to severe traffic concentration. Each switch node opens an external interface to exchange messages with vehicles proceeding to the intersection from the pre-assigned branch as well as switches the received messages via the internal interfaces, accessing two shared channels according to slot-based MAC. Within each synchronized slot, channel probing and switching can efficiently deal with channel errors. The simulation result shows that the proposed scheme improves the delivery ratio by up to 13 % for the channel error rate range as well as up to 8.1 % for the given network load distribution.

1 Introduction

Nowadays, empowered by the penetration of in-vehicle telematics devices and the evergrowing deployment of vehicular networks, wireless vehicular communications have become an important priority for car manufactures[1]. On the vehicular network, information retrieval is one of the most promising applications[2]. For example, a driver may want to know the current traffic condition of specific road segments that lie on the way to its destination, query several shops to decide where to go, and check parking lot availability[3]. Such queries must be delivered to the destinations and the result is sent back to the query issuer.

A message proceeds to its destination according to a routing protocol each vehicle cooperatively runs. As each vehicle can move only along the road segment and static nodes, such as gas station and traffic light, are generally placed on the roadside[2], the message delivery path must trail the actual road layout. In this delivery path, intersection areas are important for the delay and reliability of message transmissions. Particularly, the carry and forward strategy is preferred to cope with disconnection in the sparse network part. Here, when a vehicle node can't find a receiver, it stores the message in its buffer not discarding the

---

* This research was supported by the MKE, Korea, under the ITRC support program supervised by the IITA. (IITA-2009-C1090-0902-0040).
** Corresponding author.

C.S. Hong et al. (Eds.): APNOMS 2009, LNCS 5787, pp. 523–526, 2009.

message and thus tolerating an increased delay, until it enters the range of a new receiver. This behavior further increases the complexity of message routing near the intersection area.

In this regard, this paper is to design and measure the performance of a reliable message switching scheme around the intersection area for vehicular network, taking advantage of multiple channels. It is possible to create multiple channels in a cell in many available wireless protocols such as IEEE 802.11 series, Zigbee, and WirelessHart. Existing researches have also pointed out that multiple network interfaces does not cost too much[4]. Moreover, TDMA style access can be employed for predictable channel access, short delay, and reliability, while the probing-based access can efficiently reduce the effect of the channel error inherent in the wireless frequency channel.

2   Wireless Switch Design

Fig. 1 shows the basic idea of this paper. At the intersection, four static switch nodes from A to D are installed at each corner as shown in Fig. 1(a). The number of switch nodes is equal to the number of branches, whether they cross on the same plane or in different layers. For the sake of speeding up the transit time in the intersection area, each switch node opens an external interface to receive and send messages with other vehicles, while internally switching messages to an appropriate direction via the 2 additional channels not exposed to the vehicle. While the external interface must follow the standard vehicular communication protocol, we can define a new protocol for the internal access. For internal message exchange, this paper suggests that each node be connected to dual frequency channels which run slot-based MAC as shown in Fig 1. (b).

A vehicle having messages, whether they are created at the node or received from the other vehicles, checks if it can reach a switch node, when it approaches an intersection. If so, it sends its message directly to the switch node without

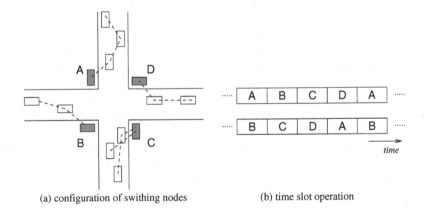

(a) configuration of swithing nodes   (b) time slot operation

Fig. 1. Basic concept

contacting any other vehicles. Otherwise, it sends to the vehicle in its forward. The buffer space can be assumed to have no limitation, but it is desirable to discard a message which stayed in the buffer too much so as to obviate undue delay for the subsequent messages. The slots of two channels are synchronized, and two slots of the same time instance must be assigned to different nodes. How to assign slots is not our concern.

At the beginning of each slot, two nodes probe the channel to their destinations, for example, by RTS/CTS handshake. The destination responds if the channel is good. Meanwhile, other nodes can hear this procedure through the shared channel, so they can cooperatively react to channel errors. Let's assume that slot i is originally allocated to A on channel 1 as well as B on channel 2, namely, $< A, B >$. Table 1 shows the probing result and corresponding actions. As shown in row 1, A on channel 1 and also B on channel 2 are both in good state, A and B send as scheduled. In row 2, every channel status is good except the one for B on channel 2. If we switch $< A, B >$ to $< B, A >$, both nodes can successfully send their messages. Otherwise, only A can succeed. In this case, the channel switch can save one transmission loss. Row 8 describes the situation that A is good only on channel 2 while B on channel 1. Switching channels saves 2 transmissions that might fail on ordinary schedule.

Table 1. Channel status and transmission

No.	Ch1–D1	Ch2–D2	Ch1–D2	Ch2–D1	Ch1	Ch2	save
1	Good	Good	X	X	D1	D2	0
2	Good	Bad	Good	Good	D2	D1	1
3	Good	Bad	Good	Bad	D1	–	0
4	Good	Bad	Bad	X	D1	–	0
5	Bad	Good	Good	Good	D2	D1	1
6	Bad	Good	Good	Bad	–	D2	0
7	Bad	Good	Bad	X	–	D2	0
8	Bad	Bad	Good	Good	D2	D1	2
9	Bad	Bad	Good	Bad	D2	–	1
10	Bad	Bad	Bad	Good	–	D1	1
11	Bad	Bad	Bad	Bad	–	–	0

X : don't care

3   Performance Measurement

We evaluate the performance of our scheme via simulation using SMPL, which provides a simple and robust discrete event trace library similar to ns-2 event scheduler[5]. The main performance metric is the delivery ratio, while the measurement result for the access delay is omitted due to space limitation. Fig. 2 shows the delivery ratio according to the error rate with the network load fixed to 1.0. At the higher error rate, we can achieve larger improvement, as there are more cases channel switching is possible. When the channel error probability is

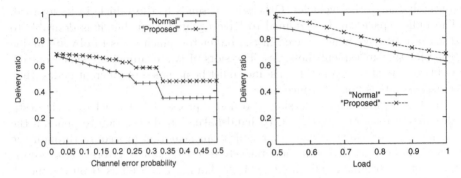

Fig. 2. Delivery ratio vs. error rate   **Fig. 3.** Delivery ratio vs. load

larger than 0.4, the proposed scheme can improve the delivery ratio by up to 13.0 %. In addition, Fig. 3 plots the delivery ratio according to the offered load for the fixed channel error probability of 0.1. The two curves show the constant gap for all load range, while the gap is 8.1 % at maximum, indicating that the early discard more affects the delivery ratio at the higher load. It can be additionally pointed out by our experiment that the improved delivery ratio doesn't affect the transit time.

4   Conclusion

This paper has proposed and measured the performance of a message switch scheme for an intersection area in vehicular telematics network. Installed at each corner of an intersection, switch nodes cooperatively exchange messages according to the current channel status, speeding up the intersection transit time and improving the delivery ratio. The simulation result shows that the proposed scheme can improve the delivery ratio by up to 13.0 % for the given range of channel error probability, and 8.1 % for the given range of network load. In short, the wireless switch can be installed at the intersections, which have a lot of traffic especially in the urban area, to enhance the communication quality and reliability.

References

1. Lee, J., Park, G., Kim, H., Yang, Y., Kim, P.: A telematics service system based on the Linux cluster. In: Shi, Y., van Albada, G.D., Dongarra, J., Sloot, P.M.A. (eds.) ICCS 2007. LNCS, vol. 4490, pp. 660–667. Springer, Heidelberg (2007)
2. Lochert, C., Scheuermann, B., Wewetzer, C., Luebke, A., Mauve, M.: Data aggregation and roadside unit placement for a VANET traffic information system. In: ACM VANET, pp. 58–65 (2008)
3. Zhao, J., Arnold, T., Zhang, Y., Cao, G.: Extending drive-thru data access by vehicle-to-vehicle relay. In: ACM VANET, pp. 30–39 (2008)
4. Naumov, V., Gross, T.: Connectivity-aware routing (CAR) in vehicular ad hoc networks. In: IEEE INFOCOM, pp. 1919–1927 (2007)
5. MacDougall, M.: Simulating Computer Systems: Techniques and Tools. MIT Press, Cambridge (1987)

Exploiting Network Distance Based Euclidean Coordinates for the One Hop Relay Selection*

Sanghwan Lee

Kookmin University, Seoul, Korea
sanghwan@kookmin.ac.kr

Abstract. The one hop relay selection problem is to find a node that can relay traffic from a source to a destination. The NAT traversal scheme in Skype VoIP networks is such an example. To find the optimal relay node requires a large amount of measurement traffic. Thus, a more scalable way to find a *good* relay node is crucial for the success of relaying. In this paper, we investigate the applicability of network distance based Euclidean coordinates for the one hop relay selection problem. Through a number of experiments with real measurement data sets, we show that the Euclidean coordinate system achieves close to the optimal performance in terms of delay.

1 Introduction

Recently, peer to peer technologies are adapted to a broad range of applications such as VoIP and real time multimedia streaming. One commercially successful example is Skype, a VoIP service. Among many features that incur the success of Skype is the NAT traversal functionality. To be specific, when two hosts behind NAT cannot directly communicate with each other, Skype finds a peer node with a public IP address to serve as a relay node between the two hosts. Choosing a relay node and relaying the voice traffic are transparent to the users. One of the main issues in the one hop relay selection problem is how to find a node that minimizes the overall delay on the one hop path.

The one hop relay selection has been an interesting research issue during the last decade [1,2]. One trivial way is to measure the network distances from both nodes to all the candidates (e.g. all the hosts in a Skype p2p network) and choose the one that minimizes the overall delay. Since this trivial method does not scale, more scalable methods are needed.

In this paper, we investigate the applicability of the network distance based Euclidean coordinates in the one hop relay selection problem to reduce the amount of measurement traffic. In the Euclidean coordinate embedding schemes (such as [3,4]), each host has a set of coordinates in a Euclidean space such that the Euclidean distance between two hosts approximates the actual network distance between them. We quantitatively evaluate how well the Euclidean embedding scheme works in selecting a relay node in a peer to peer network.

---

* This work was supported in part by research program 2007 of Kookmin University in Korea and by Seoul R&BD Program NT070100.

C.S. Hong et al. (Eds.): APNOMS 2009, LNCS 5787, pp. 527–530, 2009.

1.1   Background

The one hop relay finding problem has been intensively studied recently. [1] proposes RON that enables overlay nodes to find a one hop relay node in a small size (50-100) overlay network. Each node measures the network distances to all the other nodes and finds an optimal one hop relay node to a destination. The high measurement traffic of RON motivates other approaches such as [5], which reduces the measurement traffic by exploiting tomography. [2] exploits the rich measurement information provided by Akamai to find a one hop relay node.

The basic idea of Euclidean Coordinate Embedding is to assign coordinates to each host in such a way that the Euclidean distances in the virtual Euclidean space approximate the actual network distance such as round trip time and one way delay. Landmark based schemes such as GNP [3] and Virtual Landmarks [4] require a set of special nodes called landmarks. On the other hand, distributed schemes such as Vivaldi [6] do not require the landmarks, but they suffer from instabilities of the coordinates when nodes join and leave frequently.

2   Relay Selection with Euclidean Coordinate Embedding

In this section, we describe several relay selection methods and investigate the performance of Euclidean Embedding on the relay selection problem. Before we look at the specific schemes, we first define the relay selection problem as follows. Let P be the set of all peers in a peer to peer network. For any pair of nodes $x, y \in P$ and a subset $Q \subset P$, we find another node $p \in Q$ that minimizes the sum of delays in the overlay links of $x - p$ and $p - y$.

In a general selection scheme, a subset Q of nodes of P is chosen and the two nodes x and y measure the network distances to all the nodes in Q. Finally, x and y select the node that results in the smallest one hop network distance. The size of the chosen subset Q determines the amount of measurement traffic. We present three relay selection schemes, which only differ in how to choose the candidate subset from the entire node set of P.

- *Optimal* : We consider $Q = P$. It should be noted that this scheme finds the optimal relay node.
- *K-Random* : We select a random set of nodes of size K from the set P.
- *K-Euclidean Embedding* : We compute the one hop relay distance for all the nodes in P in the Euclidean space and select the K nodes that show the K smallest one hop relay distance in the Euclidean space. It should be noted that no distance measurement is needed at this stage.

Optimal scheme can find the optimal one hop relay path. However, if the number of nodes in P increases, the scheme is not scalable. *K-Random* scheme may reduce the measurement nodes, but the chosen node may not be the optimal one since there is no guidance in selecting the candidate subset. *K-Euclidean Embedding* exploits the estimated distances among the hosts and finds a candidate subset Q that is likely to contain the nodes that result in small one hop relay distances.

It is likely that the one hop relay path results in a longer network distance than the direct path. However, several papers show that sometimes the one hop relay paths have smaller network distances [1,2]. To quantitatively evaluate the performance of the three schemes, we define the notion of *relative penalty*. Let $d(x, y)$ be the network distance between nodes x and y. The relative penalty $Rp(p, x, y)$ of the relay node p for x and y is defined as follows.

$$Rp(p, x, y) = \frac{d(x, p) + d(p, y) - d(x, y)}{d(x, y)} \qquad (1)$$

If the relative penalty is negative, then the one hop relay actually provides a shorter network distance than the direct path.

In the evaluation, we use two real network distance measurement data sets collected from many sources. "K462" is the RTT matrix of 462 nodes measured by King [7] method. "PL" is the RTT matrix of 148 Planetlab nodes. For the *K-Euclidean Embedding* scheme, we use both GNP and Virtual Landmark schemes to assign coordinates to the nodes. We vary K for *K-Euclidean Embedding* with 2, 6, and 10. For all the src-dst pairs, we find the one hop relay node with the three schemes. We compute the relative penalty for each src-dst pair and the cumulative distributions of the relative penalties for all the src-dst pairs.

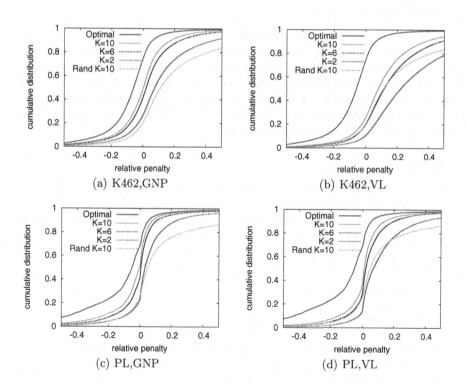

Fig. 1. The cumulative distribution of relative penalties

Fig. 1(a) shows the cumulative distribution of relative penalty of various schemes. "Optimal" shows the result of *Optimal* scheme. "Rand K=10" shows the result of *K-Random* scheme with $K = 10$. "K=2" "K=6" and "K=10" show the results of *K-Euclidean Embedding* scheme with $K = 2, 6, 10$ respectively. As can be seen in the figure, *Optimal* scheme shows the best performance. For more than 75% of the src-dst pairs, the relative penalty is less than 0, which means that the chosen one hop relay node provides a shorter distance than the direct distance. The performance of *K-Euclidean Embedding* increases as K increases. It should be noted that *K-Euclidean Embedding* with $K = 2$ shows much better performance than *K-Random* with $K = 10$.

However, the result of *K-Euclidean Embedding* with Virtual Landmark embedding scheme (VL) is a little worse than that of GNP (See Fig. 1(b)). The reason is because the estimation accuracy of Virtual Landmark scheme is usually lower than that of GNP so that the chosen candidate subset may not contain good relay nodes. Nevertheless, with $K = 10$, *K-Euclidean Embedding* shows higher performance than *K-Random*. The result with Planetlab data set (PL) is similar to that of K462 (See Fig. 1(c) and Fig. 1(d)). In summary, through various experiments, we show that the network distance based Euclidean coordinates are useful to select the one hop relay nodes.

3   Conclusion

In this paper, we evaluate the applicability of network distance based Euclidean coordinates on the one hop relay selection problem. We present three one hop relay selection schemes and compare the relative penalties of the three schemes. Through experiments with various data sets and embedding methods, we show that the relative penalties of *K-Euclidean Embedding* are smaller than that of *K-random*, which indicates the applicability of Euclidean coordinate embedding schemes in the one hop relay selection problem.

References

1. Andersen, D.G., Balakrishnan, H., Kaashoek, M., Morris, R.: Resilient overlay networks. In: Proc. 18th ACM SOSP, Banff, Canada (October 2001)
2. Ao-Jan, S., Choffnes, D.R., Kuzmanovic, A., Bustamante, E.F.: Drafting behind akamai. In: Proceedings of ACM SIGCOMM 2006, Pisa, Italy (September 2006)
3. Ng, T.E., Zhang, H.: Predicting Internet network distance with coordinates-based approaches. In: Proc. IEEE INFOCOM, New York (June 2002)
4. Tang, L., Crovella, M.: Virtual landmarks for the Internet. In: Proceedings of the Internet Measurement Conference(IMC), Miami, Florida (October 2003)
5. Chen, Y., Bindel, D., Katz, R.H.: Tomography-based overlay network monitoring. In: Proc. of IMC 2003, Miami, Florida (October 2003)
6. Dabek, F., Cox, R., Kaashoek, F., Morris, R.: Vivaldi: A decentralized network coordinate system. In: Proc. of SIGCOMM 2004, Portland, OR (August 2004)
7. Gummadi, K.P., Saroiu, S., Gribble, S.D.: King: Estimating latency between arbitrary Internet end hosts. In: Proc. of IMW 2002, Marseille, France (November 2002)

Architecture of IP Based Future Heterogeneous Mobile Network Using Network Discovery for Seamless Handoff

Ihsan Ul Haq[1], Khawaja M. Yahya[1], James M. Irvine[2], and Tariq M. Jadoon[3]

[1] Dept of CSE, NWFP University of Engineering and Technology, Peshawar Pakistan
{ihsan,yahya.khawaja}@nwfpuet.edu.pk
[2] University of Strathclyde, UK
j.m.irvine@strath.ac.uk
[3] Dept of Computer Science, LUMS, Lahore, Pakistan
jadoon@lums.edu.pk

Abstract. This article presents an architecture of all-IP Based Future Heterogeneous Mobile Communication System (FHMCS). In this architecture Mobile Virtual Network Operator (MVNO) network has been proposed as an integral part of FHMCS that will facilitate the Mobile Station/User Equipment (MS/UE) in Network Discovery. The Network Discovery is achieved by using Mobile Network Information Repository (MNIR), embedded at the MVNO site. The network discovery will help reduce handoff latency, decreasing the probability of packet loss and handoff failures. In this article the population and updation of MNIR is presented. The network information will only be available to legitimate users of all Mobile Network Operators having a single Service Level Agreement (SLA) with the MVNO.

Keywords: Seamless communication, MVNO, FHMCS, MNIR.

1 Introduction

The integration and internetworking of existing Radio Access Technologies (RATs) is a logical step to provide service continuity and better user experience in FHMCS. This involves several challenges such as mobility management, security, QoS and cost efficiency. In a heterogeneous environment, mobility management enables the system to locate roaming terminals in order to deliver data packets (i.e., location management) and maintain connections with them when moving into a new network (i.e., handoff management). For better user experience, seamless handover should be provided while the user moves from the network of one access technology to another and from the domain on one Mobile Network Operator (MNO) to another. Seamless handover means that the user does not perceive any disruption in service or quality during the handover. Although several Mobile IPv6-based mobility protocols as well as interworking architectures have been proposed in the literature, they cannot guarantee seamless roaming, especially for real-time applications. Moreover, mobility management protocols are designed for specific needs, for example, the purpose of IPv6-based mobility schemes consists of managing users roaming while ignoring access network

C.S. Hong et al. (Eds.): APNOMS 2009, LNCS 5787, pp. 531–535, 2009.

discovery [1, 2]. This article presents an architecture for FHMCS having the capability to use network discovery to reduce the inter-technology handover latency for seamless roaming in a heterogeneous environment.

Broadly speaking, the handover delay, TMIPHandoff, in mobile IP networks is the sum of TMovementDetection, the movement detection delay (defined as the time interval necessary to detect change of location, that is to realize that a MS/UE is no more in the coverage area of the existing network) and THandoffComp the time interval that elapses between the detection of the MS/UE migration and the reception of the first data packet in the new cell. In wide area networks, TMIPHandoff can be in the neighborhood of seconds; thus, suitable enhancements are required, to reduce TMovementDetection and so THandoffComp [3, 4]. Our proposals reduce TMovementDetection and so THandoffComp for Mobile IP based FHMCS. This article:

a) Proposes an architecture of FHMCS including MVNO network. The MVNO contains network information embedded in MNIR.
b) Presents a mechanism to populate and update the MNIR.
c) Uses the above information to reduce the time interval necessary to detect migration or availability of another network without keeping all interfaces active perpetually.

2   Related Work

Network Discovery: Suk Yu Hui and Kai Hau Yeung propose network discovery by processing the signals from different wireless systems [5]. The major problem of this approach is that the MS/UE needs to process signals which are in a different frequency band and for this purpose the MS/UE has to keep all its interfaces on. This incurs significant battery load.

[6] presents the architecture for internetworking between WLAN and GPRS network and proposes that a dynamic network discovery procedure is needed in order to detect any access network in range of the MS/UE location. The article has emphasized the importance/requirement for network discovery but has not proposed any dynamic network discovery procedure.

This Basic Access Signaling (BAS) has been proposed for RAN discovery in Multimedia Integrated Network by Radio Access Innovation (MIRAI) project [7]. In this mechanism, the MS/UE initiates the call, i.e., the MS/UE sends a packet including its location information to a Resource Manager (RM) on the network. The RM uses the location information of the MS/UE to send a list of available RANs to the MS/UE. The MS/UE then selects a RAN from the list and connects to it. However, there is no mechanism proposed to populate and update the RM because the network conditions may change at any time and the RM will thus provide wrong network information to the MS/UE resulting is handover failures.

In our earlier work, we proposed a Network Discovery Approach [8] by establishing a Network Information Repository (NIR) and populating the NIR by a Reporting Agents which are normal MS/UEs. The RAs collect the information about Network Elements in a domain and send it to the NIR (e.g. if a specific network element is attached/detached or becomes operational/ non-operational its information is reported to the NIR). The NIR was proposed to be a part of Core Network of PMLN. In this proposed mechanism, every mobile network needs to have its own repository. Further, from business perspective it seems difficult – technically and commercially – for an MNO to provide network information about other networks, and they will favour their own networks. A way to achieve roaming among networks is by using bilateral Service Level Agreements (SLAs), but the increasing number of wireless networks and service providers makes it impractical for network operators to have SLAs with all other operators. So in our present proposed solution, an MNIR has been proposed at the single location (the MVNO) that presents as excellent solution for inter-technology handovers with a single SLA.

Movement Detection: For L3 handoff to occur in mobile networks using MIPv6 for mobility management, the MS/UE should find that it has changed its location and has to connect to a new link to maintain existing connection(s). In prior work, the only means by which a Mobile Node is made aware of its change of location is through the Movement Detection Algorithms (MDAs) like Lazy Cell Switching (LCS), Early Cell switching (EyCS), Eager cell switching ErCS), Enhanced Lazy Cell Switching (ELCS) and prefix matching [4,9].

Fast handover minimizes the delay associated with movement detection. To achieve this goal, fast handover is designed to allow mobile node to anticipate their IP layer mobility [10]. However the anticipation also adds to delay and if anticipation is successful, the movement detection time is reduced but if the anticipation is not successful the decision about movement detection and premature forwarding of data by the Previous Access Router (PAR) to the New Access Router (NAR) may be wasted. Moreover it may result in handover failure.

3   Proposed Architecture of FHMCS

The core of this article is to present architecture of FHMCS. In this architecture, a novel role of the MVNO has been proposed. Our proposed architecture incorporates the existing Mobile Virtual Network Operator (MVNO), third party operators, which are providing mobile communication services in some part of the world including UK, Germany, France, USA etc but does not have there own licensed frequency allocation of radio spectrum. An MVNO is an entity or a Center of Excellence that works independently of the Mobile Network Operator (MNO) and functions as a switch less reseller. In normal scenario it buys minutes wholesale from an MNO and retails them to their customers. We have proposed a redesigned picture of MVNO to be used in all IP based FHMCS. We have proposed a Mobile Network Information Repository embedded in MVNO and reachable to all MNOs having SLA with MVNO. Thus, any legitimate user of any MNO having SLA with MVNO will be able to handover the session to other MNO networks ensuring seamless roaming and service continuity in a heterogeneous environment.

The proposed architecture of FHMCS including MVNO is depicted in Figure 1. The model has been proposed keeping in view the authentication model proposed by Internet Engineering Task Force (IETF) so that all the MS/UEs should be authenticated before being authorized to access the MNIR. The model consists of Network Access Server (NAS), a Backend AAA server, Home AAA Server (like HSSS) and MNIR. The model may have a Billing Server for billing. In this model, we assume that MS/UE has a preconfigured security association with either a Backend or Home AAA server, being the users of either MVNO or MNO respectively.

The MS/UE request network related information from MNIR by sending a message to Network Access Server (NAS). The request includes identity of MS/UE and its location obtained from GPS. The NAS sends this to the Backend AAA server which checks if the MS/UE belongs to the MVNO's domain based on MS/UE's identity. If the Backend AAA server does not have association with MS/UE, it locates the Home AAA server on the corresponding MNO's network.

After successful completing the authentication process, the MVNO authorizes the MS/UE and sends its request for network related information to MNIR. The MNIR after processing the request forwards a list of available networks, their IP address, type along with QoS parameter to the MS/UE and it uses it as a neighbor list. Thus the MS/UE has the list of available mobile networks at a specific location along with its type (RAT) and QoS parameters. The beauty of our proposed architecture is that the MVNO can still work like a switchless reseller. As shown in Figure 1, the user using WLAN does not have any authorization to use WiMax resources but the MNIR provides the neighbor list and the backend AAA server at MVNO authenticate and authorize the user to use the WiMax, as MVNO has an SLA with both.

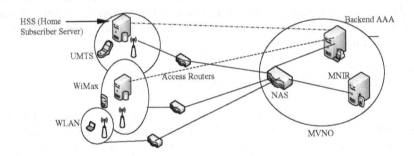

Fig. 1. Architecture of FHMCS

4 Population of MNIR

The success of our proposed model lies in the mechanism that provides the accurate network information to the MS/UEs. The process of population/updating the MNIR is as follows:

During Service Level Agreement (SLA) with MVNO, the MNO provides the IP addresses of all interfaces of the Access Router (AR) attached to different RATs, the type of RAT attached to that interface along with the coordinates (location) of the coverage area. This network related information is initially populated in the MNIR. The MNOs normally have Network Management Centers (NMC) which will note any fault in the network to update this information. For small mobile networks like WLAN without an NMC, the MNIR generates test signals. In response to the test signal the AR after checking the specified network responds to the MNIR about the status of the network. In this manner the MNIR is populated and updated.

In our earlier approach about the network discovery, RAs populate the NIR but a rouge RA may provide wrong information continuously to the NIR thus enabling the whole of the network discovery approach to fail. In our present approach such attacks are not possible.

5 Conclusion

We present a novel architecture capable of Network Discovery. The MVNO's network has been proposed as an integral part of FHMCS which will facilitate the integration and internetworking of different MNOs. The user can move from the network of one RAT to another and from the domain of one MNO to another seamlessly, resulting in a truly user-centric approach. Also, the MS/UE has to search one or a small subset of RATs depending upon the neighbor list, and so can select the best RAT in a heterogeneous situation. This will reduce the battery drain.

References

1. Kim, Y.K., Prasad, R.: 4G Roadmap and Emerging Communication Technologies. Artech House Universal Communication Series (2005)
2. Makaya, C., Pierre, S.: Enhanced fast handoff scheme for heterogeneous wireless networks. Computer Networks, 2016–2029 (January 2008)
3. Fikouras, N.A., El Malki, K., Cvetkovic, S.R., Kraner, M.: Performance Analysis of Mobile IP Handoffs. In: Asia Pacific Microwave Conference, November 30- December 3, vol. 3, pp. 770–773 (1999)
4. Blefari-Melazzi, N., Femminella, M., Fugini, F.: A Layer 3 Movement Detection Algorithm Driving Handovers in Mobile IP. Journal of Wireless Networks 11(3), 223–233 (2005)
5. Hui, S.Y., Yeung, K.H.: Challenges in the Migration to 4G Mobile Systems. IEEE Communications Magazine, 54–59 (December 2003)
6. Ramos, R.C., Gonzalez Perez, L.F.: Quasi Mobile IP-based Architecture for Seamless Interworking between W LAN and GPRS Networks. In: 2nd ICEEE, September 2005, pp. 455–458 (2005)
7. Wu, G., Mizuno, M.: MIRAI Architecture for Heterogeneous Network. IEEE Communications Magazine 40, 126–134 (2002)
8. Yaqub, R., Haq, I.U., Yahya, K.M.: Architecture Supporting Network Discovery in Future Heterogeneous Networks. In: 12th IEEE International Multitopic Conference, December 23-24, pp. 313–317 (2008)
9. Blefari-Melazzi, N., Femminella, M., Fugini, F.: Movement Detection in IP Heterogeneous Wireless Networks. In: 5th EPMCC, April 22-25, pp. 121–125 (2003)
10. Davies, J.: Understanding IPv6, 2nd edn., p. 74. Microsoft Press, Washington (2008)

Analytical Model of the Iub Interface Carrying HSDPA Traffic in the UMTS Network

Maciej Stasiak and Piotr Zwierzykowski

Poznan University of Technology, Chair of Communications and Computer Networks
ul. Polanka 3, Poznań 60965, Poland
piotr.zwierzykowski@put.poznan.pl

Abstract. The analytical model proposed in the paper can be directly applied for modeling and dimensioning of the Iub interface in the UMTS network servicing the mixture of the Release 99 and HSDPA traffic streams. The presented model can be used for the blocking probability and the throughput determination for particular traffic classes carried by the Iub interface. The proposed scheme can be applicable for a cost-effective radio resource management in 3G mobile networks and can be easily applied to network capacity calculations.

Keywords: analytical model, Iub interface, UMTS, HSDPA.

1 Introduction

Cellular network operators define, on the basis of service level agreement, a set of the key performance indicator parameters that serve as determinants in the process of network dimensioning and optimization. The dimensioning process for the system should make it possible to determine such a capacity of individual elements of the system that will secure - with the assumed load of the system - a pre-defined level of grade of service. With dimensioning the UMTS system, the most characteristic constraints are: the radio interface and the Iub interface. When the radio interface is a constraint, then, in order to increase the capacity, access technology should be changed or subsequent branches of the system should be added (another NodeB). If, however, the constraint on the capacity of the system results from the capacity of the Iub interface, then a decision to add other nodes can be financially unfounded, having its roots in incomplete or incorrect analysis of the system. This means that in any analysis of the system, a model that corresponds to the Iub interface should be also included. Several papers have been devoted to traffic modelling in cellular systems with the WCDMA radio interface, e.g. [1]. A model of the Iub interface carrying a mixture of Release 99 (R99) and HSDPA traffic with compression was presented by the authors in [2]. In that model it was assumed that the level of compression was the same for all traffic classes. In the proposed model we assume that different classes are compressed to a different degree. The paper has been divided into 4 sections. Section 2 presents an analytical model applied to blocking probability determination for R99 and HSDPA traffic classes. The following section includes the results obtained in the study and the final section sums up the discussion.

C.S. Hong et al. (Eds.): APNOMS 2009, LNCS 5787, pp. 536–539, 2009.

2   Analytical Model of the System

The Iub interface in UMTS network can be treated as the full-availability group (FAG) carrying a mixture of multi-rate traffic streams with and without compression property. Let us assume that the total capacity of the FAG is equal to V Basic Bandwidth Units (BBUs). The group is offered $M = M_k + M_{nk}$ independent classes of Erlang traffic classes: M_k classes whose calls can change requirements while being serviced and M_{nk} classes that do not change their demands in their service time. It was assumed that class i traffic stream is characterized by the Erlang traffic streams with intensity A_i. The demanded resources in the group for servicing particular classes can be treated as a call demanding an integer number of BBUs [1]. The value of BBU, i.e. R_{BBU}, is calculated as the greatest common divisor of all resources demanded by traffic classes offered to the system. The occupancy distribution can be expressed by the modified Kaufman-Roberts recursion presented in [2]:

$$n\left[P_n\right]_V = \sum_{i=1}^{M_{nk}} A_i t_i \left[P_{n-t_i}\right]_V + \sum_{j=1}^{M_k} A_j t_{j,\min} \left[P_{n-t_{j,\min}}\right]_V, \tag{1}$$

where $[P_n]_V$ is the probability of state n BBUs being busy, t_i is the number of BBUs required by a class i call: $t_i = \lfloor R_i/R_{BBU} \rfloor$, and ($R_i$ is the amount of resources demanded by class i call in *kbps*) and $t_{j,\min} = \lfloor R_{j,\min}/R_{BBU} \rfloor$ is the minimum number of BBUs required by class j call in the condition of maximum compression. Formula (1) describes the system in the condition of maximum compression. Such an approach is indispensable to determine the blocking probabilities E_i for a class i call in the system with compression:

$$E_i = \begin{cases} \sum\limits_{i=V-t_i+1}^{V} [P_n]_V & \text{for} \quad i \in \mathbb{M}_{nk}, \\ \sum\limits_{i=V-t_{i,\min}+1}^{V} [P_n]_V & \text{for} \quad i \in \mathbb{M}_k, \end{cases} \tag{2}$$

where V is the total capacity of the group and is expressed in BBUs ($V = \lfloor \frac{V_{Iub}}{R_{BBU}} \rfloor$, where V_{Iub} is the physical capacity of group in kbps).

The average number of class i calls serviced in the state n can be determined on the basis of the formula [3]:

$$y_i(n) = \begin{cases} A_i [P_{n-t_i}]_V / [P_n]_V & \text{for} \quad n \le V, \\ 0 & \text{for} \quad n > V. \end{cases} \tag{3}$$

Let as assume now that the full-availability group services a mixture of different multi-rate traffic streams with compression property [2]. This means that in the traffic mixture there are such calls in which a change in demands caused by the overload system follows unevenly.

The measure of a possible change of requirements is *maximum compression coefficient* that determines the ratio of the maximum demands to minimum demands for a given traffic class: $K_{j,\max} = \frac{t_{j,\max}}{t_{j,\min}}$, where $t_{j,\max}$ and $t_{j,\min}$ denote respectively the maximum and the minimum number of basic bandwidth units (BBUs) demanded by a call of class j. In the model we assumed that all classes can undergo compression to a different degree.

We assume that the real system operates in such a way as to guarantee the maximum use of the resources, i.e. a call of compressed class always tend to occupy free resources and decrease its maximum demands in the least possible way. The measure of the degree of compression in state n is the quotient $\frac{V-Y^{nk}(n)}{n-Y^{nk}(n)}$, where $V - Y^{nk}(n)$ expresses the amount of resources available for calls with compression and $n - Y^{nk}(n)$ is the number of BBUs occupied by calls with compression.

Let us consider now the average number of busy BBUs in the system occupied by class j calls with compression:

$$Y_j^k = \sum_{n=0}^{V} y_j(n) \left[\xi_{k,j}(n) t_{j,\min}\right] [P_n]_V, \tag{4}$$

where $\xi_{k,j}(n)$ is the compression coefficient in the current model:

$$\xi_{k,j}(n) = \begin{cases} K_{j,\max} & \text{for} & \frac{V-Y^{nk}(n)}{n-Y^{nk}(n)} \geq K_{j,\max}, \\ \frac{V-Y^{nk}(n)}{n-Y^{nk}(n)} & \text{for} & 1 \leq \frac{V-Y^{nk}(n)}{n-Y^{nk}(n)} < K_{j,\max}. \end{cases} \tag{5}$$

In Formula (5), the parameter $Y^{nk}(n)$ is the average number of busy BBUs in state n occupied by calls without compression, and $Y^{nk}(n) = \sum_{i=1}^{M_{nk}} y_i(n) t_i$.

3   Numerical Study

The proposed analytical model of the Iub interface is an approximate one. Thus, the results of the analytical calculations of the Iub interface were compared with the results of the simulation experiments. In the study, it was assumed that: R_{BBU} was equal to 1 kbps, the considered Iub interface carried traffic from three radio sectors; and a physical capacity of Iub in the downlink direction was equal to: $V_{Iub} = 3 \times 7\ 200$ BBUs $\cong 21\ 000$ BBUs. The study was carried out for users demanding a set of following services in the downlink direction: Speech ($t_{1,\min} = 16$ BBUs, $K_{1,\max} = 1$); Real-time Gaming ($t_{2,\min} = 10$ BBUs, $K_{2,\max} = 20$); Mobile TV ($t_{3,\min} = 64$ BBUs, $K_{3,\max} = 10$) and Web Browsing ($t_{4,\min} = 500$ BBUs, $K_{4,\max} = 14$). It was also assumed that the services were offered in the following proportions: $A_1 t_1 : A_2 t_2 : A_3 t_3 : A_4 t_4 = 1 : 3 : 5 : 10$.

Figure 1 presents the influence of traffic offered per BBU in the Iub interface on the average carried traffic by Iub. It can be noticed that the exponential dependence characterizes the plots corresponding to the traffic classes with compression. The linear relation between the compression coefficient and the average carried traffic (see Eq. (4)) explains the similar character of the curves. The results confirm strong dependence between the throughput and the load of the system – the more overloaded system the lower value of throughput. The value of

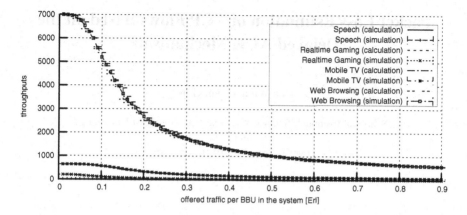

Fig. 1. Average carried traffic for particular classes carried by the Iub interface

throughput for all traffic classes from the value of 0.9 traffic offered per BBU decreases very slowly and these values are not presented in the Figure.

4   Conclusions

This paper presents a new analytical model with uneven compression that finds its application in modeling the Iub interface in the UMTS network carrying a mixture of different multi-rate Release 99 and HSDPA traffic classes, such as background (i.e. Web Browsing) and interactive (i.e. Mobile TV) traffic classes. The method can be used for determination of the blocking probability and the average throughput for calls of each traffic class carried by the interface.

It is worth emphasizing, that the main advantages of the proposed method are simplicity and accuracy which determine the usefulness from the engineering perspective. The presented methodology is limited to multi-rate Poisson traffic and, in future works, the authors intend to extend the presented method by adding Bernoulli and Pascal traffic as well.

References

1. Stasiak, M., Wiśniewski, A., Zwierzykowski, P., Głąbowski, M.: Blocking probability calculation for cellular systems with WCDMA radio interface servicing PCT1 and PCT2 multirate traffic. IEICE Transactions on Communications E92-B(4), 1156–1165 (2009)
2. Stasiak, M., Zwierzykowski, P., Wiewióra, J., Parniewicz, D.: An Analytical Model of Traffic Compression in the UMTS network. In: Bradley, J.T. (ed.) EPEW 2009. LNCS, vol. 5652, pp. 79–93. Springer, Heidelberg (2009)
3. Kaufman, J.: Blocking in a shared resource environment. IEEE Transactions on Communications 29(10), 1474–1481 (1981)
4. Holma, H., Toskala, A.: HSDPA/HSUPA for UMTS: High Speed Radio Access for Mobile Communications. Wiley and Sons, Chichester (2006)

Packet Loss Estimation of TCP Flows Based on the Delayed ACK Mechanism*

Hua Wu and Jian Gong

School of Computer Science & Engineering, Southeast University
Nanjing, China
{hwu,jgong}@njnet.edu.cn

Abstract. Based on the Delayed ACK mechanism in TCP protocol, a new method is proposed to estimate the packet loss based on the variation of the rate between the bidirectional TCP packet numbers. This method made it possible to estimate the TCP packet loss based on the export of Netflow quickly.

Keywords: Congestion window (CWND), packet lose, delayed ACK, TCP, Netflow.

1 Introduction

Packet loss is one of the most important basic metrics in the network SLA (Service Level Agreement), but the measurement of packet loss is a difficult problem. There are two kinds of methods to measure packet loss: active method [1] and passive method.

Paul Barford[2] have compared the active and passive methods, it is found that current methods for active probing for packet loss suffer from high variance inherent in standard sampling techniques and from effects of end-host interfaces loss. The level of agreement between passive measures and active measures is quite low. In order to get the feelings of the network users, passive method should be used.

This paper proposes a packet loss estimation model base on the TCP Delayed ACK(acknowledgment) mechanism proposed in rfc1122. Since almost all of the systems in the Internet use the delayed ACK mechanism to improve the performance of TCP transfers, the ration between the bidirectional packet numbers of the TCP flow reflects the packet loss of the TCP flow. Most of the network equipments provide the Netflow[4] data, this method made it possible to estimate the TCP packet loss based on the output of the Netflow, and the real time packet loss estimation of transport layer that reflect the user feelings is possible.

---

* Supported by the State Scientific and Technological Support Plan Project under Grant No. 2008BAH37B04; the National Grand Fundamental Research 973 Program of China under Grant No. 2009CB320505; the 2008 Natural Science Fundamental Program of Jiangsu Province under Grant No. BK2008288.

C.S. Hong et al. (Eds.): APNOMS 2009, LNCS 5787, pp. 540–543, 2009.

2 Packet Loss Estimation Based on the Delayed ACK Mechanism

2.1 The TCP Delayed ACK Mechanism

In the implementation of TCP protocol, delayed ACK is used to improve the TCP transfer performance. A host that is receiving a flow of TCP data packets can increase efficiency in both the Internet and the hosts by sending few than one ACK packet per data packet received.

Because a packet loss event will cause the change of the behavior pattern of the end points, investigation of the ration between the bidirectional packet numbers give clues of network packet loss.

We define two metrics as follows:

$Rate_{data/ack}$: The number rate of the data packets and the ACK packets of a TCP bulk transfer flow.

$Rate_{drop\_data/data}$: The number rate between the lost data packets and all the data packets of a TCP bulk transfer flow. This is the metric we want to estimate.

If there are no drop evens, $Rate_{data/ack}$ is 2:1.But actually, $Rate_{data/ack}$ is between 1 and 2, because when a drop event happens, $Rate_{data/ack}$ will change from 2:1 to 1:1 until the lost packets is arrived by retransmission. Compared to the packet loss, the packet numbers of data packets and ACK packets is easy to obtain, the export format of Netflow have the statistical value of the bidirectional packet number, it is easy to obtain $Rate_{data/ack}$, the important thing is to investigate the relationship between $Rate_{data/ack}$ and $Rate_{drop\_data/data}$.

2.2 Relationship between $Rate_{data/ack}$ and $Rate_{drop\_data/data}$

The main control algorithms of TCP are slow start, congestion avoidance and fast retransmit. Actually the slow start period is not slow, CWND(Congestion window) have an exponential growth and the time is relatively short, it is a detection process to achieve the equilibrium state. We do not consider this period in the bulk data transfer.

After a lost event is detected, according to the Fast Retransmit/Fast Recovery algorithms[rfc2581], after receiving 3 duplicate ACKs, TCP performs a retransmission of what appears to be the missing packet. CWND will decrease to half of the current value and the connection change to the congestion avoidance statue. This kind of transmission will happen when the TCP bulk transfer performance is limited by the network bandwidth. And the random drop event is ignored. Fig. 1 is such a transfer process and one cycle of the stable transfer statue. This cycle begins from a drop event and ends at the next drop event.

During this cycle in Fig.1, suppose the drop event happens when CWND is w, after a drop event, CWND is decreased to half of w. before the drop event, the receive host send one ACK for every two data packets, after the drop event, it sent one ACK for every data packet, until the lost packet was received, then the rate change to 2:1.

If the flow is in a stable state and long enough, the packet rate and the drop rate in one cycle can be deemed as the packet rate and drop rate of this flow, that is $Rate_{data/ack}$ and $Rate_{drop\_data/data}$. There are three stages in one cycle, we call the three stages as sustainable sending, drop event processing and window recovery.

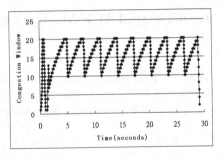
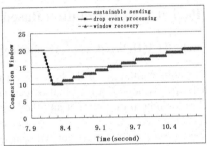

Fig. 1. CWND change pattern of the stable bulk data transfer and one cycle of the stable transfer statue

Since the receiver will send ACK every two packets, whether w is even or odd when the drop event happens will result in different results. Because of the page limit, we will not discuss that situations in detail and only give the results in Table 1.

Table 1. Packet numbers in differnet stages

		CWND is even when drop event happens	CWND is odd when drop event happens
Data packets of one cycle	Sustainable sending	w	w-1
	Drop event processing	3	4
	Window recovery	$\frac{3}{4}w^2 + \frac{3w}{2}$	$\frac{3}{4}w^2 + 2w + \frac{1}{4}$
ACK packets of one cycle	Sustainable sending	$w / 2$	$(w-1)/2$
	Drop event processing	$3 + \frac{w}{2}$	$(w-1)/2 - 4$
	Window recovery	$\frac{3}{8}w^2 + \frac{3}{4}w$	$\frac{3}{8}w^2 + w + \frac{5}{8}$
$Rate_{data/ack}$		$\dfrac{\frac{3}{4}w^2 + \frac{5}{2}w + 3}{\frac{3}{8}w^2 + \frac{7}{4}w + 3}$,	$\dfrac{\frac{3}{4}w^2 + 3w + \frac{13}{4}}{\frac{3}{8}w^2 + 2w + \frac{29}{8}}$
$Rate_{drop\_data/data}$		$\dfrac{1}{\frac{3}{4}w^2 + \frac{5}{2}w + 3}$	$\dfrac{1}{\frac{3}{4}w^2 + 3w + \frac{13}{4}}$

2.3 Simulation Results

In order to verify the accuracy of the above analysis, we use ns2 to validate. We change the bandwidth at bottleneck from 0.1Mb/s to 2 Mb/s at interval of 0.1Mb/s, at the same time, change the delay from 10ms to 20ms at every bottleneck bandwidth. The simulation results show that, no matter how the link parameters changes, when

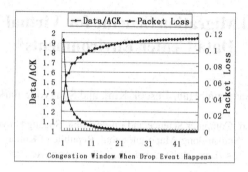

Fig. 2. Relationship between Rate$_{data/ack}$ and Rate$_{drop\_data/data}$

delayed ACK mechanism is used ,in such a stable transfer statue, Rate$_{data/ack}$ and Rate$_{drop\_data/data}$ are decided by CWND when the packet is dropped. The experiment result is showed in Fig. 2.

3   Conclusions

This paper proposes a method to estimate the packet loss easily based on the TCP delayed ACK mechanism. The export of Netflow can be used to estimated the packet loss . But two important things should be noticed when this method is applied. One is to find the suitable flow records in Netflow export, the flow should be a TCP bulk data transfer flow which is long enough to ignore the slow start period. Another thing is the estimation result will have errors because of the bandwidth variations of the flow, a correction method should be found to make the result more accurately.

References

1. Sommers, J., Barford, P., Duffield, N., Ron, A.: A geometric approach to improving active packet loss measurement. IEEE/ACM Transactions on Networking 16(2), 307–320 (2008)
2. Barford, P., Sommers, J.: Comparing Probe- and Router-Based Packet-Loss Measurement. IEEE Internet Computing 8(5), 50–56 (2004)

A QoS Based Migration Scheme for Virtual Machines in Data Center Environments*

Saeyoung Han, Jinseok Kim, and Sungyong Park

Dept. of Computer Science and Engineering, Sogang University
Shinsu-dong, Mapo-gu, Seoul, Republic of Korea
{syhan,sabin,parksy}@sogang.ac.kr

Abstract. We propose a QoS based migration system for virtual machines in data center environments. In order for our QoS based migration scheme to ensure QoS requirements for virtual machines, it estimates the QoS during migrations for each virtual machine, and selects the virtual machine that is most likely not to violate its QoS requirement. Moreover, it improves the overall system utilization by choosing the destination server for migration to balance resource usages.

1 Introduction

There have been studies about migration techniques in various domains, but most of them cannot be used directly in a data center environment, due to the differences in their system environments or the workloads [1][2]. Sandpiper was proposed as a study of virtual machine migration scheme, which used a greedy algorithm such that the most loaded virtual server, hosted in the most loaded server, is migrated into the least loaded server [3]. Sandpiper, however, does not consider any QoS aspects, so temporal QoS violation can be occurred during migrations [4]. Therefore, we propose a QoS based migration scheme that minimize the QoS violations during migrations of virtual machines.

When virtual machines requires additional resources, the QoS based migration scheme decides to migrate, estimates the service qualities of them during the migration, selects the virtual machines to migrate based on the estimation, and finally determines destinations so as to balance the usages of the different resources. By simulation, we validate that the QoS based migration system ensures the QoS requirements of virtual machines during their migrations, and improves system utilization by using resources efficiently.

The reset of this paper is organized as follows. Section 2 briefly discusses the proposed QoS based migration scheme and section 3 presents its performance results. Section 4 finalizes this paper.

---

* This Work was partially supported by the IT R&D program of MKE/KEIT. [2009-F-039-01, Development of Technology Base for Trustworthy Computing].

C.S. Hong et al. (Eds.): APNOMS 2009, LNCS 5787, pp. 544–547, 2009.

2 QoS-Based Migration Scheme

The QoS based migration scheme ensures minimum QoS not to experience performance degradation during migrations for all services in a data center. In a data center, service users can request a certain level of service quality, and the operators of the data center can easily characterize the services by contracting a Service Level Agreement (SLA) with each other. When select destinations for migrations, we consider the requested service quality such that the service quality during the migration does not fall down below the minimum service quality requested.

The QoS based migration system consists of a migration initiator, a QoS estimator, a victim selector, and a destination selector. It receives some information as a input, such as virtual machines that requires additional resources, the demands, the resource usages, and workload characteristics. As an output, the system designates resource reallocations or migrations for virtual machines to satisfy the service level agreements.

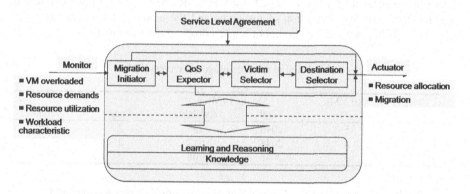

Fig. 1. The architecture of the QoS-based migration system

When a virtual machine needs additional resources, the migration initiator should decide whether to solve it with migration or simple provisioning of additional resources. If the remaining capacity of the server hosting the virtual machine cannot meet its demand, the migration initiator starts the migration process by calling the QoS estimator. Since there could be performance degradation during a migration, the victim selector should choose virtual machines hosted on the server, considering migration overhead and service quality required like follow.

$$v^* = \arg\max_v \left(\frac{\left| QoS_v^{SLA} - QoS_v^{Mig} \right|}{Mem_v} \right)$$

The victim selector picks the virtual machine v such that maximizes the difference between the estimated service quality (QoS_v^{Mig}) during a migration, and the required service level (QoS_v^{SLA}) among the servers whose QoS_v^{Mig} is better than QoS_v^{SLA}. On the other hand, the victim selector considers the allocated memory size of virtual machines to minimize the migration overhead such as migration time and the migration traffic.

Based on the selection of virtual machines and their resource demands, the destination selector should designate a destination server considering its resources, in terms of not only the remaining capacity but also the usage balances. Keeping the ratio of CPU utilization to memory utilization near around 1, the system can improve the resource utilization and reduce unused resources due to unbalanced resource usage.

3 Performance Results

To validate that the QoS based migration scheme meets the required QoS during migrations and utilizes resources evenly, we implemented the migration systems, and compared the performance with those of the migration system using a greedy algorithm (e.g., Sandpiper). Table 1 shows the initial resource usage of servers, such as the number of virtual machines, the available memory size, and the available CPU capacity. Table 2 presents the resource usages of virtual machines hosted in server 1 and some metrics used to be migrated, such as VSR and QoS. The VSR, defined in Sandpiper, means the utilization of resources and the amount of overloaded of the server. As a QoS, we use the number of request drops in these simulations.

Table 1. Initial resource usages of servers (PM)

PMID	# of VMs	Memory (MB)		CPU	
		Total	Available	Total	Available
1	5	2048	256	2	0.05
2	4	2048	512	2	0.45
3	3	2048	768	2	0.9

Table 2. Resource usages and migration metrics for virtual servers hosted on PM1

ID	CPU	Mem	VSR	QoS^{Mig}	QoS^{SLA}
1	0.45	256	0.00578	11.451	10
2	0.4	256	0.00559	15.425	20
3	0.4	512	0.00328	15.425	20
4	0.4	512	0.00328	15.425	20
5	0.3	256	0.00527	29.572	25

In the first scenario, virtual machine 4 is about to be allocated 0.1 more CPU. Since server 1(PM1) has only 0.05 CPU left, migrations are required. In Sandpiper using a greedy algorithm, virtual machine 1 is selected for migration, since it has the largest VSR. Virtual machine 1, however, has QoS of 11.451 during the migration, that violates the SLA, QoS of 10. On the other hand, the QoS based migration system chooses virtual machine 2, that ensures required QoS during the migration and has the minimum memory allocated.

Moreover, after a sequence of migrations, it is confirmed that the resource usages are various and unbalanced among the resources in Sandpiper, but the QoS based migration system keeps the balanced usages among resources.

4 Conclusion

In this paper, we propose a QoS based migration scheme that ensure to meets service requirements and to use resources efficiently. By simulations, we confirm that the QoS based migration avoids the QoS violation during migrations and the usages are balanced among different resources after migrations. In our simulations, however, we consider only CPU and memory as resources, and use limited QoS metrics, the number of request drops, based on the offline measurements. We need more studying on various QoS metrics to measure and estimate in data center environments.

References

1. Vadhiyar, S.S., Dongarra, J.J.: A performance oriented migration framework for the grid. In: Proceedings of The 3rd IEEE/ACM International Symposium on Cluster Computing and the Grid, May 2003, pp. 130–137 (2003)
2. Ruth, P., Rhee, J., Xu, D., Kennell, R., Goasguen, S.: Autonomic live adaptation of virtual computational environments in a multi-domain infrastructure. In: Proceedings of The 3rd IEEE International Conference on Autonomic Computing (June 2006)
3. Wood, T., Shenoy, P., Venkataramani, A., Yousif, M.: Black-box and gray-box strategies for virtual machine migration. In: Proceedings of the 4th USENIX Symposium on Networked Systems Design & Implementation, April 2007, pp. 229–242 (2007)
4. Clark, C., Fraser, K., Hand, S., Hansen, J.G., Jul, E., Limpach, C., Pratt, I., Warfield, A.: Live migration of virtual machines. In: Proceedings of the 2nd ACM/USENIX Symposium on Networked Systems Design and Implementation (May 2005)

Author Index